ENGINEERING ECONOMY

ENGINEERING ECONOMY

A Manager's Guide to Economic Decision Making

THIRD EDITION

AMERICAN TELEPHONE AND TELEGRAPH COMPANY

Construction Plans Department

McGraw-Hill Book Company

New York   St. Louis   San Francisco   Auckland   Bogotá
Düsseldorf   Johannesburg   London   Madrid   Mexico
Montreal   New Delhi   Panama   Paris   São Paulo
Singapore   Sydney   Tokyo   Toronto

Library of Congress Cataloging in Publication Data

American Telephone and Telegraph Company. Construction Plans Dept.
    Engineering economy.

    Updated version of the 2d ed. by American Tele-
phone and Telegraph Company, Engineering Dept.
    Includes index.
    1. Telephone—Finance.   2.   Engineering economy.
I.   American Telephone and Telegraph Company.
Construction Plans Dept.   Engineering economy.   II.   Title.
HE8783.A54   1977   384.6′3   77-6444
ISBN 0-07-001530-9

34567890 H D H D   7865432109

This book was set in Primer by University Graphics, Inc.
Printed and bound by Halliday Lithograph Corporation.

Contents

Preface

The first edition of AT&T's *Engineering Economy* was a thin green booklet, published in 1952. During the 1950s, the Bell System grew rapidly, but it was also changing. Technological advances made new services available and provided significant productivity gains. Many of these improvements—such as the conversion from manual to dial, direct dialing of long-distance calls, and mechanized billing procedures—were accomplished through the substitution of capital for labor. More comprehensive analytical techniques were obviously required to ensure that the capital was used efficiently.

In 1963, in this climate, the second edition of *Engineering Economy* was published. Over 35,000 copies of an expanded, hard-cover *Green Book*, as it came to be called, have been distributed.

Although the 1963 edition is still being used effectively, the environment in which studies are conducted has continued to change. For example:

The Bell System has adopted accelerated tax depreciation, and the tax regulations themselves have changed significantly;

Inflation is a more pervasive influence both in the economy and in our business;

The computer is now widely used as an analytical tool, with the result that a number of study approaches are now possible that would be far too complex for manual calculation.

For these and other reasons, it has become necessary to bring the *Green Book* up to date with publication of a third edition.

Perhaps the most obvious change is a modification of the title to read *Engineering Economy: A Manager's Guide to Economic Decision Making.* It is well known that engineers are concerned not merely with technical design, but also with economic design, yet the engineer is not alone in this responsibility. The concepts and techniques necessary for engineering economy studies are needed increasingly by managers in all areas of the business.

The basic purpose of this book is to provide all managers with guidelines for performing an economy study. The text will be useful directly to those managers who perform actual studies. It should also be useful as a convenient reference to those managers who may provide inputs for studies or use the study results.

The primary objective of this text is to enable the manager to identify the economic choice among alternatives involving different amounts of capital expenditures, expenses, and revenues. A secondary objective, essential to the first, is to provide the basis for a thorough understanding of the discipline of engineering economy. Continuing changes in accounting and service classifications, tax regulations, technology, valuation methods, and competitive thrusts make fundamental conceptual understanding a managerial requirement.

This book is not intended to cover all aspects of economic analysis, however. Certain more sophisticated analytical techniques were omitted in the belief that the fundamental need was for an up-to-date elementary text. The focus is on economic selection studies. Other areas of economic analysis, such as the development of cost data for pricing studies, have not been included because they are complex, specialized areas which might better be treated independently.

Neither is this text intended as a "cookbook," with standardized procedures. The many studies needed will vary enough so that a cookbook treatment would be without end.

This volume was prepared by members of the AT&T Construction Plans Department. Deep appreciation is due and tendered, however, to an Advisory Committee composed largely of operating company managers, and to a task force of people borrowed from the operating companies who developed a large part of the contents.

LOUIS M. KRAFT
Assistant Vice President
Construction Plans Department
American Telephone and Telegraph Company

ENGINEERING ECONOMY

1

ENGINEERING ECONOMY: A MANAGERIAL TOOL

THE MANAGER'S ROLE

The managers of a business are responsible for making the many decisions which determine how well the business will operate. Often, these decisions influence the company's spending patterns far into the future. It would be difficult to exaggerate the collective importance of all these decisions on the company as a whole. They are, in effect, building the business for the future.

It is vitally important that the managers of a business be aware of the economic impact that their decisions will have on the ability of the business to meet its corporate goals. In the past, much of this economic responsibility has been borne by engineers, since engineers plan most of the facilities to be used in the operation of the business. Therefore, the discipline which enables managers to measure the economic impact of their decisions has come to be called *engineering economy.*

Engineering economy is basically concerned with money—money, both as a resource which is itself in scarce supply, and as the price of other resources which reflects their scarcity of supply. Business exists in an environment which requires the prudent and efficient use of all resources, including money. An economy study of a particular problem applies the principles of engineering economy to compare alternatives in monetary terms. Thus it helps to encourage the efficient use of resources. Engineering economy is therefore an important tool for the use of all managers of a business in achieving this objective.

In the process of determining how to achieve the corporate goals, all managers must make decisions among alternative plans. For example, they may evaluate a modernization proposal against the status quo, or determine the optimum amount of equipment capacity which should be provided. These are the types of questions which the discipline of engineering economy can help all managers to answer.

This text is aimed primarily at the managers actually conducting economy studies. They should be cautioned against immediately consulting advanced chapters without reading the prior material. Engineering economy is a discipline requiring a careful building of knowledge in an orderly fashion. Chapters taken out of sequence can impair the analyst's understanding.

THE STRUCTURE OF THE BOOK

The book has three main divisions: The earliest chapters provide background information; subsequent chapters focus on specific, important cost components—those associated with a capital investment; and the final portion of the book discusses study techniques and presentation. The highlights of each of these sections are described next.

The Background Chapters

The unit of measurement in an economy study is the dollar. Although other factors may influence—even dominate—the final decision, the standard of measurement in an economy study is the monetary impact. Money, then, along with its role in the firm, is the theme of Chapters 2 through 6, which deal with accounting, finance, and costs—and particularly the cost of money.

Chapter 2: Basic Accounting Concepts

This chapter presents the basic concepts of accounting and shows how they are applied in a small business. The accounting system offers valuable information for the economic analyst because it shows the impact of monetary transactions. Yet the data in accounting reports are not always relevant to making economic decisions. Before using accounting data in an economy study, the analyst must understand just what the accounting system is measuring.

Chapter 3: Accounting and the Telephone Company

This chapter makes the transition from the basic concepts of accounting to the use of the many accounts, reports, and other data which the accounting system of a large business—the telephone company—provides.

Chapter 4: Introduction to Financing

Economy studies almost always consider the possible investment of money, known as capital. This chapter briefly covers financing, the process of acquiring this capital. It also describes some of the primary forms of financing and the costs and constraints they impose.

Chapter 5: The Mathematics of Money

The dollar is the unit of measurement in an economic analysis, yet dollars spent or received at different points in time have different values. This is because there is a cost for the use of money—a cost which is proportional to time. Chapter 5 is devoted to techniques for equating monetary transactions which occur at different times.

Chapter 6: The Nature of Costs

Costs are at the heart of an economic analysis. This chapter identifies the two important kinds of costs—capital costs and operations costs—which are relevant to studies. It explains the fact that capital costs occur because the business has funds invested in it, and that operations costs occur because the business has physical plant.

Chapters on Capital Costs

The intermediate chapters in the book are about the costs which are associated with an investment.

Plant expenditures are just as much a part of business costs as are expenditures for current operations, but there is a basic difference between the timing of these costs and the way they are supported. Current expenses are charged against current operations, and should be covered by the revenues of the period. Expenditures for plant which will provide service for a number of years are charged against operations—not at the time they are bought, but over the period the plant will provide service. Expenditures for plant are supported by invested capital until the cost of the plant is recovered from revenues through the depreciation process. The use of capital therefore imposes an obligation to earn an adequate return, and to pay the income tax burden associated with it.

Chapter 7: Capital Repayment and Depreciation

The expenditure of capital money to provide service is just as much a cost of the service as the wages of the employees who work directly to provide it. Both the need for the eventual recovery of this cost (repayment of capital) and the means of doing so (depreciation) are treated in this chapter.

Chapter 8: Return and Income Taxes

Investors do not provide the capital free; they need to be paid as encouragement to invest in any firm. The amount paid to investors, over and above the simple repayment of their investment, is profit, or return. This return then becomes the cost of getting money. Since the firm must pay corporate income taxes on any profit it makes, the return necessary for investors also has an associated income tax burden. This chapter discusses these two aspects of capital costs.

Chapter 9: Federal Income Taxes

Income taxes are a burden on the profits of the firm. However, determining just what is defined as *profit* for tax purposes is a subject of continual change. From time to time, the federal government revises its procedures for computing income taxes, and certain costs are allowed or disallowed as tax-deductible expenses. Often the forms of capital repayment used for tax purposes are different from those used on the financial reports. Under certain circumstances, tax credits are available to encourage investment when economic conditions would otherwise discourage it. Many of these income tax considerations can have substantial effects on economy studies. This chapter provides the means for including these effects by modifying the basic concept outlined in Chapter 8.

Chapter 10: Inflation and the Cost of Capital

The results of economy studies are only as good as the forecasts of the future. Investors, too, are forecasting the future, and including their expectations in their requirement for return. As a result, managers must understand the phenomenon of inflation in economy studies. This chapter discusses the relationships between inflation and the cost of capital, and it provides guidelines for including the effects of inflation in economy studies.

Chapters on Techniques

The final chapters of the book are intended to provide guidelines for structuring an actual economic analysis, to present some observations about the significance of study results, and to offer recommendations for specific problem areas.

Chapter 11: Identifying Alternatives

This chapter begins the discussion of how to perform a study, and the first and very important step is to identify the alternatives to be studied. Of course, any study is limited to the alternatives known to be available. Some techniques are presented to help ensure that all viable alternatives are considered, and that they are comparable.

Chapter 12: Economy Study Techniques

After identifying all suitable alternatives and making certain that they are comparable, the analyst's principal goal is to determine which plan achieves service objectives most economically.

Chapter 12 presents three compatible techniques for finding the most economic alternative. Two of these techniques, the present worth of expenditures (PWE) and present worth of annualized costs ($PWAC$), are quite similar. They are based on the concept that to remain healthy the firm must recover the additional costs imposed by any alternative. One of those costs includes paying the investors for the use of their money. In essence, these two techniques demonstrate that the most economic alternative is the one which results in the least increase in the firm's requirement for revenue or "revenue requirements." A third technique, internal rate of return ($IROR$), when carefully applied to the differences between alternatives, is actually making the same economic evaluation, but from a different perspective. This technique is based upon maximizing the benefit, or net present value (NPV), which an alternative will produce.

These statements of economic objectives are harmonious because they are simply different ways to measure the efficiency of the firm's economic process. For example, it would be equally valid to appraise the operation of a machine by measuring

- the amount of input required to produce a given output (revenue requirements)
- the amount of output resulting from a given input (net present value)
- the ratio of output to input (efficiency, or rate of return)

Chapter 13: Retirement, Replacement, and Other Economy Study Considerations

The special considerations which often occur in replacement studies are discussed in this chapter. Many of these considerations also occur in other studies too. The fact that what is done now cannot influence the past is the basis of the concept of sunk costs, which is covered in detail. The effects which salvage or salvage foregone have on the retaining of plant are also discussed extensively.

Chapter 14: Dispersion of Retirements

Throughout most of this book, it is assumed that investments have a known or explicit life. Actually, the mortality data provided by years of experience indicate that investments in large groups of plant items are retired over a period of time. This chapter provides the tools for recognizing this dispersion of retirement in economy studies.

Chapter 15: Leasing versus Owning

An analysis of whether to own or to lease requires comparisons of alternative ways to finance an asset. Difficulty arises because these alternatives would have different financial impacts on the firm. This chapter develops three progressively sophisticated ways to analyze the owning versus leasing question. The financial aspects of leasing mean that owning versus leasing studies are different from other economy studies. Therefore, there are special techniques and special considerations which apply. This chapter illustrates how the engineering economy concepts are combined with the financial concepts in the owning versus leasing study.

Chapter 16: Presenting and Evaluating an Economy Study

Even when the results of an economic study are found, the decision about which plan to choose is by no means certain. The presentation of study results is often critical to implementing the recommendations of the study. This final chapter provides guidelines for material to be presented and for the manner of its presentation, as well as the necessary supporting documentation. It includes a checklist which is useful in the preparation of a study as well as in the analysis of a study done by others.

2

BASIC ACCOUNTING CONCEPTS

INTRODUCTION

The American Institute of Certified Public Accountants defines *accounting* as follows:*

> Accounting is the art of recording, classifying and summarizing in a significant manner and in terms of money, transactions and events which are, in part at least, of a financial character and interpreting the results thereof.

In simpler terms, accounting is a systematic way to keep track of the flow of cash through time in order to communicate financial information about an economic entity—usually a business. Accountants develop this information by recording, classifying, and summarizing a firm's financial events; they then produce the records and reports which summarize this firm's financial position and its operations.

The principal objective of the economic analyst also is to keep track of the flow of cash in a business and to apply this information in making new financial decisions. Accounting is therefore a basic tool for an engineering economist to use.

Limitations of Accounting

Accounting is basic to business, because it is a system for measuring business results. Nevertheless, there are limits to what accounting can measure—and also to the accuracy with which it measures.

The standard for all accounting measurements is money. However, a business is a complex operation that includes many factors, such as policies and people's morale, which cannot be expressed in monetary terms. Furthermore, even those events that can be measured in terms of money cannot always be expressed with accuracy. This aspect of accounting will become more apparent later. Even at this stage, though, it is clear that some accounting figures are not exact. One reason is that reports are needed and therefore prepared periodically, when only approximate figures may be available for certain items—in fact, precise data may not be available until a long time afterward.

This lack of precision does not actually weaken the importance of accounting, since

---

Accounting Terminology Bulletin No. 1.

an approximation that is available today can be acted upon now. This immediate action might prevent the disaster which would have been inevitable had action been delayed until more accurate figures were available.

Controller, or Comptroller

The chief accounting officer in a business firm is usually designated the *controller,* or *comptroller.* The two terms are used synonymously, with the choice a matter of preference or style.

As each of the designations *controller* or *comptroller* suggests, the primary use of accounting data is to help in *controlling* business operations. The comptroller (controller) oversees the work of the accounting staff and is part of the management team charged with the responsibility for setting objectives and operating the business in such a way as to satisfy its goals.

Information for Whom?

Financial information is of interest to many people both inside and outside a business. Besides preparing the financial data, the accountant analyzes the data and, in doing so, looks for significant trends and meaningful relationships. The accountant then reports any important indications to other managers of the enterprise for their guidance.

Managers are very much interested in accounting reports, since these reports are a definitive picture of the condition of the business. As administrators of that business, the managers have many responsibilities, but a major concern is that the business make money and remain solvent. Accounting reports show the earnings and solvency situation and help the managers to assess results of past decisions and actions. This kind of assessment, in turn, helps them to plan future actions and to judge their consequences.

Current or prospective investors or creditors of a business are particularly interested in the information in financial reports. Investors are concerned with the safety of their investment and the prospects for a return. Creditors look for assurance that they will be paid, or that they can safely extend additional credit. Other groups, too, seek financial reports for appraisal and analysis. These groups include regulatory and governmental agencies, labor unions, writers for business journals, and credit agencies.

What Are the Reports?

Of the many reports the accounting process produces, two are best known: the *balance sheet* and the *income statement.* A balance sheet gives an instantaneous view of the financial position of a business on a particular day—usually the last day of the year. An income statement shows the flow of income—the moving view—during the period between two balance sheets (usually, during 12 months of the year). Sometimes it is called a *profit and loss* statement, since it shows how successfully the business is operating.

Nearly everyone has seen examples of these two reports, because they are widely distributed. Some are even published in newspapers. Corporate annual reports, which include a balance sheet and an income statement, are furnished to stockholders and are often made available to employees and others.

However, a third major accounting report is needed to show the complete financial picture of a business. This is the *funds-flow,* or *cash-flow,* statement, which is not nearly so well known or so widely distributed as the balance sheet and income statement. A funds-flow statement summarizes the financial activities of a business, resulting from both its operation and changes in its financial condition during an accounting period.

This chapter describes the balance sheet and the income statement, which the economic analyst must be able to interpret. It also describes those basic elements of the accounting process which the analyst must understand. Chapter 4 covers the funds-flow statement.

The balance sheet is presented here first because it is a convenient starting point to the understanding of basic accounting principles.

THE BALANCE SHEET

The balance sheet shows the financial position of a business on a particular date. It is always prepared annually—usually at the end of the year; but it may also be prepared monthly, or at other intervals, as well. Table 2-1 is a relatively simple balance sheet, showing the financial position of a small appliance repair business on December 31.

The heading shows three items: (1) the name of the business, (2) the name of the statement, and (3) the date of the balance sheet. Below the heading is the body of the statement, listing Assets on one side and Equities on the other. The term *assets* means "what the business owns." The term *equities* means "claims against the assets"; it is used to cover both *liabilities* and *owner's equity,* since both are claims against the assets. *Liabilities* are what is owed to a creditor other than the owner; *owner's equity* is what is owed to the owner.

Dual-Aspect Principle

Equality between the total assets and the total equities is a fundamental characteristic of the balance sheet, and is the reason for its name. The two dollar totals are equal

TABLE 2-1
Appliance Experts Company
BALANCE SHEET
December 31, ——

Assets			Equities		
Current Assets			Current Liabilities		
Cash	$ 7,440		Accounts Payable	$8,620	
Accounts Receivable	2,000		Interest Payable	40	
Inventory	1,715		Total, Current		$ 8,660
Total, Current		$11,155	Other Liabilities		
			Note Payable		5,000
Fixed Assets			Owner's Equity		
Land		6,000	A. Fixer, Capital		20,050
Building	$15,000				
Less:					
Accumulated Dep.	50	14,950			
Furniture and Fixtures	1,620				
Less:					
Accumulated Dep.	15	1,605			
		$33,710			$33,710

because the balance sheet shows two views of a business entity. One view shows the things a business owns, its assets. These have been obtained by expenditures of money supplied by creditors and owners. The other view shows what the business owes; these are its liabilities and the owner's equity. This concept of looking at a business from the two opposing aspects is known in accounting as the *dual-aspect* principle.

Basic Accounting Equation

The agreement of total assets with total equities leads to a basic accounting equation:

$$\text{Assets} = \text{equities}$$
$$\text{Assets} = \text{liabilities} + \text{owner's equity}$$

The equation for the Appliance Experts Company states the dual-aspect principle:

$$\$33,710 = \$13,660 + \$20,050$$

The balance sheet, then, is simply a detailed statement of the basic equation. This statement lists the several different kinds of assets which the business owns, as well as the different kinds of liabilities and the owner's equity.

The term *net worth* can be substituted for the term *owner's equity*. If the amount of liabilities is subtracted from the amount of assets, the remainder represents the net worth. Individuals who prepare a personal balance sheet could also develop their personal net worth. To state it another way, the net worth of an individual (or a business) is the difference between what the individual (or business) owns and what the individual (or business) owes.

Balance Sheet, a Statement of Position

The American Institute of Certified Public Accountants recommends that the name *statement of financial position* be used instead of *balance sheet* to avoid any misconception that there is something good or important about having the two totals numerically agree. Identifying the statement as showing only "financial position" emphasizes that it does not show everything about the financial condition of the business. It is important to understand that other financial statements are needed in order to get a picture of the overall condition of the firm. However, the term *balance sheet* is used in this text because it is more familiar and it is still widely used elsewhere.

A Breakdown of the Balance Sheet

Headings under Assets

Each individual asset on the balance sheet can be listed separately. Usually, however, the things a company owns are grouped under two or more headings. One heading is *Current Assets*, which includes cash and other items that can reasonably be expected to be converted into cash in the near future—usually one year. One item under Current Assets is *Accounts Receivable*. These are amounts which customers owe plus other amounts that will be received within the usual one-year period. Often this item is adjusted on the basis of experience to allow for the inevitable uncollectibles. Another item under Current Assets is *Inventory*. This consists of merchandise on hand for direct sale, or materials to be either processed for direct sale or consumed in the production of goods and services that are for sale.

A second group of assets is *Fixed Assets*. These are tangible, relatively long-lived items which the company owns and uses—either to produce goods or to furnish services. Fixed assets are offset by the accumulated depreciation which applies to them, however.

Another group of assets that is often found on balance sheets is *Other Assets*. Examples are investments in other companies or in long-term securities. Some companies may include Prepaid Expenses and Deferred Charges under Other Assets, whereas other companies may list Prepaid Expenses under Current Assets. Other assets may also include nonphysical, but valuable, things such as leases, licenses, and franchises which the business owns. Any of these might be listed as a separate asset, depending on the importance the company attaches to it.

Headings under Equities

Equities were defined earlier as "claims against the assets of a business." This group includes *Liabilities*, which represent claims against the business by others than owners, and *Owners' Equity*, which represents claims by the owners.

One heading under Liabilities is *Current Liabilities*, which are obligations due—usually within one year. They include: *Accounts Payable*, or payments due to the suppliers of goods and services to the company; *Advance Billings*, which is money billed to customers in advance; *Dividends* payable to owners and *Interest Accrued*—that is, accumulated—to outstanding debt holders for interest payments that will be made shortly; *Tax Liability*, an estimate of tax payments that are owed to governmental entities; *Accrued Expenses*, representing amounts known to be owed but for which a bill has not yet been received; *Deferred Income*, a liability of the company to furnish a future service for payment already received. A second heading under Liabilities is *Other Liabilities*. These are long-term claims, such as bonds or notes, which will mature after the coming year. Such bonds and notes may also be listed under the heading *Debt*.

The title *Owners' Equity*, to show claims of owners, varies depending on the form of the organization. For example, partnerships use the caption *Partners' Equity*, and each partner's claim is shown. Owners of a corporation are stockholders, or shareholders; their equity is listed under *Stockholders'* or *Shareholders' Equity*. Each of the different classes of stock may be listed separately. The owners' equity will also reflect any gain—profit—or loss from the operation of the business. In a small business, these amounts are added to, or subtracted from, the owner's capital. However, gains or losses are stated separately on the balance sheet of a corporation. The portion of gain that is not paid in dividends appears on the balance sheet as *Retained* or *Reinvested Earnings*. Losses may also be listed as *Deficits*.

The Effects of Business Transactions on the Balance Sheet

Accountants record all the events pertaining to the business which can be expressed in dollars. Each such event is known as a *transaction;* accounting, then, is the process that analyzes, records, and summarizes the transactions of a business. The effect of every transaction is reflected on the balance sheet, even though the balance sheet itself does not have a direct record of it. However, every transaction is recorded on auxiliary records; periodically, these auxiliary records are summarized to produce the balance sheet and other financial reports.

A good way to illustrate the effects of transactions on the balance sheet is to observe the results of each change as a new, small business is established. The illustration that follows shows the development of Appliance Experts Company and its balance sheet. For simplicity, events are assumed to occur during one month.

Mr. A. M. Fixer started this business by opening a bank account in the name of

Appliance Experts Company, and he deposited $20,000 of his own money on December 1. After the heading, the December 1 balance sheet would show these items:

Assets		Equities	
Cash	$20,000	A. Fixer, Capital	$20,000

Borrowing, a Liability

To make more cash available, on December 2 Mr. Fixer borrowed $5,000 from the bank on a note. Cash is increased by $5,000, but there is an offsetting liability to the bank. The December 2 balance sheet would therefore look like this:

Assets		Equities	
Cash	$25,000	Note Payable	$ 5,000
		Owner's Equity	
		A. Fixer, Capital	$20,000
	$25,000		$25,000

Cash Transaction

On December 3, Mr. Fixer bought a parcel of land for a shop site for $9,000, and he paid cash. In this transaction, one asset—an amount of cash—is exchanged for another asset—land. Total assets are unchanged, and, likewise, total equities. The December 3 balance sheet shows this change in assets:

Assets		Equities	
Cash	$16,000	Note Payable	$ 5,000
Land	$ 9,000	Owner's Equity	
		A. Fixer, Capital	$20,000
	$25,000		$25,000

Accounts Payable

Mr. Fixer learned of a small building that was to be moved (to make way for a new apartment structure). For $15,000 he was able to buy this building and have it moved to the recently acquired lot. Mr. Fixer made a down payment of $7,500, with the remainder due in three months. This transaction occurred on December 4. The following balance sheet reflects the acquisition of the new asset. The building is shown at cost, $15,000 (just as the land was also shown at cost), even though it was not fully paid for. To buy the building, cash was reduced by $7,500; and Appliance Experts assumed accounts payable of $7,500. Both the assets and the equities increased, yet they remain equal; and the owner's equity is unchanged.

Assets		Equities	
Cash	$ 8,500	Liabilities	
Land	$ 9,000	Accounts Payable	$ 7,500
Building	$15,000	Note Payable	$ 5,000
		Owner's Equity	
		A. Fixer, Capital	20,000
	$32,500		$32,500

Furniture and Fixtures

On December 10, the company acquired shelving, work benches, test equipment, and a cash register—all at a cost of $1,620, with payment due in 90 days. Again the new assets are recorded at cost, and the amount owed is added to accounts payable:

Assets		Equities	
Cash	$ 8,500	Liabilities	
Land	9,000	Accounts Payable	$ 9,120
Building	15,000	Note Payable	5,000
Furniture and Fixtures	1,620	Owner's Equity	
		A. Fixer, Capital	20,000
	$34,120		$34,120

Accounts Receivable

By December 11, Mr. Fixer realized that his lot was larger than necessary; and since the adjacent owner wanted more land, he sold a portion of the lot for $3,000. This piece was sold at cost, so there is no profit or loss; and the purchaser agreed to pay in 90 days. A new asset, Accounts Receivable, now appears:

Assets		Equities	
Cash	$ 8,500	Liabilities	
Accounts Receivable	3,000	Accounts Payable	$ 9,120
		Note Payable	5,000
Land	6,000	Owner's Equity	
Building	15,000	A. Fixer, Capital	$20,000
Furniture and Fixtures	1,620		
	$34,120		$34,120

Inventory

Repair parts in the amount of $2,000 are purchased December 20, for cash. The stock of these repair parts appears as Inventory on the new balance sheet:

Assets		Equities	
Cash	$ 6,500	Liabilities	
Accounts Receivable	3,000	Accounts Payable	$ 9,120
Inventory	2,000	Note Payable	5,000
Land	6,000	Owner's Equity	
Building	15,000	A. Fixer, Capital	20,000
Furniture and Fixtures	1,620		
	$34,120		$34,120

Reflecting Sales

Completion of the first few repair jobs by December 29 is a milestone in progress, resulting in receipts of $440. The parts which were used cost $285—clearly a gain of $155 at that moment. Table 2-2—the following balance sheet for December 29—reveals a change of $440 in cash to reflect the receipts and shows the addition of the $155 gain to the owner's equity:

TABLE 2-2
BALANCE SHEET
December 29, 19____

Assets		Equities	
Cash	$ 6,940	Liabilities	
Accounts Receivable	3,000	Accounts Payable	$ 9,120
Inventory	1,715	Note Payable	5,000
Land	6,000	Owner's Equity	
Building	15,000	A. Fixer, Capital	20,155
Furniture and Fixtures	1,620		
	$34,275		$34,275

The completion of repairs therefore produced revenue for Appliance Experts. This introduces a new accounting term, the *realization principle:* Revenue is realized when goods or services are furnished for cash or for some other acceptable consideration. What this means is that revenue is counted during the accounting period in which it is realized (month, quarter, or year), even though it may be reflected in accounts receivable rather than in cash.

A Recapitulation

A number of liberties were taken and simplifications made in this hypothetical case to show the effect of transactions on a balance sheet. The important point to remember, however, is that the accountant does not actually record transactions by changing the balance sheet; yet each transaction does have an effect on the assets and equities. In all cases, though, assets continue to equal equities.

In the example, some transactions produced an exchange of one asset for another but no change in the total assets or liabilities. Other events increased assets and increased equities (claims) by an equal amount.

Also in the example, gain is added to the owner's equity. This treatment of gain is customary in small businesses, which may have only one owner—or only a few owners. Corporations, which may be owned by many stockholders, show gain as a separate item under Shareholders' Equity.

Gain in corporations is identified as *net income.* It consists of two parts: one part may be paid out in *dividends;* the other part, which is not paid out in dividends, is frequently labeled *retained earnings.* Gain from the operation of a business adds to the owners' equity, and any loss reduces it.

Gain Reduced by Expenses

To return to the Appliance Experts Company, if no further receipts are assumed during the month of December, the December 29 balance sheet—Table 2-2—might be assumed to show the end-of-month (and year) position of the business. Not quite, because the real-life situation is not so simple. On the sample balance sheet, an increase in the owner's equity from a gain was shown deliberately to illustrate the way an owner benefits from successful operation of a business. However, this was unrealistic and premature, because no expenses of operation had been accounted for up to that time. Business expenses that are recognized during the month will tend to reduce the gain—or could even eliminate it entirely.

Business expenses introduce the *accrual concept.* The standard accounting procedure is to record expenses incurred during the accounting period (month, quarter, or year) whether or not the actual payment has been made. Under the accrual concept, expenses for a given period will be accrued (accumulated) and shown on the appropriate financial statements. This is done even though the expenses may be reflected by an increase in accounts payable rather than by an outlay of cash.

Expenses may also be reflected by decreases in asset accounts. One of these expenses is the $285 worth of inventory parts, which has already been accounted for. The $285 represents an allocation of a portion of the total inventory expenditure to the accounting period when the parts were actually used. Thus, only the $285 worth of parts used was counted as a cost—and not the whole $2,000 cash inventory, because the remainder is still available for use in later periods.

Tallying Expenses

Expenses that any business might have would include utilities, taxes, wages, insurance, operating supplies, interest on loans, and depreciation of fixed assets. So far, only the cost for repair parts used has been included in the balance sheet of Appliance Experts. The company would receive bills for most of the other expenses. However, two are incurred and accrued even though no bill is rendered: The first of these is depreciation; the second is interest on loans.

Depreciation is a major topic in engineering economy and is covered in detail in Chapter 7. Depreciation is another example of an expense which is reflected by a decrease in an asset account. Its treatment is simplified in the Appliance Experts example by the existence of only a few assets. It is a periodic allocation to business expense of a measured portion of the original cost of long-lived fixed assets other than land.

Mr. Fixer decided in favor of an equal allowance for depreciation during each month of the assumed life of his tangible assets. Appliance Experts listed two depreciable assets, a building and its furniture and fixtures. Mr. Fixer assumed a life of 25 years for the building and 9 years for the furniture and fixtures, with zero salvage value. He therefore decided to allocate the cost over 300 months for the building and 108 months for the furniture and fixtures. By dividing the original cost of the assets by the months of life, the monthly depreciation was figured at $50 for the building and $15 for furniture and fixtures.

The expense of interest accrued on the loan can also be computed. The $5,000 two-year note carries an interest rate of 9.6% per year, and the interest can be calculated as a cost of $40 per month on the $5,000 loan ($5,000 × 0.096 × $\frac{1}{12}$).

These expense items, depreciation of $65 ($50 + $15) and the interest of $40 for the month, resulted in an accrued expense of $105. Allowance for these expenses reduces the previously counted gain, leaving a $50 gain at the end of the month.

Allowance for additional expenses during the period would further reduce the gain, and could even create a loss situation. The end-of-month balance sheet, Table 2-1, is reproduced on page 16. A comparison with the December 29 balance sheet, Table 2-2 (shown earlier in this section), shows the treatment and effect of accounting for depreciation and the interest obligation, as well as the increase in owner's equity. As the last transactions of the year, the buyer of a portion of the land paid $1,000 of the $3,000 owed and Mr. Fixer made a $500 payment on the furniture and fixtures. These transactions result in the final end-of-month balance sheet.

Limitations of the Balance Sheet

An important point to remember is that the balance sheet is a statement of position at a particular time; it is not a complete statement of the financial condition of a business

TABLE 2-1
(Reproduced from beginning of this chapter)
Appliance Experts Company
BALANCE SHEET
December 31, 19____

Assets			Equities		
Current Assets			Current Liabilities		
Cash	$ 7,440		Accounts Payable	$8,620	
Accounts Receivable	2,000		Interest Payable	40	
			Total Current		$8,660
Inventory	1,715				
Total, Current		$11,155	Other Liabilities		
			Note Payable		5,000
Fixed Assets			Owner's Equity		
Land		6,000	A. Fixer, Capital		20,050
Building	$15,000				
Less:					
Accumulated Dep.	50	14,950			
Furniture and Fixtures	1,620				
Less:					
Accumulated Dep.	15	1,605			
		$33,710			$33,710

entity. While the balance sheet for Appliance Experts was being developed, for example, the operation increased the owner's equity by $50. However, inspection of Table 2-1, the December 31 statement, does not show this increase. Furthermore, by looking at the owner's equity alone, there is no way of knowing what the owner's actual contribution was—$5,000, $20,000, or what. The only way to see changes in the owner's equity is to compare one balance sheet with an earlier balance sheet, or to refer to an income report.

Something that the balance sheet does show is that the current assets ($11,155) are only slightly greater than the current liabilities ($8,660). Yet it does not show whether or not this is serious. In other words, the company must be able to meet its liabilities when they become due. The hope is that the ongoing operation of the business will make that possible. However, the balance sheet does not furnish this information.

AUXILIARY RECORDS

The point has been made that the major accounting reports are prepared from the accounting records. All transactions, in fact, are recorded in a *journal*, which resembles a business diary of financial events. Periodically, the entries in this journal are assigned to the appropriate asset and equity records, called the *ledger accounts*. This process and the records themselves are described next. The ledger accounts are described first, because transactions recorded in the ledger accounts are summarized on the balance sheet.

Ledger Accounts

Transactions Recorded in Accounts

If a balance sheet were prepared after each transaction, this balance sheet would confirm the change that each event produced. The process would also demonstrate that assets and equities continue to be equal. Accounting is not done that way, though, because that procedure is neither a practical nor a satisfactory one. Instead, the dollar value of a transaction is entered in a record form known as an *account*. In a small business, an account is established and maintained for each asset and for each liability. In a large business, accounts are established for groups of similar assets and for groups of similar liabilities.

In a small business, each account is usually on a separate sheet or page, and the sheets are kept in a binder known as a *ledger*. In a larger entity, accounting is mechanized or computerized. Financial reports—balance sheets, income statements, and others—can be prepared periodically from entries in the ledger accounts.

A "T" Account

The diagram below illustrates the simplest form of account, called a "T" account because it looks like the letter *T:*

Title

Left or debit side	Right or credit side

Calling the left side the "debit" side and the right side the "credit" side is an arbitrary but fixed accounting rule. Accountants use the words *debit* and *credit* as verbs. Making an entry on the left side is therefore known as *debiting,* or *to debit.* Likewise, recording an entry on the right side is known as *crediting,* or *to credit.*

In accounting, the terms *debit* and *credit*—often abbreviated as "Dr." and "Cr."— have no other meaning than "left" and "right." In everyday usage, these words have other connotations. For example, a credit is regarded as something good, a debit as something bad. Not so in accounting: The terms are strictly neutral in value. The accountant also considers the word *charge* as synonymous with *debit.*

Asset Accounts

Increases in an asset account are entered on the left, or debit, side, whereas decreases are entered on the right, or credit, side. This fixed rule for any asset account is illustrated below:

Any Asset Account

Debit	Credit
(Increases)	(Decreases)

In Table 2-1, which is the balance sheet for Appliance Experts on December 31, cash was the first item listed under current assets. This amount reflects the summary of changes in cash; each cash transaction is actually recorded as an increase or a decrease in the cash account. Table 2-3 illustrates:

TABLE 2-3
CASH

12-1	(Deposit)	20,000	12-3	(Land purch.)	9,000
12-2	(Loan)	5,000	12-4	(Building)	7,500
12-29	(Receipts)	440	12-20	(Stock)	2,000
12-30	(Land Sale)	1,000	12-30	(Furn. & Fix.)	500
		26,440			19,000
19__					
Jan. 1					
New Balance		7,440			

When the debit entries (increases) and credit entries (decreases) are added, and the smaller number is subtracted from the larger, there is a left-side balance of $7,440. This is called a *debit balance,* and it is the actual source of the amount of cash listed on the balance sheet.

As in many accounting records, neither the dollar sign ($) nor the terms *debit* and *credit* are used. They are not needed in accounting because transactions are understood to be in the unit of money. Also, the locations of debit and credit are fixed.

Equity Accounts

Rules governing increases and decreases in equity accounts are just the reverse of the rules for asset accounts. Consequently, increases in an equity account are entered on the credit (right) side, and decreases are entered on the debit (left) side. For example:

Any Equity Account

Debit	Credit
(Decreases)	(Increases)

Just the opposite of asset accounts, equity accounts have credit (right) balances. Table 2-4 below illustrates the activity in Accounts Payable for Appliance Experts:

TABLE 2-4
ACCOUNTS PAYABLE

12-30 (Paid on F&F)	500	12-4 (Bal. on Bldg.—3 months)	7,500
		12-10 (F&F—due 90 days)	1,620
	500		9,120
		19__	
		Jan. 1	
		New Balance	8,620

The amount of the account balance is entered on the balance sheet under Accounts Payable, a liability and one of the main headings on the equity side.

Debits = Credits

Every transaction is recorded by debiting one or more accounts and, at the same time, crediting one or more other accounts so that total debits are equal to total credits. The procedure is named *double-entry bookkeeping,* since every transaction requires a double entry—at least one debit and at least one credit. Because debits and credits for each transaction are equal in the ledger accounts, the debits total equals the credits total. If not, an error has been made.

The two accounts which were illustrated do not fully prove the principle of equality in debits and credits. To do this, it would be necessary to review the activity on the balance sheet and relate this activity to the asset and equity accounts that would be affected.

A BUSINESS DIARY

The approach to accounting in this chapter might be considered as "backing into accounting." The effect of transactions on the balance sheet was illustrated first. Because transactions influence the balance sheet through ledger accounts, the effect of transactions on the ledger accounts was covered next. However, transactions are not entered on the ledger accounts first, either. The day-to-day transactions are chronologically compiled in a book called a *journal,* which is a diary of the firm's financial events. Although various documents may be used to notify the accountant of a transaction, the initial accounting entry, then, is in the journal. Entries in the ledger accounts are made later, from the journal entries.

Table 2-5 illustrates some typical journal entries of some transactions of the Appliance Experts Company. To record a transaction, the name of the account to be debited is listed first, with the dollar amount placed in the left-hand money column. The account to be credited is entered below the debit entry, and the amount appears in the right-hand money column. However, the dollar sign ($) is not used in the ledger.

Table 2-5 shows that each entry is briefly described. The column marked "L" refers to the ledger account in which the entry is to be made. Although each ledger account is assigned a number, the assigned account number is recorded in the journal only after the transaction has been listed in the ledger. This process of making ledger entries from the journal is called *posting.* In the example, events through December 10 have been "posted" to the ledger, because these events have account numbers; but subsequent events have not yet been "posted." Normally, posting is not done after each transaction; it can be done daily, weekly, or whenever it is convenient.

OTHER DETAILS

The examples of relatively simple ledger and journal entries show how events are recorded during Appliance Experts' first month of operation. They also illustrate the elements of the accounting process. Some procedures have been simplified, and some steps in the monthly and yearly summaries of the accounts have been omitted. In a real-life situation, many additional accounts would be needed to keep track of the various kinds of purchases, expenses, and revenues. In fact, there is no limit to the proliferation of accounts and subaccounts that can be adopted to facilitate meaningful summaries of financial information for use in the administration of a business.

There are differences in style, however. For example, the practice of handwritten entries in a small operation would have to give way in a large company to specially designed mechanical recording equipment. Even more sophisticated techniques might be used, with entries recorded on magnetic tape.

TABLE 2-5
GENERAL JOURNAL

Date	Account titles and explanation	L	Debit	Credit
19__				
Dec. 1	Cash	100	20,000	
	A. Fixer, Capital	150		20,000
	Cash investment in business			
2	Cash	100	5,000	
	Note Payable	140		5,000
	Two-year note from bank @ 9.6%			
3	Land	101	9,000	
	Cash	100		9,000
	Purchased land for store site			
4	Building	102	15,000	
	Cash	100		7,500
	Accounts Payable	125		7,500
	To buy and relocate building: one-half down, balance in 3 months			
10	Furniture and Fixtures	103	1,620	
	Accounts Payable	125		1,620
	Shelving benches, test equipment, cash register			
11	Accounts Receivable		3,000	
	Land			3,000
	Sell portion of land at cost, due 90 days			
20	Inventory		2,000	
	Cash			2,000
	Purchased merchandise			

THE INCOME STATEMENT

In contrast to the balance sheet, which shows the financial position at a single point in time, the income statement shows the course of business operations over a period of time. It is also a record of those events that affect the owners' equity during the period between two balance sheets.

Another name for the income statement is *profit and loss* statement, with *profit* used to mean "gain."

Earlier in this chapter, the term *gain* was also used to refer to the amount by which the receipts for the Appliance Experts Company exceeded the costs of goods sold, and selected expense items were shown to reduce the gain. Although "gain" is one of the dictionary definitions for *profit*, accountants prefer another definition: *net income* for a specified time period. For accounting purposes, and for use in this text, *net income* is defined as "revenue less expenses for a stated time interval." Income is measured periodically, and it is shown on an income statement. This report shows revenues and expenses, and it also displays the resulting income or loss. The terms *revenue* and *expense* need to be discussed further.

Revenue

It is important to remember that revenue increases the owners' equity in a business, and that receipts are not exactly revenue. But what exactly is *revenue?*

Under the realization principle, payment or promise of payment is the revenue which a business receives in exchange for goods or services. According to the realization principle, too, revenue may be represented by cash payment or by an account receivable. In either case, the revenue is counted in the period being measured. This may be a month or another interval. The actual receipt of cash in a subsequent accounting period is not revenue for that period, but simply a substitution of one asset (cash) for another (accounts receivable).

The accounting system for a business will furnish one or more revenue accounts, with appropriate labels such as these: *sales, commissions,* or *fees.* Revenue accounts are really subdivisions of the owners' equity account; and, under the rules of debit and credit, revenue is to be credited to the proper revenue account, because it represents an increase of the owners' equity. The offsetting debit entry (since debits equal credits) is made to cash and/or accounts receivable.

It is important to distinguish here between *revenue* and *cash.* Revenue may be represented by cash, or by an increase in accounts receivable. Cash may be increased by the collection of an account receivable, by borrowing, by investment by the owners, or by the receipt of revenue.

Expense

Expense has an impact on a business which is opposite from that of revenue: Expense reduces the owners' equity. It is the cost of business operation for a stated time period that is incurred with the expectation of generating revenue. Whether or not revenue is actually realized during the period, though, is a separate matter. Nevertheless, this does not change the obligation to allocate the expense of operations to the right time period (the accrual concept).

An *accrual* is simply the gradual accumulation, by the assignment to specific accounting periods, of portions of a cash transaction which has occurred in the past or is expected to occur in the future. Accruals are recorded as adjusting entries on the worksheets used in preparing financial statements. Expense accounts are debited to record each accrued expense item. This debiting of an account for expense in effect reduces the owners' equity. Furthermore, an expense debit must be offset by an equal credit to an asset account such as cash, or to a liability account such as accounts payable or depreciation reserve. In short, the same distinction is necessary between expenses and expenditures as was made earlier between revenue and cash receipts.

Measuring Income

Net income has been defined as "revenues less expenses for a given time period." To find income, then, the total estimated expense must be subtracted from the total estimated revenue amount. Assuming that revenue exceeds expense, net income will be positive. If expense exceeds revenue, however, net income will be negative. The several steps that are taken in the accounting process to accumulate the revenues and expenses in order to develop the measurement of net income will not be described here, since knowledge of those technical bookkeeping procedures is not necessary for understanding accounting.

It is important to realize that the net income shown on the income statement is not an amount that can be determined with absolute accuracy. The reason for this limitation is the problem of exactly matching the expenses incurred or accrued with the revenue realized for a given accounting period. Revenue amounts, for example, may be known fairly accurately; but the allowance for uncollectibles in the accounts receivable has to be estimated. Expenses may be even more of an estimate. The firm's telephone bill is one example. The billing period seldom coincides with the calendar month. The basic service is billed in advance, whereas toll charges are billed in arrears. On a practical basis, the bill would probably be treated as an expense for the month in which it is received. Advertising expense is even more difficult to associate with the revenue it generates. Perhaps no revenue results, or it may not occur for many periods.

One of the most difficult expense items to deal with is depreciation expense on fixed assets. The primary reason is that the lives of such assets are unknown. The best that can be done is to estimate the expected life on the basis of any available experience and/ or informed judgment. Anticipated salvage can be estimated in the same way.

Accountants recognize the limit of the accuracy by which expenses can be matched against revenues by referring to "measurement" of income rather than "determination" of income. The word *determine,* then, expresses an accuracy that is not possible in accounting. *Measurement* means simply "to estimate or appraise by a criterion."

The typical auditor's opinion, in fact, recognizes the limit to the accuracy of accounting by stating in part: "In our opinion, the accompanying balance sheet and statement of income present fairly the financial position at December 31, 19__, and the results of operations for the year then ended of the ____ Company, in conformity with generally accepted accounting principles applied on a consistent basis."

To sum up, it is necessary to distinguish between expenses and expenditures of cash, because they are not necessarily the same, just as revenue and cash receipts may not be the same. The relationship is as follows:

- Receipts and the application of the realization principle result in revenue.

- Expenditures and the application of the accrual concept result in expense.

The total receipts minus expenditures therefore represents the flow of cash to an engineering economist. However, the total revenue minus expenses represents net income to an accountant.

Example of an Income Statement

Table 2-6 is a sample income statement for the Appliance Experts Company during its first month of operation.

Headings on an Income Statement

Most income statements furnish considerable detail, especially when they are prepared for management's use. The amount of detail varies for different types of businesses and for different users.

TABLE 2-6
Appliance Experts Company
INCOME STATEMENT
Year Ended December 31, 19___

Sales		$440
Beginning Inventory	—	
Purchases	$2,000	
	$2,000	
Less: Ending Inventory	1,715	
Cost of Goods Sold		285
Gross Profit		$155
Expenses		
Depreciation:		
Building	50	
Furniture and Fixtures	15	
Interest	40	
Total Expenses		105
Net Income		$ 50

The following headings are found on most income statements, and in this order:

1. *Sales or Revenue.* This may be subdivided to list different types.

2. *Cost of Goods Sold.* The cost of merchandise is shown for merchandising; the cost of the manufacturing process is shown for manufacturing entities.

3. *Gross Profit* (or *Gross Margin* or *Net Sales*). This is the difference between sales and the cost of goods sold. It is labeled "Total Revenue" when it is a summation of several revenue types, and the cost of goods sold does not apply.

4. *Operating Expenses.* Often, operating expenses are separated into such groups as Selling Expense, Operating Expense, and Administrative Expense, with a further breakdown within these groups.

5. *Operating Profit or Revenue.* This is the amount remaining after operating expenses are subtracted from gross profit.

6. *Other Revenues.* This includes revenue unrelated to the principal purpose of the business.

7. *Other Expenses.* These are expenses not related to the principal purpose of the business.

8. *Profit before Taxes* (or *Revenue before Taxes*). This is the total of items 5 and 6, less item 7.

9. *Provision for Taxes.*

10. *Profit or Income before Interest Deductions.*

11. *Interest Deductions.* These are stated separately when long-term debt exists.

12. *Net Income or Loss.*

Taxes

The format for the income statement provides for taxes, even though no tax liability was shown in Table 2-6 for Appliance Experts. Various levies could be effective, including real estate or income tax, and one or more business taxes such as gross receipts, mercantile, or use tax. Taxes were ignored in this illustration for the sake of simplicity. Nevertheless, they are another expense which would further reduce net income. In a small business, federal income tax is not ordinarily levied on the business itself but on the owners as individuals. However, the income of corporations is subject to federal income tax. Corporate income taxes are a very complex subject, which is treated in Chapters 8 and 9.

STATEMENT OF RETAINED INCOME

Quite often, a statement of owners' equity is included with the income statement. For a corporation, there is a statement of retained income or reinvested earnings. Table 2-7 is a statement of owner's equity for the Appliance Experts Company:

TABLE 2-7
Appliance Experts Company
STATEMENT OF OWNER'S EQUITY
Year Ended December 31, 19__

A. Fixer, Capital Dec. 1, 19__		$20,000
Add: Net income for year	$50	
Less: Withdrawals	—	
Increase in Capital		50
A. Fixer, Capital Dec. 31,19__		$20,050

Withdrawals, a new heading, is used to record any amounts the owner has taken from the business. The owner of an unincorporated business invests time and money in the expectation of earning a profit and does not ordinarily receive a salary or interest on the effort and money invested. A drawing account is used to record the owner's withdrawal of cash, merchandise, or other assets. As the statement of owner's equity shows, the withdrawals are not an expense of the business, but a reduction of the owner's equity.

Withdrawals in a small business can be compared with a payment of dividends to the owners of an incorporated business. All the net income is gain to the owner of a small business whether or not it is withdrawn. In a corporation, if net income is left in the business, it is regarded as retained or reinvested earnings.

SUMMARY

Accounting is able to furnish financial information about an economic entity—usually a business—by recording, classifying, and summarizing its financial events. Two major accounting reports supplying this information are the *balance sheet* and the *income statement*. The balance sheet applies the dual-aspect principle to show the things a business owns (assets) at a particular time balanced against what a business owes (equities) at the same time. The income statement gives a measure of income (or loss) resulting from the operation of the business over a specified period of time between two balance sheets. For a given period, it matches the expenses accrued against the revenues realized to show the resultant gain or loss.

These two reports, and others as well, are prepared from the accounting records. All transactions are recorded in a journal, which is like a business diary of financial events. The transactions are posted from the journal to the proper asset and equity accounts—the so-called ledger accounts. When transactions are recorded, however, a double-entry system is followed in which one or more accounts are debited and one or more other accounts are credited an equal amount. In this way, accounting performs its role of communicating financial information.

Engineering economy, of course, needs the information which accounting communicates. However, the economy study often requires knowledge of the actual cash expenditures and the receipts resulting from a specific decision. The conventions of accounting complicate the task of interpreting accounting information for economy studies. Examples are the accrual concept applied to expenses, and the realization principle applied to revenue. At the very least, the engineering economist must be aware of these concepts and how they may influence information which accounting supplies.

3

ACCOUNTING AND THE TELEPHONE COMPANY

INTRODUCTION

Bell System accounting procedures follow the general concepts introduced in Chapter 2. However, they also conform to the particular requirements mandated for telephone companies under the jurisdiction of the Federal Communications Commission (FCC). The FCC specifies what is known as a *Uniform System of Accounts* for telephone companies, in Part 31 of its Rules and Regulations. Not only does Part 31 prescribe the names and numbers of accounts which telephone companies must use, but it also outlines the content of each account and gives instructions for keeping the records.

Overall, there are about 200 designated accounts which must be used on a company's balance sheet, its income statement, and various other reports. These accounts can be grouped into eight classes:

Asset accounts	100.1–139
Liability accounts	150–181
Plant accounts	201–277
Income accounts	300–343
Retained Earnings accounts	400–416
Operating Revenue accounts	500–530
Operating Expense accounts	602–677
Clearing accounts	702–732

The Asset accounts are actually composites of other accounts; they show the gross plant in service. The Liability accounts and Income accounts are also composites of other accounts. Clearing accounts are just convenient places to accumulate certain costs of business operation until they can be properly distributed to particular accounts (usually, at the end of the month).

Under the Uniform System of Accounts, telephone companies are also permitted to set up any subdivisions of prescribed accounts which they may need for making specific allocations or for internal control and analysis. However, these subdivisions must preserve the integrity of the designated accounts, and the FCC must be informed of their nature and purpose. Accordingly, the Bell System has standardized some nine hundred subaccounts to be used in accounting work.

All these specified accounts are maintained in the Bell System accounting records, and they appear in the report which AT&T submits annually to the FCC. This report is prepared in a book of forms the FCC has designated as the *Annual Report Form M*. It has over 100 printed pages, with entry space for each of the prescribed accounts. Besides the financial and accounting data, the Annual Report Form M also includes information about officers, the board of directors, and certain stockholders. It includes a Plant Summary and a Message Summary, as well as benefits and employment data.

The telephone companies also publish and distribute their own annual reports, containing the financial and accounting data which AT&T submitted to the FCC. These published company reports are somewhat streamlined and simplified, compared with the itemized account data on the Report Form M. Nevertheless, they follow the same general arrangements provided in Form M. On their balance sheets, for example, Telephone Plant In Service is listed first under Assets, and Capital Stock is listed first under Liabilities. This sequence is typical for utilities. Nonutilities customarily list Current Assets and Current Liabilities first.

The annual reports to stockholders and to regulatory bodies are really summaries of a sizable group of reports which the companies prepare for internal management and control. These reports are known as Operating, or Monthly, Reports—a series of numbered reports summarizing the record of company transactions and other pertinent data for an accounting period (usually, one month). Each company uses standardized report forms; in fact, some multistate companies use a different report form for each state area. The Long Lines Department also prepares its own reports, and it too uses its own standard forms.

These company reports are of course quite helpful to managers. The Monthly Report No. 1, called the *Summary of Reports*, is particularly useful and is widely distributed to many managers. Traditionally, it is colored blue and consists of two sheets. Sheet 1 is a summary income statement, and Sheet 2 is a summary balance sheet. The *No. 1 Report*, as it is often called, not only lists financial data for the current month and year to date but also compares current data with the data from certain previous periods. Furthermore, it shows selected averages and percentages which may be useful for analyzing certain results.

The details for the information on the No. 1 Report actually appear on one of the other numbered series of reports (after No. 1). Although these detailed reports are not distributed regularly to all managers, they are available to any manager who needs such detailed information.

This chapter describes the major accounting entries, most of them appearing on the No. 1 Report. It also explains how managers can make best use of this detailed accounting information. The hope is that this chapter will illustrate how the principles explained in Chapter 2, and applied there to a small business, apply equally to a big business, although with some modifications.

BALANCE SHEET ACCOUNTS

Sheet 2 is the balance sheet portion of the Monthly Report No. 1, summarizing the several asset and liability accounts. Table 3-1 is an example of Sheet 2, showing a skeleton listing of assets and liabilities. It is important to keep in mind that details about these accounts appear in other numbered reports. For example, Monthly Report No. 2 is an expanded balance sheet, and other reports contain details about particular assets and liabilities.

Assets and liabilities will be discussed separately—first, assets.

TABLE 3-1

FORM S.N. 152(2) (1-76)	SUMMARY OF REPORTS			MONTHLY REPORT NO. 1
	SAMPLE TELEPHONE COMPANY		MARCH 1976	Sheet 2

	BALANCE SHEET	a At End of Month	b Increase over Last Month	c Amount (Increase over Dec. 31 Last)	d %	
1	Telephone Plant	4,621,821	23,471	54,083	1.2	1
2	Less – Depreciation and Amortization Reserves	919,245	6,075	14,469	1.6	2
3	Total Telephone Plant less Reserves	3,702,576	17,396	39,615	1.1	3
4	Miscellaneous Physical Property	7,676	116	201	2.7	4
5	Investments in Subsidiaries	–	–	–	–	5
6	Other Investments and Funds	1,482	–	–	–	6
7	Cash and Temporary Cash Investments	11,083	1,519*	404*	3.5*	7
8	Accounts Receivable	205,236	17,080*	7,577*	3.6*	8
9	Material and Supplies	16,519	152	56	.3	9
10	Prepaid Accounts and Deferred Charges	21,310	669*	4,414	26.1	10
11	Total Assets	3,965,883	1,603*	36,305	.9	11
12	Capital Stock – par value ($ per share)	1,386,767	–	–	–	12
13	Premium on Capital Stock	20	–	–	–	13
14	Other Capital	–	–	–	–	14
15	Retained Earnings Reserved	–	–	–	–	15
16	Unappropriated Retained Earnings	460,147	3,351	8,167	1.8	16
17	Less – Capital Stock Expense	–	–	–	–	17
18	Total Capital Stock & Retained Earnings	1,846,934	3,351	8,167	.4	18
19	Funded Debt	1,130,000	–	–	–	19
20	Advances From Affiliated Companies	–	–	20,000*		20
21	Other Long-Term Debt	–	–	–	–	21
22	Notes	272,375	1,935	9,965*	3.5*	22
23	Current and Accrued Liabilities	326,350	17,390*	28,882	9.7	23
24	Deferred Credits	390,225	10,501	29,222	8.1	24
25	Total Liabilities and Capital	3,965,883	1,603*	36,305	.9	25
26	Total Capital Obligations, Prem., and Retained Earnings	3,249,309	5,286	21,798*	.7*	26
27	% Debt Obligations to Total	43.16% %	x x x	x x x	x x	27

	COMPANY TELEPHONES	e At End of Month	f Increase over Last Month	g Increase over Dec. 31 Last	Percent of Total: h End of This Month	i Dec. 31 Last	
28	Total Company Telephones	7,535,424	34,176	89,063	x x	x x	28
29	Main	3,689,612	6,691	23,007	49.0	49.3	29
30	Residence Extension	2,424,024	22,266	53,129	32.2	31.8	30
31	Other	1,421,788	5,219	12,927	18.8	18.9	31

	GAIN IN CO. TELEPHONES (Excl. Purchases, Sales & Adj.)	This Month	k Increase over Last Month	l Increase over Same Month Last Year	m This Year to Date	n Increase over Same Period Last Year	
32	Inward Tel. Movement	160,009	37,210	34,768	416,313	55,148	32
33	Outward Tel. Movement	125,676	30,089	18,001	327,599	18,383	33
34	Net Gain	34,381	7,996	16,896	88,947	37,083	34
35	Employees – End of Month	39,222	234*	2,005*	x x x	x x x	35
36	Salary and Wage Payments	52,893	3,308	5,961	153,902	12,214	36

	CALLING RATE AND TOLL MESSAGES	o This Month	p Same Month Last Year	q This Year to Date	r Same Period Last Year	
37	Av. Daily Calling Rate per Tel. — Total	5.15	5.25	5.22	5.25	37
38	Excl. Residence Extension	7.58	7.54	7.67	7.54	38
39	Toll Messages Originating	45,812,708	40,327,724	131,245,952	119,445,816	39

	INDEPENDENT CO. TOLL SETTLEMENTS	s This Month	t Same Month Last Year	u This Year To Date	v Same Period Last Year	
40	Intrastate	10,384	7,130	29,185	23,385	40
41	Interstate	5,915	4,576	16,943	14,285	41

	CONSTRUCTION AND NEW MONEY	w This Month	x This Year to Date	y Amount (Increase over Same Period Last Year)	z Percent	
42	Construction Expenditures	36,832	98,064	2,506*	2.5*	42
43	New Money Requirements	1,935	29,965*	82,600*	x x x	43

( ) Denotes negative amount.

Date Issued April 15, 1976

Asset Accounts

Telephone Plant

The first entry on Table 3-1 is Telephone Plant. The amount of money listed there is the original cost of long-lived (generally more than one year) physical property that is now being used, or will be used in the future, to furnish communication services. Four groups are included in this entry. (The accounts themselves are summarized in Monthly Report No. 2, shown in Table 3A-3, Appendix 3A, at the end of this chapter.) By far the largest amount is for Account 100.1, or the cost of Telephone Plant in Service. This is the property actually dedicated to providing service. A second group is for Account 100.2, or the cost of Telephone Plant Under Construction. This group represents the money spent on plant that is not yet in service at the date of the balance sheet. The third group is for Account 100.3, or the cost of Property Held for Future Telephone Use, according to a plan that anticipates this use. The last group is for Account 100.4, or the cost of the Telephone Plant Acquisition Adjustment. This group is used when dedicated telephone plant is purchased outside the System at a price different from that of the net plant. It represents any amount by which the purchase price differs from the original cost, less the appropriate depreciation.

The sum of these four groups, then, makes up the entry for Telephone Plant listed on the No. 1 Report and therefore represents all applicable costs required to provide the plant. Included are costs for material, labor for installation and construction, the use of motor vehicles, engineering, plant administration, and supply expense. Also included are capitalized amounts of interest, taxes, relief and pensions, and other general expenses. The ratio of the total installed cost for some type of plant (including the appropriate increments) to the material cost will range from 120% to over 200%.

DETAILED LISTINGS FOR TELEPHONE PLANT   The detailed amounts for Telephone Plant in Service are recorded in separate accounts and subaccounts. For the designated accounts, these amounts are on the Monthly Report No. 2, which is a detailed balance sheet. The Quarterly Report No. 2A lists changes in telephone plant. Other details about plant are on the Quarterly Report No. 2B, which shows telephone plant purchased and sold; and the Quarterly Report No. 2C shows details about telephone plant under construction and property held for future telephone use. Examples of some of these reports (the No. 2 Report and the No. 2A Report) are in Appendix 3A, at the end of this chapter.

THE CONTINUING PROPERTY RECORD   Even though the continuing property record is not actually a part of the balance sheet accounts, it is germane to a discussion of telephone plant because it shows the detailed quantities and locations of the plant which the dollars listed on the balance sheet have bought. This collection of information varies in depth depending upon the kind of plant, and it is a requirement of the Uniform System of Accounts. According to this System,

> the continuing property record shall be arranged in conformity with the plant accounts prescribed in this System of accounts. . . . The record shall contain such detailed description and classification of property record units as will permit their ready identification and verification. They shall be maintained in such manner as will meet the following basic objectives:
>
> 1. An inventory of property record units which may be readily spot-checked for proof of physical existence.
>
> 2. The association of cost with such property record units to assure accurate accounting for retirements.
>
> 3. The determination of dates of installation and removal of plant retired to provide data for use in connection with depreciation studies.

However, even if the FCC did not require the continuing property record, accounting and engineering people would have used it, or something like it, in their work. For one reason, this record is an inventory which can be used in the appraisals of plant required for rate cases, sales of plant, and property tax assessments, or other studies. It also provides the details needed for developing the average unit costs and *broad-gauge* costs used in estimating the cost of construction projects. These average unit costs are also helpful in separation studies required for rate cases, revenue settlements with other companies, and various rental arrangements. (*Separation studies* refer to the apportionment of revenues due different telephone companies for calls across state lines and for certain calls within a single state.) In fact, without this kind of record, the analyst could not perform many of the costing and pricing functions required in economy studies.

In general, the continuing property record for land, buildings, and central office equipment is primarily a record of actual dollars spent for each parcel of land, for each building, and for each central office. However, the engineering records of these transactions are detailed enough so that the cost of a property can be analyzed by segments and subunits of plant. This kind of analysis may be necessary for retirements, rentals, sales, property taxes, separation studies, depreciation studies, and various types of cost studies. These engineering records may be in various forms, such as completion reports, priced-out "B" specifications, floor layouts, or miscellaneous drawings and diagrams. The Plant Operations, Engineering, and Accounting Departments also maintain inventories of supplies and equipment, as well as location records for cable, wire, and poles.

For station equipment and outside plant, a summary continuing property record is generally maintained for each state or subdivision of a state. These and the other continuing property records are kept in the Accounting Department, but they are based on sources originating in other departments. For example, quantities of outside plant items placed in service are reported to Accounting on material reports which are prepared by the construction forces. Quantities of station apparatus, central office equipment, and other materials are shown on billing statements and summaries from Western Electric. Engineering and other time reports also provide the basis for construction charges for central office equipment, station equipment, and other outside plant. Plant operations and engineering forces report removals of plant from service as well, so that the appropriate amount of money will be subtracted from the affected accounts—also, so that the quantities shown on the continuing property record can be reduced to reflect retirements and to show those items remaining in service.

Of course, checks and audits are made of the many transactions occurring daily, not only to protect the integrity of the accounts, but also to protect the integrity of the continuing property records.

Depreciation Reserve

Line 2 of Sheet 2 of the No. 1 Report is for *Depreciation Reserve*. This is defined as the accumulation of a periodic allocation to business expense of the original cost for long-lived fixed assets other than land.

Four kinds of dollar amounts are included in the entry Depreciation Reserve: (1) the amount allocated each month for depreciation; (2) any gross salvage realized upon retirement of plant; (3) the original cost of plant retired; and (4) the cost to remove and dispose of retired plant. Any entries into the account for items (1) and (2) increase the account balance, whereas entries for items (3) and (4) decrease it.

Even though Depreciation Reserve is listed among the asset accounts, it is actually Account 171. This account number classifies it as a liability account, under the heading Deferred Credits and Reserves. This classification of Depreciation Reserve as a liability is an historic one, and a few companies (in other industries) list Depreciation Reserve on the equities side of the balance sheet. Before 1952, the Bell System, too, carried Depreciation Reserve as a liability on the No. 1 Report. However, this practice is obsolete. The current accounting practice is to consider the Depreciation Reserve as a *contra-asset* account—meaning "the opposite of an asset"—and it is shown as a deduction from the original cost of fixed assets.

There are several different methods of computing depreciation. However, Part 31 of the FCC Rules and Regulations requires that depreciation allocations be computed on what is called a *straight-line basis*—meaning equal annual amounts—over the estimated service life of each class of depreciable telephone plant.

The term *Reserve* itself is somewhat misleading, because it might connote the idea of a sum of money actually set aside or reserved, and this is not the case. To avoid any misunderstanding, the term *Accumulated Depreciation* is preferred, and it is now used in Bell System published financial statements such as the Annual Report. Of course, the monthly depreciation accrual does result in an equivalent amount of cash on hand, which is actually the partial recovery of capital. If the company had no need for additional long-term capital, it could logically return the recovered capital to investors and retire a portion of the outstanding securities. This does not happen, though, since additional fixed assets are invariably required. The accumulated depreciation is therefore regularly applied to the acquisition of new assets, so that the capital previously provided by owners or lenders is merely transferred from an old asset to a new one. This transfer of capital also minimizes the amount of new capital the company would otherwise require.

Chapter 4 discusses further how money accumulated for depreciation pays for additional assets. Even at this point, however, readers can see that the balance in the Depreciation Reserve account represents the amount of the original plant investment that has been recovered by regular allocations to operating expense.

Because it was originally considered a liability account, the Depreciation Reserve account will have a credit (right-side) balance. Monthly allocations for depreciation expenses are credited to this account, and so are amounts for salvage. Both credits of course increase the balance.

Salvage may be in the form of cash actually received for junked material, or it may represent the original cost for material which was judged to be reusable. The cost of the material itself might then be placed in the Materials and Supplies account, pending reuse; or, if it is held for sale to others, this cost might be placed in the Miscellaneous Physical Property account.

The retirement of plant and the related expenditure to physically remove it are also reflected in the Depreciation Reserve account. When depreciable telephone plant is retired from service, the original cost (which is determined from the continuing property record) is deducted from both the plant account (Buildings, Central Office Equipment, Pole Lines, etc.) and the Depreciation Reserve account. There is a credit to the plant account and a concurrent debit to the Reserve. The cost of removal is also debited to the Reserve account. This total may include amounts for engineering, labor, transportation, and other costs which occur in dismantling, removing, and disposing of retired plant.

Telephone Plant Less Reserves

Line 3 of Sheet 2 of the Monthly Report No. 1 is Total Telephone Plant less Reserves. This entry is the amount remaining after the Depreciation Reserve balance is subtracted from the sum of the balances in the plant accounts (the Total Telephone Plant amount). This difference is commonly called *net plant*. The dollars for net plant are the amount of the original plant investment which is not yet recovered by depreciation accruals.

Net plant, however, is not a very good approximation of the value of the business at any one time. This is because depreciation is not intended to be a measure of value. Another reason why net plant does not represent value is that because of inflation, the amounts for Total Telephone Plant and for Depreciation Reserve represent dollars of widely different values or purchasing power. As a result, the recovered investment dollars may not be enough to furnish new assets that are equivalent to the old. At the time of this writing, there is no generally accepted way of really measuring value. The various governmental agencies, the investment community, and industry in general are all involved in a continuing search for a solution.

The Other Investment Accounts

Entries for the other investment accounts are actually composites of several asset accounts: (1) Miscellaneous Physical Property; (2) Investments in Subsidiaries; and (3) Other Investments and Funds. (These entries are on lines 4, 5, and 6 of Sheet 2 of the No. 1 Report—Table 3A-2 in Appendix 3A at the end of this chapter.)

The account Investments in Subsidiaries is used to record any investments in affiliated companies when the reporting company holds a controlling interest. Any investments in a company representing a minority or noncontrolling interest are recorded in the Other Investments and Funds account, along with any advances to such companies. The Miscellaneous Physical Property account is used to record an investment in physical property such as land, buildings, and equipment owned by the company but not used to furnish communication services. Also included is the investment in telephone plant which has been retired from service but is being held for sale.

Current Assets

The Current Assets accounts (lines 7, 8, and 9 of Sheet 2 of the No. 1 Report—Table 3A-2 in Appendix 3A at the end of this chapter) include cash and certain assets that will shortly be converted into cash. Also included is the amount represented by Material and Supplies on hand to be used in day-to-day operations.

In the operating telephone companies, accounts receivable are the largest part of current assets. Receivables are primarily customer billings, but they also include amounts of interest or other payments due the company currently or within one year. As these due amounts are collected or received, the balance in accounts receivable is reduced and the cash balance increased.

Bell System companies keep necessary quantities of cash on hand and deposited in demand deposits (checking accounts) to meet payrolls, bills, and other obligations promptly. Any cash over the foreseen needs is invested in short-term securities.

The Material and Supplies account is used to record the investment in telephone plant material and in supplies that are on hand for a construction program or for maintenance requirements. Station apparatus on hand for installation does not appear in this account, however. To reduce record keeping, station apparatus is carried in the appropriate plant account from the time it is purchased until it is disposed of and retired from service. (This practice is referred to as *cradle to grave* accounting.)

Prepaid Accounts and Deferred Charges

The entry on line 10 of Sheet 2 of the No. 1 Report is a composite of several accounts used to record amounts paid in advance for rents, taxes, insurance, directory expenses, and other items. For example, if space is leased for a 12-month period and $6,000 is paid in advance, the full amount is charged (debited) to the Prepaid Rent account. Then, to properly allocate the lease expense over its duration, $500 is credited to that account each month and debited to rent expense.

The account Other Deferred Charges is widely used for recording preliminary engineering costs incurred in order to determine whether prospective construction projects are feasible. Such charges are later transferred to a plant account if construction is authorized. If it is not, they are transferred to the account Miscellaneous Income Charges. Still another use of the Other Deferred Charges account is to temporarily record expenditures whose accounting classifications have not yet been determined.

The account Other Deferred Charges also includes the account Discount on Long-Term Debt. When a bond issue is sold, the company usually—although not always—receives less than the face amount. This difference (discount), along with the inevitable expenses of every bond issue, is listed under Discount on Long-Term Debt. The amount for each bond issue is then amortized over its life by debits to expense.

Equity and Liability Accounts

Capital Stock

The balance in the Capital Stock account represents the book amount for stock certificates issued by the company. The *book amount* means the par value of any stock which has a par value; or, for stock with no par value, it means the amount properly authorized to be included in the account. Subaccounts are used to record information about different classes of stock according to par value. This information might include whether the stock is common or preferred, what the voting rights are, and what the conditions are for retiring the bonds. A separate account is used to record the amounts received from the sale of stock if these amounts exceed the par or stated value. There are additional accounts to record the amounts of stock for which legally enforceable subscriptions have been received, and to record installment payments under a stock purchase plan. As these subscriptions are paid up and certificates issued, the appropriate dollar amounts are transferred to the proper capital stock subaccounts.

Retained Earnings

The Retained Earnings sum which appears on the balance sheet is a record of accumulated earnings from the business to date that have been retained and reinvested. There are two classifications: Retained Earnings Reserved and Unappropriated Retained Earnings. In the Bell System, most retained earnings are unappropriated—meaning, they are not reserved but are available for investment in the business. Only a minor part of retained earnings is presently reserved; and if there is a reservation, it usually represents a revenue amount that has already been collected but is subject to refund resulting from rate case proceedings or litigation.

Long-Term Debt

Debt (borrowed money) that matures (must be repaid) more than one year from its date of issue is recorded in the appropriate account under Long-Term Debt. Funded Debt includes the face amount of outstanding bonds and debentures. (These and other forms of debt are discussed in Chapter 4.)

Operating companies list advances from AT&T as Advances from Affiliated Companies if permanent financing is expected to take care of the repayment. The Other Long-Term Debt account is used to record other borrowings, like notes, which mature more than one year from the date of issue.

There is a balance sheet account for listing the amount of premiums (excess over face value) received from the sale of a bond issue. The premium is amortized over the life of the bond issue by monthly credits to expense. (An expense credit is a reduction.) The account Premium on Long-Term Debt appears under Deferred Credits, which will be discussed shortly. (The method for recording Discount on Long-Term Debt was previously noted under Prepaid Accounts and Deferred Charges.)

Current Liabilities

The entry Current Liabilities covers accounts for recording payments due to others which have not actually been paid out at the date of the balance sheet. These liabilities consist of short-term notes (maturity less than one year), accounts payable, customers' deposits, accrued interest, dividends and rent, advance billings and payments, and other miscellaneous amounts. It is the present accounting practice to also include the account Accrued Liabilities Not Due under Current Liabilities.

Accounts Payable is primarily for materials, supplies, and service from Western Electric and others. Payrolls due employees are also included, as well as interest and rents due on a monthly basis. Taxes withheld from the employees' payroll are carried under Accounts Payable, too.

Cash deposits received from customers as security for payment for service are carried as a current liability. Many telephone services are billed in advance, but these advance billed amounts will be applied to the revenue of a future period. Advance billings are carried as a current liability until the period when the service is actually furnished and the revenue accounted for.

Accrued Liabilities Not Due

Accounts under Accrued Liabilities Not Due cover taxes, interest, dividends, and rents. These are amounts which have been recognized as expenses of the current or a recent accounting period, even though payments are not yet due. The accounts are credited as the expenses are accrued. As payments are made, however, they are debited to reduce the balance. This is an example of the accrual concept, under which revenues and expenses are matched for an accounting period so that income can be measured.

Deferred Credits and Reserves

Liabilities of a known amount which are payable at some future time are recorded under Deferred Credits and Reserves. So are certain reserves for possible contingencies. The reserve for depreciation is actually a part of this group, even though it is carried as a contra-asset account under Assets. Most of the money under deferred credits consists of unamortized income tax credits and accumulated deferred income taxes; these funds have been invested in telephone plant.

An income tax credit resulting from an investment is amortized over the life of the plant that created it. The amount remaining to be amortized is carried in the Other Deferred Credits account. However, there is a separate account for accruing deferred income taxes resulting from the use of depreciation methods on the tax return that are different from the methods used in the financial reports. (More rapid depreciation accruals, called *accelerated tax depreciation*, are allowed on the tax return, whereas the equal annual depreciation amounts—that is, straight-line depreciation—are specified for the financial reports.) The deferred tax amounts are carried in that account until the future periods when the tax deductions from accelerated depreciation decline. Small amounts in Deferred Credits represent liabilities whose final allocation has not yet been determined. The account Other Deferred Credits is also used to accumulate depreciation on miscellaneous physical property.

Monies accumulated for employee pensions and death benefits are recorded in the so-called Provident Reserve account. As these accumulations are transferred to the trust funds, the account is debited. Subaccounts are similarly used to carry amounts involved in the employees' savings plans. The Employment Stabilization Reserve account may be used to set up a reserve for termination allowances when employees are laid off because of lack of work.

INCOME STATEMENT ACCOUNTS

Table 3-2 is an example of the Earnings and Expenses portion (Sheet 1) of the Monthly Report No. 1. It is really a summary of many accounts, details of which appear on other numbered reports. For example, further detail for Operating Revenues is on the No. 4 Report, and for Operating Expenses on the No. 5 Report. The total from each of those reports becomes a line at the top of Monthly Report No. 3, the Income Statement. The Report No. 3 also shows some detail about taxes, other income, interest charges, and dividends. Examples of these reports are in Appendix 3A at the end of the chapter. The items found on Sheet 1 of the No. 1 Report are covered in the next sections.

Operating Revenues

The accounts under Operating Revenues are designed to show the amount of money the companies are entitled to receive for communication services during the period covered by the report. Revenues are classified as follows:

1. *Local Service:* Local service revenue includes local exchange charges from all classes of customer service and local private-line services.

2. *Toll Service:* Toll service revenue is the sum of message tolls, Wide-Area Toll Service (WATS), and toll private-line service.

3. *Miscellaneous*
   a. Commissions for billing or collecting tolls on telegraph, cable, or radio messages.
   b. Directory advertising and sales.
   c. Rents for operating telephone plant rented to others. (Rent from miscellaneous physical property is listed under Other Income; it is not operating revenue.)
   d. General services and licenses. (AT&T receives such revenue; the operating companies usually do not.)
   e. Other miscellaneous amounts.

4. Less: A provision for uncollectible revenue.

Toll revenues and also—to a small extent—local private-line revenues are separated into intra state (meaning "within *a* state") and interstate (meaning "between states") amounts. The reporting company lists as interstate its share of revenues from service that crosses a state line(s). For example, a message originating in New York and terminating in Denver uses the communication facilities of the New York Telephone Company, the AT&T Long Lines Department, and the Mountain Bell Telephone Company. Each company would be due a share of the revenue which is proportionate to the investment (and associated expenses) it has contributed. Similar principles hold when revenue is shared with independent telephone companies. Separation of revenue among Bell System companies is accomplished through the Division of Revenue procedures.

A certain portion of billed revenues will not be collectible, however. To avoid overstating revenue and, therefore, net income, the gross revenues for a given period are

TABLE 3-2

FORM S.N. 152(1) (1-76)　　　　　　　　　　　MONTHLY REPORT No.1
　　　　　　　　　　　　　　　　　　　　　　　　　Sheet 1

SUMMARY OF REPORTS
SAMPLE TELEPHONE COMPANY　　　　　MARCH 1976

EARNINGS AND EXPENSES	a This Month	b Increase over Last Month	c Increase over Same Month Last Year	d This Year to Date	Increase over Same Period Last Year	
					e Amount	f Percent
1 Local Service Revenues	93,948	3,143	8,894	273,987	18,464	7.2
2 Gross Toll Service Revenues	67,521	6,355	11,038	190,857	23,527	14.1
3 Less — Ind. Co. Toll Settlements	16,299	1,755	4,592	46,128	8,457	22.4
4 Net Toll Service Revenues	51,222	4,600	6,446	144,729	15,069	11.6
5 Misc. Operating Revenues	4,283	90*	463	13,153	912	7.5
6 Less — Uncollect. Oper. Revenues	899	29*	6	2,795	266	10.5
7 Total Operating Revenues	148,533	7,682	15,797	429,073	34,179	8.7
8 Maintenance	29,514	6,743	6,162	77,992	9,183	13.3
9 Depreciation & Amortization	19,718	123	1,149	58,806	3,360	6.1
10 Traffic Expenses	8,959	7*	6*	27,291	138	.5
11 Commercial Dept.	4,874	499	486	13,856	591	4.5
12 Marketing — Sales Expenses	3,310	191	658	9,684	1,748	22.0
13 Marketing — Other Expenses	2,836	421	673	7,448	1,329	21.7
14 Operating Rents	617	120*	95*	2,106	66*	3.0*
15 Executive and Legal Depts.	403	39*	18	1,279	152	13.5
16 Accounting Dept.	3,635	113	214	10,602	244	2.4
17 Prov. for Serv. Pens. & Death Bens.	9,173	30*	1,111	27,556	3,275	13.5
18 Other Employees' Benefits	4,938	40	1,244	15,030	3,689	32.5
19 General Services & Licenses	1,987	961*	332	7,017	1,530	27.9
20 Other General Expenses	6,910	1,171	1,623	18,279	2,480	15.7
21 Less — Exp. Charged Const.	3,350	388	829	8,908	1,284	16.8
22 Total Operating Expenses	93,522	7,756	12,741	268,037	26,369	10.9
23 Net Operating Revenues	55,031	74*	3,056	161,036	7,810	5.1
24 Federal Income Taxes-Operating	12,441	242*	715	35,726	1,462	4.3
25 State, Local and Other Oper. Taxes	18,996	291	964	56,114	2,925	5.5
26 Total Operating Taxes	31,436	49	1,678	91,840	4,387	5.0
27 Operating Income	23,595	123*	1,378	69,196	3,423	5.2
28 Other Income	482	91*	193*	1,598	558*	25.9*
29 Misc. Deductions from Income	24	11	122*	334	137*	29.1*
30 Income before Interest Deductions	24,054	224*	1,307	70,460	3,001	4.4
31 Interest Deductions	7,422	56	219	22,418	255	1.2
32 Income after Interest Deductions	16,632	280*	1,087	48,042	2,746	6.1
33 Extraordinary & Delayed Items-Net.	9	23	9	6*	16	74.1
34 Net Income	16,641	257*	1,097	48,036	2,761	6.1
35 Dividends (accrual basis)	13,290	—	578	39,870	1,733	4.5
36 Balance	3,351	257*	519	8,167	1,028	14.4

AVERAGES	g This Month	h Same Month Last Year	i This Year to Date	j Same Period Last Year	k% Increase i over j
37 Av. Tel. Plant (Excl. Plant under Const.)	4,543,990	4,264,837	4,518,541	4,242,357	6.5
38 Av. Cap. Oblig., Prem., & Retained Earnings	3,238,349	3,116,351	3,251,361	3,123,192	4.1
39 Av. No. of Company Telephones	7,518,336	7,209,928	7,489,087	7,192,147	4.1
40 Av. Tel. Plant per Av. Co. Telephone	604.39	591.52	603.35	589.86	xxx

PERCENTAGES (Annual Basis)	l This Month	m Last Month	n Same Month Last Year	o This Year to Date	p Same Period Last Year
41 Total Oper. Revenues to Average Tel. Plant	39.23	37.43	37.26	37.98	37.14
42 Total Oper. Expenses to Total Oper. Revenues	62.96	60.88	60.85	62.47	61.20
43 Maintenance to Average Tel. Plant	7.79	6.05	6.57	6.90	6.49
44 Depreciation (Accts. 608, 702, 704) to Av. Depr. Tel. Plant	5.39	5.39	5.41	5.39	5.41
45 Income* to Av. Cap. Oblig., Prem., & Retained Earnings	8.89	8.94	8.73	8.64	8.61
46 Pre-Fed. Tax Income* to Av. Cap. Oblig, Prem. & Ret. Earn.	13.48	13.62	13.21	12.99	12.97
47 Income to Av. Common Equity Capital	10.82	11.01	10.30	10.43	10.01
48 Post-Tax Interest Coverage	3.26	3.32	3.24	3.17	3.13

* Income before Interest Deductions less interest charges not related to Capital Obligations.

( ) Denotes negative amount.

reduced by an allowance for estimated uncollectibles. This estimate is based on a continuing analysis of the uncollectibles during recent months.

Revenue amounts on an income statement are for services which the company furnishes during the period covered by the statement, regardless of when the customer was, or will be, billed—or will pay. Local exchange service is generally billed in advance, whereas toll service is billed in arrears. The task of customer billing is so tremendous that the load is distributed by dividing the calendar month into ten or more billing periods. For example, one customer receives a bill dated March 1, which covers local service for the period March 1 to March 31 and toll service for the period February

1 to February 28. Another customer receives a bill dated March 3, and the charges will include advance billing for local services from March 3 to April 2 and for toll charges from February 3 to March 2. And so on. Since amounts due from federal government agencies are always billed in arrears, the monthly bill to such an agency, dated March 6, would cover both local and toll charges for the period February 6 to March 5.

A given subscriber's bill may represent revenues which span three different months. The customers generally pay their bills sometime after they receive them. Although their payments result in the actual receipt of cash by the firm, they have little to do with the amount of revenue that is assigned to an accounting period.

Despite the apparent difficulties, the revenues for a given month can be estimated and accrued in the accounts, and subsequent adjustments can be made as required to keep the revenue amounts in the right period of time. This matching of revenues with time is important with large amounts. The reason is that if the revenues for a service are not booked in the same month or year as the expenses for furnishing that service, the level of earnings might be distorted in each accounting period.

Operating Expenses

Because Operating Expenses on the income statement includes many types of expenses, many accounts are needed so that managers and others can keep track of what is going on in the company. In general, operating expenses are outlays covering the cost of operating and maintaining plant, providing for its depreciation, marketing the services offered, and administering the overall policies and activities of the business. The major classifications of expenses are described next.

Current Maintenance

Ordinary repairs and rearrangements and changes of plant in service are recorded separately for the various kinds of telephone plant. The costs of plant testing, commercial power, and shop repairs are also important kinds of current maintenance expense. Maintenance training expenses are recorded here, too.

Depreciation

In general, expenditures for telephone plant with an estimated life of more than one year are included in the plant accounts and are carried as an asset on the balance sheet. Such expenditures represent prepaid costs; and depreciation accounting is the procedure for assigning this kind of cost to each accounting period during the life of the plant. Amounts are therefore added to the depreciation reserve each month and simultaneously charged to expense to represent an allocation of the cost of plant previously constructed and placed in service. The rate of depreciation depends upon the estimated life of the particular plant and the estimated net salvage when the plant is removed from service. The life of the plant is affected by wear and tear, decay, the action of the elements, casualty, inadequacy, obsolescence, changes in the art, and public requirements. The accountants accrue depreciation expense monthly for each class of plant, using depreciation rates developed by depreciation engineers.

Traffic Expenses

The wages of traffic operators is the largest single entry under Traffic Expenses. Additional entries include amounts for traffic supervision, customer instruction, employment, training, lunch room, and house service. General Traffic Supervision includes subaccounts for the costs of network design, with a provision to transfer appropriate charges to construction accounts.

Commercial Department

Expenses for the payroll and other costs for general commercial administration, advertising, and public telephone commissions, as well as all directory costs, are recorded as Commercial Expenses. Expenses for local commercial operations, sales, and connecting-company relations are charged to the Commercial Department if they are performed there. In some organizations, certain of these functions may be in a separate Marketing Department.

Marketing Department

Historically, the functions now assigned as a marketing responsibility were part of the Commercial Department, and the Uniform System of Accounts continues to list only Commercial Expenses. With the use of subaccounts, however, the items can be separated and appropriately charged as either a commercial or a marketing function. Certain items are identified under Commercial Expenses as subject to alternate assignment to the Marketing Department. Additional items which might be treated this way are responsibilities for rate and tariff, customer relations, and forecast and development activities. In either department, Commercial or Marketing, some of the expenses may properly be transferred to construction.

Operating Rents

This account is used to record rents paid by the company for building space, conduit space, pole-line attachment space, and elements of telephone plant not provided for in other accounts. It is also used to record payments for the rental of telephone plant items. In contrast, the rental of a complete telephone plant, a complete exchange, or a complete toll system would not be charged to the Operating Rents account but to the income account called *Rent for Lease of Operating Property.*

Executive and Legal Departments

The account called *Executive and Legal Departments* includes the pay, office, traveling, and other expenses for the following Executive Department employees:

Chairman of the Board of Directors

President

Vice President

Secretary

Assistants authorized to act for officers

Managers in the General and Division Office, and staff forces

On the No. 1. Report, Law Department expenses are included with Executive Department expenses to make up one entry. However, there are separate accounts and subaccounts for the two departments.

Accounting Department

Several subaccounts are used to show the pay and other expenses for the subdivisions of accounting. The pay and expenses of the Vice President and the Comptroller are included. Sheet 2 of the Monthly Report No. 5 shows some of the subaccount listings.

Relief and Pensions

The Uniform System of Accounts specifies Relief and Pensions as the only expense account for employee benefits. Nevertheless, the current list of Bell System standard accounts has over 50 subaccounts which are used to classify and record the costs related to the many employee benefits, including the provision for pensions.

In general, the Relief and Pensions account is used to accumulate the following: all company contributions to trust funds intended solely for service pensions and death benefits; pay and expenses of the Benefit and Medical Departments; accident, sickness, death, and other disability benefits and supplementary pensions paid by the company; group insurance premiums (including life and medical expense policies); company contributions to the Bell System Savings Plan for Salaried Employees; and other miscellaneous but related expenses.

All of these costs are debited (added) to the account, which is also credited (reduced) by appropriate amounts transferred to construction accounts and billed to others (primarily custom work). The Monthly Report No. 5 (Sheet 1, line 43, in the sample) shows the debit balance left in the account after the transfer. To help make the analysis of employee expenses more meaningful, the No. 1 Report listing of pension and benefit costs includes the amount previously transferred to construction. The total is subdivided as described next.

The account Provision for Service Pensions and Death Benefits (in the No. 1 Report) includes only the company contributions to the trust funds. These contributions are based on actuarial studies made by the AT&T Company to determine the amounts that must be accrued to pay mandatory pensions and death benefits. Each associated company deposits amounts representing a certain percentage of the regular employees' payroll with the appointed funds trustees. The deposited funds are invested by the trustees in various securities and other investments.

The account Other Employees' Benefits, also in the No. 1 Report, covers all other expenses associated with employee benefits.

General Services and Licenses

The account General Services and Licenses is used by the operating companies, and also by the Long Lines Department, to record payments made to the AT&T Company for services rendered under agreements known as *license contracts*.

Under these contracts, the AT&T Company agrees: (1) to maintain arrangements for manufacturing telephones and related equipment under patents it owns or controls related equipment which the licensee can buy at reasonable prices; (2) to conduct research in telephony and to make the benefits of this research available to the licensee; and (3) to furnish advice and assistance on almost all phases of the licensee's business. The AT&T Company also agrees to maintain connections between the licensee's telephone system and the systems of the other associated companies of the Bell System, and to provide for the joint use of certain rights-of-way and facilities.

Such services can best be handled on a centralized basis. They include the extensive research and development work which Bell Telephone Laboratories, Inc., performs for the AT&T Company; they also include advice and assistance on matters related to engineering, commercial, traffic, accounting, legal, and financial services.

Supplementary and standard supply contracts are related to the license contracts, but are distinctly separate from General Services and Licenses payments. The *supplementary agreements* cover the sharing of revenues derived by the AT&T Company and the other companies of the Bell System from interstate and foreign services. (The sharing arrangement was described earlier, under Operating Revenues.) *Standard supply contracts* are other agreements which the AT&T Company, as well as each of the license contract companies, has with Western Electric. Under these agreements, Western Electric agrees to either manufacture or purchase materials, to sell such materials to the company, to maintain stocks at distributing points, to prepare equipment specifications, to perform installations of materials, and to repair or dispose of used materials which the company has returned. These agreements and contracts contribute significantly to the unified Bell System objective of furnishing widespread, essentially universal, and reasonably priced communication services and facilities.

Other General Expenses

The entry Other General Expenses on the No. 1 Report is a consolidation of numerous expense items from the No. 5 Report called *Operating Expenses*. A large amount represents engineering costs that cannot be attributed to a construction or other expense account. Engineering costs for studies of a general nature are included here, too. Other costs are for Treasury Department, personnel, security, insurance, accidents and damages not charged to other accounts, and certain publicity and information expenses. For example, the costs of publications for employees and for stockholders are lumped here. The Other General Expenses account also includes membership fees and dues in trade, technical, and professional associations that cannot be allocated to individual departments.

Expenses Charged Construction

The classification Expenses Charged Construction provides a way to transfer to the cost of construction part of the administrative costs initially accounted for as expenses. These transferred costs are primarily a portion of the pay and expenses (including relief and pensions) for the Executive, Accounting, Treasury, Law, and General Departments. Because pay and expenses for these departments are considered to be incurred in connection with the overall administration and operation of the business, some portion of these expenses is logically related to construction.

It would be impractical to attempt to identify and classify the many separate work operations and expense items as either expense or construction. This difficulty is resolved by assigning such costs to construction in the same proportion that all other wages and salaries are assigned to construction. For example, if 18% of the monthly payroll (excluding the payroll for the general administrative offices) is spent for construction, 18% of the general office salaries and expenses is also charged to construction.

Net Operating Revenues

The amount of revenue remaining after the allowance for operating expenses is *Net Operating Revenues*. This amount naturally plays a part (along with other factors) in calculating operating income taxes.

Operating Taxes

Two simple entries on the Monthly Report No. 1, for federal income tax and for other operating taxes, cover an array of accounts and detailed procedures. Some of these items are identified in the next two classifications. First, however, other real but unseen taxes must be mentioned.

Some governmental jurisdictions levy sales and use taxes which are paid on material purchases for telephone company use. Under the accounting system, such taxes become part of the cost of material and are accounted for in a construction or an expense account. For example, taxes on gasoline and vehicle license fees (a form of tax) are accounted for as motor vehicle expenses, which are then distributed to construction and expense accounts. Taxes incurred on property while under construction become part of the construction cost. Some portion of payroll-related taxes, such as Social Security, is also prorated to construction. Although the total of these various levies is a significant amount of tax expense, none of it appears under the classification Operating Taxes.

Federal Income Taxes—Operating

The entry for Federal Income Taxes—Operating represents the federal income tax expense for the current accounting period. The amount is the net balance of debits and credits in several accounts and subaccounts for the following records: the charges and amortization of credits related to investment tax credits (to be explained shortly); deferred federal income taxes from the use of accelerated tax depreciation; and the actual operating federal income tax payment.

The calculation of corporate federal income tax is complicated and, of course, is a specialized field. It is not a simple matter of subtracting expenses and taxes (other than federal income taxes) from revenues and then applying a tax rate to the difference. The federal tax code outlines the procedure, and it defines the revenue and the deductions. As a result, the income subject to federal income taxes is not identical with the income before taxes as it appears in the company's accounts.

The calculation of the federal income tax owed has been further complicated by the introduction of *investment tax credits* and the use of accelerated depreciation. According to the Internal Revenue Code, a tax credit can be taken for a percentage of specified investments in new plant facilities. This credit reduces the amount that would otherwise be paid out for income tax.

The associated companies have elected to amortize investment credits to income. These credits are therefore added to the Other Deferred Credits account (a liability), and the money representing them is available (less amortization) to invest in the business for more plant construction. Each credit is amortized during each accounting period over the average life of the plant that was the basis for it. The amounts for amortization are taken into income by crediting them to the subaccount Investment Credits—Net. A separate subaccount is then debited with the realized investment credits. In this way, the credits for each accounting period are offset by the amount of amortization.

Accounting procedures covering income tax deferrals resulting from the use of *accelerated depreciation* are similar to those outlined for investment tax credit. Naturally, different accounts are used, and there are some other minor variations. It is important to note that under the provisions of the tax law, the companies may use accelerated depreciation for income tax purposes. However, because this tax depreciation is different from the straight-line depreciation provided in the company accounts, there is a resulting tax differential. This difference is deferred and credited to the liability account Accumulated Deferred Income Taxes—Accelerated Tax Depreciation. Deferred amounts are not amortized, but are used when required for future tax payments. This occurs when the tax depreciation from accelerated depreciation for any year's eligible depreciable plant becomes less in the current year than it would be with straight-line depreciation. The Uniform System of Accounts specifies that detailed records must be maintained so that the deferred tax amounts can be shown by year of purchase and separately for each class of eligible depreciable plant. Until the deferred income tax amount is used for tax payments, it is invested in additional plant.

A separate account is provided for recording the calculated federal income tax payment accrued for the accounting period. The single-line entry on the Monthly Report No. 1 is the net liability for operating federal income tax for the reporting period. It is the sum of net charges to investment tax credit, actual operating federal income tax, and deferred income tax. Further detail appears on the Monthly Report No. 3. The entry for federal income taxes on the No. 1 Report is the amount of liability for the period, which is not necessarily the amount paid.

State, Local, and Other Operating Taxes

This catchall entry—State, Local, and Other Operating Taxes—contains all operating taxes other than federal income tax. These include property, gross receipts, franchise, capital stock, Social Security, and unemployment taxes and the tax effect of state and local income taxes. (State and local income taxes in some cases may also be subject to deferral and amortization treatment similar to federal income tax arrangements.) Federal income tax is a well-known burden on business expense. However, a fact that is often overlooked is that the total of other taxes can exceed the federal obligation, with the result that the sum for all taxes is substantial.

Other Income

Amounts under Other Income include (1) dividends received; (2) interest received from cash deposits, advances, and short-term investment; (3) interest charged to construction; and (4) the net of rentals and nontax expenses from miscellaneous physical property owned by the company. Dividend income is received from the ownership of the stock of affiliated and subsidiary companies, and it is a major item of income to the AT&T Company. Interest is received because all cash beyond the immediate needs is placed in interest-bearing deposits or other liquid securities. Subsidiary companies also pay interest on amounts advanced to them.

Interest charged to construction warrants further explanation. In the discussion of the entry Telephone Plant in the balance sheet accounts, capitalized interest was identified as one of the costs of plant construction. This interest charge recognizes that the return on capital money tied up during the construction period is a part of the cost of the completed plant. The amount of interest charged to construction is added to the plant construction accounts and is simultaneously credited as interest income. The interest amounts charged and added to construction costs will be recovered over the life of the plant through the depreciation accruals. However, income credited from interest during construction is excluded from taxable income for income tax calculations.

Miscellaneous Deductions from Income

The account Miscellaneous Deductions from Income covers certain charitable and other contributions, expenses of trustees administering trusts for outstanding debt issues, and an allowance for uncollectible amounts recorded under Other Income. It also covers all taxes related to Other Income, such as income, property, and gross receipts taxes.

This account also reflects the impact of the deferral and amortization of income tax amounts resulting from the use of accelerated depreciation on miscellaneous physical property.

Interest Deductions

The total interest paid on money borrowed from various sources is listed under Interest Deductions. Separate accounts are used to record different interest obligations for outstanding bonds, bank loans, commercial paper, and miscellaneous interest costs. Interest is also paid on customer deposits; it is recorded in a miscellaneous interest account. Interest costs are accrued monthly—even though, as with bond interest, payment may be made only semiannually.

Extraordinary and Delayed Items—Net

Income or losses resulting from nonrecurring transactions that are not customary business activities of the company are considered extraordinary income credits or losses. However, a single entry shows the net financial result of unusual transactions that may be recorded individually in one of several special accounts. For example, profit or loss on the sale of land is an extraordinary item. When AT&T disposed of its investment in the Communications Satellite Corporation, the gain was treated as an extraordinary income credit. Delayed amounts generally represent major adjustments, or corrections, for entries recorded in operating accounts in an earlier period. Major adjustments are those which would seriously distort the operating-income result for the current period if the adjustments were reflected in operating income or expense accounts. Income tax effects of both extraordinary and delayed items are debited or credited to a single tax account. The net debit or credit balance of these accounts is listed on the No. 1 Report.

Net Income

Net income is important because it represents the money the business has earned. Earnings belong to the stockholders, and they may be distributed to them in the form of dividends or may be retained by the company and reinvested in the business. In practice, companies prefer to pay out a portion of earnings and reinvest the remainder. Such a dividend policy is possible only with adequate earnings, coupled with the company's ability to supplement retained earnings with sufficient external financing to meet its needs. Chapter 4 treats the problem of financing.

Dividends

In the Bell System, dividends are declared and paid quarterly. Dividends paid by the operating companies and Western Electric are an important part of the income of the AT&T Company.

In turn, the AT&T Company has a long and outstanding record of paying dividends to its stockholders. During the last several years, the amount paid out in dividends on all classes of the AT&T Company stock has averaged approximately two-thirds of consolidated Bell System earnings.

Balance

The balance remaining after dividends are paid is transferred to the Retained Earnings account on the balance sheet. Retained earnings are available for reinvestment in the business or for future payment of dividends. By law, corporations cannot pay out more in dividends than is available from net income for the period unless there is a positive balance in retained earnings which can be used to make up the deficiency. This is how the Bell System managed to pay dividends during the depression of the 1930s and shortly after World War II as well.

OTHER DATA ON THE MONTHLY REPORT NO. 1

The lower portions of Sheets 1 and 2 of the No. 1 Report show significant data about the business. Certain computed annual averages are listed on Sheet 1, since these are more appropriate to use in evaluating operating results than the end-of-period figures. Several ratios are calculated and listed for selected indicators. Sheet 2 shows statistical data relating to telephones in service, as well as inward and outward movement. Information is included about employees, wage and salary payments, calling rates, and construction expenditures.

CLEARING ACCOUNTS

Clearing accounts are a special group of expense accounts which do not appear on the No. 1 Report but are an important accounting tool nevertheless. They provide a place to accumulate certain costs of business operations which should be charged to several different accounts. Nearly all these costs are distributed at the end of the month to their proper final accounts. Any minor uncleared balances are then included on the balance sheet as deferred items.

Part 31 of the FCC Rules and Regulations contains the following instructions for clearing accounts:

Instructions for Clearing Accounts

31.7-70 Purpose for clearing accounts.

The clearing accounts (702 to 707, inclusive) are provided as a medium for the distribution of certain items which affect more than one account and which cannot be appropriately allocated as they are incurred.

Items included in these accounts are:

702 Vehicle and other work equipment expense

704 Supply expense

705 Engineering expense

706 Plant supervision expense

707 House service expense

Motor vehicle expense is a good example. An installation or construction vehicle is used for both construction (capital expenditure) and repair (maintenance expense) of telephone plant. Vehicle or other work equipment expense is accumulated during the month in one clearing account and then prorated to final plant and expense accounts. The bulk of the expense is allocated depending on the distribution of the time of people using the vehicles or equipment during the month. For certain vehicles, the allocations are based on periodic studies of use.

Vehicle expense includes not only repair and operations costs (such as for gas, oil, and tires) but also the depreciation expense and costs for garage and other applicable rent. To the extent that vehicle expense includes some rent and is prorated to construction, the plant accounts also contain an element of rent expense.

Similar principles apply to the remaining clearing accounts. Some further information about the distribution of the amounts in the clearing accounts is shown on the Quarterly Report No. 19. (Appendix 3A at the end of the chapter has an example of this report.)

USE OF ACCOUNTING DATA

Overall, this book presents techniques for conducting economic evaluation studies of alternative courses of action. A basic requirement for such studies is valid cost information for elements of the alternatives being investigated. Inasmuch as accountants record all the costs of a business, accounting records and reports are essential for cost information. In consulting accounting data, however, the analyst must be discriminating, select appropriate information, and then apply it properly to study work.

This chapter is an overview of the nature of accounts and the major items included in each account. It should therefore help the analyst understand the financial impact of various transactions, as well as the ways these transactions are recorded. However, the analyst cannot apply the concepts explained here directly, to determine the current accounting of a particular transaction. The main reason for this is that the purpose of accounting is to keep a record of the past; the analyst, on the other hand, needs to estimate the future. Consideration of the future often involves different methods, changed procedures, or both. Although costs incurred in the past may serve as a guide, they may not be directly applicable in estimating for the future. Furthermore, recorded costs often include a portion of overhead or other loadings (for the Executive, Legal, and Treasury Departments) that may be inappropriate for the specific problem the analyst has at hand.

To summarize, accounting records can be useful so long as the user understands their nature. A most important point for the analyst to keep in mind is that the result of economic analysis will be reflected in future accounting records; however, past accounting records and procedures should not be permitted to control the future.

Appendix 3A, which follows, contains examples of the chief accounting reports with which the analyst should be familiar.

APPENDIX 3A

Examples of Reports

REPORTS INCLUDED

Reproduced in this appendix are some examples of the accounting reports that will be particularly helpful to the manager in analyzing the current financial results and position of the company. Although most reports are issued monthly, some are issued only quarterly. Annual supplements are also issued for certain reports.

A list of the reports included here follows:

Report	Description
No. 1	Summary of Reports
No. 2	Balance Sheet
No. 2A	Changes in Telephone Plant Accounts
No. 3	Income Statement
No. 4	Operating Revenues
No. 5	Operating Expenses
No. 6	Depreciation Reserve
No. 19	Clearing and Certain Other Accounts
No. 21	Construction Expenditures

Data on the Monthly Report No. 1 shown in Tables 3A-1 and 3A-2 are for a different company than data on the reports illustrated in Tables 3-1 and 3-2 in the body of the chapter. As a result, there are some minor differences between the two examples. However, all the reports in this appendix are for the same company and month, and many of the entries can be traced through several reports.

TABLE 3A-1

FORM S.N. 152(1) (1-76)

SUMMARY OF REPORTS

SAMPLE TELEPHONE COMPANY   MARCH 1976

EARNINGS AND EXPENSES	a This Month	b Increase over Last Month	c Increase over Same Month Last Year	d This Year to Date	e Amount	f Percent	
1 Local Service Revenues	58 315	1 223	4 379	171 434	10 181	6.3	1
2 Gross Toll Service Revenues	60 601	6 920	10 787	169 765	28 422	20.1	2
3 Less — Ind. Co. Toll Settlements	11 088	191	2 019	32 680	4 505	16.0	3
4 Net Toll Service Revenues	49 513	6 729	8 768	137 085	23 917	21.1	4
5 Misc. Operating Revenues	5 485	294	718	15 886	2 216	16.2	5
6 Less — Uncollect. Oper. Revenues	573	6	(37)	1 734	(57)	(3.2)	6
7 Total Operating Revenues	112 740	8 240	13 902	322 671	36 371	12.7	7
8 Maintenance	24 700	3 490	4 549	68 984	9 246	15.5	8
9 Depreciation & Amortization	14 096	94	984	42 028	2 770	7.1	9
10 Traffic Expenses	7 531	399	642	22 363	1 214	5.7	10
11 Commercial Dept.	4 026	345	733	11 572	1 478	14.6	11
12 Marketing — Sales Expenses	1 906	36	511	5 629	1 514	36.8	12
13 Marketing — Other Expenses	3 529	62	616	10 174	1 525	17.6	13
14 Operating Rents	619	(8)	213	1 929	214	12.5	14
15 Executive and Legal Depts.	199	15	(121)	639	(336)	(34.5)	15
16 Accounting Dept.	3 402	89	350	10 195	887	9.5	16
17 Prov. for Serv. Pens. & Death Bens.	7 155	(5)	1 096	21 461	3 292	18.1	17
18 Other Employees' Benefits	3 195	4	691	9 541	2 020	26.9	18
19 General Services & Licenses	1 506	(744)	250	5 343	1 177	28.2	19
20 Other General Expenses	4 327	750	946	11 448	1 116	10.8	20
21 Less — Exp. Charged Const.	2 716	543	802	7 276	1 420	24.2	21
22 Total Operating Expenses	73 475	3 984	10 658	214 030	24 697	13.0	22
23 Net Operating Revenues	39 265	4 256	3 244	108 641	11 674	12.0	23
24 Federal Income Taxes-Operating	9 894	1 551	926	26 201	3 379	14.8	24
25 State, Local and Other Oper. Taxes	9 644	555	743	27 725	2 404	9.5	25
26 Total Operating Taxes	19 538	2 106	1 669	53 926	5 783	12.0	26
27 Operating Income	19 727	2 150	1 575	54 715	5 891	12.1	27
28 Other Income	566	43	(242)	1 588	(314)	(16.5)	28
29 Misc. Deductions from Income	58	41	(75)	94	(116)	(54.9)	29
30 Income before Interest Deductions	20 235	2 152	1 408	56 209	5 693	11.3	30
31 Interest Deductions	6 856	(44)	393	20 733	1 038	5.3	31
32 Income after Interest Deductions	13 379	2 196	1 015	35 476	4 655	15.1	32
33 Extraordinary & Delayed Items-Net	2	2	2	(4)	(4)		33
34 Net Income	13 381	2 198	1 017	35 472	4 651	15.1	34
35 Dividends (accrual basis)	10 053	(1)	1 115	30 160	3 347	12.5	35
36 Balance	3 328	2 199	(98)	5 312	1 304	32.5	36

AVERAGES	g This Month	h Same Month Last Year	i This Year to Date	j Same Period Last Year	k% Increase i over j	
37 Av. Tel. Plant (Excl. Plant under Const.)	3 660 041	3 384 381	3 640 566	3 365 378	8.2	37
38 Av. Cap. Oblig., Prem., & Retained Earnings	2 590 101	2 452 978	2 600 987	2 454 892	6.0	38
39 Av. No. of Company Telephones	7 189	7 002	7 171	6 991	2.6	39
40 Av. Tel. Plant per Av. Co. Telephone	590.11	483.33	507.68	481.41	xxx	40

PERCENTAGES (Annual Basis)	l This Month	m Last Month	n Same Month Last Year	o This Year to Date	p Same Period Last Year	
41 Total Oper. Revenues to Average Tel. Plant	36.96	34.46	35.04	35.45	34.03	41
42 Total Oper. Expenses to Total Oper. Revenues	65.17	66.50	63.56	66.33	66.13	42
43 Maintenance to Average Tel. Plant	8.10	6.99	7.14	7.58	7.10	43
44 Depreciation (Accts. 608, 702, 704) to Av. Depr. Tel. Plant	4.79	4.78	4.82	4.78	4.84	44
45 Income* to Av. Cap. Oblig., Prem. & Retained Earnings	9.34	8.30	9.18	8.61	8.20	45
46 Pre-Fed. Tax Income* to Av. Cap. Oblig, Prem. & Ret. Earn.	13.90	12.14	13.61	12.62	11.93	46
47 Income to Av. Common Equity Capital	10.90	9.12	10.77	9.64	8.96	47
48 Post-Tax Interest Coverage	2.98	2.64	2.93	2.73	2.58	48

* Income before Interest Deductions less interest charges not related to Capital Obligations.

( ) Denotes negative amount.

Sheet 1 of the Monthly Report No. 1 presents summaries of pertinent items found in more detail on other reports. It is a summary income statement, plus several important averages and percentages.

TABLE 3A-2

FORM S.N. 152(2) (1-76)	SUMMARY OF REPORTS	MONTHLY REPORT NO. 1

SAMPLE TELEPHONE COMPANY   MARCH 1976   Sheet 2

	BALANCE SHEET	a At End of Month	b Increase over Last Month	Increase over Dec. 31 Last		
				c Amount	d %	
1	Telephone Plant	3 763 646	28 369	69 241	1.9	1
2	Less — Depreciation and Amortization Reserves	806 375	4 432	15 315	1.9	2
3	Total Telephone Plant less Reserves	2 957 271	23 937	53 926	1.9	3
4	Miscellaneous Physical Property	1 467	2	(118)	(7.5)	4
5	Investments in Subsidiaries					5
6	Other Investments and Funds	298				6
7	Cash and Temporary Cash Investments	14 672	(1 171)	(3 275)	(18.2)	7
8	Accounts Receivable	149 652	(5 131)	(1 167)	(.8)	8
9	Material and Supplies	16 496	82	1 029	6.7	9
10	Prepaid Accounts and Deferred Charges	36 214	5 396	5 816	19.1	10
11	Total Assets	3 176 070	23 115	56 211	1.8	11
12	Capital Stock — par value ($ per share)	1 040 000				12
13	Premium on Capital Stock	676				13
14	Other Capital					14
15	Retained Earnings Reserved					15
16	Unappropriated Retained Earnings	434 116	3 328	5 312	1.2	16
17	Less — Capital Stock Expense					17
18	Total Capital Stock & Retained Earnings	1 474 792	3 328	5 312	.4	18
19	Funded Debt	1 080 000				19
20	Advances From Affiliated Companies	34 500	7 000	(1 000)	(2.8)	20
21	Other Long-Term Debt					21
22	Notes	21 400		(200)	(.9)	22
23	Current and Accrued Liabilities	195 509	2 684	22 921	13.3	23
24	Deferred Credits	369 869	10 103	29 178	8.6	24
25	Total Liabilities and Capital	3 176 070	23 115	56 211	1.8	25
26	Total Capital Obligations, Prem., and Retained Earnings	2 610 692	10 328	4 112	.2	26
27	% Debt Obligations to Total	43.51 %	X X X	X X X	X X	27

	COMPANY TELEPHONES	At End of Month	f Increase over Last Month	g Increase over Dec. 31 Last	Percent of Total		
					h End of This Month	i Dec. 31 Last	
28	Total Company Telephones	7 196 067	13 819	50 445	X X	X X	28
29	Main	3 580 120	5 532	17 428	49.7	49.9	29
30	Residence Extension	2 366 764	9 897	27 189	32.9	32.7	30
31	Other	1 249 183	(1 610)	5 828	17.4	17.4	31

	GAIN IN CO. TELEPHONES (Excl. Purchases, Sales & Adj.)	This Month	k Increase over Last Month	l Increase over Same Month Last Year	m This Year to Date	n Increase over Same Period Last Year	
32	Inward Tel. Movement	99 952	17 279	12 025	266 541	13 141	32
33	Outward Tel. Movement	82 785	20 049	10 534	214 978	(3 590)	33
34	Net Gain	17 167	(2 770)	1 491	51 563	16 731	34
35	Employees — End of Month	32 890	(41)	(465)	X X X	X X X	35
36	Salary and Wage Payments	42 706 403	2 588 840	5 851 458	124 929 763	13 136 802	36

	CALLING RATE AND TOLL MESSAGES	o This Month	p Same Month Last Year	q This Year to Date	r Same Period Last Year	
37	Av. Daily Calling Rate per Tel. { Total	5.07	4.92	5.08	4.86	37
38	Excl. Residence Extension	7.55	7.28	7.57	7.14	38
39	Toll Messages Originating	53 575 562	48 737 324	153 228 130	143 993 795	39

	INDEPENDENT CO. TOLL SETTLEMENTS	s This Month	t Same Month Last Year	u This Year To Date	v Same Period Last Year	
40	Intrastate	6 913	4 647	20 387	15 334	40
41	Interstate	4 175	4 422	12 293	12 841	41

	CONSTRUCTION AND NEW MONEY	w This Month	x This Year to Date	Increase over Same Period Last Year		
				y Amount	z Percent	
42	Construction Expenditures	37 818	94 081	4 334	4.8	42
43	New Money Requirements	7 000	(1 200)	(25 425)	X X X	43

( ) Denotes negative amount.

Date Issued April 19, 1976

- - - - - - - -

Sheet 2 of the No. 1 Report is a summary balance sheet, and it also includes several other items.

TABLE 3A-3

FORM S.N. 153 (1-77)
This Copy for

BALANCE SHEET

SAMPLE TELEPHONE COMPANY   MARCH 1976

		a At End of Month	b Increase over Last Month	c Increase over December 31 last	
1	Organization, Franchises & Patent Rights (201, 202, 203)				1
2	Land (211)	27 810 584	6 520	5 632	2
3	Buildings (212)	334 906 285	916 954	2 804 584	3
4	Central Office Equipment (221)	1 249 487 165	10 523 186	24 063 992	4
5	Station Apparatus (231)	345 586 883	1 750 368	4 446 782	5
6	Station Connections (232)	392 734 843	2 744 388	7 860 573	6
7	Large Private Branch Exchanges (234)	141 401 314	472 371	883 002	7
8	Pole Lines (241)	108 093 065	279 650	854 986	8
9	Cable (242)	774 200 953	4 695 285	12 464 672	9
10	Aerial Wire (243)	10 653 187	(60 496)	(224 641)	10
11	Underground Conduit (244)	199 421 923	1 464 883	2 858 442	11
12	Furniture and Office Equipment (261)	36 563 581	781 784	942 299	12
13	Vehicles and Other Work Equipment (264)	47 362 033	642 179	789 635	13
14	Telephone Plant Acquired Less Plant Sold (276, 277). .				14
15	**Total Telephone Plant In Service (100.1)**	3 668 221 822	24 217 075	57 749 963	15
16	Telephone Plant Under Construction (100.2).	91 581 333	4 321 041	11 690 312	16
17	Property Held for Future Telephone Use (100.3)	3 843 331	(168 964)	(198 914)	17
18	Telephone Plant Acquisition Adjustment (100.4)				18
19	**Total Telephone Plant**	3 763 646 486	28 369 152	69 241 333	19
20	Less—Depreciation Reserve (171)	806 375 071	4 431 583	15 314 580	20
21	—Amortization Reserve (172)				21
22	**Total Telephone Plant less Reserves**	2 957 271 415	23 937 569	53 926 753	22
23	Investments In, and Advances to Affiliated Cos. (101)				23
24	Other Investments and Funds (102, 136, 137)	298 065	16	(48)	24
25	Miscellaneous Physical Property (103)	1 467 247	2 442	(118 143)	25
26	**Total**	1 765 312	2 459	(118 191)	26
27	Cash (113).	13 867 286	(890 650)	(3 388 856)	27
28	Special Cash Deposits (114)	804 643	(280 649)	114 274	28
29	Temporary Cash Investments (116)				29
30	Material and Supplies (122)	16 495 920	81 868	1 028 782	30
31	Other Current Assets (115, 117, 118, 120, 121, 123)	149 651 368	(5 131 245)	(1 167 635)	31
32	**Total Current Assets**	180 819 219	(6 220 676)	(3 413 434)	32
33	Subscriptions to Funded Debt (127)				33
34	Prepaid Accts. & Deferred Chgs (129 to 133, 135, 138, 139)	36 214 410	5 396 192	5 816 454	34
35	**Total Assets**	3 176 070 358	23 115 544	56 211 580	35
36	Capital Stock, Com. (150-01) (Auth. $. . . . .)	1 040 000 000			36
37	Capital Stock, Pref. (150-02) (Auth. $ . . . .)				37
38	Premium on Capital Stock (152)	676 388			38
39	Capital Stock Subscribed & Stock Installments (153) . . .				39
40	Other Capital (179)				40
41	Retained Earnings Reserved (180)				41
42	Unappropriated Retained Earnings (181).	434 115 722	(16 778 936)	5 311 500	42
43	Less—Capital Stock Expense (134.2)				43
44	**Total Capital Stock and Retained Earnings**	1 474 792 111	(16 778 936)	5 311 500	44
45	Funded Debt (154.1)	1 080 000 000			45
46	Funded Debt Subscribed (154.2).				46
47	Advances from Affiliated Companies (156)	34 500 000	7 000 000	(1 000 000)	47
48	Other Long-Term Debt (157)				48
49	Notes Payable — Coml. Paper & Banks (158.2-07, 08). .	21 400 000		(200 000)	49
50	Current Liabilities (158 to 165; except 158.2-07, 08). . .	131 364 267	7 699 997	7 795 447	50
51	Accrued Liabilities (166, 167)	64 144 861	15 091 531	15 125 619	51
52	Deferred Credits (168, 174, 176.1, 176.2)	368 682 429	10 115 290	29 181 666	52
53	Miscellaneous Reserves (169, 170, 173).	1 186 688	(12 337)	(2 654)	53
54	**Total Liabilities and Capital**	3 176 070 358	23 115 544	56 211 580	54

( ) Denotes negative amount.

Date issued   April 19, 1976   . . . .
                                                Vice President and Comptroller

The Monthly Report No. 2 contains more detail on major asset and liability accounts than is included on the No. 1 Report. Certain additional information is referenced. Beginning with this report, the entries include account numbers in parentheses.

TABLE 3A-4a

FORM S.N. 281-A(1) (3-69)
This Copy for

QUARTERLY REPORT No. **2A.**
Sheet 1.

SAMPLE TELEPHONE COMPANY
CHANGES IN TELEPHONE PLANT ACCOUNTS

MARCH 1976

		a Plant Added	*b* Plant Retired	*c* Net Increase	*d* Total at End of Period	
1	Organization (201)					1
2	Franchises (202)					2
3	Patent Rights (203)					3
4	Land (211)	6 086	454	5 632	27 810 584	4
5	Buildings (212)	3 297 466	492 881	2 804 584	334 906 285	5
6	**Total Land and Buildings**	3 303 552	493 336	2 810 216	362 716 869	6
7	C.O.E. (221) — Manual.	12 111	1 502 227	(1 490 116)	15 420 384	7
8	— Panel	65 560	5 722	59 837	26 986 225	8
9	— Step-by-Step	2 278 319	726 939	1 551 380	148 603 235	9
10	— Crossbar	6 791 512	1 013 152	5 778 359	458 022 984	10
11	— Circuit	12 659 104	523 115	12 135 989	345 083 532	11
12	— Radio.	498 823	81 526	417 297	10 419 617	12
13	— Electronic	5 863 779	252 535	5 611 243	244 951 186	13
14	**Total Cent. Office Equipment**	28 169 211	4 105 219	24 063 992	1 249 487 165	14
15	Sta. App. (231 — Teletypewriter	937 418	552 111	385 306	25 301 361	15
16	— Tele. & Misc.	9 605 194	5 540 573	4 064 620	315 841 581	16
17	— Radio	18 429	21 574	(3 145)	4 443 939	17
18	Total Station Apparatus	10 561 042	6 114 260	4 446 782	345 586 883	18
19	Sta. Conn. (232) — Teletypewriter . . .	204 082	64 860	139 222	4 309 014	19
20	— Tele. & Misc. . . .	16 833 662	9 120 723	7 712 938	388 040 184	20
21	— Radio	29 467	21 054	8 412	385 645	21
22	Total Station Connections	17 067 212	9 206 638	7 860 573	392 734 843	22
23	Large Priv. Br. Exchs. 234)	3 586 242	2 703 239	883 002	141 401 314	23
24	**Total Station Equipment**	31 214 497	18 024 138	13 190 358	879 723 041	24
25	Pole Lines (241)	1 530 676	675 689	854 986	108 093 065	25
26	U.G. Conduit (244)	2 918 330	59 887	2 858 442	199 421 923	26
27	**Total Outside Plant Structures**	4 449 006	735 576	3 713 429	307 514 988	27
28	Aerial Cable (242.1) — Exch.—Bldg. . .	2 222 216	270 388	1 951 828	85 627 663	28
29	—Other .	7 293 354	1 635 574	5 657 779	331 991 937	29
30	— Toll	112 976	154 143	(41 166)	27 354 513	30
31	Total Aerial Cable	9 628 547	2 060 106	7 568 441	444 974 113	31
32	U.G. Cable (242.2) — Exchange	4 242 820	360 393	3 882 427	234 337 586	32
33	— Toll	(192 431)	16 870	(209 302)	24 139 043	33
34	Total Underground Cable	4 050 389	377 264	3 673 125	258 476 629	34
35	Buried Cable (242.3) — Exchange . . .	1 341 813	125 191	1 216 621	64 718 943	35
36	— Toll	7 070	593	6 477	5 417 308	36
37	Total Buried Cable	1 348 883	125 784	1 223 099	70 136 252	37
38	Submarine Cable (242.4) — Exchange .	6		6	194 698	38
39	— Toll				419 260	39
40	Total Submarine Cable	6		6	613 958	40
41	Total Exchange Cable	15 100 211	2 391 547	12 708 663	716 870 828	41
42	Total Toll Cable	(72 384)	171 607	(243 991)	57 330 125	42
43	**Total Cable**	15 027 826	2 563 154	12 464 672	774 200 953	43
44	Aerial Wire (243) — Exchange	252 591	465 421	(212 830)	10 595 966	44
45	— Toll		11 811	(11 811)	57 220	45
46	**Total Aerial Wire**	252 591	477 232	(224 641)	10 653 187	46
47	**Total Outside Plant (L 27 + 43 + 46)**	19 729 424	3 775 964	15 953 460	1 092 369 130	47

( ) Denotes negative amount.

TABLE 3A-4b

FORM S.N. 281-A (2) (1-67)
This Copy for

QUARTERLY REPORT No. 2A.
Sheet 2.

SAMPLE TELEPHONE COMPANY

CHANGES IN TELEPHONE PLANT ACCOUNTS   MARCH 1976

Printed in U.S.A.

	a Plant Added	b Plant Retired	c Net Increase	d Total at End of Period	
	This Year to Date				
Furniture & Office Equip. (261).					
1 Storeroom Furn. & Office Equip.	8 068	6 205	1 862	570 879	1
2 Other Furn. & Office Equip.	180 037	129 179	50 858	21 388 757	2
3 Computer and AMA Systems	875 632	(13 946)	889 578	14 603 944	3
4 Total Furniture & Office Equip. . . .	1 063 738	121 438	942 299	36 563 581	4
Vehicles and Other Work Equip. (264).					
5 Motor Vehicles	743 690	157 597	586 093	35 988 139	5
6 Garage & Motor Vehicle Shop Equip. .	17 221	1 872	15 349	657 596	6
7 Special Tools & Work Equip.					7
8 Other Shop Equipment.				28 187	8
9 Other Tools and Work Equip.	223 008	39 002	184 005	10 163 994	9
10 Storeroom Work Equipment	6 099	1 912	4 186	524 115	10
11 Total Vehicles & Other Work Equip..	990 019	200 384	789 635	47 362 033	11
12 Total General Equipment.	2 053 757	321 822	1 731 935	83 925 614	12
13 Tel. Plant Acquired (276)		xxxx			13
14 Tel. Plant Sold (277)	xxxx				14
15 Total Tel. Plant in Ser. (100.1) . . .	84 470 443	26 720 480	57 749 963	3 668 221 822	15
16 Tel. Plt. under Construction (100.2) .	11 704 997	14 685	11 690 312	91 581 333	16
17 Prop. Held for Future Tel. Use (100.3)	(198 941)		(198 941)	3 843 331	17
18 Tel. Plt. Acquisition Adj. (100.4). . .					18
19 Total Telephone Plant	95 976 499	26 735 165	69 241 333	3 763 646 486	19

( ) Denotes negative amount.

Date issued   April 22, 1976

....
Vice President and Comptroller

The Quarterly Report No. 2A lists changes for the 100.1, 100.2, 100.3, and 100.4 investment accounts for the year and by plant accounts. The totals for these accounts appear on Sheet 2 of the No. 2A Report. These totals also appear on the No. 2 Report, and their sum is carried through to the No. 1 Report, Sheet 2. (Accounts 100.2 and 100.3 are further analyzed on Report No. 2C, which is not included.)

TABLE 3A-5

FORM S.N. 154 (1-77)

MONTHLY REPORT NO. 3

INCOME STATEMENT

SAMPLE TELEPHONE COMPANY MARCH 1976

		a This Month	b Increase over Last Month	c This Year to Date	d Increase over Same Period Last Year	
1	Operating Revenues (300)	112 740 924	8 240 904	322 671 528	36 371 369	1
2	Operating Expenses (301)	73 475 493	3 984 630	214 030 348	24 696 976	2
3	Net Operating Revenues	39 265 431	4 256 274	108 641 180	11 674 393	3
4	Investment Credits Realized (304-01)	2 554 666	43 111	9 010 367	5 603 511	4
5	Amort. of Invest. Credits – Cr. (304-02) . . .	444 433	10 666	1 297 517	409 916	5
6	Investment Credits – Net (304).	2 110 233	32 445	7 712 850	5 193 594	6
7	Federal Income Taxes – Operating (306) . .	2 677 993	1 824 532	3 088 827	(3 596 990)	7
	Other Operating Taxes (307) –					
8	Property Taxes (01)	784 582	24 285	2 308 322	247 991	8
	State & Local Income Taxes (02) –					
9	Current Period (12).	1 119 683	409 001	2 572 737	428 903	9
10	Deferred (22)	1 168 649	(107 594)	3 524 152	547 652	10
11	Inc. Cr. and Chgs. Prior Defrls (32) . . .	57 332	(39 593)	172 389	172 389	11
12	Gross Receipts Taxes (03)	3 646 000	267 500	10 421 100	1 139 000	12
13	Capital Stock Taxes (04)	1 122 500		3 367 500	231 300	13
14	S.S. State Unemployment Taxes (315) . . .	190 225	115	570 545	(14 262)	14
15	Other Taxes (06)	94 136	12 716	256 509	4 616	15
16	Total State and Local Taxes (L 8 Thru 15) .	8 068 444	645 617	22 848 477	2 412 810	16
17	Other S.S. Taxes (115, 215,75, 85, & 95) . .	1 575 818	(89 844)	4 876 285	(9 410)	17
18	Total (307) (L. 16 + 17)	9 644 262	555 773	27 724 763	2 403 400	18
19	Operating FIT Deferred (308.1, 308.2)* . . .	5 341 975	(486 006)	16 109 171	2 492 371	19
20	Inc. Cr. Rsltg. Frm. Prior Dfrls. of FIT (309)*.	236 634	(179 363)	709 865	709 865	20
21	Operating Income	19 727 602	2 150 166	54 715 433	5 891 882	21
22	Dividend Income (312).					22
23	Interest Earned (313-01)	3 804	455	12 388	(243 690)	23
24	Interest Charged Construction (313-02)	562 557	42 497	1 577 269	(69 515)	24
25	Sink. & Oth. Funds & Misc. Phys. Prop. (314,315)	(378)	104	(1 741)	(835)	25
26	Miscellaneous Income (316)	1	1	11	(129)	26
27	Total Other Income.	565 984	43 058	1 587 927	(314 170)	27
28	Miscellaneous Income Charges (323).	116 017	86 638	177 690	22 154	28
29	Federal Income Taxes-Nonoperating (326) *. . .	(50 600)	(37 500)	(77 600)	(119 400)	29
30	Other Nonoperating Taxes (327) *.	(7 135)	(7 903)	(5 334)	(17 991)	30
31	Total Miscellaneous Deductions	58 282	41 235	94 756	(115 236)	31
32	Income before Interest Deductions	20 235 304	2 151 990	56 208 604	5 692 948	32
33	Interest on Funded Debt (335)	6 541 666		19 625 000	2 187 500	33
	Other Interest Deductions (336) –					
34	Advances – From A.T.&T. Co. (11)	109 968	(15 114)	419 225	381 510	34
35	Commercial Paper (07)	24 624	(45 545)	167 910	(1 149 165)	35
36	Banks (08).	68 260	6 287	196 801	(465 178)	36
37	Misc. – Related to Cap. Obligations (19). .					37
38	– Unrelated to Cap. Obligations (29).	84 044	9 079	243 088	76 297	38
39	Amort. of Disc. on LT Debt (338)	28 467		85 403	8 345	39
40	Release of Prem. on LT Debt – Cr. (339) .	4 777		14 332		40
41	Other Fixed Charges (340)	3 697	447	10 197	(402)	41
42	Total Interest Deductions.	6 855 952	(44 846)	20 733 294	1 038 906	42
43	Income after Interest Deductions	13 379 352	2 196 837	35 475 310	4 654 042	43
44	Extraordinary Income Credits (360).					44
45	Delayed Income Credits (365).					45
46	Extraordinary Income Charges (370)	(3 411)	(3 436)	8 209	8 209	46
47	Delayed Income Charges (375)					47
48	Inc. Tax eff. of Ext. & Dlyd. Items-Net (380)	1 700	1 700	(4 400)	(4 400)	48
49	Net Income	13 381 063	2 198 573	35 471 500	4 650 233	49
50	Dividends Declared–Pref. Stk. (416-01)					50
51	–Com. Stk. (416-02) . . .	30 160 000	30 160 000	30 160 000	3 347 500	51

( ) Denotes negative amount

*See Supplement to Monthly Report No. 3 for details of specific subaccounts.

. . . .

Vice President and Comptroller

The Income Statement is a detailed breakdown of information appearing on Sheet 1 of the No. 1 Report. Line 1, Operating Revenues, is derived from the No. 4 Report. Line 2, Operating Expenses, is from the No. 5 Report.

TABLE 3A-6a

FORM S.N. 155 (1) (1-77)

OPERATING REVENUES

MONTHLY REPORT NO. 4

Sheet 1

SAMPLE TELEPHONE COMPANY   MARCH 1976

		a This Month	b Increase Over Last Month	c This Year to Date	d Increase over Same Period Last Year	
	LOCAL SERVICE REVENUES					
	SUBSCRIBERS' STATION REVENUES:					
1	Monthly Charges (500-01)	44 951 675	38 594	134 472 218	6 373 371	1
2	Message Charges (500-02)	8 150 234	598 206	22 365 745	1 195 518	2
3	Non-recurring Charges (500-03)	2 973 860	443 984	8 173 305	2 186 115	3
	DATA-PHONE 50					
4	Monthly Charges (500-14)					4
5	Non-recurring Charges (500-34)					5
6	Total DATA-PHONE 50 (500-04)					6
	PICTUREPHONE					
7	Monthly Charges (500-15)	413	104	1 030	(48)	7
8	Local Msg. Charges (500-25)		(5)	11	(16)	8
9	Non-recurring Charges (500-35)					9
10	Total PICTUREPHONE (500-05)	413	98	1 041	(64)	10
11	Subscribers' Station Revenues (500)	56 076 183	1 080 884	165 012 310	9 754 940	11
12	Public Telephone Revenues (501)	1 521 077	133 721	4 266 247	256 890	12
13	Service Stations (503)	1 518	(4)	4 571	148	13
	LOCAL PRIVATE LINE SERVICES:					
14	Telephone – Inter (504-11)	1 715	864	4 668	(1 208)	14
15	– Intra (504-21)	160 244	6 679	471 374	48 671	15
16	Teletypewriter – Inter (504-12)	107 331	(5 209)	335 941	(38 179)	16
17	– Intra (504-22)					17
18	DATAPHONE Digital Svc. – Inter (504-13) .					18
19	– Intra (504-23) .					19
20	Other Telegraph – Inter (504-14)	164	48	263	169	20
21	– Intra (504-24)					21
	PROGRAM TRANSMISSION					
22	Audio – Inter (504-118)	14 983	5 463	32 256	62	22
23	– Intra (504-218)	10 638	(1 060)	33 365	5 575	23
24	Television – Inter (504-128)	6 461	25	24 652	7 971	24
25	– Intra (504-228)	454	(965)	3 102	(2 003)	25
26	Other Services – Inter (504-19)	7 148	1 583	18 284	5 677	26
27	– Intra (504-29)	407 458	1 011	1 226 916	142 193	27
28	Total (504) – Inter	30 307	7 935	79 860	12 501	28
29	– Intra	686 291	503	2 070 963	156 427	29
30	Total (504) (L28+29)	716 598	8 438	2 150 824	168 928	30
	OTHER LOCAL SERVICE REVENUES					
31	Directory Assistance (506-01)					31
32	Other Services (506-09)					32
33	Total (506)					33
	TOTAL LOCAL SERVICE REVENUES					
34	– Inter (Line 28)	30 307	7 935	79 860	12 501	34
35	– Intra (L11+12+13+29+33)	58 285 070	1 215 104	171 354 093	10 168 4C6	35
36	**Total Local Service Revenues (L34+35)** . . .	58 315 377	1 223 039	171 433 953	10 180 908	36

( ) Denotes negative amount

This report is a detailed breakdown of information about the entry Operating Revenues, which appears on line 1 of the Income Statement, Monthly Report No. 3.

TABLE 3A-6b

FORM S.N. 155 (2) (1-76)

OPERATING REVENUES

SAMPLE TELEPHONE COMPANY

MONTHLY REPORT NO. 4
Sheet 2

MARCH 1976

	a This Month	b Increase over Last Month	c This Year to Date	d Increase over Same Period Last Year	
TOLL SERVICE REVENUES					
MESSAGE TOLLS:					
1 Tel. & Misc. – Inter (510-11)	17 447 992	1 554 855	50 293 758	5 259 865	1
2 – Intra (510-21)	23 101 304	4 835 190	60 725 203	13 637 475	2
3 DATA-PHONE 50 – Inter (510-14) .				(10)	3
4 – Intra (510-24)					4
5 Total – Inter	17 447 992	1 554 855	50 293 758	5 259 855	5
6 – Intra	23 101 304	4 835 190	60 725 203	13 637 475	6
7 Total (510)	40 549 297	6 390 046	111 018 962	18 897 330	7
WIDE AREA TELECOMM. SERVICES:					
8 WATS – Inter (511-11)	2 908 152	75 302	8 199 991	1 360 370	8
9 – Intra (511-21)	1 577 454	78 833	4 557 366	1 213 809	9
10 Toll Calling Plans (511-03)					10
11 Total – Inter	2 908 152	75 302	8 199 991	1 360 370	11
12 – Intra	1 577 454	78 833	4 557 366	1 213 809	12
13 Total (511)	4 485 606	154 135	12 757 357	2 574 179	13
TOLL PRIVATE LINE SERVICES:					
14 Telephone – Inter (512-11)	696 704	72 166	2 003 039	499 127	14
15 – Intra (512-21)	1 301 863	(51 543)	3 998 438	753 639	15
16 Teletypewriter – Inter (512-12)	140 506	15 190	399 869	32 670	16
17 – Intra (512-22)	28 698	(883)	91 843	7 624	17
18 DATA-PHONE Digital Svc. – Inter (512-13)	10 356	2 672	24 057	22 566	18
19 – Intra (512-23)					19
Other Telegraph –					
20 Tel. Grd. Svc. – Inter (512-114)	7 711	886	22 159	8 408	20
21 – Intra (512-214)	366	12	765	477	21
22 Teleg. Grd. Svc. – Inter (512-124) ..	10	279	(259)	(265)	22
23 – Intra (512-224) ..	(163)	(413)	(162)	(157)	23
Multipurpose Wideband –					
24 Telpak – Inter (512-117)	1 123 440	119 977	3 224 714	380 070	24
25 – Intra (512-217)	293 150	(8 722)	920 228	62 895	25
26 Other – Inter (512-127)	40 135	7 633	107 448	10 418	26
27 – Intra (512-227)	(228)	(260)	27	(147)	27
Program Transmission –					
28 Audio – Inter (512-118)	37 006	(850)	123 914	31 943	28
29 – Intra (512-218)	2 033	(3 482)	11 805	6 929	29
30 Television – Inter (512-128)	153 778	21 066	427 518	58 341	30
31 – Intra (512-228)	9 355	1 858	16 854	15 647	31
Other Services –					
32 Tel. Grd. Svc. – Inter (512-119)	329 284	46 389	938 132	310 257	32
33 – Intra (512-219)	165 862	(39 788)	579 641	74 634	33
34 Teleg. Grd. Svc. – Inter (512-129) ..	20 637	3 683	57 614	16 293	34
35 – Intra (512-229) ..	116 422	(2 428)	360 279	153 264	35
36 Total – Inter	2 559 567	289 091	7 328 205	1 369 828	36
37 – Intra	1 917 359	(105 650)	5 979 721	1 074 807	37
38 Total (512)	4 476 926	183 440	13 307 926	2 444 635	38
OTHER TOLL SERVICE REVENUES					
39 Directory Assistance (516-01)...........					39
40 Other Services (516-09)................					40
41 Total (516)					41
TOTAL TOLL SERVICE REVENUES					
42 – Inter (L5+11+36)	22 915 711	1 919 248	65 821 954	7 990 053	42
43 – Intra (L6+12+37+41)	26 596 118	4 808 373	71 262 291	15 926 092	43
44 **Total Toll Service Revenues (L42 + 43)** .	49 511 830	6 727 621	137 084 246	23 916 145	44

( ) Denotes negative amount

TABLE 3A-6c

FORM S.N. 155 (3) (1-77)

OPERATING REVENUES

SAMPLE TELEPHONE COMPANY

MARCH 1976

	a This Month	b Increase over Last Month	c This Year to Date	d Increase over Same Period Last Year	
TOTAL MISCELLANEOUS REVENUES					
1 Telegraph Commissions (521)	2 468	78	7 147	(1 056)	1
2 Directory Advertising and Sales (523) . .	4 723 044	276 600	13 593 822	1 757 774	2
3 Rent Revenues (524)	591 147	9 733	1 753 175	349 483	3
4 Revs. from Genl. Services & Lic. (525) .					4
5 Other Operating Revenues (526)	168 460	7 716	532 021	110 377	5
6 Total Miscellaneous Revenues	5 485 121	294 128	15 886 166	2 216 579	6
TOTAL OPERATING REVENUES — BEFORE UNCOLLECTIBLES (500 to 526, INCL)					
7 Total — Inter (Sh.1, L 34 + Sh.2, L42) . .	22 946 018	1 927 183	65 901 815	8 002 554	7
8 — Intra (Sh.1, L35 +Sh2,L43+Sh.3,L6'	90 366 310	6 317 606	258 502 550	28 311 078	8
9 Total	113 312 328	8 244 790	324 404 365	36 313 633	9
TOTAL UNCOLLECTIBLE OPERATING REVENUES — DR					
10 Inter — Bell DR. (530-01)	237 920	8 273	702 321	(95 298)	10
11 Inter — Other (530-02)	(9 851)	(16 113)	4 042	(62 546)	11
12 Intra — Bell (530-03)	331 157	2 157	1 010 197	106 165	12
13 Intra-Other (530-04)	12 177	9 568	16 277	(6 058)	13
14 Total (530)	571 403	3 885	1 732 837	(57 736)	14
TOTAL OPERATING REVENUES					
15 Interstate (L7 — (L10+L11))	22 717 949	1 935 023	65 195 452	8 160 398	15
16 Intrastate (L8 — (L12 + L13)).	90 022 975	6 305 880	257 476 076	28 210 970	16
17 Total Operating Revenues	112 740 924	8 240 904	322 671 528	36 371 369	17

MONTHLY AVERAGE REVENUES PER TELEPHONE

	a This Month	b Same Mo. Last Yr.	c This Year to Date	d Same Per. Last Yr.	
LOCAL SERVICE REVENUES					
18 Inter	.004	.003	.004	.003	18
19 Intra	8.107	7.699	7.965	7.686	19
20 Total Local	8.111	7.702	7.969	7.689	20
TOLL SERVICE REVENUES					
21 Inter	3.188	2.872	3.059	2.758	21
22 Intra.	3.699	2.947	3.313	2.638	22
23 Total Toll	6.887	5.819	6.372	5.396	23
24 TOTAL MISCELLANOUS REVENUES .	.763	.681	.738	.651	24
25 Total (L20 + 23 + 24)	15.761	14.202	15.079	13.736	25
UNCOLLECTIBLE OPER. REVENUES					
26 Inter	.032	.044	.033	.041	26
27 Intra	.047	.043	.048	.044	27
28 Total Uncollectibles — Dr — . . .	.079	.087	.081	.085	28
TOTAL OPERATING REVENUES					
29 Inter	3.160	2.831	3.031	2.720	29
30 Intra	12.522	11.284	11.968	10.931	30
31 Total Operating Revenues	15.682	14.115	14.999	13.651	31

( ) Denotes negative amount.

Date Issued   April 19, 1976

. . . .
Vice President and Comptroller

TABLE 3A-7a

FORM S.N. 156 (1) (1-76)   MONTHLY REPORT No. 5.
                                                            Sheet 1

SAMPLE TELEPHONE COMPANY
OPERATING EXPENSES

MARCH 1976

	a — This Month	b — Increase over Last Month	c — This Year to Date	d — Increase over Same Period Last Year	
1 Repairs of Outside Plant (602.1 to .8)	3 850 871	561 901	10 915 795	1 923 321	1
2 Test Desk Work (603)	2 885 210	329 610	8 238 020	1 477 926	2
3 Repairs of C.O. Equipment (604)	7 544 979	983 546	21 206 997	2 237 396	3
4 Repairs of Station Equipment (605)	9 013 178	1 349 574	24 947 318	3 888 876	4
5 Repairs of Building & Grounds (606)	878 993	182 360	2 239 092	(268 024)	5
6 Maintaining Transmission Power (610) ...	346 541	39 315	994 569	(4)	6
7 Other Maintenance Expenses (612)	180 120	43 124	442 448	(13 198)	7
8 **Total Maintenance Expenses**	24 699 996	3 489 433	68 984 243	9 246 293	8
9 Depreciation (608)	14 096 073	93 683	42 028 166	2 770 779	9
10 Extraordinary Retirements (609)					10
11 Amort. of Intangible Property (613)					11
12 Amort. of Tel. Plant Acquis. Adj. (614) ..					12
13 **Total Depreciation & Amort. Exps.**	14 096 073	93 683	42 028 166	2 770 779	13
14 General Traffic Supervision (621)	1 164 220	133 764	3 290 133	283 048	14
15 Serv. Inspection & Customer Inst. (622) ..	295 482	39 639	841 761	(13 897)	15
16 Operators' Wages (624)	5 458 607	113 641	16 577 952	856 118	16
17 Rest and Lunch Rooms (626)	44 317	(4 935)	145 395	(5 394)	17
18 Opers.' Employment & Training (627) ...	99 793	18 172	265 132	(7 353)	18
19 C.O. Stationery and Printing (629)	113 826	17 895	315 899	(84 416)	19
20 C.O. House Service (630)	96 697	(3 043)	294 361	27 745	20
21 Miscellaneous C.O. Expenses (631)	251 506	77 059	622 341	155 368	21
22 Public Telephone Expenses (632)	7 695	6 299	13 623	2 934	22
23 Other Traffic Expenses (633)				(10)	23
24 Joint Traffic Expenses – Dr. (634)					24
25 Joint Traffic Expenses – Cr. (635)	1 237	235	3 255	463	25
26 **Total Traffic Expenses**	7 530 909	398 257	22 363 345	1 213 680	26
27 General Coml. Administration (640)	1 262 493	57 740	3 717 868	916 329	27
28 Advertising (642)...................	410 475	47 361	930 863	281 214	28
29 Sales Expense (643)	1 123 335	(2 463)	3 321 975	646 027	29
30 Connecting Company Relations (644)....	49 200	5 572	135 729	(4 161)	30
31 Local Commercial Operations (645)	3 545 688	326 091	10 161 396	1 432 757	31
32 Public Telephone Commissions (648)	467 866	29 152	1 299 894	131 453	32
33 Directory Expenses (649)	2 601 299	(20 068)	7 807 056	1 115 739	33
34 Other Commercial Expenses (650)	100	(139)	(352)	(3 546)	34
35 **Total Coml. & Marketing Expenses**	9 460 461	443 247	27 374 433	4 515 814	35
36 **Total Commercial Dept. Expenses**	4 025 745	345 178	11 571 666	1 477 552	36
37 **Marketing – Sales Expenses**	1 905 873	36 051	5 629 222	1 514 015	37
38 **Marketing – Other Expenses**	3 528 842	62 018	10 173 544	1 524 246	38
39 **Total Genl. Office Sal. & Expenses**	6 811 002	380 413	19 927 993	1 197 631	39
40 Insurance (668)	19 048	5 024	46 136	19 318	40
41 Accidents and Damages (669)	305 814	259 854	383 348	186 495	41
42 Operating Rents (671)	619 240	(6 998)	1 928 927	214 108	42
43 Relief and Pensions (672)	8 513 569	(432 764)	26 207 057	4 200 941	43
44 Tel. Franchise Requirements (673)	7 833	(18)	23 577	(1 198)	44
45 General Services & Licenses (674)	1 506 136	(743 486)	5 343 206	1 176 949	45
46 Other Expenses (675).	603 910	234 368	1 293 924	202 969	46
47 Tel. Franchise Requirements-Cr. (676) ..	7 833	(18)	23 577	(1 198)	47
48 Expenses Charged Const. – Cr. (677)	690 668	136 405	1 850 434	248 004	48
49 **Total Other Operating Expenses**.......	10 877 049	(820 406)	33 352 166	5 752 777	49
50 **Total Operating Expenses (301)**	73 475 493	3 984 630	214 030 348	24 696 976	50

( ) Denotes negative amount.

This report is a detailed breakdown of information about the entry Operating Expenses, which appears on line 2 of the Income Statement, Monthly Report No. 3.

TABLE 3A-7b

FORM S.N. I56 (2) (7-76)

SAMPLE TELEPHONE COMPANY

OPERATING EXPENSES

MONTHLY REPORT NO. 5
Sheet 2

March 1976

	a This Month	b Increase over Last Month	c This Year to Date	d Increase over Same Period Last Year	
1 Salaries (11)	74 913	19 015	222 137	(368 152)	1
2 Other Expenses (16)	23 849	6 601	63 039	(66 098)	2
3 Executive Department (661)	98 763	25 617	285 177	(434 250)	3
4 Salaries (11)	1 210 531	107 055	3 521 090	64 081	4
5 Machine Cost (12)	30 037	(2 502)	94 053	(15 166)	5
6 Postage (14)	457 632	(11 751)	1 419 924	321 834	6
7 Printing and Stationery (15)	35 397	6 314	103 491	(3 902)	7
8 Other Expenses (16)	125 332	13 514	347 816	(26 303)	8
9 Transfers — Cr. (19)	31 405	4 653	97 859	4 528	9
10 Total Operations (662-01)	1 827 525	107 977	5 388 516	336 016	10
11 Salaries (21)	332 669	20 717	963 487	155 327	11
12 Machine Cost (22)	493 792	(42 201)	1 500 119	217 377	12
13 Postage (24)	527	16	1 546	427	13
14 Printing and Stationery (25)	193 086	45 788	520 444	16 398	14
15 Other Expenses (26)	46 064	11 434	125 953	(12 657)	15
16 Transfer — Cr. (29)	291 856	43 472	738 245	117 885	16
17 Total EDP (662-02)	774 284	(7 715)	2 373 305	258 988	17
18 Salaries (31)	541 105	(12 010)	1 628 555	235 567	18
19 Machine Cost (32)	5 322	(1 872)	17 837	16 067	19
20 Postage (34)	75	(52)	330	251	20
21 Printing and Stationery (35)	11 946	6 004	35 728	5 143	21
22 Other Expenses (36)	36 516	(6 099)	122 597	42 025	22
23 Transfers — Cr. (39)	106 594	(8 895)	322 579	78 520	23
24 Total Data Systems (662-03)	488 371	(5 134)	1 482 469	220 534	24
25 Salaries (41)	281 854	(6 184)	856 365	51 306	25
26 Machine Cost (42)				(13)	26
27 Postage (44)	46	(185)	661	(564)	27
28 Printing and Stationery (45)	5 119	1 820	14 786	2 575	28
29 Other Expenses (46)	30 720	(1 307)	90 995	(2 738)	29
30 Transfers — Cr. (49)	5 880	344	12 469	(20 653)	30
31 Total Gen'l Acctg. (662-04)	311 859	(6 200)	950 338	71 219	31
32 Accounting Department (662)	3 402 041	88 926	10 194 631	886 758	32
33 Salaries (11)	81 772	11 510	229 418	23 152	33
34 Other Expenses (16)	47 157	(7 263)	149 167	15 926	34
35 Treasury Department (663)	128 929	4 247	378 586	39 078	35
36 Salaries 11, 21 and 31	77 303	(4 001)	237 938	39 308	36
37 Other Expenses 16, 26, 29, 36 and 39	14 950	322	47 873	5 109	37
38 Attorney's Fees and Costs 17, 27 and 37	7 923	(6 731)	67 807	53 952	38
39 Law Department (664)	100 176	(10 409)	353 619	98 369	39
40 Information (01) — Salaries (11)	213 713	786	638 271	96 452	40
41 — Other Exps. (16)	87 439	8 790	226 920	49 833	41
42 General Security (02) — Salaries (21)	161 745	3 739	478 165	27 752	42
43 — Other Exps. (26)	16 708	1 813	46 819	5 209	43
44 Personnel (03) — Salaries (31)	723 259	21 515	2 140 562	222 071	44
45 — Other Exps. (36)	162 169	27 098	457 855	171 509	45
46 Planning (04) — Salaries (41)	15 220	(1 003)	46 674	16 543	46
47 — Other Exps. (46)	1 518	559	3 578	(244)	47
48 General Services (05) — Salaries (51)	186 751	5 362	551 690	110 716	48
49 — Other Exps. (56)	82 818	24 351	194 577	(94 005)	49
50 Other (06) and (08) — Salaries (61) and (81)	99 939	14 283	277 654	48 007	50
51 — Other Exps. (66), (67) and (86)	41 021	4 783	144 754	34 904	51
52 Engineering (09)	1 288 784	159 951	3 508 454	(81 075)	52
53 Other Gen'l Off. Sal. & Expenses (665)	3 081 091	272 032	8 715 978	607 675	53
54 Total Gen'l Off. Sal. Expenses	6 811 002	380 413	19 927 993	1 197 631	54

— Denotes negative amount.

TABLE 3A-7c

FORM S.N. 156 (3) (1-72)
This Copy for

OPERATING EXPENSES

MONTHLY REPORT No. **5.**
Sheet 3

SAMPLE TELEPHONE COMPANY

MONTHLY AVERAGE PER TELEPHONE

MARCH 1976

		a This Month	b Same Mo. Last Yr.	c This Year to Date	d Same Per. Last Yr.	
1	Total Maintenance Expenses	3.436	2.878	3.206	2.848	1
2	Total Depreciation & Amort. Expenses	1.960	1.872	1.954	1.872	2
3	Total Traffic Expenses.	1.048	.984	1.040	1.009	3
4	Total Commercial Dept. Expenses . . .	.560	.470	.538	.481	4
5	Total Marketing Dept. Expenses	.756	.615	.735	.609	5
6	Executive Department	.014	.034	.013	.034	6
7	Accounting Department.	.472	.436	.474	.444	7
8	Treasury Department.	.018	.018	.018	.016	8
9	Law Department	.014	.012	.016	.012	9
10	Other Gen'l Off. Sal. & Expenses	.429	.362	.405	.387	10
11	Total Gen'l Office Salaries & Expenses .	.947	.862	.926	.893	11
12	Total Other Operating Expenses	1.153	1.290	1.550	1.316	12
13	Total Operating Expenses.	10.220	8.971	9.949	9.028	13

— Denotes negative amount.

Date issued   April 19, 1976

- -
Vice President and Comptroller

TABLE 3A-8a

FORM S.N. 281-C (1) (3-64)

QUARTERLY REPORT No. 6.
Sheet 1

SAMPLE TELEPHONE COMPANY
DEPRECIATION RESERVE

MARCH 1976

		a This Quarter	b This Year to Date	c Same Period Last Year	
1	Balance at Beginning of Period................		791 060 490	765 335 069	1
	Credits —				
2	Charged Depreciation Expense (608)		42 028 166	39 257 386	2
3	Charged Extraordinary Retirements (609)				3
4	Charged Clearing Accounts (702, 704)..........		1 115 859	1 070 860	4
5	Other Credits..............................		157 796	34 204	5
6	Total Credits........................		43 301 822	40 362 451	6
	Debits —				
7	Net Charge for Plant Retired		27 987 241	29 111 048	7
8	Other Debits				8
9	Total Debits		27 987 241	29 111 048	9
10	Net Increase in Reserve.....................		15 314 580	11 251 402	10
11	Balance at End of Period....................		806 375 071	776 586 472	11
12	Average Depreciable Tel. Plant (Accts. 212 thru 264).....................		3 608 770 955	3 336 092 505	12
13	% Depreciation (608, 702, 704) to Avg. Depr. Tel. Plant (Annual Basis).......... %		4.78 %	4.84 %	13
14	Total Depreciable Tel. Plant (Accts. 212 thru 264)—End of Period	X X X	3 640 411 237	3 365 592 262	14
15	% Depreciation Reserve (171) to Total Depr'ble Tel. Plant—End of Period........	X X X	22.15 %	23.07 %	15

( ) Denotes negative amount.

Lines 2 to 9 on Sheet 1 summarize the credits and debits to the Depreciation Reserve account. Sheet 2 provides the details for entries to the Depreciation Reserve; the net is carried on line 10 of Sheet 1. The amount charged to Depreciation Reserve (column *g*, Sheet 2) is equal to the book cost, plus the cost of removal, less salvage. The amounts in column *g* equal the sum of columns *a* through *d* less the sum of columns *e* and *f*.

FORM S.N. 281-C (2) (3-68)

TABLE 3A-8b

SAMPLE TELEPHONE COMPANY

DEPRECIATION RESERVE

QUARTERLY REPORT No. 6. Sheet 2

MARCH 1976

NET CHARGE TO RESERVE FOR PLANT RETIRED
This Year to Date

| Classes of Depreciable Plant | a — Charge to Reserve for Plant Sold with Traffic (Sheet 3 Col. d) | Entries to Reserve for Other Plant Retired | | | | | g — Net Charge to Reserve for Plant Retired (Cols. a thru d minus Cols. e and f) |
| | | Book Cost | | d — Cost of Removal | Salvage, Insurance and Miscellaneous Adjustments | | |
		b — Plant Sold without Traffic to other Consolidated Bell System Companies	c — Other		e — Plant Sold without Traffic to other Consolidated Bell System Companies	f — *Other	
1 Buildings			592 406	183 993		73 470	702 929
2 C.O.E. — Manual			1 502 227	52 434		(32 714)	1 587 376
3 — Panel			5 722	544 930		50 881	499 770
4 — Step-by-Step			726 939	305 876		349 711	683 103
5 — Crossbar		33 019	1 013 152	247 988		161 701	1 099 440
6 — Circuit			490 095	232 731	20 489	301 552	433 804
7 — Radio			81 526	6 291		9 403	78 413
8 — Electronic			252 535	62 991		208 103	107 423
9 Station Apparatus — Teletypewriter		12 484	539 626	x x x	10 200	2 748	539 162
10 — Telephone & Misc.		563 996	4 976 576	x x x	453 085	125 740	4 961 748
11 — Radio			21 574	x x x		137	21 436
12 Station Connections			9 206 638	1 764 999		9 485	10 962 152
13 Large Priv. Branch Exchanges			2 703 239	263 944		1 048 834	1 918 349
14 Pole Lines			675 689	288 896		82 123	882 462
15 Aerial Cable — Exchange			1 905 962	473 325		286 997	2 092 291
16 — Toll			154 143	39 932		250 814	(56 739)
17 Underground Cable — Exchange			360 393	85 235		76 119	369 510
18 — Toll			16 870	19 923		39 347	(2 553)
19 Buried Cable — Exchange			125 191	6 713		3 962	127 942
20 — Toll			593	207		202	597
21 Submarine Cable — Exchange				126			126
22 — Toll				(26)			(26)
23 Aerial Wire — Exchange			465 421	127 278		6 771	585 928
24 — Toll			11 811	(3 414)		1 513	6 883
25 Underground Conduit			59 884	45 322			105 210
26 Furniture & Office Equipment			135 384			26 668	108 716
27 Computer & AMA Systems			(13 946)	43		1	(13 904)
28 Motor Vehicles			157 597			14 032	143 564
29 Other Work Equipment			42 787			667	42 119
30 Total		609 501	26 210 048	4 749 745	483 775	3 098 278	27 987 241

* Insurance and Miscellaneous Adjustments included in this column are detailed on Sheet 3. ( ) Denotes negative amount.

Date issued April 22, 1976

_____
Vice President and Comptroller

TABLE 3A-9a

FORM S.N. 172(1) (1-76)

SAMPLE TELEPHONE COMPANY
CLEARING AND CERTAIN OTHER ACCOUNTS

QUARTERLY REPORT No. **19.**
Sheet 1

March 1976

	a This Quarter	b This Year To Date	Increase Over Same Period Last Year		
			c Amount	d Percent	
MOTOR VEHICLE AND GARAGE EXPENSE					
1 Balance Beginning of Year	X X X	(1 987)	X X X	X X	1
2 Charges–Rents & Repairs of Rented Quarters (18)		468 047	71 344	17.98	2
3 —Rented Motor Vehicles (22)		262 352	79 318	43.33	3
4 —Other Expenses		4 445 174	348 465	8.51	4
5 —Total		5 175 575	499 128	10.67	5
6 Clearance (detailed on Sheets 3 and 4)		5 149 598	X X X	X X	6
7 Balance End of Period (702-01)	X X X	23 989	X X X	X X	7
SPECIAL TOOL & WORK EQUIPMENT EXPENSE					
8 Balance Beginning of Year	X X X		X X X	X X	8
9 Charges		(17)	(17)		9
10 Clearance (detailed on Sheet 3)		(17)	X X X	X X	10
11 Balance End of Period (702-03)	X X X		X X X	X X	11
OTHER SHOP EXPENSE					
12 Balance Beginning of Year	X X X	(1 448)	X X X	X X	12
13 Charges		458	7	1.61	13
14 Clearance (detailed on Sheet 3)		304	X X X	X X	14
15 Balance End of Period (702-04)	X X X	(1 295)	X X X	X X	15
OTHER TOOL & WORK EQUIPMENT EXPENSE					
16 Balance Beginning of Year	X X X	5 164	X X X	X X	16
17 Charges		1 143 503	207 368	22.15	17
18 Clearance (detailed on Sheet 3)		1 128 981	X X X	X X	18
19 Balance End of Period (702-05)	X X X	19 686	X X X	X X	19
PLANT SUPV. EXP.-PLANT SUPERINTENDENCE					
20 Balance Beginning of Year	X X X	(1 042)	X X X	X X	20
21 Charges–Salaries (11)		2 444 845	260 193	11.91	21
22 —Rents & Repairs of Rented Quarters (15)		58 223	7 034	13.74	22
23 —Other Expenses (16)		584 547	35 448	6.46	23
24 —Total		3 087 616	302 675	10.87	24
25 Clearance (detailed on Sheet 3)		3 213 096	X X X	X X	25
26 Balance End of Period (706-01)	X X X	(36 522)	X X X	X X	26
MISCELLANEOUS EXP. OF PLANT LABOR					
27 Balance Beginning of Year	X X X	340	X X X	X X	27
28 Charges–Rents & Repairs of Rented Quarters (01)		165 909	33 907	25.69	28
29 —Other Expenses (02)		2 313 640	174 788	8.17	29
30 —Total		2 479 549	208 695	9.19	30
31 Clearance (detailed on Sheet 3)		2 458 160	X X X	X X	31
32 Balance End of Period (709)	X X X	21 728	X X X	X X	32
HOUSE SERVICE EXPENSE					
33 Balance Beginning of Year	X X X		X X X	X X	33
34 Charges		6 000 722	199 462	3.44	34
35 Clearance (detailed on Sheets 3 and 4)		5 837 184	X X X	X X	35
36 Balance End of Period (707)	X X X	163 588	X X X	X X	36
SUPPLY EXPENSE					
37 Balance Beginning of Year	X X X	(1 186)	X X X	X X	37
38 Charges–Salaries (11)		993 946	363 934	57.77	38
39 —Rents & Repairs of Rented Quarters (15)		204 729	37 965	22.77	39
40 —Other Expenses (16)		763 455	218 395	40.77	40
41 —Total		1 962 130	620 295	46.23	41
42 Clearance (detailed on Sheet 3)		2 000 704	X X X	X X	42
43 Balance End of Period (704)	X X X	(39 760)	X X X	X X	43
44 Total Material for Supply Expense Clearance . .	X X X	34 681 406	X X X	X X	44
45 % Supply Expense Clearance to Total Material (Line 42 Col. b ÷ Line 44, Col. b)	X X X	5.77 %	X X X	X X	45

( ) Denotes negative amount.

Totals may not balance due to reporting to nearest dollar.

The clearing accounts (the 700 series) are a temporary catchall for charges that are not made directly to final accounts. However, these charges are cleared to the final reporting codes or accounts monthly, by the application of average hourly rates to the hours summarized from time reports or else on a percentage basis—whichever is appropriate.

Sheets 1 and 2 list the various clearing accounts for such items as vehicles, tools, supplies, and engineering. Sheet 3 analyzes the clearances shown on Sheets 1 and 2 in these four groups: Construction, Depreciation Reserve, Maintenance, and Other. Sheet 4 then analyzes some of the clearances shown under column d (Other) of Sheet 3.

TABLE 3A-9b

FORM S. N. 172(2) (2-77)

QUARTERLY REPORT No. 19.
Sheet 2

SAMPLE TELEPHONE COMPANY
CLEARING AND CERTAIN OTHER ACCOUNTS MARCH 1976

		a This Quarter			b This Year To Date	Increase Over Same Period Last Year				
						c Amount			d Per Cent	
	PLANT LABOR									
1	Balance Beginning of Year	×	×	×		×	×	×	× ×	1
2	Charges (Prior to Transfers to Accts. 712 & 713) .				56 955 928	6 218 032			12.26	2
3	Clearance and Transfers {to Accts. other than 712 and 713*.				48 768 206	×	×	×	× ×	3
4	to Account 712				8 190 566	×	×	×	× ×	4
5	to Account 713				(2 844)	×	×	×	× ×	5
6	Balance End of Period (710)	×	×	×		×	×	×	× ×	6
	PLANT LABOR – ANNUAL									
7	Balance Beginning of Year	×	×	×	25	×	×	×	× ×	7
8	Charges (line 4)				8 190 566	402 763			5.17	8
9	Clearance and Transfers {to Accounts other than 713......*				10 328 335	×	×	×	× ×	9
10	to Account 713					×	×	×	× ×	10
11	Balance End of Period (712)	×	×	×	(2 137 742)	×	×	×	× ×	11
	PLANT LABOR – VARIABLE									
12	Balance Beginning of Year	×	×	×	(1 952)	×	×	×	× ×	12
13	Charges (lines 5 and 10)				(2 844)	(689 899)			(100.41)	13
14	Clearance ....*				(17 953)	×	×	×	× ×	14
15	Balance End of Period (713)	×	×	×	13 156	×	×	×	× ×	15
	ENGINEERING EXPENSE									
16	Balance Beginning of Year	×	×	×	71 142	×	×	×	× ×	16
17	Charges—Salaries (11)				8 983 770	863 821			10.64	17
18	—Rents & Repairs of Rented Quarters (15)				190 417	91 469			92.44	18
19	—Other Expenses (16)				1 270 088	(16 322)			(1.27)	19
20	—Total				10 444 276	938 967			9.88	20
21	Clearance—to Accts. other than 139-23.........#				8 198 460	×	×	×	× ×	21
22	—to Account 139-23				2 891 455	×	×	×	× ×	22
23	Balance End of Period (705)	×	×	×	(574 497)	×	×	×	× ×	23
	PLANT SUPV. EXPENSE – PLANT ENGINEERING									
24	Balance Beginning of Year	×	×	×	4 053	×	×	×	× ×	24
25	Charges—Salaries (21)				4 065 417	174 286			4.48	25
26	—Rents & Repairs of Rented Quarters (25)				73 015	30 333			71.07	26
27	—Other Expenses (26)				835 389	358 516			75.18	27
28	—Total				4 973 821	563 136			12.77	28
29	Clearance—to Accts. other than 139-33 ....#				3 408 737	×	×	×	× ×	29
30	—to Account 139-33				1 444 367	×	×	×	× ×	30
31	Balance End of Period (706-02)	×	×	×	124 770	×	×	×	× ×	31
	PORTION OF NET. DESIGN, BUS. SVCS. DESIGN, NET. ADM. WAGES, BUS. SVCS. FAC. ADM. WAGES, NET. ADM. DATA PROCESSING, BUS. SVCS. FAC. ADM. DATA PROCESSING AND FORECAST AND DEVEL. TRANS. TO CONST.									
32	Traf Exp (621-20Cr,621-40Cr,624-220Cr,624-250Cr)#				441 143	132 328			42.85	32
33	Genl. Com'l Admin. (640-190 CR. and 390 CR.) #									33
	OTHER DEFERRED CHARGES – ENGINEERING – PROJECTS									
34	Balance Beginning of Year	×	×	×	3 357 649	×	×	×	× ×	34
35	Charges				2 003 033	549 781			37.83	35
36	Clearance ....#				1 495 111	×	×	×	× ×	36
37	Balance End of Period (139-13)	×	×	×	3 865 571	×	×	×	× ×	37
	OTHER DEFERRED CHARGES – ENGINEERING – UNCLASSIFIED PRODUCTIVE TIME									
38	Balance Beginning of Year	×	×	×	(293)	×	×	×	× ×	38
39	Charges (lines 22 and 30)				4 335 822	104 322			2.47	39
40	Clearance—General Engineering Portion ....#				2 886 116	×	×	×	× ×	40
41	—Plant Engineering Portion ....#				1 314 796	×	×	×	× ×	41
42	Balance End of Period (139-23 and 33)	×	×	×	134 616	×	×	×	× ×	42
	PERCENT OF NET. DESIGN, BUS. SVCS. DESIGN, NET. ADM. WAGES, BUS. SVCS. FAC. ADM. WAGES, NET. ADM. DATA PROCESSING, BUS. SVCS. FAC. ADM. DATA PROCESSING AND FORECAST AND DEVEL. TRANS. TO CONST.									
						e This Year To Date			f Same Period Last Year	
43	From Traf Exp(621-21,25,26,27,41,45,46&47,624-122,222,322,422,522,622,722,125,225&325,631-37&38)					30.9 %			24.5%	43
44	From General Com'l Admin.– (640-191, 196, 391 and 396)..					%			%	44

Totals may not balance due to reporting to nearest dollar.   * Detailed on Sheet 3
( ) Denotes negative amount.   # Detailed on Sheets 3 and 4.

TABLE 3A-9c

SAMPLE TELEPHONE COMPANY

CLEARING AND CERTAIN OTHER ACCOUNTS

MARCH 1975

ANALYSIS OF CLEARANCES (Sheets 1 and 2, Col. b)

#		Construction	Depreciation Reserve	Maintenance	Other
	VEHICLE, SPECIAL TOOL, AND SHOP EXPENSE				
1	Motor Vehicle and Garage Expense ....(702-01)	1 445 121	201 719	2 365 838	• 1 136 918
2	Special Tool and Work Equipment Expense.......(702-03)	105	12	182	(17)
3	Other Shop Expense............................(702-04)				
4	**SUPPLY EXPENSE**...............................(704)	1 185 275		810 760	• 4 669
5	**HOUSE SERVICE EXPENSE**........................(707)	30 826	3 961	3 236 894	• 2 565 500
	ENGINEERING				
6	Clearances of Current Charges — General (Sheet 2, lines 21 and 40)..(705)	2 177 252	288 883	2 935 666	• 5 682 773
7	Plant (Sheet 2, lines 29 and 41)..(706-02)	2 244 367	267 500	902 860	• 1 308 805
8	Traf. (Sh.2,L.32)(621-20,621-40,624-220,&624-250)	171 086	X	X	• 270 057
9	Com'l (Sheet 2. line 33) (640-190 & 390)		X	X	•
10	Charges—Engineering—Clearances from Acct. — General 139-13 (705)	691 886	35 155	67 534	• 27 510
11	139-13 Other Deferred — Plant 139-13 (706-02)	463 448	13 866	32 250	• 15 392
12	Traffic 139-13 (621-20 & 621-40)	145 586	X	X	• 2 481
13	Projects (Sheet 2, line 36) — Com'l 139-13 (640-190 & 390)		X	X	•
14	Other Clearances Based on Engineering Costs — General Expenses	410 499	X	X	309
15	Social Security Taxes	320 695	X	X	1 061
16	Relief and Pensions	1 196 960	X	X	4 774
17	Total (Lines 6 through 16)	7 821 782	605 405	3 938 311	7 313 166
	PLANT LABOR & CLEARANCES Based Thereon				
18	Plant Labor (Sheet 2, Lines 3, 9 and 14)..(710, 712, 713)	13 689 670	1 912 669	39 362 940	4 113 306
19	Plant Superintendence ....(706-01)	787 732	106 451	2 184 920	43 992
20	Other Tool and Work Equip. Exp...(702-05)	379 202	50 777	672 147	26 854
21	Miscellaneous Exp. of Plant Labor (709)	619 286	83 314	1 718 607	36 952
22	Related Clearances — General Expenses	1 439 934	X	X	(309)
23	Social Security Taxes	872 328	X	X	22 119
24	Relief and Pensions	4 228 423	X	X	96 308
25	Total (Lines 19 through 24)	8 326 907	240 543	4 575 675	225 917
26	% Clearances to Plant Labor Costs (Line 25 ÷ Line 18)	60.8 %	X	X	X

* See Sheet 4 for Analysis.

Totals may not balance due to reporting to nearest dollar.

( ) Denotes negative amount.

Line 8 Col A includes $33,653 transferred from Acct 624-220

FORM S. N. 172(4) (2-77)

TABLE 3A-9d

SAMPLE TELEPHONE COMPANY

CLEARING AND CERTAIN OTHER ACCOUNTS

ANALYSIS OF CERTAIN CLEARANCES ON SHEET 3 COLUMN D

MARCH 1976

Line		a Motor Vehicle and Garage Expense (Sheet 3, Line 1)	b House Service Expense (Sheet 3, Line 5)
1	Rest and Lunch Rooms (626)	x	152 709
2	Central Office House Service (630)	x	287 594
3	Other Traffic Accounts	x	99 243
4	Total Traffic Accounts (Lines 1 through 3)	49 587	539 547
5	Total Commercial Accounts	279 809	354 297
6	Clearing Accounts	632 397	1 120 810
7	Other	175 124	550 845
8	Total (Lines 1 through 7; also Sheet 3, Column d, Lines 1 and 5, respectively)	1 136 918	2 565 500

Line			c Acct. 139-13 Other Def. Chgs.—Engineering—Projects	d Acct. 665-09 Gen'l Ofce. Exp.—Engineering	e Acct. 675 Other Expenses	f Other
9	Clearances of Current Charges—Engineering	General (Sheet 3, Line 6)	1 202 672	2 692 066		1 788 034
10		Plant (Sheet 3, Line 7)	530 303	773 484		5 017
11		Traffic (Sheet 3, Line 8)	270 057	x	x	
12		Commercial (Sheet 3, Line 9)	x	x	x	
13	Clearances from Acct. 139-13 Other Deferred Charges—Engineering—Projects	General (Sheet 3, Line 10)	x	27 510	x	
14		Plant (Sheet 3, Line 11)	x	15 392	x	2 481
15		Traffic (Sheet 3, Line 12)	x	x	x	
16		Commercial (Sheet 3, Line 13)	x	x	x	
17	Total (Lines 9 through 16)		2 003 033	3 508 454		1 795 532

MISCELLANEOUS DATA

Line		g This Year To Date	h Same Period Last Year
18	% of General Office Salaries and Expenses (Accounts 661 through 665, except Accounts 662-58,* 664-17, 664-02, 665-06, and 665-09) Transferred to Construction	18.4 %	17.3 %

Totals may not balance due to reporting to nearest dollar.

( ) Denotes negative amount.

*Account 662-58 .... $ 6 299 958

Account 621-27 & 621-47 $ 1 427 127   (included in Line 9, Column f)

Date issued   April 26, 1976

....
Vice President and Comptroller

FORM S.N. 195 (8-75)

TABLE 3A-10

CONSTRUCTION EXPENDITURES AND RETIREMENTS

SAMPLE TELEPHONE COMPANY

MONTHLY REPORT No. 21

MARCH 1976

	a) Specific Estimates	b) Routine Estimates	c) Gross Const. Expenditures (a + b)	d) Intra-Company Reused Material	e) Inter-Company* Reused Material	f) Plant Retired This Year to Date	
	Construction Expenditures — This Year to Date						
1 Land	16 989	7 558	24 547	XXX	XXX	454	1
2 Buildings	1 847 174	597 387	2 444 562			492 881	2
3 Total Land and Buildings	1 864 163	604 946	2 469 109			493 336	3
4 Manual	(33 933)	10 884	(23 048)	1 166		1 502 227	4
5 Panel	64 314	51 957	116 272	10 280		5 722	5
6 Step-by-Step	2 108 372	395 477	2 503 850	318 673		726 939	6
7 Crossbar	5 835 832	583 433	6 419 265	143 443	5 581	1 013 152	7
8 Circuit	10 516 718	976 743	11 493 462	112 649	103 563	501 027	8
9 Radio	391 004	66 485	457 490	53 126		81 526	9
10 Electronic	17 960 283	357 445	18 317 728	67 451		252 535	10
11 Total Central Office Equipment	36 842 592	2 442 427	39 285 019	706 791	109 144	4 083 131	11
12 Station Apparatus	8 842	10 195 336	10 204 178	2 284	12 552	5 539 414	12
12A Base Shipment Adjustment	XXX	XXX	XXX	XXX	XXX	XXX	12A
13 Station Connections	8 721	17 087 720	7 096 441	31 495		9 206 638	13
14 Large Private Branch Exchanges	76 546	3 284 697	3 361 243	828 337	4 227	2 706 992	14
15 Total Station Equipment φ	94 109	30 567 754	30 661 863	862 137	16 779	17 453 045	15
16 Pole Lines	261 112	1 281 708	1 542 820	467		675 689	16
17 Underground Conduit	1 783 434	463 222	2 246 656	868		59 887	17
18 Total Outside Plant Structures	2 044 547	1 744 930	3 789 477	1 336		735 576	18
19 Aerial Cable – Building	206 060	2 086 327	2 292 387	31 450		270 388	19
20 – Other	3 570 843	4 454 269	8 025 112	30 774		1 635 574	20
21 Underground Cable	4 218 828	1 073 952	5 292 781	44 572		360 393	21
22 Buried Cable	299 878	1 094 045	1 393 923	3 793		125 191	22
23 Submarine Cable							23
24 Aerial Wire	1 958	252 163	254 122	752		465 421	24
25 Total Exchange Lines	8 297 569	8 960 757	17 258 327	111 343		2 856 969	25
26 Aerial Cable	97 860	46 769	144 629	1 159		154 143	26
27 Underground Cable	559 823	46 210	606 034	1 506		16 870	27
28 Buried Cable	16 564	6 699	23 264			593	28
29 Submarine Cable							29
30 Aerial Wire						11 811	30
31 Total Toll Lines	674 248	99 679	773 928	2 665		183 418	31
32 Total Outside Plant (L18 + 25 + 31)	11 016 365	10 805 368	21 821 733	115 345		3 775 964	32
33 Furniture and Office Equipment	774 604	(237 647)	536 957	1 403		121 438	33
34 Vehicles & Other Work Equipment	4 581	987 354	991 935			200 384	34
35 Total General Equipment	779 185	749 706	1 528 892	1 403		321 822	35
36 Grand Total – This Year to Date φ	50 596 417	45 170 203	95 766 620	1 685 677	125 923	26 127 300	36

* Material purchased from a Consolidated Bell System Company.

( ) Denotes negative amount.

φ Lines 15 and 36 will not be additive across the columns if an entry appears on line 12A.

....
Vice President and Comptroller

Listed on this Monthly Report No. 21 are the construction expenditures by the types of estimates (specific and routine) and the retirements. These data are reflected in the proper investment accounts—100.1, 100.2, or 100.3—listed on the No. 2 Report series.

4

INTRODUCTION TO FINANCING

INTRODUCTION

This chapter is an overview of financing, which is the activity of raising funds for a business. There are discussions of why funds are required, how they are obtained and in what forms, and how the investors who provide much of the money measure the firm's performance. The subject of financing is of particular interest to the student of engineering economy, because the concepts of financing are the basis for the discipline of engineering economy.

NEED FOR FUNDS

The flow of cash is vital to a business. In fact, some writers have compared it with the flow of blood in a living creature. Certainly a business could not exist for long if the flow of funds ceased.

Businesses need funds for three primary purposes:

1. To pay current operating expenditures

2. To make the necessary capital expenditures for replacing old plant and providing for growth

3. To meet obligations to those who have invested in the business

Operating expenditures, which require cash, include wages, salaries, employee benefits, provision for pensions, supplies, rents, insurance, advertising, and various taxes. Capital expenditures are needed to buy the land, buildings, outside plant, and equipment which the business needs in order to provide service. Obligations to investors include the payment of interest and principal to lenders and bondholders, the payment of dividends to stockholders, and the preservation of the value of the stockholders' investment.

Sources of Funds

Fundamentally, there are just two ways in which funds can enter the business. The first is revenue. If the company is to remain financially healthy, revenues must be adequate to cover the current operating expenditures, to provide the capital needed to replace plant as it wears out, and to meet the obligations to investors. The portion of revenue which is not required for these needs can be retained by the company and reinvested in the business. In this way, the revenues of the business itself are a source of funds for investment in the business. Revenues, then, are the source of *internally generated funds*.

The second way funds enter the business is from existing or new investors who buy new stock or bonds which the company sells. Because this second source comes from outside the current business operation, these funds are called *externally generated funds*.

A very small amount of funds also enters the business from the sale of plant or property that the business no longer wants or needs. This source is so small in a business like a telephone company that it is insignificant to the overall financing of the business.

All these sources were used in establishing and operating the Appliance Experts Company, the small business discussed in Chapter 2. Chapter 2 developed the balance sheet and income statement for that business, which are reproduced here as Figures 4-1 and 4-2. However, those reports do not describe the actual flow of funds in the business very well.

THE FUNDS-FLOW STATEMENT

As Chapter 2 pointed out, a third major accounting report is needed to show the complete picture, since the balance sheets and income statements do not always disclose the complete financial activity of a business. As a result, the Accounting Principles Board (APB) of the American Institute of Certified Public Accountants (AICPA) issued its Opinion No. 19 requiring companies to include a *Funds-Flow Statement* in annual reports.

FIGURE 4-1
**Appliance Experts Company
BALANCE SHEET**
December 31, 19__

Assets			Equities		
Current Assets			Current Liabilities		
Cash	$ 7,440		Accounts Payable	$8,620	
Accounts Receivable	2,000		Interest Payable	40	
Inventory	1,715		Total, Current		$ 8,660
Total, Current		$11,155	Other Liabilities		
			Note Payable		5,000
Fixed Assets			Owner's Equity		
Land		6,000	A. Fixer, Capital		20,050
Building	$15,000				
Less: Accumulated Dep.	50	14,950			
Furniture and Fixtures	1,620				
Less: Accumulated Dep.	15	1,605			
		$33,710			$33,710

FIGURE 4-2
Appliance Experts Company
INCOME STATEMENT
Year Ended December 31, 19___

Sales		$440
Beginning Inventory	—	
Purchases	$2,000	
	$2,000	
Less: Ending Inventory	1,715	
Cost of Goods Sold		285
Gross Profit		$155
Expenses		
Depreciation:		
Building	50	
Furniture and Fixtures	15	
Interest	40	
Total Expenses		105
Net Income		$ 50

Purpose of the Funds-Flow Statement

Funds-flow statements are intended to summarize the financial and investment activities of a company resulting from the operation of the business and from changes in its financial condition during an accounting period. In its opinion, therefore, the APB further recommended that the title "Funds-Flow Statement" be changed to "Statement of Changes in Financial Position." This title change has in fact been adopted by AT&T in its annual report.

Chapter 2 also mentioned that the title "Statement of Financial Position" was suggested in place of the title "Balance Sheet." The statement of changes in financial position would then account for the changes in the balance sheet accounts from one year to the next. Since the associated Income Statement can be used to show the necessary details of expenses, the details of funds flow need not be included in the funds-flow statement. Net income from the income statement therefore becomes the starting point for funds flow.

Sources and Application of Funds

A clearly presented funds-flow statement is important to both investors and managers because it helps them to recognize changes in the financial position. The funds flowing from operations, sale of stock, and borrowing are accumulated as sources. The outflows for dividends, debt repayment, and fixed assets are listed as applications. Changes in the balance sheet accounts furnish additional data for developing funds flow; and these data are supplemented by the disclosure of gross—rather than net—amounts of offsetting or related transactions. The sources and applications of funds can be described as follows:

Sources	*Applications*
Increases in Liabilities	Decreases in Liabilities
Decreases in Assets	Increases in Assets
Increases in Owners' Equity	Decreases in Owners' Equity

The three categories listed—assets, liabilities, and owners' equity—are the elements which make up the basic accounting equation:

$$\text{Assets} = \text{liabilities} + \text{owners' equity}$$

Usually, assets increase from one balance sheet to the next, thus representing an application of funds. Maintaining the equality of the balance sheet requires an equal increase in sources—either liabilities or owners' equity, or both. In the less frequent instance, assets might be reduced (a source of funds) if, for example, one asset such as a marketable security is conveyed to a creditor to satisfy a debt. The reduction in liabilities would then be a funds application.

An Example of a Funds-Flow Statement

The funds-flow statement for the Appliance Experts Company can be developed from a review of the company's transactions in Chapter 2. Figure 4-3 is the Statement of Changes in Financial Position (Funds-Flow Statement) for that company. It illustrates how this report helps give a more complete picture of that company's financial position. Since it is a new company, the Balance Sheet (Statement of Financial Position)—Figure 4-1—also represents the changes, because the previous balance sheet accounts are zero.

However, in Figure 4-3 the flow of funds is greater than the changes in either the balance sheet accounts or the income statement would indicate. A review of the formation of the Appliance Experts Company will confirm that $25,620 was actually paid out for land, building, and furniture and fixtures; and that a portion of the land was subsequently sold for $3,000. The accountant preparing the funds-flow statement knows of these transactions and properly includes them. However, the income statement and balance sheet do not illustrate transactions like this $3,000 sale of land.

FIGURE 4-3
Appliance Experts Company
STATEMENT OF CHANGES IN FINANCIAL POSITION
(FUNDS-FLOW STATEMENT)
Year Ended December 31, 19__

Sources of Funds

From Operations:		
Net Income	$50	
Depreciation	65	$ 115
From Financing:		
Long-Term Borrowing		5,000
Owner's Capital Contribution		20,000
From Sale of Land		3,000
		$28,115

Application of Funds

Acquisition of Fixed Assets (Plant)	$25,620
Change in Working Capital	2,495
	$28,115

The APB opinion also recommends showing the details of the changes in the accounts that make up working capital. One way of doing this is to include a table listing changes in the current accounts, as follows:

Change in working capital is accounted for by:	
Increase in Current Assets	
Cash	$ 7,440
Accounts Receivable	2,000
Inventory	1,715
	$11,155
Less—Increase in Current Liabilities	
Accounts Payable	$ 8,620
Interest Payable	40
	$ 8,660
Change in Working Capital	$ 2,495

Another way to display the details of working capital is to include changes in the current accounts in the funds-flow statement. When this is done, however, the statement is regarded as a cash-flow statement. Figure 4-4 shows the cash-flow format.

A decrease in working capital is a source of funds, whereas an increase is an application of funds. Since working capital is a combination of assets and liabilities, a firm may consistently choose to list working capital either as a source or as an application. If it is listed as a source and working capital actually increases, the increase would be shown in parentheses to indicate a negative source (in reality, an application).

Most economy studies can assume that there is no difference in working capital between alternative courses of action. As a result, the funds-flow statement and the cash-flow statement can usually be considered as identical.

Strictly interpreted, the cash-flow statement in Figure 4-4 is not a true representation of the cash transactions which occurred in the one month of this company's life. This is because it includes as a source the $3,000 from the sale of land, and $2,000 of that has yet to be actually received. When the cash-flow statement is prepared over a full year, that type of transaction will be completed; the statement can then be considered representative of the actual cash flow for the year.

OBTAINING FUNDS

This small appliance repair business obtained funds from the two major sources for larger businesses. Furthermore, it got some funds from the sale of a fixed asset, which is generally a negligible source for a larger business. Because Appliance Experts had been in business for only one month, the internally generated funds were quite small. In larger businesses, this source is much greater; and in some businesses, it provides all the funds required.

Most companies begin with external financing, and many must also go outside the business to get additional funds. Mr. Fixer did not own the business and was therefore external to it, until he put up the $20,000. Then the business needed more money, so it got a bank loan. Therefore, Appliance Experts has both forms of external financing—equity, or ownership, and debt.

Even though the Appliance Experts Company has all these elements, it is not very

FIGURE 4-4
Appliance Experts Company
CASH-FLOW STATEMENT
Year Ended December 31, 19___

Source of Funds

From Operations:		
Net Income	$50	
Depreciation	65	$ 115
Increase in Payables		8,660
From Long-Term Borrowing		5,000
From Owner's Capital Contribution		20,000
From Sale of Land		3,000
		$36,775

Application of Funds

Increase in Cash	$ 7,440
Increase in Receivables	2,000
Increase in Inventory	1,715
Acquisition of Fixed Assets	25,620
	$36,775

representative of a larger firm. Besides, it did not have any income tax, which would change the picture somewhat. For the remainder of this chapter, then, the financial reports of the mythical Old Glory Telephone Company will be used because they are more illustrative of a large company. Figure 4-5 is a condensed income statement; Figure 4-6 is a condensed balance sheet as of December 31, 1976 and also shows the data from the previous balance sheet; and Figure 4-7 is a funds-flow statement. Figure 4-7 shows that Old Glory had several internal sources, which were not available to the small appliance repair business.

Internally Generated Funds

As pointed out earlier, revenues are the actual source of internally generated funds. However, because the amount of internal funding is equal to revenues less the expenses which actually require cash payment, this amount is also equal to those expenses which do not require current cash payments plus the net income. Therefore, internally generated funds are made up of depreciation, net income, and certain tax effects. These elements are described in the following sections. Figure 4-8 illustrates that their ultimate source is revenue.

Depreciation

As Chapters 2 and 3 explained, the cost of telephone plant is charged to an asset account at the time the plant is installed. Then, each year of the plant's service life, a portion of its cost is charged against that year's revenues. This charge, called *deprecia-*

tion, is designed to provide for the recovery of capital invested in plant as that plant is used up.

In theory, depreciation accruals could actually be repaid to the investors, and in some ventures this is done. However, in a business which requires substantial amounts of money each year for construction, there would be no point in repaying the investors an amount equal to the depreciation accrual and then going to the capital market for that much more in new funds. Instead, depreciation accruals are reinvested in the business, and these accruals provide funds for the purchase of new plant. Therefore, $140,000 of depreciation expense is shown on the funds-flow statement of the Old Glory Telephone Company as a source of funds.

In a sense, the reinvestment of depreciation represents a recycling of capital. Money invested in plant which is growing older and reaching the end of its usefulness is simply transferred to new plant. Chapter 7 will discuss depreciation further.

Net Income

The net income from the income statement is another source of internally generated funds. However, not all of the net income can be reinvested in the business. Earnings which are not paid out to the owners as dividends are generally referred to as *retained earnings,* or *reinvested earnings.* The portion retained is a source of capital for the company. These retained earnings actually represent an increase in the investment in the company by the owners (the stockholders).

The funds-flow statement for the Old Glory Telephone Company shows net income as a source of funds, and dividend payments as an application. However, the net effect is the same as if the difference between them—or the reinvested earnings—were shown as a source and the payment of dividends were not even considered.

FIGURE 4-5
Old Glory Telephone Company
CONDENSED INCOME STATEMENT
Year 1976

Operating Revenues		$810,000
Maintenance Expense	$175,000	
Depreciation Expense	140,000	
Other Operating Expense	235,000	
Total Operating Expense		550,000
Net Operating Revenues		$260,000
Income Taxes		
Current	$ 19,000	
Deferred	38,000	
Investment Credit—Net	10,000	
Total Operating Taxes		67,000
Operating Income		$193,000
Interest		68,000
Net Income		$125,000
Dividends		82,000
Retained Earnings		$ 43,000

FIGURE 4-6
Old Glory Telephone Company
CONDENSED BALANCE SHEET

	12/31/76	12/31/75
Assets		
Plant in Service	$2,875,000	$2,665,000
Less: Accumulated Depreciation	719,000	665,000
Net Plant	2,156,000	2,000,000
Current Assets	169,000	169,000
Total Assets	$2,325,000	$2,169,000
Liabilities and Capital		
Equity		
Common Stock (25,000 Shares)	$ 625,000	$ 625,000
Retained Earnings	425,000	382,000
Total Equity	1,050,000	1,007,000
Debt		
Long-Term	890,000	790,000
Notes Payable	85,000	120,000
Total Debt	975,000	910,000
Current Liabilities	150,000	150,000
Deferred Credits		
Accumulated Deferred Income Taxes	108,000	70,000
Unamortized Investment Tax Credits	42,000	32,000
Total Deferred Credits	150,000	102,000
Total Liabilities and Capital	$2,325,000	$2,169,000

Tax Effects

The funds-flow statement for the Old Glory Telephone Company also lists two tax-related effects as sources of funds. These tax effects are described in detail in Chapter 9, and therefore they are mentioned only briefly here. It is important to realize that even though these tax effects contribute significantly to internally generated funds, government tax programs are subject to change. If they should change enough so that these funds are not available, these funds would have to be raised externally .

DEFERRED TAXES   Federal and some state tax regulations permit the use of depreciation methods for tax purposes which allow greater amounts of depreciation expense in the early years of the life of an item of plant than in the later years. This method is known as *accelerated tax depreciation*. Because of it, tax payments are less in the early years and more in later years than they would be if straight-line depreciation were used for both book and tax purposes. Tax regulations applicable to the Bell System generally require that the financial reports show total operating taxes as the taxes which would have been paid if straight-line depreciation had been used. This means that some of the taxes shown on the income statement are not currently due but have been deferred to a later time. The company has the use of those funds until they are needed to offset increased taxes in the future due to reductions in tax depreciation.

INVESTMENT TAX CREDITS—NET Because the government tries to encourage the investment of money in capital goods, it provides tax incentives to taxpayers who do this. The major incentive is known as an *investment tax credit*. Unlike accelerated depreciation, which is a deferral of taxes, an investment tax credit is an actual reduction. Under the tax regulations applicable to the Bell System, however, the reduction in taxes is not allowed to increase net income at the time it occurs. Instead, it is gradually released, or amortized, to net income over the life of the plant that produced it. Until the reduced taxes have been completely amortized, those funds are also available for the company's use.

Externally Generated Funds

Many companies cannot generate enough funds internally to meet their total capital requirements. These companies must then get funds from outside the business. For example, the Old Glory Company's funds-flow statement showed internally generated funds of $313,000 and total requirements of $513,000. The difference, $200,000, must be made up through external financing.

External financing is accomplished by issuing stock, or by borrowing—usually by issuing bonds. Stock issues are classified as *equity financing*, since stock represents ownership in the company. Bond issues are classified as *debt financing*, because bondholders do not own any part of the company; they merely loan money to the

FIGURE 4-7
Old Glory Telephone Company
STATEMENT OF CHANGES IN FINANCIAL POSITION
(Funds-Flow Statement)
1976

Sources of Funds

Internal Sources		
Net Income		$125,000
Depreciation		140,000
Deferred Income Taxes		38,000
Investment Tax Credits—Net		10,000
Total from Internal Sources		$313,000
External Sources		
Sale of Bonds		200,000
		$513,000

Applications of Funds

Telephone Plant		$296,000
Dividends		82,000
Repayment of Debt		
Long-Term	$100,000	
Short-Term	35,000	
		135,000
		$513,000

FIGURE 4-8
Revenue Creates Internal Financing

FIGURE 4-8
Revenue Creates Internal Financing

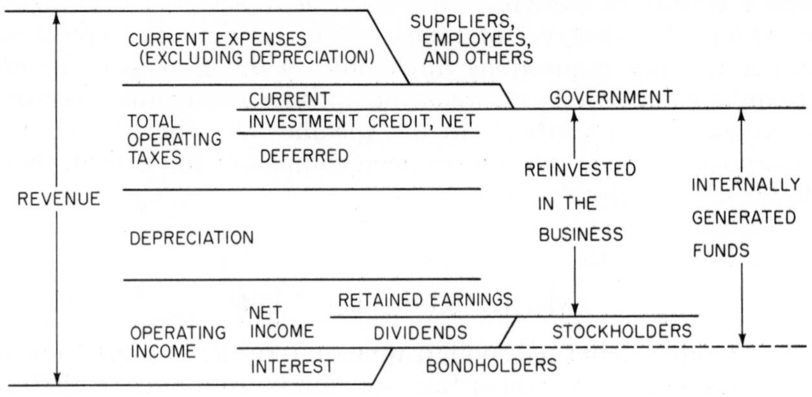

company, and they are therefore its creditors. Equity and debt financing are discussed separately in the next sections.

Equity Financing

Ownership in a corporation is represented by shares of stock. A buyer of stock owns the corporation with the other stockholders in proportion to the number of stock shares the individual owns, compared with the total number of shares issued. Stock ownership is not restricted to individuals, however. The stock of many corporations is owned by partnerships, institutions, and other corporations. Stock may be issued in the form of common stock or preferred stock, or both.

COMMON STOCK   Common stock is the lowest-ranking capital of a corporation in terms of its claim on assets and earnings. Dividends on common stock are paid only after all other obligations have been met; and if the business should be liquidated, the claims of common stockholders would be the last to be recognized. If the business prospers, common stockholders will benefit. If it fails, they may lose their entire investment. Of all the investors in a corporation, then, the common stockholders bear the most risk, since their future return depends on the success of the business.

Privileges and rights associated with ownership of common stock are shared in common with all other stockholders. Usually, one of these rights is the right to vote at meetings of the corporation's stockholders. All shares of common stock also share equally in dividends which the board of directors declares. Dividends on common stock are not guaranteed but are subject to the availability of earnings, and they are paid at the discretion of the directors.

Common stock is usually bought and sold through stockbrokers and others authorized to conduct these transactions. Investors and financial analysts follow the market value of common stock very closely. The market price of a stock is what investors are willing to pay for it; and the price is a reflection of the company's performance and prospects, of general business conditions, and of the alternative investment opportunities which investors may have.

Two other values associated with stock are *book value* and *par value. Book value* is the amount of ownership each common share represents in the property of the company that is not pledged to other owners or lenders. If new common stock is sold for a price below this value, the book value after the sale is less, and each existing stockholder's share of ownership has been diluted or reduced. If new stock can be sold for a price higher than book value, each existing share represents a larger portion of the total—a situation which is much more favorable. *Par value* is just an arbitrary value (sometimes zero) set by the corporation. Although it has little significance to investors, it may be of some interest to the corporation. Some states do not permit a no-par stock to be issued; and in some cases, an annual capital stock tax may be based on the par value.

PREFERRED STOCK   Preferred stock has some of the features of both debt and common stock. It resembles debt in the sense that it generally has a fixed life, that the rate of the dividend payment to the holder is generally fixed, and that preferred stockholders usually do not have the power to vote or to participate in the management of the firm. However, preferred stock is considered to be equity for accounting and tax purposes; and it is like common stock in that the failure to pay the preferred dividend generally cannot force the company into bankruptcy. There are many varieties of preferred stock, and different classes may be issued to furnish various features—particularly in matters related to dividends, conversion, and redemption.

Dividends on preferred stock are ordinarily a fixed percentage of a stated or par value per share. In other words, the fixed dividend will not increase, even though the earnings are good, whereas the common stock dividend could increase. However, it is important to remember that preferred dividends must be paid before any dividends can be declared on common stock; and if the business should be liquidated, preferred stock has priority over common stock for distribution of the proceeds.

Most preferred stock is of the *cumulative* type. This means that if periodic dividends are not paid, these amounts will accrue and must be paid in total before any dividend can be paid on common stock. Preferred stock may also be of a *participating* type. This kind of stock receives the fixed dividend, of course; but it may also receive further dividends after the declared dividends have been distributed to the common shares.

Corporations sometimes stipulate that a preferred issue may be converted to common stock, in which case it is called *convertible preferred*. The conversion privilege will be at a stated ratio of common shares in exchange for each share or shares of preferred. Some preferred stock can be redeemed at the call of the corporation in accordance with terms specified at the time it is issued. Quite often, early redemption requires that a premium be paid to the shareholder; but the premium amount usually reduces with time and shortly disappears. The company can also specify that preferred issues be retired by the establishment of a special account for the purpose, called a *sinking fund*.

The variety of features associated with preferred stock gives a corporation flexibility in its financing and makes it possible to attract a range of investor types under many different market conditions. For example, in the early 1970s, AT&T issued preferred stock for the first time in decades. By doing so, the System was able to get capital through equity financing during periods when market conditions were not favorable for issuing common stock. The several preferred issues include combinations of the features just described, except that none has the participation feature.

Debt Financing

The chief characteristic of debt capital is that the company contracts to pay a specific rate of interest, and also to repay the principal at some stated maturity date. Failure to do either of these could result in bankruptcy.

In addition to issuing stock, the Bell System has financed its operations with both short- and long-term borrowing. These are explained in the next sections.

LONG-TERM BORROWING—BONDS   Although banks could provide long-term loans, the most significant corporate mechanism for long-term borrowing is to issue bonds.

A *bond* represents debt; the buyer of a corporate bond has loaned the firm an amount of money. In return for this loan, the corporation issuing the bonds promises to make periodic interest payments and to pay back a specific amount at the end of a stated period of years. These features clearly distinguish bonds from stock. Bonds obligate the corporation to fixed interest payments and ultimate redemption. Stocks, representing ownership, do not obligate the corporation to the payment of dividends, and presumably they are issued for the life of the corporation.

Like preferred stock, bonds may have a variety of features involving type, life, interest rate, redemption, and conversion. Normally, bonds are issued in denominations of $1,000, although denominations of $10,000 may also be issued for the convenience of institutional investors. Bond prices are expressed as a percentage, with a $1,000 bond being 100 (% is implied). Gradations of ⅛ of a point are observed in the marketplace, with a price of 98⅛ meaning $981.25 and a price of 100¾ meaning $1,007.50.

Corporations issue two general bond types, *mortgage bonds* and *debenture bonds*. *Mortgage bonds* represent a lien on a company's physical properties. Since physical property is the security for the loan, mortgage bonds somewhat reduce management's freedom of action toward that property. Part or all of the assets may be pledged, depending on the size of the issue. Some of the Bell Companies have issued mortgage bonds.

Debenture bonds are distinguished from mortgage bonds in that they represent a claim on the company's general assets, but they are not secured by real property or by any other collateral. The only security backing the debenture is the credit and earning power of the company. To market a debenture, then, requires that the company have a sustained record of regular earnings that is likely to continue—even during an economic downturn. Utilities usually meet this criterion better than manufacturing concerns, since their customer demand is not so volatile. Most of the Bell System debt, in fact, is in debentures.

Convertible debentures are bonds which generally are convertible into common stock at a fixed price per share. (The mechanism for conversion is similar to that for convertible preferred stock.) The convertible debenture remains a debt instrument until it is converted. Upon conversion, the debt disappears, and it is replaced with common stock. The stated life of this kind of debenture is typically 20 years or so, but generally it is converted into the company's common stock before maturity.

INTEREST COST AND YIELD OF BONDS   Borrowing money through the sale of bonds is an agreement to pay a specified amount of interest stated as a percentage of the face value, usually $1,000. In other words, the company must pay a price for the use of the bondholders' funds. However, the company seldom gets exactly the face value for the sale of bonds. Therefore, the interest cost of those funds, as a percentage of the amount of money received, is usually different from the specified rate, which is called the *coupon rate*. A recent Bell Company issue illustrates this effect.

The company decided to issue debenture-type bonds with a 37-year life and an 8⅝% coupon rate, and asked for competitive bids. A group (or syndicate) of investment bankers won this security issue with a bid of 98.904. In accepting this bid, the company received $989.04 for each $1,000 face value bond. Because the annual interest payment on each bond will be $86.25 ($1,000 × 8⅝%), the current annual interest cost to the company will be 8.72% ($86.25 ÷ $989.04). The actual interest cost over the life of the bond will be still higher, because the company must pay back the entire $1,000 to retire each bond at maturity.

The price the syndicate bid was related to the price which it felt it could get for the resale of the bonds to investors. The syndicate estimated that the buyers would pay $997.23 for each bond, and it therefore would make $8.19 ($997.23 − $989.04) on each bond it sold. The actual buyers of the bonds would then pay $997.23 to get $86.25 yearly for 37 years; at the end of that time, they would then get $1,000. These terms mean that the buyer will receive 8.65% ($86.25 ÷ $997.23) of the investment as interest each year.

At the time the investment is made, that percentage is called the *current yield*. The current yield of a bond is the annual interest divided by the *current* bond price. Those investors who are interested in current income evaluate bonds based upon their current yields. More often, though, bonds are evaluated upon the basis of their *yield to maturity*. Yield-to-maturity calculations will recognize that the $1,000 to be received at maturity is different from the price, a fact which will change the yield somewhat. Such calculations involve concepts and techniques related to the time value of money, which is covered in Chapter 5. Most people use prepared tables which list yields for bonds as a function of the time to maturity, the price, and the coupon rate. In this example, because the bonds do not mature for 37 years, and because the price is quite close to the face value, the yield to maturity is approximately equal to the current yield of 8.65%. In other cases, the two may be quite different.

SHORT-TERM DEBT   Short-term debt has a maturity of less than one year and is another source of external debt financing. The two principal sources of short-term funds are bank loans and *commercial paper*. Bank loans are simply funds borrowed from commercial banks. *Commercial paper* is a short-term (3 to 270 days), unsecured promissory note sold to commercial-paper dealers, who, in turn, sell them to investors. These investors are generally other corporations with a temporary excess of cash.

Other External Sources

LEASING   The basic reason for getting external funds is to buy assets. A financial alternative to borrowing the capital funds to buy these assets would be to borrow the assets directly. Leasing is this kind of alternative. More specifically, a *lease* is a contractual arrangement under which the owner of the property (the lessor) allows another party (the lessee) to use the property over a specified period of time and for a specified sum of money.

The difference between lessors, bondholders, and shareholders is the relative priority which each has on the revenues of the company—or, in the case of insolvency, on the property itself. Theoretically at least, each one is paid a return commensurate with the risk assumed. Legally, the position of the lessor appears to be about equal with that of the bondholder. The only practical difference between a bondholder and a lessor is that the bondholder lends capital to be turned into an asset, whereas the lessor lends the asset itself. Of course, the risk to the common stockholder is increased by leases just as it would be with additional debt.

Leases can be arranged for almost any asset the company needs, on either a short- or a long-term basis. Although leasing furnishes the use of an asset without the external financing that outright purchase would require, it does obligate the company to a series of lease payments. These are much like the payments which are necessary to repay a loan. The question of leasing versus owning is sufficiently extensive for it to be treated separately, in Chapter 15.

ADVANCES FROM AT&T   Advances from AT&T to associated companies are short-term loans and are one of the techniques the Bell System has used in its overall financing strategy. Before 1969, the bulk of the associated companies' short-term debt was based on advances. Since that time, however, the Bell Companies are more likely to obtain short-term loans from local banks and to issue commercial paper, and the short-term debt arising from advances is now relatively small.

CORPORATE FUNDS FLOW

Figure 4-9 illustrates how a company gets and uses funds; it also indicates which financial reports show what happens to these funds. Whether derived from an internal or an external source, the funds go into the cash drawer and lose their identity. It is then the accountants' task to classify and record the flows in the proper categories so they can be summarized later and presented in financial reports. The income statement shows what is left as net income, to be used in the business after all expenses are deducted from revenues. The funds-flow statement, on the other hand, shows all sources of funds, internal and external, and how these funds are used.

CAPITAL STRUCTURE

The preceding material has explained the various forms of debt and equity financing. All the capital which has been invested in a business was obtained, originally, from one or both of those two forms of financing. Generally, a company will use both debt and equity types of capital. In fact, a company usually tries to find an optimum balance

FIGURE 4-9
Corporate Funds Flow

between the two, depending on the relative degree of risk, cost, availability, and other factors. The mix between debt and equity capital is called the firm's *capital structure.*

Debt Ratio

The capital structure of a business is determined by the proportion of debt-type capital in the total capital of the business. This proportion is called the *debt ratio,* and it is one of the fundamental financial indicators. The debt ratio is the amount of debt capital divided by the amount of total capital. At the end of 1976, the Old Glory Telephone Company's debt ratio was (see Figure 4-6):

$$\text{Debt ratio} = \frac{\text{debt capital}}{\text{total capital}} = \frac{\$975,000}{\$1,050,000 + \$975,000}$$

Debt ratio = 0.481, or 48.1%

The debt ratio measures the amount of the firm's capital which is subject to the contractual return requirement. Too much debt increases the possibility that the bondholders may not receive their interest and principal payments if earnings decline. The result could be bankruptcy. Too much debt also increases the risk to the stockholders because, if earnings decline, they lose first.

The prudent use of debt can help the stockholders, however. Debt costs less than equity financing, because loaning money with a contract for its repayment is less risky than simply giving the firm money and hoping for a return. Therefore, if the business prospers, the stockholders realize a greater proportionate increase in their earnings if there is more debt than if there is less debt. This phenomenon is known as *financial leverage.* Most homeowners are familiar with the effect of financial leverage, even though they might not recognize the term. When a buyer gets a mortgage on a home, any change in the value of the home flows into the homeowner's equity. If the home increases in value, the equity increases by a larger percentage than the total value of the home, and the homeowner is well pleased. Of course, if the value decreases, the equity will decrease by a greater percentage, and then the homeowner is very unhappy.

The amount of debt which a company can prudently carry depends on many factors. They include the basic riskiness of the business, the need to maintain a good credit rating, and the ability to borrow additional funds when equity capital may not be available. These last two factors are particularly important to firms which have large external financing requirements.

Assuming that other factors are similar for companies in the same industry or providing similar products or services, the firms with higher debt ratios will be more risky than those with lower debt ratios. Firms with reliable demand for their product or service, such as utilities, are able to have higher debt ratios than firms with more volatile demand. This is so because the lower risk of revenue decline offsets the higher risk of larger debt obligations.

Interest Coverage

The debt ratio measures the amount of debt in the firm's capital structure, but it does not indicate much about the firm's present ability to handle that much debt. For example, a firm may have a relatively high debt ratio; but if the debt is mostly old bonds with a very low interest rate, the burden of that much debt is not so great as the debt ratio alone would indicate. The ratio of income to the interest payments is a very important indicator of the amount of debt that a firm can prudently carry in its capital structure. This ratio is sometimes called Times Interest Earned (TIE), and many financial analysts believe that it is more important than the debt ratio. There are several varieties of this type of ratio. Some are found before taxes have been deducted, and some include other fixed charges in addition to interest. The most frequently used ratio is the result of dividing the post-tax current operating income by the current interest payments. Referring to the Old Glory Telephone Company's income statement (Figure 4-5), the interest coverage, as given by times interest earned, is:

$$\text{TIE} = \frac{\text{operating income}}{\text{interest}} = \frac{\$193,000}{\$68,000}$$

$$\text{TIE} = 2.84$$

Interest coverage is like the debt ratio in that it is a risk indicator. However, it is a more current indicator than the debt ratio. Financial analysts generally use this number just as it appears above. However, the reciprocal of interest coverage offers some additional insight. The reciprocal is the percentage of the total return which is subject to the contractual obligation of debt. Since $\frac{1}{2.84}$ is 35.2%, then 35.2% of the current total return is required for the interest on the 48.1% of the total capital which is debt.

BOND RATINGS

Because potential investors are usually wary of risky investments, there is a need to evaluate the relative risks of investing in corporate bonds. The large financial houses maintain research staffs for this purpose, and some publish material reporting their conclusions. Over the years, two well-known firms have developed a bond-rating service which the investment community and many others rely on. These firms, Moody's Investors Service, Inc., and Standard & Poor's Corporation, routinely rate public bond issues of significant size, and they closely agree in their opinions. Bond ratings are designated by a letter code. Moody's ratings range from Aaa, the highest, down to C, the lowest. Standard & Poor's (S&P) ratings are coded similarly. A partial listing of the two ratings of bonds follows:

S&P	Moody's
AAA	Aaa
AA	Aa
A	A
BBB	Baa
BB	Ba
B	B

Essentially, the ratings are just a relative measure of the safety of the interest and principal. Quantifiable data are an important part of the rating process, but so are subjective judgments about such factors as the quality of management and the expectations of profits. Interest coverage and the debt ratio are the most important of the quantifiable data. The levels of these factors which qualify a firm for a specific rating are variables which change with the economic conditions.

Bonds with an AAA or Aaa rating are the highest grade and provide the maximum protection. A rating of AA or Aa is expressed as only slightly lower in degree than the highest rating; the marginally lower rating may reflect somewhat less safety or protec-

TABLE 4-1
MOODY'S BOND YIELDS AND RISK*

Rating	% Yield Jan. 1974	% Yield July 1974	
Baa	8.58	9.55	Higher Risk and Yield
Aaa	7.83	8.72	Lowest Risk and Yield

*Source: *Economic Report to the President,* February 1975.

tion. In general, these top two rating groups represent high-grade issues. Investors who buy lower-rated bonds expect a higher face rate of interest, or they will buy only at a lower price (at a discount). In this way, they can realize a yield commensurate with their perception of the risk they are assuming. Table 4-1 illustrates the correlation between risk and yield, by showing bond-yield averages compared with Moody's ratings at two selected points in time.

This table indicates that investors required about 10% more return for assuming the greater risk of bonds rated Baa than the risk of those rated Aaa. Another point of interest is that the level of yield acceptable to investors will change depending on the prevailing economic conditions. For example, the rate of inflation or the supply and demand for capital will affect the required yield. Table 4-1 shows that investors required almost a full percentage point more in July 1974 than in January 1974. Chapter 10 discusses the relationships between inflation and the cost of both debt and equity capital.

Those companies which receive the highest bond ratings can get debt capital at a lower interest cost than firms with a poorer rating. More important, companies with the highest rating have maximum access to the debt markets. In fact, in an unsettled economic climate, firms rated at the lower end of the scale may be unable to get financing at any reasonable cost. For these reasons, preservation of a high bond rating is an important objective.

COST OF CAPITAL

Both current investors in a business and those prospective investors deciding on a possible investment are very much interested in the performance of that business. Investors will be willing to provide capital for a business venture only if they can expect that the money will eventually be returned with a profit. To put it another way, the business can attract the funds it needs only if it earns a fair return on the capital invested in it. From the point of view of the firm, then, the capital is obtained at some cost—the *cost of capital.* This cost of capital is synonymous with the rate of profit that the investors require. It has two elements, representing the two types of capital in a business: the cost of debt, and the cost of equity.

Cost of Debt

Debt investors who buy specific bond issues can measure their rate of profit or return by the yield of the particular bond. The firm can also measure the cost of this specific bond issue in a similar way. The interest cost derived from the terms of current bond issues is the cost of new, or *incremental,* debt. The average cost of all existing debt, or *embedded* debt, is the total amount of interest as a percentage of the average amount of debt capital. For the Old Glory Telephone Company, the embedded cost of debt is computed thus:

$$\text{Average debt capital} = \frac{\$910,000 + \$975,000}{2} = \$942,500$$

$$\text{Embedded cost of debt} = \frac{\text{interest}}{\text{average debt capital}}$$

$$= \frac{\$68,000}{\$942,500} = 7.2\%$$

Cost of Equity

Because there is no contractual obligation of the firm to pay a specific amount to most owners of stock, the incremental cost of equity is very difficult to determine. The rate of earnings on existing equity capital is, however, a measure of how profitable the business *was* for its owners. The ratio of net income to average equity capital is an indicator of this performance. This ratio is called the *return on equity* (ROE). For the Old Glory Telephone Company, it is:

$$\text{Average equity capital} = \frac{\$1,050,000 + \$1,007,000}{2}$$

$$= \$1,028,500$$

$$\text{Return on equity (ROE)} = \frac{\text{net income}}{\text{average equity capital}}$$

$$= \frac{\$125,000}{\$1,028,500} = 12.2\%$$

To potential investors, the 12.2% is not very meaningful by itself. Investors compare a company's return on equity with that expected on other equity investments of comparable risk. Just how investors make these comparisons is a matter of conjecture. There are many ways, and all are at least partially subjective. However they do it, the required return on equity will be a function of current economic conditions, the riskiness of the business, and other factors.

Cost of Total Capital

The capital structure of a business is a composite of both debt and equity capital. Therefore, the return on total capital is also a composite of the cost of debt and the return on equity. This embedded composite is a frequently used measure of the profitability of public utilities. It shows how well the firm is earning on all capital, both debt and equity. Because it is the result of past operations, this ratio is usually known simply as the *rate of return*. It is equal to operating income, which includes both interest and net income, divided by the average total capital. For Old Glory, the rate of return for 1976 was, then:

$$\text{Average capital} = \text{average debt} + \text{average equity}$$

$$= \frac{\$910,000 + \$975,000}{2} + \frac{\$1,007,000 + \$1,050,000}{2}$$

$$= \$1,971,000$$

$$\text{Rate of return} = \frac{\text{operating income}}{\text{average capital}}$$

$$= \frac{\$193,000}{\$1,971,000} = 9.8\%$$

The rate of return on total capital can also be found by compositing the cost of debt and the return on equity in proportion to the debt and equity in the capital structure. The proportion of debt is, of course, the debt ratio; the proportion of equity is, then, 1 (100%) minus the debt ratio. Table 4-2 illustrates this relationship for the Old Glory Telephone Company.

The rate of return on total capital is the profitability indicator which regulatory commissions generally use to control public utilities. It is considered an indicator of how well the utility is able to compete for the capital it needs in both the stock and the bond markets.

Incremental Cost of Capital

Table 4-2 indicates the rate of return that the Old Glory Telephone Company achieved during 1976 on funds obtained in the past. However, it does not reflect the cost of capital that the firm will have to pay if it needs to get new capital. In many respects, the cost of new capital, or the *incremental cost of capital,** is far more important to the business. This is so because it reflects how the firm will be expected to perform in the future. The managers of the firm can control only the future activities of the firm; the past is now out of their control.

The incremental cost of capital is much more difficult to determine than the current rate of return. Any way it is done is partly subjective, and assumptions must be made about the future. The interest cost of the most recent bond issues is often a starting point. Because the next bond issues will probably have an interest cost approximately the same as the last one, that cost is usually considered to be the incremental cost of debt.

The incremental cost of equity is another story. Many ways have been tried, and it is always a difficult estimate. Because stock is a more risky investment than bonds, the incremental cost of equity should be greater than the incremental cost of debt. But how much greater depends on many factors and is different for different businesses. Although other theories may be equally valid, one theory is that the risk premium is about one-half the cost of debt. On that basis, if the interest cost of the latest bond issue of the Old Glory Telephone Company were 9%, the incremental cost of equity might be 13.5% (9% × 1.5). Assuming that the 48% debt ratio is maintained, the incremental cost of capital is then determined as in Table 4-3.

TABLE 4-2
RATE OF RETURN

Capital	Proportion	×	Return	=	Portion of total return
Debt	0.481		7.2%		3.5%
Equity	1 − 0.481		12.2%		6.3%
Total	1.00				9.8%

*Also called *marginal cost of capital.*

TABLE 4-3
INCREMENTAL COST OF CAPITAL*

Capital	Proportion	×	Cost	
Debt	0.48		9.0%	4.3%
Equity	1 − 0.48		13.5%	7.0%
Total	1.00			11.3%

*When embedded cost of debt is used in this table, the result is called *embedded cost of capital.*

The significance of Table 4-3 is that as the managers of the Old Glory Telephone Company make decisions affecting the future, they must recognize that the cost of the capital they must work with is represented by 11.3%; it is not the 9.8%, since that represents the past.

OTHER FINANCIAL RATIOS

In their efforts to measure and compare the performance of businesses, investors and financial analysts use many indicators in addition to the ratios presented so far. One of the most important of these is *earnings per share,* or EPS. The Old Glory balance sheet shows that there are 25,000 shares of stock outstanding. The income statement shows earnings, or net income, of $125,000. Therefore:

$$\text{Earnings per share (EPS)} = \frac{\text{earnings}}{\text{average number of common shares}}$$

$$= \frac{\$125,000}{25,000} = \$5.00$$

The $5.00 by itself, or in relationship to an EPS number for another company, is almost meaningless, because the EPS figure does not tell whether the $5.00 was earned on a $30 investment or a $100 investment. However, EPS becomes meaningful when it is combined with another important relationship, the *book value per share.* Book value has already been mentioned as the amount of ownership each share represents in the property of the company that is not pledged to others. This amount is the common stock plus the retained earnings, or the total equity capital shown on the balance sheet. For Old Glory, it is:

$$\text{Average equity capital} = \frac{\$1,050,000 + \$1,007,000}{2} = \$1,028,500$$

$$\text{Book value per share} = \frac{\text{average equity capital}}{\text{average number of shares}}$$

$$= \frac{\$1,028,500}{25,000} = \$41.14$$

When the EPS is divided by the book value per share, it is the return on equity (ROE), which has already been shown to be a meaningful indicator.

$$\text{ROE} = \frac{\text{EPS}}{\text{book value}} = \frac{\$5.00}{\$41.14} = 12.2\%$$

Another way to impart meaning to EPS numbers is to consider the growth in earnings per share. In general, those stocks with a higher growth in EPS are more attractive to investors than stocks with a lower growth in EPS. However, the amount of growth in EPS which makes a stock attractive depends on several factors. For those companies which pay out all earnings in dividends, and are therefore attractive to investors seeking income, any positive growth in EPS will indicate an increasing return on equity; and this pleases investors. If a company retained all earnings, it would be attractive to investors seeking long-term gains; and the growth in EPS would have to be greater than the desirable return on equity to reflect an increasing return on equity. Most companies fall somewhere in between these two extremes, and a satisfactory growth in EPS is, then, somewhat less than the desirable return on equity.

Investors and financial analysts also compute numerous other ratios for a variety of reasons. One of these, the *price earnings ratio* (*P/E*), is frequently published in the newspapers along with the daily stock quotations, and it is widely used. The *P/E* is the current market price of the stock divided by the earnings per share.

If a market price of $50.00 is assumed for the Old Glory Stock, the *P/E* is:

$$P/E = \frac{\text{market price}}{\text{earnings per share}}$$

$$= \frac{\$50.00}{\$5.00} = 10$$

Because the market price of stock is a function of so many factors, it is difficult to establish the meaning of the *P/E*. It means different things to different investors, and it has a number of special meanings. One popular meaning is that a high *P/E* relative to other stocks implies that investors expect a high growth in earnings per share. Some investors look, then, for stocks with a high *P/E*. Other investors actively look for stocks with a low *P/E*, because they want to recover their investment as quickly as possible. So many factors influence the *P/E* that an ideal level probably does not exist.

SUMMARY

Financing is the activity of raising the funds or capital necessary to the operation of a business. Internal financing comes from the revenue resulting from the continued operation of the firm. External financing comes from stockholders who buy shares of ownership in the firm, lenders who buy the firm's bonds, and lessors who loan property to the firm. In general, a familiarity with the flow of funds or cash through the firm is necessary to the understanding of financing. An accounting report, the funds-flow statement, describes this flow of funds or cash in the business. Like the income statement, it usually covers a period of one year; but it includes the cash received from and paid to investors, and the gross expenditures for new plant as well as net income.

Because many firms have a continuing need for external sources of capital, corporate financial managers must constantly make very important decisions about their companies' financing needs. For one thing, many different types of securities can be issued which may have significant impact on the success of the business, as well as its attractiveness to investors. It is very important that financial decisions be made keeping in mind the various measures which investors use to evaluate the firm. Some of these measures are extremely important, such as debt ratio, interest coverage, bond rating, rate of return, and earnings per share.

One measure, the rate of return, actually reflects the performance of the firm in meeting the cost of funds or capital raised through financing. This cost of capital is a cost which the firm must meet if it is to continue to attract the capital necessary to keep its plant modern and efficient and to provide for growing customer demand. Each firm's cost of capital is determined in the marketplace. Its securities are evaluated there based on their risk, earnings potential, and other factors, compared with those of alternative investment opportunities. Although the primary objective of this chapter has been to introduce the subject of financing, the chapters that follow will demonstrate that the cost of capital has a vital role in engineering economics. It is the price which must be paid for the resource of capital, and therefore it is just as much a part of the engineering economy study as the price paid for any other resource.

5

MATHEMATICS OF MONEY

INTRODUCTION

Money can "work" for its owner, by earning more money. Money can be placed in a bank savings account, where it will earn interest. Money can be used to buy things like land, gold, jewelry, or stamps in the hope that these will increase in value and that their eventual sale will bring in more money than was paid out originally. Money can also be used to buy stocks or bonds in anticipation of the payment of dividends or interest, or in anticipation of their eventual sale for an increased price. The mathematics of money is based on the fact that all money works and therefore has potential earning power.

This earning power of money can also be viewed as a cost of using money. The bank which pays interest to a depositor, for example, is paying for the privilege of using the depositor's money. Likewise, the business which sells stocks or bonds to an investor is paying for the use of the investor's money with dividends, interest, and increases in value. The term *interest* is commonly used in financial writings and mathematical tables to denote the rate of money's earning power; and since the conventions of finance seem most appropriate for this chapter, the term *interest* is used here, too. However, elsewhere in this text the term *return* is used to refer to the earning power of composite debt and equity capital, and the term *interest* is reserved for the earning power, or the cost, of borrowed money (debt).

Money has earning power because it works over a period of time. Before the actual return from the investment of money can be realized, then, time must pass; the entire concept of the earning power of money can therefore be viewed as the *time value* of money. An analyst must be able to apply this concept when estimating the costs of alternative ways to accomplish a single business goal. Often, the estimated expenditures for alternative plans will occur in varying amounts and at varying times. As a result, the analyst cannot reasonably compare these sums before converting or translating them into *equivalent* amounts as of some common date. This chapter provides the background and the instruction for making such translations.

MONEY AND TIME

A good way to see how the time-value concept of money works is to compare the various effects that time and earning power have on $20.00. If $20.00 is placed in a bank paying an interest rate of 5% per year, it grows to $21.00 when the interest—$1.00—is paid at the end of one year. So, $20.00 today at 5% is equivalent to $21.00 a year from now. Or, in other words, to plan on having $21.00 one year from now, only $20.00 has to be deposited today at 5% interest.

Stated in this second way, this example illustrates one of the basic principles of the mathematics of money: what is known as *discounting*. The $20.00 is called the *discounted present value* of the $21.00 needed in one year. Because the $20.00 is the result of discounting the $21.00 at 5%, the 5% is said to be the *discount rate*.

Both the initial deposit and the earned interest left in an account have earning power because interest is *compounded*. This means that it is computed on the principal plus the accrued interest. Economy studies are always based on the use of compounding, because *all* money, both return and investment (or, in banking terms, accrued interest and principal), have time value. When the second year begins, $21.00 is in the bank; and it will earn $1.05 at 5%. At the end of the second year, then, the original $20.00 will have a value of $22.05.

The $20.00 could also be used to pay four equal annual installments if it is put into a desk drawer and $5.00 is withdrawn each year. If $20.00 is deposited in the bank at 5% instead, Table 5-1 shows that there would be $21.00 after one year, the time of the first payment. After the $5.00 payment is made, $16.00 would be left on deposit. This $16.00 would earn $0.80 interest (at 5%). At the end of the second year, there would be $16.80 from which to make the second payment. The $11.80 left would earn $0.59, so that the account would have $12.39 for the third payment at the end of the third year. The remaining $7.39 would earn $0.37. When the time comes for the fourth and final payment, there would be $7.76. After the final $5.00 payment is made, the $2.76 remaining would represent the accrued interest that the original $20.00 earned, less the four annual payments of $5.00.

Because of this earning power of money, the initial deposit could have been reduced to $17.73. Table 5-1 also shows that this sum would have allowed $5.00 to be paid out at the end of each of four years, with nothing left in the bank at the end of the fourth year.

It follows that the cost of using money is intimately associated with time and cannot be considered independently of it.

TABLE 5-1
THE EFFECT OF TIME ON THE EARNING POWER OF $20.00

Year (A)	Deposit at beginning of year (B)	Balance at beginning of year $(G_{n-1} + B)$ (C)	Interest paid at end of year $(C \times 0.05)$ (D)	Balance at end of year $(C + D)$ (E)	Withdrawn at end of year (F)	Balance at beginning of next year $(E - F)$ (G)
A sum of $20.00 earning 5% is more than enough to make four annual $5.00 withdrawals.						
1	$20.00	$20.00	$1.00	$21.00	$5.00	$16.00
2	—	16.00	0.80	16.80	5.00	11.80
3	—	11.80	0.59	12.39	5.00	7.39
4	—	7.39	0.37	7.76	5.00	2.76
A sum of $17.73 earning 5% is just right.						
1	$17.73	$17.73	$0.89	$18.62	$5.00	$13.62
2	—	13.62	0.68	14.30	5.00	9.30
3	—	9.30	0.46	9.76	5.00	4.76
4	—	4.76	0.24	5.00	5.00	0.00

Equivalence

The mathematics of money distinguishes between the terms *equivalent* and *equal:* the concept of *equivalence* requires the use of an interest rate; the concept of *equality* does not. The $20.00 in the desk drawer is *equal* to the four $5.00 withdrawals, but this $20.00 is also *equivalent* to the four $5.00 payments at an interest rate of zero. The $17.73 is certainly not *equal* to the four $5.00 withdrawals; but when it is placed in a 5% interest-bearing account, it is *equivalent* to the four $5.00 payments at 5%.

The term *equivalent* always implies that the time-value-of-money concept is applied at some interest rate. At this given interest rate, an amount of money has an infinite range of equivalent and potential values, over infinite points in time, even though it actually exists at only one point in time. If an amount of money is to have a precise meaning, then, it must be fixed in time as well as in amount. Before comparing different amounts of money in an economy study, the analyst must first express them all in terms of a common date or some equivalent period of time. Mathematical formulas and tables are available to translate an amount at any particular date into an equivalent amount at any other date.

There are many different kinds of translations. A frequent example is to translate a single amount of money into an equivalent amount as of either an earlier or a later date. This is done by finding the *present worth* (PW) or the *future worth* (FW) of the money on this date.

A single amount of money can also be translated into an equivalent *annuity* (A). An annuity is a series of *uniform* amounts of money occurring each year over a period of years. The annuity equivalent to an initial amount is called the *annuity from a present amount* (A/P); the annuity equivalent to a final amount is called the *annuity for a future amount* (A/F). Translations of single amounts into annuities are very common in business and in everyday affairs as well. Parents who want to determine how much money to put aside each year for a child's college education, for example, may calculate the annuity they must save for that future amount.

Conversely, an annuity can be translated into an equivalent single amount as of either an earlier or a later date. As one example, a future series of uniform annual expense savings can be restated as an equivalent single amount of money at the present time. This translation enables the analyst to compare the single equivalent savings amount with the cost for items of plant which would be spent right now in an alternative plan. The single amount that is equivalent to a future annuity is called the *present worth of an annuity* (P/A). The single amount that is equivalent to a past annuity is called the *future worth of an annuity* (F/A).

A single amount of money can also be found which is equivalent to any nonuniform series. Although there are many kinds of nonuniform series, two special types occur quite often. One is called an *arithmetic,* or linear, *progression.* This is a series which changes by a *constant amount* called a *gradient* (G). An example of an arithmetic progression might be maintenance costs on equipment which requires more attention as it ages.

The other special type of nonuniform series is called a *geometric progression.* This is a series which changes at a *constant rate.* A familiar example of this kind of change is the effect of inflation or deflation.

Use of Time-Value Factors

Making a translation from one dollar amount to an equivalent could be a very tedious process indeed, requiring many multiplications and additions. As one example, many calculations were needed to construct Table 5-1—and that was just a simple translation of an initial amount into an annuity. However, so-called *time-value factors* have been developed which will save long arithmetic computations and will make possible practically any conceivable conversion. There are several factors, of course; and the right one to use for a given translation depends on the particular conditions being studied.

For most translations which managers are required to make, payments can be assumed to occur at discrete points in time—usually, once a year. The factors which are used in the mathematics of money to make these translations are called *periodic* (often *annual*) *compounding factors*. The time-value factors for periodic compounding are therefore used most frequently; and, of these, the six time-value factors used for translations between single amounts and uniform series are the most common.

The basic forms for these factors are listed here, with i for interest rate and N for the number of periods:

- *The Future Worth (F) of a Present Amount (P)*
$$(F/P, i\%, N) = (1 + i)^N$$
- *The Present Worth (P) of a Future Amount (F)*
$$(P/F, i\%, N) = \frac{1}{(1 + i)^N}$$
- *The Future Worth (F) of an Annuity (A)*
$$(F/A, i\%, N) = \frac{(1 + i)^N - 1}{i}$$
- *An Annuity (A) for a Future Amount (F)*
$$(A/F, i\%, N) = \frac{i}{(1 + i)^N - 1}$$
- *The Present Worth (P) of an Annuity (A)*
$$(P/A, i\%, N) = \frac{(1 + i)^N - 1}{i(1 + i)^N}$$
- *An Annuity (A) from a Present Amount (P)*
$$(A/P, i\%, N) = \frac{i(1 + i)^N}{(1 + i)^N - 1}$$

Appendix A at the back of the book contains tabulated values for these factors for interest rates from 1 to 20% and for periods from 1 to 75. Appendix 5A at the end of the chapter contains the mathematical derivation of these factors and others discussed in this chapter.

Before considering how the first six factors are applied, however, some general background is helpful.

Notation

The notation used in this book for the time-value factors has been endorsed as standard by the Engineering Economy Division, American Society for Engineering Education. However, readers may already be familiar with other notation in other popular texts and handbooks. Table 5-2 therefore compares different sets of notation for the six most common time-value factors.

In this book, the basic form of the notation used for all time-value factors always consists of a ratio of two letters, representing two amounts of money (for example, P/A or F/P), plus an interest rate ($i\%$) and the number of periods (N). The whole factor is written like this:

$$(P/A, i\%, N)$$

or

$$(F/P, i\%, N)$$

Sometimes, in a general discussion of a factor, the text will find it convenient to drop the interest rate and the period in the notation. For example, the present worth (P) of a known annuity (A) is as follows:

$$P = A(P/A)$$

Or, the future worth (F) of a known present amount (P) is:

$$F = P(F/P)$$

Forms such as $PW(\ )$ and $FW(\ )$ are also used. These are called *operators* because they represent some operation—the present worth (PW) or future worth (FW) of whatever is inside the parentheses.

TABLE 5-2

A COMPARISON OF NOTATIONS USED IN DIFFERENT TEXTBOOKS

Engineering Economy 3rd Ed. AT&T Co	Engineering Economy 2nd Ed. AT&T Co	Engineering Economy 1st Ed. AT&T Co	Compound Interest And Annuity Tables[1] Kent	Principles Of Engineering Economy[2] Grant & Ireson
Future Worth Of A Present Amount (F/P)	Future Worth Of A Present Amount (f/p)	Present Worth To Future Worth (s)	Compound Amount Of 1 (s)	Compound Amount Of A Single Payment (caf')
Present Worth Of A Future Amount (P/F)	Present Worth Of A Future Amount (p/f)	Future Worth To Present Worth $(v^{\overline{n}})$	Present Value Of 1 $(v^{\overline{n}})$	Present Worth Of A Single Payment (pwf')
Future Worth Of An Annuity (F/A)	Future Worth Of An Annuity (f/a)	Annuity To Future Worth $(s\overline{n})$	Compound Amount Of 1 Per Annum $(s\overline{n})$	Compound Amount Of A Uniform Series (caf)
Annuity For A Future Amount (A/F)	Annuity For A Future Amount (a/f)	Future Worth To Annuity $(1/s\overline{n})$	Accumulation (Sinking Fund) Factor $(1/s\overline{n})$	Sinking Fund Factor (sff)
Present Worth Of An Annuity (P/A)	Present Worth Of An Annuity (p/a)	Annuity To Present Worth $(a\overline{n})$	Present Value Of 1 Per Annum $(a\overline{n})$	Present Worth Of A Uniform Series (pwf)
Annuity From A Present Amount (A/P)	Annuity From A Present Amount (a/p)	Present Worth To Annuity $(1/a\overline{n})$	Amortization Factor (Annuity That 1 Will Purchase) $(1/a\overline{n})$	Capital Recovery Factor (crf)

1. *Compound Interest And Annuity Tables*, Kent, McGraw-Hill Book Company, Inc., New York And London, 1926.

2. *Principles Of Engineering Economy*, Eugene L. Grant And W. Grant Ireson, Fourth Edition, The Ronald Press Company, New York, 1960.

In general, it is well to keep in mind that the notation used for time-value factors is always consistent with the mathematical operation required. The symbols can also be adapted to computer operations and are easily read as the abbreviations of the operations they symbolize.

Time-Line Diagrams

In doing economy studies, managers typically have trouble keeping all the relationships straight unless they illustrate the conditions. Two types of time-line diagrams help to illustrate the relationships. The cash-flow diagram describes the amount and timing of each cash amount, and the balance diagram includes the effect of interest. They are explained below.

Cash-Flow Diagrams

Cash-flow diagrams appear throughout this text because they are a particularly good way to organize one's thoughts before trying to solve a complex problem. This diagram is simply a horizontal line that is marked off to represent periods of time. Upward arrows represent the receipts of cash, and downward arrows represent expenditures— outflows. (A handy way to remember the direction is to think of "*down* and *out*.") Borrowing money from the bank and repaying it in four installments would therefore be shown like this:

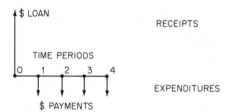

Of course, to interpret a cash-flow diagram, it is essential to know whose vantage point is being shown. For example, the cash-flow diagram above is from the borrower's viewpoint. The lender's diagram for this same transaction would look like this:

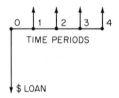

These two diagrams may be used to discuss two other important points:

- First, since a loan is repaid with payments that include interest, a debt is eventually satisfied by payments which are equivalent to the loan at that rate of interest. To find the loan payments, then, the banker translates the loan into an equivalent annuity at that interest rate over the period of the loan.

- Second, in tables of annuity factors, the annuity begins one full period later than its present equivalent amount, but the future equivalent of an annuity coincides with the final payment.

These points will become clear with experience in applying the time-value factors.

Balance Diagrams

A balance diagram is also a useful time-line diagram, because it displays the effect of interest on the cash flow. In fact, it provides the same kind of information as that in Table 5-1, except that it is in a graphic form. Like a cash-flow diagram, it can be constructed from either the borrower's or the lender's point of view. The balance diagram which follows is from the viewpoint of the depositor of the $17.73 in Table 5-1:

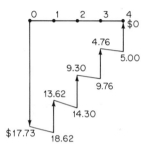

For consistency with the cash-flow diagram, *down*ward arrows represent the *out*flow of money—the deposit or investment; the *up*ward arrows represent the *in*flows—the withdrawals. The balance diagram illustrated here goes to zero because it is constructed at 5%, and the series of $5.00 withdrawals is equivalent to the $17.73 deposit. A similar 5% diagram for the $20.00 deposit would not go to zero, because four $5.00 withdrawals are not equivalent to the $20.00 deposit at 5%.

General Application

Figure 5-1 illustrates how the six most common time-value factors are applied. In each case, the upper cash flow is equivalent to the cash flow beneath it when the proper factor (indicated by the broad arrow) is applied. Conversely, the lower cash flow is also equivalent to the cash flow just above it—again, when the proper factor is applied.

FIGURE 5-1
Equivalence: For Single Amounts and Uniform Series

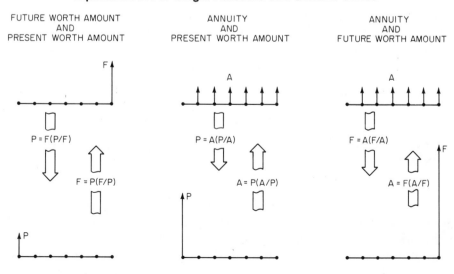

FUTURE WORTH AMOUNT AND PRESENT WORTH AMOUNT

ANNUITY AND PRESENT WORTH AMOUNT

ANNUITY AND FUTURE WORTH AMOUNT

$P = F(P/F)$

$F = P(F/P)$

$P = A(P/A)$

$A = P(A/P)$

$F = A(F/A)$

$A = F(A/F)$

Therefore, the relationship in each upper cash-flow diagram is the reciprocal of the relationship in the cash-flow diagram just beneath it. The factors are therefore reciprocals also.

$$(F/P) = \frac{1}{(P/F)} \qquad (F/A) = \frac{1}{(A/F)} \qquad (P/A) = \frac{1}{(A/P)}$$

This relationship is useful because it means that only three time-value factors actually need to be tabulated in order to handle six operations. It also illustrates the algebraic nature of the notation.

Another algebraic relationship is that if two time-value factors are multiplied, the product is a third time-value factor. For example:

$$(A/F) \times (P/A) = (P/F)$$

Particular Applications

As Figure 5-1 showed, each time-value factor makes a single conversion. The numbered arrows in Figure 5-2 illustrate how $1,000 at the end of year 8 can be transformed into various amounts and series, all equivalent if the time value of money is assumed to be 10% compounded annually. The appropriate table in Appendix A at the back of the book is used to determine each time-value factor.

- Arrow 1 shows that the future worth of the $1,000 at the end of 20 years, or 12 years after year 8, is $P(F/P, 10\%, 12)$; this is $1,000(3.138), or $3,138.

- Arrow 2 shows that an annuity of 7 payments which would accumulate to that future worth of $3,138 is $F(A/F, 10\%, 7)$; this is $3,138(0.10541), or $331 per year.

- Arrow 3 shows that the present worth at the beginning of that 7-year annuity of $331 per year is $A(P/A, 10\%, 7)$; this is $331(4.868), or $1,611.

- Arrow 4 shows that the present worth of that $1,611, 13 years earlier, is $P(P/F, 10\%, 13)$; this is $1,611(0.2897), or $466.

- Arrow 5 shows that an 8-year annuity which could be paid from the $466 is $P(A/P, 10\%, 8)$; this is $466(0.18744), or $87.44 per year.

- Arrow 6 shows that the future worth at the end of that $87.44 annuity with 8 payments is $A(F/A, 10\%, 8)$; this is $87.44(11.436), or the original $1,000.

In practice, these problems must be first set up. Arrow 4 found the present worth of a future single amount. If that amount were a single receipt for salvage in N years, the problem might be written like this:

$$PW(\text{salvage})$$

The operation is therefore:

$$PW(\text{salvage}) = \text{salvage} \times (P/F, i\%, N)$$

Or, for Arrow 3 the problem is to find the present worth of an annuity. If the annuity were uniform annual maintenance costs, the operation is like this:

$$PW(\text{maintenance}) = \text{maintenance/year } (P/A, i\%, N)$$

So far, one time-value factor has been used to make a single conversion. Often, two or more factors are required. For example, suppose the problem is to find the present worth right now of a 5-year annuity which is deferred—that is, which does not begin for 3 years. (Another example is the combination of Arrows 3 and 4 in Figure 5-2.) Two

FIGURE 5-2
An Example of Equivalence

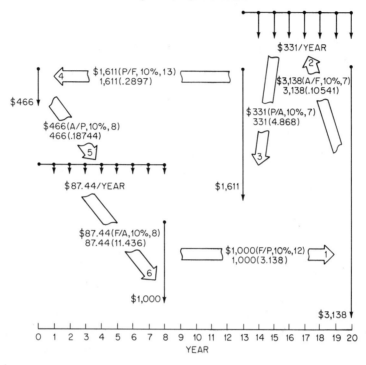

factors are now required. One factor is needed to get the present worth of the annuity 3
years from now (at the beginning of the annuity, or 1 year before the first payment).
Then a second factor is needed to get the present worth of that amount. The combined
operation is:

$$PW(\text{deferred } A) = A(P/A, i\%, 5)(P/F, i\%, 3)$$

An equally good procedure is to use one factor to find the present worth of the annuity as
if it lasted for the full 8 years, and then subtract the present worth of the nonexistent
first three payments using a second factor. Or:

$$PW(\text{deferred } A) = A(P/A, i\%, 8) - A(P/A, i\%, 3)$$

Sometimes experienced analysts use a common shortcut and abbreviate this as:

$$PW(\text{deferred } A) = A(P/A, i\%, 8 - 3)$$

This shortcut is fine, too, except that it is easily misunderstood as:

$$PW(\text{deferred } A) = A(P/A, i\%, 5)$$

which is *not* correct.

   Finding the future worth of a deferred annuity does not require the use of two factors.
For example, suppose the problem is to find the future value of a series of annual $100
costs in 20 years if this series does not start until 5 years from now. In this case, the
problem is simply to find the future value of a 15-year annuity. Or:

$$FW(\text{deferred } A) = A(F/A, i\%, 15)$$

A simple time-value diagram would show this very clearly:

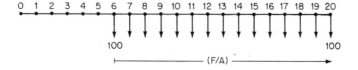

The future value for the first 5 years is zero, because there were no costs.

Some Special Properties

A knowledge of some special properties of the time-value factors will help the reader to understand their real meaning. One of these is that the (A/P) factor over an infinite number of periods is simply the interest rate:

$$(A/P, i\%, \infty) = i$$

Therefore, if a sum of money is deposited in a bank, the interest can be withdrawn each period forever. For example, in a recession or an economic depression, a mortgage holder, rather than foreclosing, may allow the borrower to pay just the interest on a loan. By accepting just the interest, the mortgage holder is preserving the value of the principal, all the while getting a return on it.

The reciprocal equation also demonstrates an important point:

$$(P/A, i\%, \infty) = \frac{1}{i}$$

Therefore, the present worth of a series of payments which continues forever is one payment divided by the interest rate. In other words, an *infinite* series still has a *finite* equivalent present value.

All the factors are mathematically related. One important relationship is that the (P/A) factor is the sum of each (P/F) factor over the time of the annuity. In other words,

$$(P/A, i\%, N) = \sum_{n=1}^{N} (P/F, i\%, n)$$

The small n represents each period, and the capital N represents the number of periods. So, the present worth of an annuity over N periods is also the sum of the present worths of all the future amounts—from the period of the first amount to the period of the last amount.

Another relationship that is sometimes very helpful for interpreting the effect of interest is:

$$(A/P) = (A/F) + i$$

The two time-value factors here are often given special names: The (A/F) factor is known as the *sinking fund,* or the *annuity depreciation* factor; the (A/P) factor is known as the *amortization,* or the *capital recovery* factor. Their relationship illustrates that capital recovery (A/P) consists of the sum of repayment (A/F) and return (i), and it is basic to the understanding of engineering economy.

NONUNIFORM SERIES

In the previous section, time-value factors were applied to uniform series. However, as pointed out earlier, any nonuniform series can also be translated into an equivalent single amount. The procedure is to apply the appropriate single-payment factor—either (P/F) or (F/P)—to each amount in the series for the required number of periods, and then to add up all the results. For example, to find the present worth of a nonuniform series of rearrangement costs, the operation is as follows:

$$PW(\text{Rearrangement}) = \sum_{n=1}^{N} R_n (P/F, i\%, n)$$

It is also possible to find the annuity equivalent to a nonuniform series. The procedure is to first find the present worth (or future worth) of the series, as above, and then to multiply by the appropriate (A/P) [or (A/F)] factor. This can become a rather tedious procedure, and there are shortcuts for two special kinds of nonuniform series—arithmetic and geometric progressions.

Arithmetic Progression

An arithmetic progression was defined earlier as a nonuniform series which changes by a constant amount. The same procedures which are used for any nonuniform series can also be applied to an arithmetic progression. To save long arithmetic computations, though, three new factors can be used—the so-called *gradient factors*. They are listed here:

- *The Future Worth (F) of a Linear Gradient (G)*

$$(F/G, i\%, N) = \frac{1}{i}\left[\frac{(1 + i)^N - 1}{i} - N\right]$$

- *The Present Worth (P) of a Linear Gradient (G)*

$$(P/G, i\%, N) = \frac{1}{i}\left[\frac{(1 + i)^N - 1}{i(1 + i)^N} - \frac{N}{(1 + i)^N}\right]$$

- *An Annuity (A) from a Linear Gradient (G)*

$$(A/G, i\%, N) = \frac{1}{i} - \frac{N}{(1 + i)^N - 1}$$

Appendix A at the back of the book includes tabulated values for these three factors, as well as for the previous six.

General Application

Figure 5-3 illustrates how an arithmetic progression can be translated into an equivalent single sum or annuity. As each of the cash-flow diagrams shows, the arithmetic progression is broken down into a uniform series plus or minus a changing series. The

FIGURE 5-3
Equivalence-Arithmetic Progressions

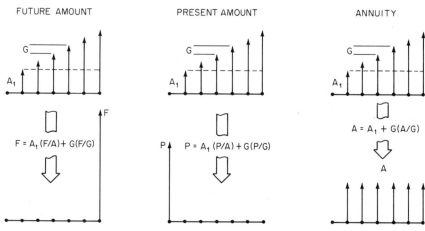

FUTURE AMOUNT

$F = A_1 (F/A) + G(F/G)$

PRESENT AMOUNT

$P = A_1 (P/A) + G(P/G)$

ANNUITY

$A = A_1 + G(A/G)$

single amounts (P or F) equivalent to the arithmetic progression are both the sum of two products: the product of the uniform portion and a uniform series factor; and the product of the gradient portion and a gradient factor. When an arithmetic progression is converted into an annuity, however, the uniform part of the progression is already in the form of an annuity. Therefore, the equivalent annuity is equal to the uniform part—that is, the base payment—plus the product of the gradient and the (A/G) factor.

The conventions for the gradient factor are slightly different from those for the other time-value factors. Each of the upper cash-flow diagrams in Figure 5-3 consists of six payments. The gradient factor operates on one payment fewer than the entire series—five payments, here. This means that the present worth is found one period before the first amount but *two periods before* the first gradient.

Particular Applications

The numbered arrows in Figure 5-4 illustrate how an arithmetic progression which begins at $500 and increases by $100 each year for 5 years can be translated into equivalent single amounts and annuities at 10%.

The translations for this increasing progression are as follows:

- Arrow 1 shows that the equivalent present worth of this progression is:

$$P = \$500(3.791) + \$100(6.862)$$
$$P = \$1,895 + \$687 = \$2,582$$

- Arrow 2 shows that the equivalent future worth of this progression is:

$$F = \$500(6.105) + \$100(11.05)$$
$$F = \$3,053 + \$1,105 = \$4,158$$

- Arrow 3 shows that the equivalent annuity for this progression is:

$$A = \$500 + \$100(1.810) = \$681 \text{ per year}$$

This uniform series can also be translated to another equivalent 5-year arithmetic progression, which starts at $1,000 and decreases by a constant amount each year.

- Arrow 4 shows that this gradient is found by solving for G:

$$A = A_1 + G(A/G, 10\%, 5)$$
$$\$681 = \$1,000 + G(1.810)$$
$$G = -\$319/1.810 = -\$176.24$$

Two more translations of the decreasing progression are also illustrated:

- Arrow 5 shows that the future worth is:

$$F = A_1(F/A, 10\%, 5) + G(F/G, 10\%, 5)$$
$$F = \$1,000(6.105) - \$176.24(11.05)$$
$$F = \$6,105 - \$1,947 = \$4,158$$

- Arrow 6 shows that the present worth is:

$$P = A_1(P/A, 10\%, 5) + G(P/G, 10\%, 5)$$
$$P = \$1,000(3.791) - \$176.24(6.862)$$
$$P = \$3,791 - \$1,209 = \$2,582$$

To summarize, in converting an arithmetic progression to an equivalent single amount or annuity, the procedure is to break up the series into a uniform series plus a gradient series. The uniform portion is then multiplied by the uniform-series factor, and this

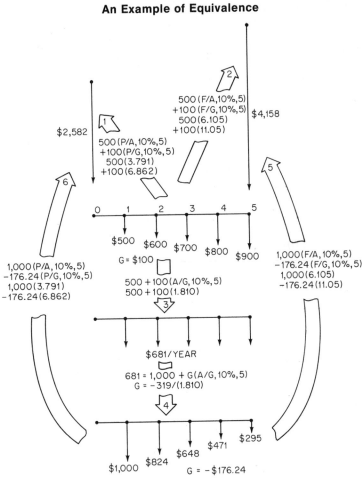

FIGURE 5-4
An Example of Equivalence

product is added to the product of the gradient portion and the gradient factor. In converting an arithmetic progression to an equivalent annuity, however, the uniform part of the series is already in the form of an annuity. Therefore, the equivalent annuity is equal to the first amount plus the product of the gradient and the (A/G) factor.

Geometric Progression

A geometric progression was defined earlier as a nonuniform series that changes by a constant rate, or percentage. The most familiar example was said to be the effect of a constant rate of inflation or deflation.

For example, suppose maintenance costs for one particular unit were $100 for the past year and they are expected to increase at 6% per year. The present worth at 12% of the maintenance costs for the next 5 years could be found as below:

Year	Maintenance cost			(P/F, 12%, N)	PW
	Prev. year × (1.06)	=	This year		
1	$100.00		$106.00	0.8929	$ 94.65
2	106.00		112.36	0.7972	89.56
3	112.36		119.10	0.7118	84.78
4	119.10		126.25	0.6355	80.23
5	126.25		133.83	0.5674	75.94

Sum of Present Worths = $425.16

However, it is not necessary to do all of this arithmetic because a mathematical shortcut can be used instead to handle both the constant percentage changes and the discounting due to time value. It is a single interest rate, which expresses the relationship between the discount rate (i) and the periodic rate of change (I). In this text, it is called the *convenience rate,* and its derivation follows.

Derivation of the Convenience-Rate Concept

If (I) represents the periodic rate of change in a series, the amount in any period of that geometric progression can always be expressed as $(1 + I)$ times the amount in the previous period. If the first amount is expressed as $(1 + I)$ times some base amount (A_0), the amount in any period (n) is

$$A_n = A_0(1 + I)^n$$

The A_0 represents the base of the geometric progression and is not one of the amounts in that series. The first amount is I percent different from the base (A_0).

The present worth of one amount in the series is found by multiplying this amount by the (P/F) factor at the appropriate discount rate and period:

$$PW(A_n) = A_n(P/F, i\%, n)$$

Earlier, the (P/F) factor was defined as

$$(P/F, i\%, n) = \frac{1}{(1 + i)^n}$$

Therefore

$$PW(A_n) = A_n \frac{1}{(1 + i)^n}$$

Since

$$A_n = A_0(1 + I)^n$$

then

$$PW(A_n) = A_0(1 + I)^n \frac{1}{(1 + i)^n} = A_0 \frac{(1 + I)^n}{(1 + i)^n}$$

The I and i are working against each other; therefore, there is some rate (i_g)—the g here represents *geometric*—which represents both the changing amount and the discounting for present value:

$$PW(A_n) = A_0(P/F, i_g\%, n)$$

Because

$$(P/F, i_g\%, n) = \frac{1}{(1 + i_g)^n}$$

then

$$PW(A_n) = A_0 \frac{1}{(1 + i_g)^n}$$

The i_g can be found as follows:

$$PW(A_n) = A_0 \frac{(1 + I)^n}{(1 + i)^n} = A_0 \frac{1}{(1 + i_g)^n}$$

$$(1 + I)(1 + i_g) = (1 + i)$$
$$(1 + I) + (1 + I)i_g = (1 + i)$$
$$(1 + I)i_g = (1 + i) - (1 + I)$$
$$i_g = \frac{i - I}{1 + I}$$

Because the (P/A) factor is a summation of a series of (P/F) factors, and because i_g does not depend on the n, this convenience-rate concept can be used to find either the present worth of a single amount or the present worth of a geometric progression. For a progression with the base A_0:

$$PW(\text{geometric progression with base } A_0) = \sum_{n=1}^{N} A_0 \frac{1}{(1 + i_g)^n}$$
$$= A_0 \sum_{n=1}^{N} \frac{1}{(1 + i_g)^n} = A_0(P/A, i_g\%, N)$$

USING THE CONVENIENCE RATE   The convenience rate can be used to convert a geometric progression directly to a single present amount or to convert a single present amount to a geometric progression. The procedure is similar to that of conversions with uniform series. The difference is that, with a geometric progression, the time-value factor at the convenience rate is applied to the base of the series; with a uniform series, the factor at the discount rate is applied to the actual cash amounts.

For the inflating maintenance expense specified at the beginning of this section, the present worth is:

$$P = \$100(P/A, i_g\%, 5)$$

where $i_g = \dfrac{i - I}{1 + I}$

$$i_g = \frac{0.12 - 0.06}{1 + 0.06} = 0.0566 = 5.66\%$$

The next section of this chapter will cover how the (P/A) factor for this nontabulated interest rate can be found or approximated. The factor needed is:

$$(P/A, 5.66\%, 5) = 4.2516$$

Therefore:

$$P = \$100(4.2516) = \$425.16$$

which is exactly the same as the tabular result.

To find the future worth of (or the annuity equivalent to) a geometric progression, however, it is necessary first to find the equivalent present worth of the series. Then the future amount (or equivalent annuity) is found, using the time-value factors at the discount rate (i).

General Application

Figure 5-5 illustrates how the convenience rate is used to convert a geometric progression directly to a present equivalent amount—or to convert a present equivalent amount directly to the base of a geometric progression.

FIGURE 5-5
Equivalence-Geometric Progressions

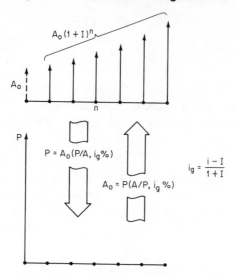

$$P = A_0(P/A, i_g\%)$$

$$A_0 = P(A/P, i_g\%)$$

$$i_g = \frac{i - I}{1 + I}$$

FIGURE 5-6
Example of Equivalence. A 5-year series, growing at 5% on a base of $500, is *equivalent* to all of these at 10%.

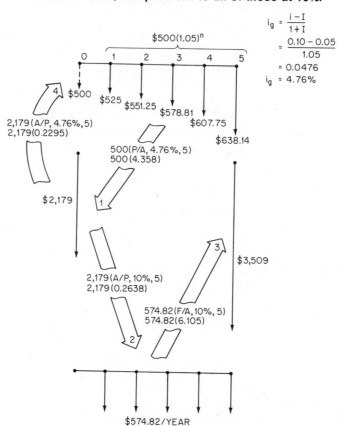

$$i_g = \frac{i - I}{1 + I}$$
$$= \frac{0.10 - 0.05}{1.05}$$
$$= 0.0476$$
$$i_g = 4.76\%$$

$500(1.05)^n$

$500

$525 $551.25 $578.81 $607.75 $638.14

2,179(A/P, 4.76%, 5)
2,179(0.2295)

$2,179

500(P/A, 4.76%, 5)
500(4.358)

$3,509

2,179(A/P, 10%, 5)
2,179(0.2638)

574.82(F/A, 10%, 5)
574.82(6.105)

$574.82/YEAR

Particular Applications

The numbered arrows in Figure 5-6 illustrate how a geometric progression growing at 5% on a $500 base can be translated into equivalent single amounts and series.

The convenience rate for a 10% discount rate would be:

$$i_g = \frac{i - I}{1 + I} = \frac{0.10 - 0.05}{1.05} = 0.0476 = 4.76\%$$

- Arrow 1 shows that this progression can be converted to a present amount using the convenience rate of 4.76%:

$$P = A_0(P/A, 4.76\%, 5)$$
$$P = \$500(4.358) = \$2,179$$

- Arrow 2 shows that the equivalent annuity can be found from that present equivalent amount, using the (A/P) factor at the discount rate:

$$A = P(A/P, 10\%, 5)$$
$$A = \$2,179(0.2638) = \$574.82/\text{year}$$

- Arrow 3 shows that the future worth can be found from the annuity, using the (F/A) factor also at the discount rate:

$$F = A(F/A, 10\%, 5)$$
$$F = \$574.82(6.105) = \$3,509$$

- Arrow 4 shows that the present amount can be converted into the original base of the geometric progression using the convenience rate of 4.76%:

$$A_0 = P(A/P, 4.76\%, 5)$$
$$A_0 = \$2,179(0.2295) = \$500$$

The convenience rate in Figure 5-6 is 4.7619% to five significant digits, and the factors were developed on a calculator. Using five significant figures here illustrates that the convenience-rate concept is exact. However, it is often satisfactory to round off the convenience rate to the nearest tabulated interest rate. Many times, the rate of change in the geometric progression is small, and the convenience rate is approximately equal to the discount rate minus the rate of change.

The convenience rate should not be considered a discount rate or a "net" time value of money. Time-value factors for discount rates are applied to future cash amounts in order to find their equivalent present value. However, the future cash amounts must be known before a discount rate can be applied. With geometric progressions, requiring forecasts of constant percentage changes in costs, it is convenient to combine the two operations. The convenience rate is simply a mathematical shortcut for doing this. In other words, when applied to the base of a geometric progression, the convenience rate will provide for both the constant percentage changes and the discounting for time value.

Furthermore, the convenience rate applies to any rate of change, positive or negative—even one that is greater than or equal to the discount rate. If the rate of change is greater than the discount rate, though, the convenience rate will be negative; the factor must then be computed using the formula, or on a calculator. If the rate of change equals the discount rate, the convenience rate is zero; the (P/F) factor is then 1.0 and the (P/A) factor is the number of periods (N) in the series.

INTERPOLATION

Tables of factors have been published for many interest rates and periods. In fact, there are handbooks of tables with factors which have more significant digits and which cover a wider range of interest rates than the tables in this text. However, factors are often needed which are not available in a table. Untabulated factors can be calculated by the use of the formulas or by interpolation between tables. The most accurate procedure is to compute the factor, because the relationship between the tables is not linear, and linearity is assumed for interpolation. For example, in Figure 5-6, the (P/A) factor was required for 4.76% interest and for a period of 5 years. On the calculator the operation is as follows:

$$(P/A, i\%, N) = \frac{(1 + i)^N - 1}{i(1 + i)^N}$$

$$(P/A, 4.76\%, 5) = \frac{(1.0476)^5 - 1}{0.0476(1.0476)^5}$$

$$= \frac{(1.2618) - 1}{(0.0476)(1.2618)} = 4.358$$

Interpolation between the 4% and 5% tables would proceed like this:

$$(P/A, 4\%, 5) = 4.452$$
$$(P/A, 5\%, 5) = 4.329$$

Assuming that the relationship is linear, the factor for 4.76% is proportionately less than the 4% factor, or 0.76/1.00 of the difference. Then:

$$(P/A, 4.76\%, 5) = 4.452 - (4.452 - 4.329)\frac{0.76}{1.00}$$

$$= 4.452 - 0.093 = 4.359$$

In this example, it is equally satisfactory to use tables or to use the calculator, because the difference between the two interest rates is small. However, in cases where interpolation is necessary over a wide range, the differences can be substantial.

EFFECTIVE AND NOMINAL RATES OF INTEREST

So far, all the applications in this chapter have considered periodic compounding, with end-of-period payments assumed. In most cases, the period is 1 year and interest is compounded annually. However, payments do not always occur annually, and interest can be compounded at many different intervals. For example, payments on home mortgages are often made monthly, and interest on bank deposits is often paid quarterly.

A bank account which has 3% interest payable quarterly is called a *12%-per-year account.* It would be more accurate to call such an interest rate a *nominal rate* of 12% per year. The actual annual interest rate which results is called the *effective rate.* These two rates are not the same. For instance, $100 deposited in an account which has 3% interest payable quarterly would actually grow to $112.55 at the end of the year and not just $112.00. What this calculation shows is that the nominal rate of 12% compounded quarterly is the same as an effective rate of 12.55% compounded annually. Furthermore, the more frequent the number of compounding periods per year, the greater the difference between the nominal and effective rates.

Before nominal rates having different compounding periods can be compared, they must first be converted to effective rates over the same period. As in the example above, a nominal rate of 12% per annum compounded quarterly is the same as:

$$\frac{12\%}{4} = 3\% \text{ per quarter}$$

This means that the future worth of $1 at 3%, compounded quarterly, for N years (or $4N$ quarters) is:

$$\$1(1.03)^{4N}$$

To find the annual effective rate i which would result from this quarterly compounding, this equation is set up:

$$\$1(1 + i)^N = \$1(1.03)^{4N}$$
$$(1 + i) = (1.03)^4$$
$$1 + i = 1.1255$$
$$i = 0.1255, \text{ or } 12.55\%, \text{ the effective rate}$$

In other words, compounding quarterly at a nominal rate of 12% per year is equivalent to compounding annually at 12.55%.

The relationship above can be restated as a general formula for finding the effective rate:

$$1 + i = \left(1 + \frac{r}{M}\right)^M$$

where r = the nominal rate per period
$\quad\ i$ = the effective rate per period
$\quad M$ = the number of compounding intervals per period

Conversion of Rates

Examples (A) and (B) illustrate the use of the formula to solve for i or r—whichever is not known.

Example A. What effective annual rate corresponds to a nominal rate of 8% per annum compounded monthly?

Solution A.

$$1 + i = \left(1 + \frac{r}{M}\right)^M$$
$$= \left(1 + \frac{0.08}{12}\right)^{12}$$
$$= 1.006667^{12}$$
$$1 + i = 1.0829$$
$$i = 0.0829 = 8.29\%, \text{ the effective annual rate}$$

Example B. If the effective rate is 8%, what is the nominal rate under monthly compounding?

Solution B. This equation can be solved for r:

$$1 + i = \left(1 + \frac{r}{M}\right)^M$$
$$1.08 = \left(1 + \frac{r}{12}\right)^{12}$$
$$\sqrt[12]{1.08} = 1 + \frac{r}{12}$$
$$r = 12\left[\sqrt[12]{1.08} - 1\right] = 0.0772 = 7.72\%, \text{ the nominal rate}$$

Applications of Effective Rates

The importance of an effective rate is that it permits the use of periodic compounded time-value factors in situations where the payments do not occur annually. It is particularly useful for determining the present worth of an annuity when the payments are not annual. Examples C and D illustrate.

Example C. Find the present worth of a 15-year series of $500 costs starting now and occurring every 18 months at 10% per year.

Solution C. The period (M) is 18 months, or 1.5 years.
The nominal rate for this period is:

$$r = 10\% \times 1.5 = 15\%$$

The effective rate for the 18-month period can then be found from this formula:

$$1 + i = \left(1 + \frac{r}{M}\right)^M$$

$$1 + i = \left(1 + \frac{0.15}{1.5}\right)^{1.5}$$

$$i = (1.10)^{1.5} - 1$$

$$i = 15.37\% \text{ every 18 months}$$

There is an initial cost of $500; and there are ten 18-month periods in 15 years with a $500 cost at the end of each. Because the convention of the (P/A) factor does not provide for the initial amount, the present worth is found as follows:

$$P = A + A(P/A, 15.37\%, 10)$$
$$P = \$500 + \$500(P/A, 15.37\%, 10)$$

Interpolation between the 15% and 20% tables gives these values:

$$(P/A, 15\%, 10) = 5.019$$
$$(P/A, 20\%, 10) = 4.192$$

Between the 20% and 15% there is a difference of 5%; and between 15% and 15.37% there is a difference of 0.37%. Therefore, the factor is:

$$(P/A, 15.37\%, 10) = 5.019 - (5.019 - 4.192)\frac{0.37}{5.00}$$

$$= 4.958$$

The present worth of this series is then:

$$P = \$500 + \$500(4.958) = \$2,979$$

Example D. Find the present worth of a 5-year series of $400 costs occurring every 6 months, at 10% per year. The first payment is in 6 months.
This time, the first step is to find the rate which, when compounded every 6 months (or $M = 2$), will result in 10% compounded annually. That rate is r/M, or the nominal rate divided by the number of periods.
The formula for the effective rate is:

$$1 + i = \left(1 + \frac{r}{M}\right)^M$$

$$1.10 = \left(1 + \frac{r}{M}\right)^2$$

$$\frac{r}{M} = \sqrt{1.10} - 1$$

$$\frac{r}{M} = 4.88\% \text{ every 6 months}$$

There are ten 6-month periods in 5 years. Particular values can now be substituted in the formula for finding the present worth of an annuity:

$$P = \$400(P/A, 4.88\%, 10)$$

Interpolation between the 4% and 5% tables gives the following:

$$(P/A, 4\%, 10) = 8.111$$
$$(P/A, 5\%, 10) = 7.722$$

$$(P/A, 4.88\%, 10) = 8.111 - (8.111 - 7.722)\frac{0.88}{1.00} = 7.769$$

$$P = \$400(7.769) = \$3,108$$

CONTINUOUS COMPOUNDING OF INTEREST

Force of Interest

At a specific nominal rate of interest, the effective rate of interest gets larger as the compounding interval is shortened. Table 5-3 shows that if the nominal rate is 10%, the effective rate is more than 10.5% when interest is compounded daily.

Theoretically, compounding periods can be even shorter than daily, although it is not feasible to compute the effective rate for such short intervals on most calculators. There is a limit as to how high the effective rate can go, however.

When a nominal rate of interest (r) is compounded for M compounding intervals per period, the result is an effective periodic rate of interest (i) given by this formula:

$$1 + i = \left(1 + \frac{r}{M}\right)^M$$

Define a quantity k as:

$$k = \frac{M}{r}, \quad \text{or} \quad M = kr$$

TABLE 5-3
COMPARISONS OF NOMINAL AND EFFECTIVE RATES*

Compounding period	Nominal rate (r)	$i = \left(1 + \frac{r}{M}\right)^M - 1$	Effective rate (i)
Annually	10%	$\left(1 + \frac{0.1}{1}\right)^1 - 1$	10%
Semiannually	10%	$\left(1 + \frac{0.1}{2}\right)^2 - 1$	10.25%
Quarterly	10%	$\left(1 + \frac{0.1}{4}\right)^4 - 1$	10.3813%
Monthly	10%	$\left(1 + \frac{0.1}{12}\right)^{12} - 1$	10.4713%
Daily	10%	$\left(1 + \frac{0.1}{365}\right)^{365} - 1$	10.5156%

*Figure 5-7 illustrates these data graphically.

FIGURE 5-7

**Increases in the Annual Effective Rate with Increases in the Number of Annual
Compounding Intervals**

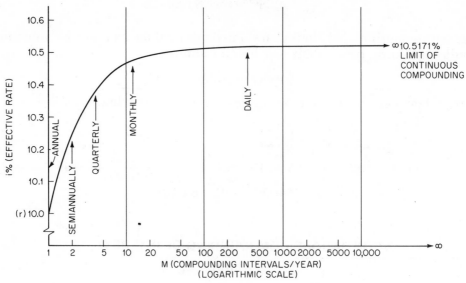

Then, as the length of the compounding interval gets shorter, M gets larger and approaches infinity; k also approaches infinity. The equation can be written as follows:

$$1 + i = \left(1 + \frac{1}{k}\right)^{kr} = \left[\left(1 + \frac{1}{k}\right)^{k}\right]^{r}$$

Mathematicians have defined the limit (the ultimate value) of the quantity $\left(1 + \frac{1}{k}\right)^{k}$, as k approaches infinity, as a special number designated e:

$$\lim_{k \to \infty} \left(1 + \frac{1}{k}\right)^{k} = e$$

As the number of compounding intervals approaches infinity, the formula then approaches a limit given by this relationship:

$$1 + i = \lim_{k \to \infty} \left(1 + \frac{1}{k}\right)^{kr} = e^{r}$$

Because the number of compounding intervals is infinite, compounding is said to be *continuous*. The nominal rate (r) for continuous compounding is called the *force of interest*. The value of e is approximately 2.71828; therefore, continuous compounding at a nominal rate of 10% would result in the maximum value for the effective rate:

$$1 + i = e^{0.10}$$
$$i = (2.71828)^{0.10} - 1$$
$$i = 10.5171\%$$

This rate is only slightly greater than the effective rate resulting from daily compounding (10.5156% on Table 5-3).

Explanation of e

The letter e represents a special number which occurs with an exponent so often, in higher mathematics, that it was convenient to develop a system of logarithms with e as the· base. This system is called the *natural,* or *Napierian, system of logarithms.* Logarithms are exponents; and all numbers can be expressed as a base number raised to some power or exponent. For example:

$$100 = 10^2$$

Here, the base 10 raised to the power of 2 is 100. The 2 is then said to be the *logarithm to the base 10* of 100. This is written as follows:

$$\log_{10}(100) = \log_{10}(10^2) = 2$$

If the base 10 is used, the subscript is usually omitted. Because the common numbering system is in multiples of 10, 10 is understood as the base for common logarithms. In the natural system of logarithms, with e as the base, a logarithm is usually written like this:

$$\log_e x = \ln x$$

The equation relating the effective and nominal rates of interest with continuous compounding is:

$$(1 + i) = e^r$$

It can also be written in the natural system of logarithms, as here:

$$\ln(1 + i) = \ln(e^r) = r$$

Natural logarithms are tabulated in handbooks, and some calculators can compute them, too. For a given effective rate (i), the natural logarithm of $(1 + i)$ will be the force of interest (r).

It is possible to determine the approximate value of e by substituting some values for k (100, 10,000, 1,000,000), as follows:

$$x = \left(1 + \frac{1}{100}\right)^{100}$$

$$\log x = 100 \log 1.01 = 0.43214$$
$$x = 2.70483$$

$$y = \left(1 + \frac{1}{10,000}\right)^{10,000}$$

$$\log y = 10,000 \log 1.0001 = 0.434273$$
$$y = 2.71815$$

$$z = \left(1 + \frac{1}{1,000,000}\right)^{1,000,000}$$

$$\log z = 1,000,000 \log 1.000001 = 0.434294$$
$$z = 2.71828$$

Here, x, y, and z represent estimated values of e. The value of e to ten decimal places is 2.7182818285.

Applications of Continuous Compounding

Savings institutions sometimes offer continuous compounding as an inducement to savers. Because the nominal interest rate which the institution can pay is regulated, compounding continuously is a device to increase the effective rate of interest for depositors.

However, continuous compounding is used for a different reason in economy studies. When cash amounts are assumed to occur at discrete points in time—usually at the end of the year—periodic time-value factors are used. When the cash amounts are better assumed to be a continuous flow, continuous compounding factors must be used.

The fact is that in industry, generally, the actual cash flow—both inward from receipts and outward for expenditures—is continuous: It goes on every business day, and hourly throughout the day. It would seem, then, that the continuous flow of cash would be a more accurate assumption for economy studies. However, even though the overall cash flows tend to be continuous, the purpose of an economy study is to analyze specific events within the firm, where the amounts may best be assumed to occur at discrete points in time. Furthermore, the estimates of cash flow for economy studies are seldom so precise that continuous compounding would increase the accuracy.

Annual compounding is used in this text because it is conceptually easier to understand and apply. Still, it is important to recognize that there are certain mathematical advantages to using continuous compounding, and it may be better for some applications.

Six time-value factors are in general use for continuous compounding of interest, where amounts of money are expected to occur as a flow throughout the year. The notation is similar to that of the periodic compounding factors, except for the bar over the amount which is a continuous flow. The bar indicates that the factor is for continuous compounding, and it also identifies which flow is continuous. The six time-value factors for continuous compounding are listed:

- *The Future Worth (F) of a Continuous Present Amount ($\overline{P}$)*

$$(F/\overline{P}, i\%, N) = \frac{e^{rN}(e^r - 1)}{re^r}$$

- *The Present Worth (P) of a Continuous Future Amount ($\overline{F}$)*

$$(P/\overline{F}, i\%, N) = \frac{e^r - 1}{re^{rN}}$$

- *A Continuous Annuity ($\overline{A}$) for a Future Amount (F)*

$$(\overline{A}/F, i\%, N) = \frac{r}{e^{rN} - 1}$$

- *The Future Worth (F) of a Continuous Annuity ($\overline{A}$)*

$$(F/\overline{A}, i\%, N) = \frac{e^{rN} - 1}{r}$$

- *The Continuous Annuity ($\overline{A}$) from a Present Amount (P)*

$$(\overline{A}/P, i\%, N) = \frac{re^{rN}}{e^{rN} - 1}$$

- *The Present Worth (P) of a Continuous Annuity ($\overline{A}$)*

$$(P/\overline{A}, i\%, N) = \frac{e^{rN} - 1}{re^{rN}}$$

Values for these six continuous compounding time-value factors are tabulated in Appendix B at the back of the book, for effective interest rates from 8% to 15% and from 1 to 75 periods. They correspond to the effective rates of the annual compounding tables in Appendix A at the back of the book.

When continuous compounding is applied, it is important to keep in mind that the continuous amount is a flow of cash, which is expressed as a rate of flow per period. For emphasis, cash-flow diagrams which include a continuous flow will show this flow as a block, rather than as the vertical arrow, which represents a discrete amount.

Figure 5-8 illustrates how the six continuous compounding time-value factors are

FIGURE 5-8
Equivalence-Continuous Compounding

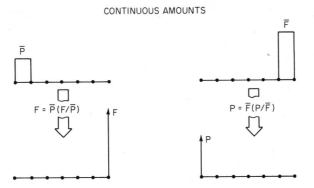

CONTINUOUS AMOUNTS

$\overline{P}$

$\overline{F}$

$F = \overline{P}(F/\overline{P})$

$P = \overline{F}(P/\overline{F})$

CONTINUOUS ANNUITIES

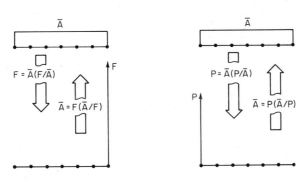

$\overline{A}$

$\overline{A}$

$F = \overline{A}(F/\overline{A})$

$\overline{A} = F(\overline{A}/F)$

$P = \overline{A}(P/\overline{A})$

$\overline{A} = P(\overline{A}/P)$

FIGURE 5-9
An Example of Equivalence-Continuous Compounding

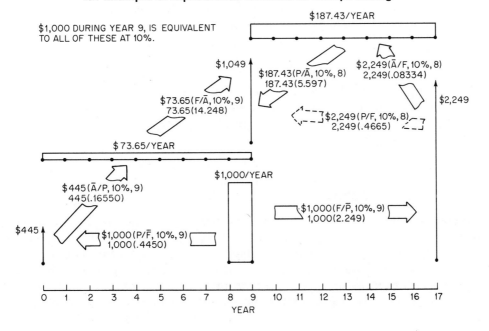

$1,000 DURING YEAR 9, IS EQUIVALENT
TO ALL OF THESE AT 10%.

$187.43/YEAR

$1,049

$187.43(P/\overline{A}, 10\%, 8)$
187.43(5.597)

$2,249(\overline{A}/F, 10\%, 8)$
2,249(.08334)

$73.65(F/\overline{A}, 10\%, 9)$
73.65(14.248)

$2,249(P/F, 10\%, 8)$
2,249(.4665)

$2,249

$73.65/YEAR

$1,000/YEAR

$445(\overline{A}/P, 10\%, 9)$
445(.16550)

$1,000(F/\overline{P}, 10\%, 9)$
1,000(2.249)

$445

$1,000(P/\overline{F}, 10\%, 9)$
1,000(.4450)

YEAR
0 1 2 3 4 5 6 7 8 9 10 11 12 13 14 15 16 17

used. One point to note here is that the continuous annuity factors are reciprocals of one another. However, the single continuous amount factors are not reciprocals.

Figure 5-9 on page 113 illustrates how the six factors are applied. A single continuous amount of $1,000 during year 9 can be transformed into various discrete amounts and continuous annuities, all equivalent at the 10% effective rate. This example also illustrates that two of the discrete amounts, which are related by the dashed broad arrow, are equivalent at 10% compounded annually.

Appendix 5A at the end of the chapter contains derivations of each of the six factors, as well as some simple examples illustrating how these factors are applied.

Notes on Continuous Compounding

It is important to remember that the reason for using continuous compounding in an economy study is to assume the continuous flow of cash rather than discrete amounts at periodic intervals. If a study includes both continuous flows and discrete amounts, the effective rates for all the factors must be the same. For this reason, tables are provided in which the effective rates for continuous compounding are equal to the effective rates for the periodic tables. This means that the force of interest (r) will usually be a non-integer number. However, many books include tables in which the force of interest is an integer. These rates may be called *continuous interest rates*, and the effective rate of equivalence is somewhat higher. The important point is that if both continuous flows and discrete flows occur in a particular problem, the factors used must have the same annual effective rate.

How these rates are interpreted affects the use of the convenience rate (i_g) for finding the present worth of a continuous geometrically changing series. With continuous compounding, if both the discount rate and the rate of change are considered as force of interest rates or nominal rates, the convenience force of interest (r_g) is simply the difference between the two:

$$r_g = r - \text{nominal rate of change}$$

To find the equivalent present worth of a continuous geometrically changing series, the factor is then found in a continuous compounding table where the *force of interest* is r_g.

However, if the discount rate and rate of change are both considered effective rates, it is necessary to return to the original formula:

$$i_g = \frac{i - I}{1 + I}$$

The factor is then found from a continuous compounding table where the *effective* rate is i_g. This second approach is most often appropriate, because rates of change derived from index numbers (see Chapter 10, on inflation) are usually effective rates rather than nominal rates.

USING THE MATHEMATICS OF MONEY

The most common use of time-value factors is to convert some amount of money at one point in time into an equivalent amount at a known interest rate at a different point in time. The procedure is to first consult a table for the proper factor. The product of this factor and the known amount of money is the equivalent unknown amount. Other applications of the time-value factors are simply variations of this procedure.

For example, to find the quarterly payments necessary to repay a $5,000 second mortgage on a house in 5 years, at 12% nominal interest per year, the procedure is as follows:

The rate per quarter = $^{12}\!/_4$ = 3%
The number of quarters in 5 years = 5 × 4 = 20

The (A/P) factor is found in the 3% table, at 20 periods:

$$(A/P, 3\%, 20) = 0.06722$$

The payment is then:

$$A = P(A/P, i\%, N)$$
$$A = \$5,000(0.06722) = \$336.10$$

The solution shows that $336.10 paid each quarter for 5 years would repay the loan.

The principles of the mathematics of money can also be used in reverse to find an interest rate which will make two known amounts equivalent. For example, the institution making the $5,000 second mortgage might require a payment of $340.00 each quarter for 5 years. The annual rate of interest these payments would represent could be found as follows.

$$P \times (A/P, i\%, N) = A$$
$$\$5,000 \times (A/P, i\%, 20) = \$340.00$$
$$(A/P, i\%, 20) = \frac{340}{5,000} = 0.0680$$

This falls between the 20-period (A/P) factors for 3% and 4%.

$$(A/P, 3\%, 20) = 0.06722$$
$$(A/P, 4\%, 20) = 0.07358$$

The quarterly compounded interest rate is found by interpolation:

$$i = 0.03 + 0.01 \times \left(\frac{0.06800 - 0.06722}{0.07358 - 0.06722}\right)$$
$$i = 0.0312, \text{ or } 3.12\% \text{ each quarter}$$

This is a nominal annual rate of:

$$3.12 \times 4 = 12.48\%$$

It is also an effective annual rate of:

$$(1.0312)^4 - 1 = 0.1308 = 13.08\%$$

The principles of the mathematics of money can be used in still another way, to find the necessary number of periods required to repay a loan. For example, the borrower might decide that $300 would be a reasonable quarterly payment, and the institution is charging 12% for a $5,000 second mortgage. How many $300 quarterly payments are required to repay the loan?

$$A = P(A/P, i\%, N)$$
$$\$300 = \$5,000 \times (A/P, 3\%, N)$$
$$(A/P, 3\%, N) = \frac{300}{5,000} = 0.0600$$

On the 3% table, 0.0600 falls between 23 and 24 quarters. This means that to keep the payments under $300, it is necessary for the loan to last for at least 24 quarters, or 6 years. The quarterly payments would then be as follows:

$$A = \$5,000(A/P, 3\%, 24)$$
$$A = \$5,000(0.05905) = \$295.25$$

Any time-value-of-money problem has four elements:

- Some amount or series of money
- An equivalent amount or series of money
- An interest rate
- A number of periods of time

The time-value factors can be used to find any one of these four elements provided each of the other three is known.

The illustrations here have been simple applications. But solving for i or N often will not be simple, because several different factors may be used to express the equivalence. These problems must be solved by trial and error, and complex problems are best left to computers.

SUMMARY

The earning power of money is the basis for the mathematics of money. Because of this earning power, an amount of money can grow in value with the passage of time, and an amount of money in the future does not have that same value today. This time-value relationship means that at a given earning power, or interest rate, an amount of money at one point in time is equivalent to another amount at some other time—or to a series of amounts over a period of time.

In economy studies, it is necessary to compare estimates of the costs of various alternatives. However, these cost estimates usually vary both in timing and in amount. Because of the concept of equivalence, they can easily be converted into equivalent amounts which can be compared.

The concept of equivalence is applied by the use of time-value factors to make the necessary translations. These factors are tabulated for a wide range of interest rates. However, they can also be calculated from mathematical formulas, or with electronic calculators which include some of the formulas as internal operations.

The most easily understood application of time-value factors is to cash amounts which are assumed to occur at discrete points in time. This assumption is adopted for most of this text and it is usually adequate for most economy studies. Time-value factors for periodic—usually annual—compounding of interest are based on this assumption. In some cases, however, a more appropriate assumption may be that cash amounts are flowing continuously throughout time, and time-value factors for continuous compounding are used.

Two rates of interest are associated with time-value factors for continuous compounding. One rate, called the *force of interest,* is the nominal rate, which compounded continuously will produce the other rate, called the *effective rate.* If both continuous factors and annual factors are mixed in an analysis, the effective rate for the continuous factors must be the same as the periodic rate of interest for the annual factors.

Various time-value factors are available, which make possible practically any translation between equivalent amounts or series of amounts. These factors apply to single amounts, periodic annuities, arithmetic progressions, geometric progressions, continuous amounts during a single year, and continuous annuities. Other series of cash amounts can be translated by the application of more than one factor or by a serial application of the single-amount factors.

PROBLEMS

These problems are based on the material covered in the chapter. In attempting to solve them, it will be helpful to draw a time-cost diagram. (Use the tables at the back of the book.)

1. Determine the equivalent values of the following amounts over the time periods specified, assuming the time-value of money is 10%.

   a. The amount at the end of 20 years which is equivalent to $100 today. (*Answer:* $672.70)

   b. The equal annual year-end payments during the last 12 of the above 20 years which are equivalent to the amount found in "a." (*Answer:* $31.46)

   c. The amount at the beginning of the 12-year period which is equivalent to the 12 equal annual payments. (*Answer:* $214.37)

   d. The minimum amount which would have been acceptable 14 years ago to accumulate to the amount found in "c" (22 years before the amount found in "c"). (*Answer:* $26.32)

   e. The minimum equal annual year-end payments, equivalent to payment "d," which would have been acceptable during the past 14 years. (*Answer:* $3.57)

   f. The payment today which is equivalent to payment "e." (*Answer:* $99.87)

2. In how many years would the following present amounts equate to the future worths shown at the specified cost-of-money rates compounded annually?

Present amount	Cost-of-money rate	Future worth	
$1,200	5%	$ 1,690	(*Answer:* 7 yr)
5,650	9%	22,430	(*Answer:* 16 yr)
310	7%	4,640	(*Answer:* 40 yr)
250	15%	4,090	(*Answer:* 20 yr)

3. Determine the interest rate which, when compounded annually over the years shown, will produce present amounts equivalent to the specified future worths.

Present amount	Years	Future worth	
$1,750	15	$7,310	(*Answer:* 10%)
400	55	1,190	(*Answer:* 2%)
4,500	5	6,020	(*Answer:* 6%)
100	26	1,190	(*Answer:* 10%)

4. A man completes his payments on an endowment policy upon reaching age 55. At this time, its value is $5,000. He may choose to receive either a single amount at age 65 or equal annual amounts from age 65 to 74. If the insurance company computes interest at 3% annually:

   a. How much would he receive as a single amount upon reaching age 65? (*Answer:* $6,720)

   b. What would the amounts be if he chooses to receive 10 equal annual amounts starting at his 65th birthday?

(Hint: The first annuity payment occurs one year after the date of the present value of the annuity.)   (*Answer:* $765)

5. Referring to problem 4, if the man's policy provided for specified equal annual payments for life starting when he reaches age 65, on what average life spans were the amounts shown below computed? (Use tables in Appendix A at the back of the book, disregarding the effect that dispersed mortality of a group has on average life.)

   $ 930/year   (*Answer:* 72 years)
       655/year   (*Answer:* 76 years)
     1,205/year   (*Answer:* 70 years)

6. What end-of-year payments are needed to accumulate to the following future amounts at the specified interest rates compounded annually?

Future amount	Years	Interest rate	
$10,000	20	12%	(*Answer:* $139)
10,000	45	2%	(*Answer:* $139)
5,000	8	10%	(*Answer:* $437)
5,000	10	3%	(*Answer:* $436)
12,000	15	8%	(*Answer:* $442)

7. Referring to the data in problem 6, what end-of-year payments would be required if final payments to the fund were to be made 5 years before the terminal date, with the balance left to accumulate to the specified future amounts?   (*Answer:* $152; $150; $938; $812; $564)

8. What is the present worth of the following series of equal annual amounts starting after the third year from today (time 0) and continuing for the number of years shown, annually compounded at a time-value rate of 8%?

Annual amounts	Years	
$ 600	17	(*Answer:* $4,345)
1,500	10	(*Answer:* $7,991)
200	32	(*Answer:* $1,816)
1,200	6	(*Answer:* $4,404)

9. Two land sites are under review for a new plant location. One, in City A, has a current property tax of $15,000 per year. This tax is expected to increase $500 per year. The other site, in City B, has taxes of $10,000 per year with an estimated increase of $2,000 per year. (*a*) How much money would have to be set aside today to provide for taxes over the next 10 years? The interest rate is 12%. (*b*) If the two sites can be bought for the same price, which site should be recommended based on the difference in taxes alone?

   (*Answer:* (*a*) City A = $ 97,702
                    City B = $108,308
             (*b*) Recommended site in City A)

10. What is the present worth at 10% interest of a 20-year geometric progression increasing at 2% per year if the first amount is $102?
                                                          (*Answer:* $994)

11a. You have a new baby. It now costs $5,000 per year to send a student to college. If costs of education are expected to increase at 4% per year, how much will each of 4 years at college cost beginning on your child's 18th birthday?

You expect your salary will increase at 5% per year until you must begin to pay for college. If during the next 17 years you make annual deposits of a constant percentage of your salary into an 8% savings account, how much must your first deposit one year from now be so that you can withdraw the annual amounts needed for college beginning on your child's 18th birthday? What is the final deposit?

(*Answer:* For college: $10,130, $10,535, $10,955, $11,395. First deposit, $756; last deposit, $1,650.)

11b. What would the first and last deposits in 11a be if you project your salary to increase at 10% per year? (*Answer:* $524; $2,408)

12. Since a nominal annual interest rate when divided by the number of compounding intervals per year is equal to the rate per interval, determine the future worths of these present amounts for the following numbers of years at the specified nominal rates:

Present amount	Years	Nominal rate	Compounding frequency	
$4,000	7	9%	Annually	(*Answer:* $7,312)
2,500	10	6%	Semiannually	(*Answer:* $4,515)
500	3	20%	Quarterly	(*Answer:* $ 898)

13. What are the effective annual rates corresponding to the following nominal rates compounded as shown below?

Nominal rate	Compounding frequency	
5%	Annually	(*Answer:* 5%)
9%	Semiannually	(*Answer:* 9.2%)
12%	Quarterly	(*Answer:* 12.55%)
6%	Monthly	(*Answer:* 6.17%)
7.7706%	Quarterly	(*Answer:* 8%)
9.5690%	Monthly	(*Answer:* 10%)
8%	Continuously	(*Answer:* 8.33%)

14. If the time value of money is 12% per year, what rate per period should be used to find the present worth of a series of amounts occurring with the frequency below?

Frequency of amounts	
Annually	(*Answer:* 12%)
Quarterly	(*Answer:* 2.87%)
Monthly	(*Answer:* 0.95%)
Every 18 months	(*Answer:* 18.53%)
Every 2 years	(*Answer:* 25.44%)

15. If the business' cost of money is 12%, what is the present worth of the following series of costs?

a. $12,000 at the end of each year for 5 years. (*Answer:* $43,260)

b. $3,000 at the end of each quarter for 5 years. (*Answer:* $45,206)

c. Continuously occurring costs at a rate of $1,000 per month for 5 years.   (*Answer:* $45,804)

16. The yield-to-maturity of a bond is the *nominal* interest rate which, when compounded semiannually (because bond interest is usually paid semiannually), will cause the purchase price to equal the present value of the future interest and principal. If you want an 8% yield-to-maturity, how much should you pay for a 4% coupon-rate, $1,000 corporate bond maturing in 3 years?   (*Answer:* $895)

Derivations and Applications of Time-Value Factors

This appendix contains the derivation of each time-value factor, as well as one or more examples of its application.

SINGLE-AMOUNT FACTORS

The Future Worth of a Present Amount (*F*/*P*)

Derivation of the (*F*/*P*) Factor

The *future worth* of a present amount is the sum, at the end of a specific period of time, that this present amount of money will accumulate to at a given rate of interest each compounding period. For example, at an 8% annual rate of interest, the future worth of $100 at the end of 4 years is:

$100.00	Present amount at time 0
8.00	Interest at 8%, at the end of year 1
108.00	Worth, end of first year
8.64	Interest at 8%, at the end of year 2
116.64	Worth, end of second year
9.33	Interest at 8%, at the end of year 3
125.97	Worth, end of third year
10.08	Interest at 8%, at the end of year 4
$136.05	Worth, end of fourth year

The worths at the end of each year were derived as follows:

Worth, end of first year $= 100 + 100(0.08) = 100(1 + 0.08)$
Worth, end of second year $= 100(1 + 0.08) + 100(1 + 0.08)(0.08)$ or
$100(1 + 0.08)(1 + 0.08) = 100(1 + 0.08)^2$
Worth, end of third year $= 100(1 + 0.08)^2 + 100(1 + 0.08)^2(0.08)$ or
$100(1 + 0.08)^2(1 + 0.08) = 100(1 + 0.08)^3$
Worth, end of fourth year $= 100(1 + 0.08)^3 + 100(1 + 0.08)^3(0.08)$ or
$100(1 + 0.08)^3(1 + 0.08) = 100(1 + 0.08)^4$

This last calculation shows that the future worth of $100 at the end of the fourth year is equal to the present amount ($100) times $(1 + 0.08)^4$. Thus:

$$\$100(1.08)^4 = \$100 \times 1.3605$$
$$= \$136.05$$

Note:

$$(1 + 0.08)^4 = 1.08 \times 1.08 \times 1.08 \times 1.08 = 1.3605$$

The *time-value factor* for the future worth of a present amount is the ratio of the future amount to the present amount. In the example, this is $136.05 divided by $100, or 1.3605, which is the same as $(1.08)^4$.

The general, or symbolized, expression of the *time-value factor* for the future worth of a present amount may therefore be developed like this:

$$\text{Worth, end of first year: } (F)_1 = P + P \times i \quad \text{or}$$
$$P + Pi = P(1 + i)$$
$$\text{Worth, end of second year: } (F)_2 = P[(1 + i) + (1 + i)i] \quad \text{or}$$
$$P[(1 + i)(1 + i)] = P(1 + i)^2$$
$$\text{Worth, end of third year: } (F)_3 = P[(1 + i)^2 + (1 + i)^2 i] \quad \text{or}$$
$$P[(1 + i)^2(1 + i)] = P(1 + i)^3$$
$$\text{Worth, end of } N \text{ years: } (F)_N = P(1 + i)^N$$

The (F/P) factor is then expressed as follows:

$$(F/P, i\%, N) = (\text{future worth of 1, } N \text{ years}) = (1 + i)^N$$

This increase in worth with the passage of time can be illustrated as follows:

$$P = \text{Present Amount}$$
$$A = \text{Interest on } P, \text{ end of year 1}$$
$$B = \text{Interest on } P + A, \text{ end of year 2}$$
$$C = \text{Interest on } P + A + B, \text{ end of year 3}$$

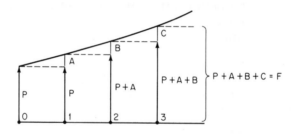

Application of the (F/P) Factor

Examples A and B illustrate how the (F/P) factor is applied.

Example A. Find what $100 now will be worth at the end of 10 years, at 8% interest.

Solution A. The equation is as follows:

$$F = P(F/P, 8\%, 10)$$
$$F = \$100(2.159) = \$215.90$$

Example B. Find the future value in 10 years of a $100 payment occurring 3 years from now if the interest rate remains 8%.

Solution B. Note that the number of periods is $10 - 3$, or 7. Therefore, the equation is:

$$F = F(F/P, 8\%, 7)$$
$$F = \$100(1.7138) = \$171.38$$

The Present Worth of a Future Amount (P/F)

Derivation of the (P/F) Factor

The *present worth* of a future amount is the amount of money at the start of a specific period of time which will accumulate to a known sum at the end of that period at a given rate of interest each compounding period.

The previous section showed that, at an 8% rate of interest, the future worth of $100 at the end of 4 years is $136.05. Conversely, the present worth of $136.05 in hand four years from now is $100 if the same 8% rate of interest is assumed.

Mathematically,

if
$$F = P(1 + i)^N$$

then
$$P = \frac{F}{(1 + i)^N}$$

The *time-value factor* for the present worth of a future amount is therefore expressed as follows:

$$(P/F, i\%, N) = (\text{present worth of 1, } N \text{ years}) = \frac{1}{(1 + i)^N}$$

Application of the (*P/F*) Factor

Example A. Find the present worth of $100 now at 8% interest.

Solution A. The equation is as follows:

$$P = \$100(P/F, 8\%, 0)$$

The (*P/F*) factor at any interest rate for 0 years is 1, so the equation becomes:

$$P = \$100 \times 1.0 = \$100$$

Example B. Find the present worth of $100 occurring 10 years from now.

Solution B. The equation is as follows:

$$P = F(P/F, 8\%, 10)$$
$$P = \$100 \times 0.4632 = \$46.32$$

UNIFORM SERIES FACTORS

The Future Worth of an Annuity (*F/A*)

Derivation of the (*F/A*) Factor

The *future worth of an annuity* for a given period of time is the sum at the end of that period of the future worths of all payments at a given rate of interest each compounding period. For example, at an 8% rate of interest, the future worth of an annuity of $100 per year for 4 years may be found as follows:

Time of payment (end of year)	Amount of payment (A)	Formula $A(1 + i)^N$	Future worth (F)
1	$100	$100(1.08)^3$	$125.97
2	$100	$100(1.08)^2$	116.64
3	$100	$100(1.08)^1$	108.00
4	$100	$100(1.08)^0$	100.00
		Sum of the future worths =	$450.61

The general, or symbolized, expression for the future worth of an annuity can therefore be developed as here:

$$F = A[(1 + i)^0 + (1 + i) + (1 + i)^2 + (1 + i)^3] \qquad (5A.1)$$

Multiplying both sides of equation (5A.1) by $(1 + i)$ gives equation (5A.2):

$$F(1 + i) = A[(1 + i)^0(1 + i) + (1 + i)(1 + i) + (1 + i)^2(1 + i)$$
$$+ (1 + i)^3(1 + i)]$$
$$= A[(1 + i) + (1 + i)^2 + (1 + i)^3 + (1 + i)^4] \qquad (5A.2)$$

Subtracting equation (5A.1) from equation (5A.2) gives this result:

$$F[(1 + i) - 1] = A[(1 + i)^4 - 1]$$
$$F = A \frac{(1 + i)^4 - 1}{i}$$

The general equation is therefore:

$$F = A \frac{(1 + i)^N - 1}{i}$$

The time-value factor for the future worth of an annuity is then expressed like this:

$$(F/A, i\%, N) = \text{(future worth of 1, } N \text{ years)}$$
$$= \frac{(1 + i)^N - 1}{i}$$

Application of the (F/A) Factor

Both Example A and Example B illustrate how to apply the (F/A) factor.

Example A. Find the future worth of a $100 annuity at 8% interest at the end of 10 years.

Solution A. The equation is as follows:

$$F = \$100(F/A, 8\%, 10)$$
$$F = \$100(14.487) = \$1,448.70$$

Example B. Find the future worth in 10 years of a $100 annuity at 8% interest, with payment starting at the end of the fourth year.

Solution B. This is actually a 7-year annuity, with the first amount occurring 4 years from now. The equation becomes:

$$F = \$100(F/A, 8\%, 7)$$
$$= \$100(8.923) = \$892.30$$

The Annuity for a Future Amount (A/F)

Derivation of the (A/F) Factor

The *annuity for a future amount* is the amount which, if it is set aside each year, will accumulate to a known sum at the end of a specific period of time at a given rate of interest each compounding period.

Earlier, the future worth of an annuity was determined by this formula:

$$F = A \frac{(1 + i)^N - 1}{i}$$

Conversely, the annuity for a future worth can be found by this formula:

$$A = F \frac{i}{(1 + i)^N - 1}$$

The time-value factor is then as follows:

$$(A/F, i\%, N) = \text{(annuity for a future amount of 1, } N \text{ years)}$$

$$= \frac{i}{(1 + i)^N - 1}$$

Application of the (A/F) Factor

Example. Find the annuity over a 10-year period which is equivalent at 8% interest to a future amount of $1,449.

Solution. The equation is:

$$A = \$1,449(A/F, 8\%, 10)$$
$$A = \$1,449(0.06903) = \$100$$

The Present Worth of an Annuity (P/A)

Derivation of the (P/A) Factor

The present worth of an annuity for a given period of time is the sum at the start of that period of the present worths of each payment in the series, at a given rate of interest each compounding period.

   This present amount is also the amount which, at a given rate of interest each compounding period, will allow specified amounts to be withdrawn for a definite period. For example, at an 8% rate of interest, the present worth of an annuity of $100 per year for 4 years may be found as follows:

Time of payment (N)	Amount of payment (A)	Formula $A(1 + i)^{-N}$	Present worth (P)
1	$100	$100(1.08)^{-1}$	$ 92.59
2	100	$100(1.08)^{-2}$	85.73
3	100	$100(1.08)^{-3}$	79.38
4	100	$100(1.08)^{-4}$	73.50
		Sum of the present value worths =	$331.20

Note: $(1 + i)^{-N}$ is the same as $1/(1 + i)^N$.

The general, or symbolized, expression for the present worth of an annuity can be developed from this example, as follows:

$$P = A\left[\frac{1}{(1+i)^1} + \frac{1}{(1+i)^2} + \frac{1}{(1+i)^3} + \frac{1}{(1+i)^4}\right] \qquad (5A.3)$$

If both sides of equation (5A.3) are multipled by $1/(1+i)$, the result is equation (5A.4);

$$P\frac{1}{(1+i)} = A\left[\frac{1}{(1+i)^2} + \frac{1}{(1+i)^3} + \frac{1}{(1+i)^4} + \frac{1}{(1+i)^5}\right] \qquad (5A.4)$$

Now equation (5A.3) can be subtracted from equation (5A.4), leaving:

$$P\left[\frac{1}{(1+i)} - 1\right] = A\left[\frac{1}{(1+i)^5} - \frac{1}{(1+i)}\right]$$

Multiplying both sides by $(1+i)$ gives this result:

$$P\left[1 - (1+i)\right] = A\left[\frac{1}{(1+i)^4} - 1\right]$$

Then,

$$P(-i) = A\left[\frac{1}{(1+i)^4} - 1\right]$$

$$P = A\left[\frac{1}{i} - \frac{1}{i(1+i)^4}\right]$$

$$P = A\left[\frac{(1+i)^4 - 1}{i(1+i)^4}\right]$$

The general equation is therefore:

$$P = A\left[\frac{(1+i)^N - 1}{i(1+i)^N}\right]$$

The time-value factor for the present worth of an annuity is then expressed like this:

$$(P/A, i\%, N) = (\text{present worth of 1, } N \text{ years}) = \frac{(1+i)^N - 1}{i(1+i)^N}$$

Application of the (*P/A*) Factor

Example A. Find the present worth of a \$100 annuity at an 8% interest rate over a 10-year period.

Solution A. The equation is:

$$P = \$100(P/A, 8\%, 10)$$
$$P = \$100(6.710) = \$671$$

Example B. Find the amount of money which must be deposited now, at 10%, to be able to withdraw \$10,000 per year for 5 years.

Solution B. This is the same type of question, but asked in a different way. The answer is found in the same way.

$$P = \$10,000(P/A, 10\%, 5)$$
$$P = \$10,000(3.791) = \$37,910$$

The Annuity from a Present Amount (*A/P*)

Derivation of the (*A/P*) Factor

The *annuity from a present amount* is the annual amount which can be withdrawn for a definite period of time from a present sum of money at a given rate of interest each compounding period.

Earlier, the present worth of an annuity was determined by this formula:

$$P = A \frac{(1 + i)^N - 1}{i(1 + i)^N}$$

Conversely, the annuity from a present amount may be found by this relationship:

$$A = P \frac{i(1 + i)^N}{(1 + i)^N - 1}$$

The time-value factor is then as follows:

$$(A/P, i\%, N) = (\text{annuity from a present amount 1, } N \text{ years})$$

$$= \frac{i(1 + i)^N}{(1 + i)^N - 1}$$

Application of the (*A/P*) Factor

Example. Determine the annuity over a 10-year period which is equivalent to a present worth of $671 at 8% interest.

Solution. The equation is:

$$A = \$671(A/P, 8\%, 10)$$
$$A = \$671(0.1490), \text{ or } \$100$$

GRADIENT FACTORS (ARITHMETIC PROGRESSIONS)

The Future Worth of a Linear Gradient (*F/G*)

The *future worth of a linear gradient* is the future worth at the end of the period of the changing portion of an arithmetic progression for a given period of time at a given rate of interest each compounding period.

Derivation of the (*F/G*) Factor

As the definition above indicates, the (*F/G*) factor will determine the future worth of only the changing portion of an arithmetic series. The future worth of the constant portion is found using the (*F/A*) factor. The constant portion is an annuity of payments equal to the first payment. The changing portion begins, then, with the second payment, as in the sketch below.

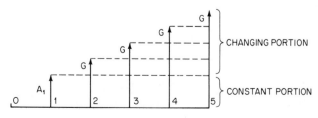

The future worth of the complete series is the sum of the two components, or:

$$F = A_1(F/A) + G(F/G)$$

The changing portion is a series of deferred annuities, each in the amount of the gradient (G). A new deferred annuity begins each year after the first. The future worth is therefore the sum of the future worths of each of those deferred annuities.

$$F = G(F/A, i\%, N-1) + G(F/A, i\%, N-2) + \cdots$$
$$+ G(F/A, i\%, 2) + G(F/A, i\%, 1)$$

Note: It is important not to confuse this notation with the (P/A) factor notation for a deferred annuity. Here, the first factor is found for $(N-1)$ periods. If $N = 5$, then the first factor is $(F/A, i\%, 4)$.

In algebraic form, the equation looks like this:

$$F = G\frac{(1+i)^{(N-1)} - 1}{i} + G\frac{(1+i)^{(N-2)} - 1}{i} + \cdots$$
$$+ G\frac{(1+i)^2 - 1}{i} + G\frac{(1+i)^1 - 1}{i}$$

Then:

$$F = \frac{G}{i}[(1+i)^{(N-1)} + (1+i)^{(N-2)} + \cdots + (1+i)^2 + (1+i)^1 - (N-1)]$$

$$F/G = \frac{1}{i}[(1+i)^{(N-1)} + (1+i)^{(N-2)} + \cdots + (1+i)^2 + (1+i)^1 + 1] - \frac{N}{i}$$

The bracketed terms in this last equation were shown to be the Future Worth of an Annuity Factor for N years in equation (5A.1).
Therefore:

$$(F/G) = \frac{1}{i}(F/A) - \frac{N}{i} = \frac{(F/A) - N}{i}$$

Using the above equation, the (F/G) can be computed from a table of (F/A) factors even if it is not tabulated itself.
Algebraically:

$$(F/G, i\%, N) = \frac{1}{i}\left[\frac{(1+i)^N - 1}{i} - N\right]$$

Application of the (F/G) Factor

Example A. Find the future worth at 10% of an 8-year arithmetic progression starting with $500 and increasing $100 each year.

Solution A. The equation is as follows:

$$F = \$500(F/A, 10\%, 8) + \$100(F/G, 10\%, 8)$$
$$F = \$500 \times (11.436) + \$100 \times (34.36) = \$9,154$$

Example B. Find the future worth at 12% of an 8-year arithmetic progression starting at $100 and increasing $100 each year.

Solution B. This problem could be solved exactly as Example A was; but because the first amount equals the gradient, it can also be solved by the following:

$$F = \$100(F/G, 12\%, 8 + 1)$$
$$F = \$100(48.13) = \$4{,}813.30$$

This is because the (F/G) factor for 9 years is the one which assumes eight gradients. With this solution, the constant portion of the series is zero.

The Present Worth of a Linear Gradient (*P/G*)

The *present worth of a linear gradient* is the present worth at the start of the period of the changing portion of an arithmetic progression for a given period of time at a given rate of interest each compounding period.

Derivation of the (*P/G*) Factor

As with the future worth of an arithmetic series, the present worth of the complete series is then the sum of two components, or:

$$P = A_1(P/A) + G(P/G)$$

Since the functional symbols for compound interest factors can be manipulated algebraically, the present worth of a linear gradient (P/G) can be found from the future worth of a gradient (F/G). Because:

$$(P/G) = (F/G) \times (P/F)$$

and the (F/G) factor was shown to be:

$$(F/G) = \frac{(F/A) - N}{i}$$

then:

$$(P/G) = \frac{(F/A) - N}{i}(P/F)$$

$$(P/G) = \frac{(F/A)(P/F) - N(P/F)}{i}$$

$$(P/G) = \frac{(P/A) - N(P/F)}{i}$$

The above formula can be used to find the (P/G) factor if it is not tabulated. Algebraically it is:

$$(P/G) = \frac{1}{i}\left[\frac{(1 + i)^N - 1}{i(1 + i)^N} - \frac{N}{(1 + i)^N}\right]$$

Application of the (*P/G*) Factor

Example A. Find the present worth at 10% of a 10-year arithmetic progression starting at $1,000 and decreasing $100 each year.

Solution A. The equation is as follows:

$$P = \$1{,}000(P/A, 10\%, 10) - \$100(P/G, 10\%, 10)$$
$$P = \$1{,}000 \times (6.144) - \$100 \times (22.891)$$
$$= \$6{,}144 - \$2{,}289 = \$3{,}855$$

Example B. Find the present worth at 12% of an 8-year arithmetic progression, starting with $100 and increasing $100 each year.

Solution B. Because the first amount is equal to the gradient, this can be found as above, or from:

$$P = \$100(P/G, 12\%, 8 + 1)(F/P, 12\%, 1)$$
$$P = \$100(17.356)(1.1200) = \$1,944$$

The (P/G) factor for 9 years has eight gradients but finds the present worth two periods prior to the first.

The Annuity for a Linear Gradient (A/G)

The *annuity for a linear gradient* is the annuity over the total period equivalent to the changing portion of an arithmetic progression for a given period of time at a given rate of interest each compounding period.

Derivation of the (A/G) Factor

The annuity equivalent to the complete series is, then, the sum of the constant portion and the annuity equivalent to the changing portion, as:

$$A = A_1 + G(A/G)$$

The annuity for a linear gradient can be derived from the future worth of a gradient (F/G) using this formula:

$$(A/G) = (F/G) \times (A/F)$$

Recall that:

$$(F/G) = \frac{(F/A) - N}{i}$$

therefore:

$$(A/G) = \frac{(F/A) - N}{i}(A/F)$$

$$(A/G) = \frac{(F/A)(A/F) - N(A/F)}{i}$$

$$(A/G) = \frac{1 - N(A/F)}{i}$$

which can be used to find the factor, or:

$$(A/G) = \frac{1}{i}\left[1 - \frac{Ni}{(1 + i)^N - 1}\right]$$

$$(A/G) = \frac{1}{i} - \frac{N}{(1 + i)^N - 1}$$

Application of the (A/F) Factor

Example. Find the annuity equivalent at 12% to a 20-year arithmetic series starting at $300 and increasing $30 a year.

Solution. The equation is as follows:

$$A = \$300 + \$30(A/G, 12\%, 20)$$
$$A = \$300 + \$30(6.020) = \$480.60$$

CONTINUOUS-SERIES FACTORS

The Future Worth of a Continuous Annuity ($F/\overline{A}$)

Derivation of the ($F/\overline{A}$) Factor

The development of the ($F/\overline{A}$) factor is really an expansion of the development of a factor for compounding intervals of less than one period. This is the (F/A) factor for annual compounding:

$$(F/A, i\%, N) = \frac{(1 + i)^N - 1}{i}$$

For quarterly amounts, where $A/4$ is the quarterly amount, the factor would be:

$$(F/A, \frac{r}{4}\%, 4N) = \frac{[1 + (r/4)]^{4N} - 1}{r/4}$$

The future worth would then be this factor times the amount of the annuity occurring each quarter:

$$F = \frac{A}{4} \frac{[1 + (r/4)]^{4N} - 1}{r/4}$$

Because the 4s cancel, the future worth is also

$$F = A \frac{[1 + (r/4)]^{4N} - 1}{r}$$

Here, A is the rate of flow per year rather than quarterly, and r is found from this equation:

$$(1 + i) = \left(1 + \frac{r}{4}\right)^4$$

Therefore, the future worth is also:

$$F = A \frac{(1 + i)^N - 1}{r}$$

In the limiting case of continuous compounding:

$$(1 + i) = e^r$$

Then the ($F/\overline{A}$) factor is:

$$(F/\overline{A}, i\%, N) = \frac{e^{rN} - 1}{r}$$

Here, the nominal rate (r) is:

$$r = \ln (1 + i)$$

Application of the ($F/\overline{A}$) Factor

Example. Find the future worth at the end of 10 years at 8% effective annual interest of a continuous flow at a rate of $100 per year.

Solution. The equation becomes:

$$F = \$100 \times (F/\overline{A}, 8\%, 10)$$
$$F = \$100(15.059) = \$1505.90$$

Earlier, the future worth of a \$100-per-year annuity at 8% was found to be \$1,448.70 when discrete payments were assumed. The difference here is not due to a difference between the effective and nominal interest rates. It happens because more money would be in the account to earn interest if the \$100 were deposited continuously throughout the year rather than at the end of each year.

A Continuous Annuity for a Future Amount ($\overline{A}$/F)

Derivation of the ($\overline{A}$/F) Factor

The future worth of a continuous annuity has already been given as this relationship:

$$(F/\overline{A}, i\%, N) = \frac{e^{rN} - 1}{r}$$

Conversely, the continuous annuity for a future amount is the reciprocal:

$$(\overline{A}/F, i\%, N) = \frac{r}{e^{rN} - 1}$$

Application of the ($\overline{A}$/F) Factor

Example. Find the continuous annuity at 8% over a 10-year period which is equivalent to a future amount of \$1,449.

Solution. The equation is as follows:

$$\overline{A} = \$1,449 \times (\overline{A}/F, 8\%, 10)$$
$$\overline{A} = \$1,449 \times (0.06641) = \$96.23$$

The Present Worth of a Continuous Annuity (P/$\overline{A}$)

Derivation of the (P/$\overline{A}$) Factor

The present worth of a continuous annuity can be developed from this observation:

$$(P/\overline{A}) = (F/\overline{A}) \times (P/F)$$

Therefore:

$$(P/\overline{A}, i\%, N) = \frac{e^{rN} - 1}{r} \times \frac{1}{(1 + i)^N}$$

Since $(1 + i) = e^r$, the (P/$\overline{A}$) factor is as follows:

$$(P/\overline{A}, i\%, N) = \frac{e^{rN} - 1}{r} \times \frac{1}{e^{rN}} = \frac{e^{rN} - 1}{re^{rN}}$$

Application of the (P/$\overline{A}$) Factor

Example. Find the present worth at 8% of a continuous \$100 per year for 10 years.

Solution. The solution is:

$$P = \$100 \times (P/\overline{A}, 8\%, 10)$$
$$P = \$100(6.975) = \$697.50$$

A Continuous Annuity from a Present Amount ($\overline{A}$/P)

Derivation of the ($\overline{A}$/P) Factor

The present worth of a continuous annuity was shown to be this:

$$(P/\overline{A}, i\%, N) = \frac{e^{rN} - 1}{re^{rN}}$$

Conversely, the continuous annuity from a present amount is as follows:

$$(\overline{A}/P, i\%, N) = \frac{re^{rN}}{e^{rN} - 1}$$

Application of the ($\overline{A}$/P) Factor

Example. Find the continuous annuity over a 10-year period equivalent to \$671 at 8%.

Solution. The solution is:

$$\overline{A} = \$671(\overline{A}/P, 8\%, 10)$$
$$\overline{A} = \$671(0.14337) = \$96.20$$

CONTINUOUS SINGLE-AMOUNT FACTORS

The Future Worth of a Continuous Present Amount ($F/\overline{P}$)

Derivation of the ($F/\overline{P}$) Factor

The $(F/\overline{P})$ factor can be developed from the observation that it is the future worth of a continuous annuity for one period which then earns for the remaining periods:

$$(F/\overline{P}, i\%, N) = (F/\overline{A}, i\%, 1) \times [F/P, i\%, (N - 1)]$$
$$(F/\overline{P}, i\%, N) = \frac{e^r - 1}{r} \times (1 + i)^{(N-1)}$$

Since $(1 + i) = e^r$:

$$(1 + i)^{(N-1)} = \frac{e^{rN}}{e^r}$$

and

$$(F/\overline{P}, i\%, N) = \frac{e^{rN}(e^r - 1)}{re^r}$$

Application of the ($F/\overline{P}$) Factor

Example. Find the future worth in 10 years at 8% of \$100 received during the first year:

Solution. The solution is as follows:

$$F = \$100 \times (F/\overline{P}, 8\%, 10)$$
$$F = \$100(2.078) = \$207.80$$

The Present Worth of a Continuous Future Amount ($P/\overline{F}$)

Derivation of the ($P/\overline{F}$) Factor

The $(P/\overline{F})$ factor can be developed by the observation that it is the present amount which will accumulate in $(N - 1)$ periods to the present worth of a one-period continuous annuity:

$$(P/\overline{F}, i\%, N) = (P/F, i\%, N - 1) \times (P/\overline{A}, i\%, 1)$$

$$(P/\overline{F}, i\%, N) = \frac{1}{(1 + i)^{(N-1)}} \times \frac{e^r - 1}{re^r}$$

since $(1 + i) = e^r$:

$$\frac{1}{(1 + i)^{(N-1)}} = \frac{e^r}{e^{rN}}$$

and

$$(P/\overline{F}, i\%, N) = \frac{e^r(e^r - 1)}{re^r e^{rN}} = \frac{e^r - 1}{re^{rN}}$$

Application of the ($P/\overline{F}$) Factor

Example. Find the present worth of $100 during the last year of a 10-year period at 8%.

Solution. The solution is as follows:

$$P = \$100 \times (P/\overline{F}, 8\%, 10)$$
$$P = \$100(0.4815) = \$48.15$$

6

THE NATURE OF COSTS

INTRODUCTION

The flow of cash is actually the lifeblood of a business enterprise. Figure 6-1 illustrates this flow of cash through the business and the various items which affect it. Assuming that revenues are equivalent to cash receipts—which is not a true assumption instantaneously but must be true over time—the revenues must be sufficient to cover all the costs of the business. Regulated businesses often expand the common definition of the term *cost* to include all the firm's obligations which must be met ultimately from revenues. Throughout this book, then, the term *cost* should be considered as synonymous with the revenues needed by the business, or the *revenue requirements*.

Using this definition, Figure 6-1 shows that there are two kinds of costs which the revenues must cover: the *plant operations costs,* which are incurred by the existence and the operation of the physical plant used to provide service; and the *capital costs,* which are incurred because the investors provide funds (capital) needed to acquire the physical plant. Figure 6-1 also illustrates the internal and external sources of funds, discussed in Chapter 4, which finance the business. Depreciation and retained earnings, as well as tax effects (tax effects were omitted from Figure 6-1 for simplicity*), are within the firm and, therefore, internal sources. External sources are, of course, funds obtained from investors.

Figure 6-1 will be a useful reference throughout this book because it presents an overall view of the business. However, this chapter emphasizes the view which is of chief interest to economic analysts: the costs of doing business. It also describes how to handle these costs in comparative studies.

CAPITAL COSTS VERSUS PLANT OPERATIONS COSTS

Because both *capital* and *plant operations* costs are of special interest to managers in conducting economy studies, it is important that the difference between these two kinds of costs be clear. *Plant operations costs* are those which are incurred because a company has a working plant. As the name implies, *capital costs* are costs associated directly with capital.

Plant operations costs depend upon the physical plant: for example, the quantities of the various kinds of plant units, how the plant is assembled, where it is located, how it

---

*Chapter 9 contains an expansion of Figure 6-1 which illustrates these income tax effects.

FIGURE 6-1
Cash Flow

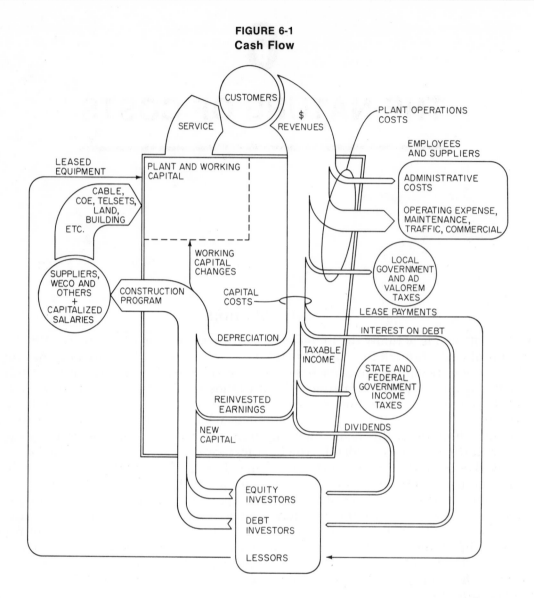

is used, how it is maintained, and the extent to which it is rearranged or relocated. However, plant operations costs are not directly related to the cost of the plant itself. One of their important characteristics is that most of them are continuous, or *recurring,* for as long as the plant remains in service; in fact, they tend to come and go with the plant. By far the greatest part of these costs is payroll costs. Examples are maintenance, traffic, and commercial expenses. However, there are also some nonrecurring, or one-time, plant operations costs.

On the other hand, capital costs are directly related to the cost of the plant. They are incurred because property is acquired and owned, and not because it physically exists. There are two forms of capital costs, too, and one is caused by the other.

The first form is the expenditure for the property and its installation, which is the initial *one-time* investment made in plant when that plant is acquired. This is known as the *first cost,* and it is a nonrecurring cost for a particular unit of plant. It is represented largely by materials, the labor of installing, and the contractor's bills if the construction work is done on contract. If this first cost is significant, and if the property is expected to be in service for more than a year, the cost is generally "capitalized." This first cost is on the left side of Figure 6-1, and it is included in the construction program. Such property is considered an asset of the company. Its original capital cost is recorded in a plant account and distributed to operating expense throughout the period when it provides service, so that a recurring annual capital cost is created.

The recurring annual cost resulting from capitalization is the second form of capital cost, and it contributes to the revenue requirement. This recurring form of capital cost is associated with long-lived property, like an electronic switching system. In contrast, the cost of small units of property, like tools, and the cost of property which is used up fast, like fuel oil and office supplies, are considered plant operations costs and are charged to expense at the time the property is purchased.

In making comparative cost studies, the analyst is interested in both forms of capital costs. The reason is that the first cost is an actual cash expenditure; and the way that cost is distributed to expense will affect income taxes, which is another actual cash expenditure.

The recurring capital costs are more difficult to estimate than first costs. They consist of three elements: *capital repayment, return on capital,* and *income taxes.* In this book, *depreciation* is considered *capital repayment;* and *return on capital* is the cost of money, or the earning power required by investors who provide the money. When return on capital is realized, an income tax liability is also incurred. *Income taxes,* therefore, are an additional capital cost. These topics are treated separately in later chapters (Chapters 7, 8, and 9).

Expensing versus Capitalizing a Cost

It was mentioned earlier that a capitalized cost is charged to operating expense throughout the life of the property. This is achieved through the process of depreciation accounting. However, the difference between expensing and capitalizing a cost is more than just an accounting convention. The cost of goods and services consumed currently is charged as a current expense and should be recovered from current revenues; no investors' funds are required. On the other hand, the funds required to purchase plant must come from investors, and not from current customers. Of course, any rational investor expects not only a return *on* the capital but also the return *of* that capital at some point. Since telephone plant generally has a limited life and a value at retirement which is less than its first cost, some means must be provided for recovering the investors' funds from the business—or else they will have been dissipated. Depreciation is the mechanism by which this is accomplished. If revenues include a component to cover depreciation charges, the customer has then repaid the firm for the capital which was consumed.

The Annual Cost of Doing Business

The analyst wants to know the total *annual cost* of doing business, which is the sum of the recurring capital costs and the plant operations costs. It is important to note that the initial capital cost by itself is not actually a part of the annual expense of doing business. However, it is the reason why there are recurring annual capital costs until the initial cost has been fully repaid.

To remain financially healthy, a business must recover all of its costs—both capital and operational—from revenues. Economic selection studies examine the costs and savings in alternative plans which affect the amount of revenue needed to run the business. Such studies are said to have a "revenue requirement" orientation. Essentially what is being measured in a comparative cost study is the amount of additional revenue required to support each proposal in order to cover all costs that the plan imposes. This means that choosing an alternative which minimizes any increase in the revenue requirement actually amounts to choosing the plan which puts the least cost burden on the customer.

Accounting Contrasted with Economy Studies

It is important to realize that recurring costs are handled differently in accounting and in economy studies. One difference is that the classification of recurring, or annual, costs into plant operations costs and capital costs does not resemble the classification appearing on the income statement.

The cost of doing business in an economy study is the sum of recurring capital costs and the plant operations costs. In the accounts, the major portion of cost is called *Operating Expenses*. (See Table 6-1, which is an example of the Monthly Report No. 5, titled Operating Expenses.) It includes depreciation (line 9), which, as stated earlier, is considered *capital repayment*—a capital cost—in this book. What is called the *Plant Operations Cost* in this book is limited to costs resulting from the physical and functional nature of the plant and the use made of it.

On the other hand, operating taxes and fixed charges (that is, interest) are not included in Operating Expenses in the accounts. Yet taxes and interest are, of course, costs of doing business, and it is merely an accounting technicality to include depreciation but not taxes and interest in the group of accounts known as Operating Expenses.

The analyst views the *return on capital* as a cost of doing business, too, because it is something that must be met out of revenues once an investment has been made. Yet this item does not even appear as a cost on the income statement. Furthermore, the return for a particular year, which is obtained from the income statement by adding interest and net income, is not properly a measure of the return on capital because the overall return depends on events that have not yet occurred.

CAPITAL COSTS

Components of First Costs

The term *first cost* designates the amount of money required to build new plant—the investment in plant. It is the total original installed cost. The first-cost estimates used in cost studies include all anticipated expenditures up to the time the project is completed and ready for use. However, any costs which are common to alternative plans are generally omitted from comparison studies.

The more important first-cost items are described below:

- *Material:* All material used in the construction of the plant; freight costs, sales taxes; and supply expense.

- *Installation:* All direct labor costs, as well as motor vehicle and incidental expense.

- *Miscellaneous Loading:* Supervision, tool expense, general expenses, social security taxes, and relief and pensions.

- *Engineering:* All engineering time and associated costs.

- *Costs Occurring during Construction Which are Added to the Plant Investment Accounts:* These costs include interest during construction, ad valorem (property) taxes, and insurance where necessary. Such costs, which are incurred before the plant is put into service, are just as much a cost of constructing the plant as the material and labor. This is clear for costs like ad valorem taxes and insurance, which are actual out-of-pocket payments. It is not so clear for interest during construction, which must be regarded as return on capital obligations deferred during the non-revenue-producing period of construction. This cost is recovered over the life of the plant through capital repayment accruals (depreciation), because it, too, is properly a part of the first cost.

TABLE 6-1A

FORM S.N. 156 (1) (1-76)

SAMPLE TELEPHONE COMPANY
OPERATING EXPENSES

MARCH 1976

#		a This Month	b Increase over Last Month	c This Year to Date	d Increase over Same Period Last Year	#
1	Repairs of Outside Plant (602.1 to .8)	3 850 971	561 901	10 915 795	1 923 321	1
2	Test Desk Work (603)	2 885 210	329 610	8 238 020	1 477 926	2
3	Repairs of C.O. Equipment (604)	7 544 979	983 546	21 206 997	2 237 396	3
4	Repairs of Station Equipment (605)	9 013 178	1 349 574	24 947 318	3 888 876	4
5	Repairs of Building & Grounds (606)	878 993	182 360	2 239 092	(268 024)	5
6	Maintaining Transmission Power (610) ...	346 541	39 315	994 569	(4)	6
7	Other Maintenance Expenses (612)	180 120	43 124	442 448	(13 198)	7
8	Total Maintenance Expenses	24 699 996	3 489 433	68 984 243	9 246 293	8
9	Depreciation (608)	14 096 073	93 683	42 028 166	2 770 779	9
10	Extraordinary Retirements (609)					10
11	Amort. of Intangible Property (613)					11
12	Amort. of Tel. Plant Acquis. Adj. (614) ..					12
13	Total Depreciation & Amort. Exps.	14 096 073	93 683	42 028 166	2 770 779	13
14	General Traffic Supervision (621)	1 164 220	133 764	3 290 133	283 048	14
15	Serv. Inspection & Customer Inst. (622) ..	295 482	39 639	841 761	(13 897)	15
16	Operators' Wages (624)	5 458 607	113 641	16 577 952	856 118	16
17	Rest and Lunch Rooms (626)	44 317	(4 935)	145 395	(5 394)	17
18	Opers.' Employment & Training (627) ...	99 793	18 172	265 132	(7 353)	18
19	C.O. Stationery and Printing (629)	113 826	17 895	315 899	(84 416)	19
20	C.O. House Service (630)	96 697	(3 043)	294 361	27 745	20
21	Miscellaneous C.O. Expenses (631)	251 506	77 059	622 341	155 368	21
22	Public Telephone Expenses (632)	7 695	6 299	13 623	2 934	22
23	Other Traffic Expenses (633)				(10)	23
24	Joint Traffic Expenses – Dr. (634)					24
25	Joint Traffic Expenses – Cr. (635)	1 237	235	3 255	463	25
26	Total Traffic Expenses	7 530 909	398 257	22 363 345	1 213 680	26
27	General Coml. Administration (640)	1 262 493	57 740	3 717 868	916 329	27
28	Advertising (642).....................	410 475	47 361	930 863	281 214	28
29	Sales Expense (643)	1 123 335	(2 463)	3 321 975	646 027	29
30	Connecting Company Relations (644)	49 200	5 572	135 729	(4 161)	30
31	Local Commercial Operations (645)	3 545 688	326 091	10 161 396	1 432 757	31
32	Public Telephone Commissions (648)	467 866	29 152	1 299 894	131 453	32
33	Directory Expenses (649)	2 601 299	(20 068)	7 807 056	1 115 739	33
34	Other Commercial Expenses (650)	100	(139)	(352)	(3 546)	34
35	Total Coml. & Marketing Expenses	9 460 461	443 247	27 374 433	4 515 814	35
36	Total Commercial Dept. Expenses	4 025 745	345 178	11 571 666	1 477 552	36
37	Marketing – Sales Expenses	1 905 873	36 051	5 629 222	1 514 015	37
38	Marketing – Other Expenses.....	3 528 842	62 018	10 173 544	1 524 246	38
39	Total Genl. Office Sal. & Expenses	6 811 002	380 413	19 927 993	1 197 631	39
40	Insurance (668)	19 048	5 024	46 136	19 318	40
41	Accidents and Damages (669)	305 814	259 854	383 348	186 495	41
42	Operating Rents (671)	619 240	(6 998)	1 928 927	214 108	42
43	Relief and Pensions (672)	8 513 569	(432 764)	26 207 057	4 200 941	43
44	Tel. Franchise Requirements (673)	7 833	(18)	23 577	(1 198)	44
45	General Services & Licenses (674)	1 506 136	(743 486)	5 343 206	1 176 949	45
46	Other Expenses (675).................	603 910	234 368	1 293 924	202 969	46
47	Tel. Franchise Requirements-Cr. (676) ..	7 833	(18)	23 577	(1 198)	47
48	Expenses Charged Const. – Cr. (677)	690 668	136 405	1 850 434	248 004	48
49	Total Other Operating Expenses......	10 877 049	(820 406)	33 352 166	5 752 777	49
50	Total Operating Expenses (301)	73 475 493	3 984 630	214 030 348	24 696 976	50

( ) Denotes negative amount.

TABLE 6-1B

FORM S.N. 156 (2) (7-76)

MONTHLY REPORT NO. 5
Sheet 2

OPERATING EXPENSES
SAMPLE TELEPHONE COMPANY

MARCH 1976

	a This Month	b Increase over Last Month	c This Year to Date	d Increase over Same Period Last Year	
1 Salaries (11)	74 913	19 015	222 137	(368 152)	1
2 Other Expenses (16)	23 849	6 601	63 039	(66 098)	2
3 Executive Department (661)	98 763	25 617	285 177	(434 250)	3
4 Salaries (11)	1 210 531	107 055	3 521 090	64 081	4
5 Machine Cost (12)	30 037	(2 502)	94 053	(15 166)	5
6 Postage (14)	457 632	(11 751)	1 419 924	321 834	6
7 Printing and Stationery (15)	35 397	6 314	103 491	(3 902)	7
8 Other Expenses (16)	125 332	13 514	347 816	(26 303)	8
9 Transfers — Cr. (19)	31 405	4 653	97 859	4 528	9
10 Total Operations (662-01)	1 827 525	107 977	5 388 516	336 016	10
11 Salaries (21)	332 669	20 717	963 487	155 327	11
12 Machine Cost (22)	493 792	(42 201)	1 500 119	217 377	12
13 Postage (24)	527	16	1 546	427	13
14 Printing and Stationery (25)	193 086	45 788	520 444	16 398	14
15 Other Expenses (26)	46 064	11 434	125 953	(12 657)	15
16 Transfer — Cr. (29)	291 856	43 472	738 245	117 885	16
17 Total EDP (662-02)	774 284	(7 715)	2 373 305	258 988	17
18 Salaries (31)	541 105	(12 010)	1 628 555	235 567	18
19 Machine Cost (32)	5 322	(1 872)	17 837	16 067	19
20 Postage (34)	75	(52)	330	251	20
21 Printing and Stationery (35)	11 946	6 004	35 728	5 143	21
22 Other Expenses (36)	36 516	(6 099)	122 597	42 025	22
23 Transfers — Cr. (39)	106 594	(8 895)	322 579	78 520	23
24 Total Data Systems (662-03)	488 371	(5 134)	1 482 469	220 534	24
25 Salaries (41)	281 854	(6 184)	856 365	51 306	25
26 Machine Cost (42)				(13)	26
27 Postage (44)	46	(185)	661	(564)	27
28 Printing and Stationery (45)	5 119	1 820	14 786	2 575	28
29 Other Expenses (46)	30 720	(1 307)	90 995	(2 738)	29
30 Transfers — Cr. (49)	5 880	344	12 469	(20 653)	30
31 Total Gen'l Acctg. (662-04)	311 859	(6 200)	950 338	71 219	31
32 Accounting Department (662)	3 402 041	88 926	10 194 631	886 758	32
33 Salaries (11)	81 772	11 510	229 418	23 152	33
34 Other Expenses (16)	47 157	(7 263)	149 167	15 926	34
35 Treasury Department (663)	128 929	4 247	378 586	39 078	35
36 Salaries 11, 21 and 31	77 303	(4 001)	237 938	39 308	36
37 Other Expenses 16, 26, 29, 36 and 39	14 950	322	47 873	5 109	37
38 Attorney's Fees and Costs 17, 27 and 37	7 923	(6 731)	67 807	53 952	38
39 Law Department (664)	100 176	(10 409)	353 619	98 369	39
40 Information (01) — Salaries (11)	213 713	786	638 271	96 452	40
41 — Other Exps. (16)	87 439	8 790	226 920	49 833	41
42 General Security (02) — Salaries (21)	161 745	3 739	478 165	27 752	42
43 — Other Exps. (26)	16 708	-1 813	46 819	5 209	43
44 Personnel (03) — Salaries (31)	723 259	21 515	2 140 562	222 071	44
45 — Other Exps. (36)	162 169	27 998	457 855	171 509	45
46 Planning (04) — Salaries (41)	15 220	(1 003)	46 674	16 543	46
47 — Other Exps. (46)	1 518	559	3 578	(244)	47
48 General Services (05) — Salaries (51)	186 751	5 362	551 690	110 716	48
49 — Other Exps. (56)	82 818	24 351	194 577	(94 005)	49
50 Other (06) and (08) — Salaries (61) and (81)	99 939	14 283	277 654	48 007	50
51 — Other Exps. (66), (67) and (86)	41 021	4 783	144 754	34 904	51
52 Engineering (09)	1 288 784	159 951	3 508 454	(81 075)	52
53 Other Gen'l Off. Sal. & Expenses (665)	3 081 091	272 032	8 715 978	607 675	53
54 Total Gen'l Off. Sal. Expenses	6 811 002	380 413	19 927 993	1 197 631	54

— Denotes negative amount.

TABLE 6-1C

FORM S.N. 156 (3) (1-72)
This Copy for

OPERATING EXPENSES

MONTHLY REPORT No. **5.**
Sheet 3

SAMPLE TELEPHONE COMPANY

MONTHLY AVERAGE PER TELEPHONE

MARCH 1976

	a This Month	b Same Mo. Last Yr.	c This Year to Date	d Same Per. Last Yr.	
1 Total Maintenance Expenses	3.436	2.878	3.206	2.848	1
2 Total Depreciation & Amort. Expenses	1.960	1.872	1.954	1.872	2
3 Total Traffic Expenses.	1.048	.984	1.040	1.009	3
4 Total Commercial Dept. Expenses . . .	.560	.470	.538	.481	4
5 Total Marketing Dept. Expenses	.756	.615	.735	.609	5
6 Executive Department	.014	.034	.013	.034	6
7 Accounting Department.	.472	.436	.474	.444	7
8 Treasury Department.	.018	.018	.018	.016	8
9 Law Department	.014	.012	.016	.012	9
10 Other Gen'l Off. Sal. & Expenses	.429	.362	.405	.387	10
11 Total Gen'l Office Salaries & Expenses .	.947	.862	.926	.893	11
12 Total Other Operating Expenses	1.513	1.290	1.550	1.316	12
13 Total Operating Expenses.	10.220	8.971	9.949	9.028	13

— Denotes negative amount.

Date issued April 19, 1976

Vice President and Comptroller

The costs of rearrangements of existing plant to accommodate new plant are not capitalized, however. These are nonrecurring costs which are classified as maintenance costs because they are similar to the recurring expenses of plant upkeep.

Costs Associated with Working Capital

Most businesses maintain a pool of investors' cash and short-term investments (principally in government securities and bank deposits), to pay bills for new construction and to meet current expenses. This pool of working capital has as much an earnings obligation as the capital invested in plant.

The usual assumption in comparative studies is that the amount of working capital is the same under each alternative and therefore can be omitted from the comparison. However, in special studies requiring total costs, all the costs associated with working capital must be included.

Obtaining First-Cost Data

The first-cost data used in economy studies must be the best possible estimate of actual first costs. The analyst must therefore be confident that such data truly represent the study situation. Very often, catalog prices, broad-gauge unit costs, and other so-called standards are mere averages that may bear no relationship whatever to any specific case and are useful only as guides in estimating.

First costs may be estimated by use of one or more of these methods:

1. A detailed pricing-out of the component parts of the physical plant requirements, such as the unit costs of building material; frames, bays, and cable racks; conduit tile, excavation, back fill, and paving; and labor requirements.

2. *Broad-gauge* costs for outside plant and central office equipment, if such costs are available. However, these figures must be used with discretion, since they are average figures and may not be pertinent to a particular situation. There are two general types of broad-gauge costs, the first being for outside plant and central office equipment in service. These are generally known as *historical costs;* they represent the original installed costs fully loaded with engineering, supervision, and other overhead. Historical costs are reasonably stable. The second type of broad-gauge costs is represented by new products now under development by the Bell Telephone Laboratories or in the early stages of production by the Western Electric Company or the general trade (outside suppliers). These are "forward-looking" or "going production" costs estimated by AT&T, Bell Telephone Laboratories, and the Western Electric Company or the general trade—whichever is involved.

3. Unit costs developed specifically for study purposes. Examples are for the following unit costs:

    • Route mile and circuit mile

    • Per square (or cubic) foot of building space

    • Central office equipment, either per frame or per line

4. Unit costs from previously developed local studies. It is desirable to review

these carefully to be certain they are truly representative of the situation under study.

5. Western Electric price lists.

6. AT&T letters describing new items under development.

Most important, the cost data an analyst uses should be specific to the exact problem at hand. Generally, these data come from the best experience, estimates, and judgment of those directly concerned. Study estimates are only as good as the experience and judgment of those who make them, and they should be tested by a standard of what is reasonable. The various cost components—that is, material, labor, and engineering—should therefore be examined separately.

OPERATIONS COSTS

Plant-related Continuing Costs

As the previous discussion pointed out, continuing, or recurring, costs of doing business fall into two general groups: those *associated with the working physical plant,* and those *associated with capital.* The remainder of this chapter deals with the plant-related continuing costs, or operations costs.

The continuing costs (usually called *expenses*) associated with working physical plant are not directly related to the amount of invested capital. Instead, expenses of plant operation are the result of the day-to-day policies and practices of operation within the company, as well as of external influences.

These expenses are largely the wage and salary costs associated with operations, purchases of supplies, and taxes other than income taxes. The usual classifications are:

Maintenance

Traffic

Commercial

Marketing

Revenue Accounting

General Administrative

Taxes Other than Income Taxes

Each one is described in the following sections.

Maintenance Expense

Superior telephone service can be provided only if the telephone equipment and plant are in top operating condition. The costs of keeping these facilities in good condition are accounted for as maintenance expenses. Included in this classification are:

- The cost of material and labor associated with the upkeep of plant. This includes the training of maintenance forces and the testing of equipment and facilities.

- The cost of rearrangements and changes of plant.

- Miscellaneous expenses like shop repairs, tool expenses, motor vehicle expenses, and the cost of house service.

- The cost of general supervision and engineering associated with maintenance work.

Although the quantities and types of equipment in service do affect maintenance costs, the broad averages found by relating the total maintenance expense to the total first cost do not necessarily hold when specific items of plant are considered. One reason is that a design objective in modernizing any type of equipment is to reduce maintenance expense. However, this reduction in expense is often achieved through higher first costs for the equipment.

Therefore, in a specific cost study the maintenance expense expressed as an annual cost is independent of the capital requirements. The type of plant, conditions of use, and test facilities must be considered, as well as all other factors that are variable in the plans under study. Sometimes, it may be appropriate to use the broad averages summarized in company reports; and, of course, maintenance expenses derived as a percentage of the investment may be helpful as a comparative check for reasonableness in detailed calculations. However, averages should be used in a cost study *only* if the study represents a cross section of the plant on which the average percentage was based—and only if, in the best judgment of the analyst, they also represent the study situation.

The Monthly Report No. 5, Operating Expenses (Table 6-1), summarizes maintenance expenditures according to broad classifications of plant. However, it is well to keep in mind that a No. 5 Report may be of only limited usefulness in a cost study because it was not prepared for specific locations or for specific types of plant within a general class.

An example is in the development of maintenance estimates for new or special items of plant. Since these are new items, or perhaps nonstandard, they cannot be analyzed from average historical data. One approach to the problem of computing an estimated annual cost for maintenance is to work with Plant Department representatives in order to derive maintenance work units for similar types of equipment. Work unit information can be useful provided the limitations are understood.

It is often possible to test whether a maintenance estimate is reasonable by equating the dollar amount to worker-hours. For example, if the dollars of maintenance expense equate to two workers full time, the analyst should ask this: "Is it reasonable that the full time of two workers will be required to operate and keep the proposed plant in service?"

Traffic Expense

Traffic operating expenses include the operators' salaries and the other costs of handling the traffic at local and toll offices. In many economy studies, traffic expenses are the same for various plans being compared and can be omitted. However, they become extremely important in studies involving alternate methods or varying efficiencies in handling traffic. In fact, these expenses can be a controlling factor in a study of such projects as the centralization of information traffic, or the consolidation of various types of traffic on Traffic Service Positions. Furthermore, in special studies which require that the total expense be considered rather than differential expenses, all pertinent traffic expenses must of course be included.

Wherever appropriate, broad averages can be derived from data found on the Monthly Report No. 5. However, as in the case of the maintenance expenditures, these data represent companywide totals and are not representative of specific study problems. Traffic Department people will help in computing specific expense estimates.

Commercial, Marketing, and Revenue Accounting Expenses

Commercial expenses primarily cover the operations of local business offices and general staff work. Marketing expenses are the costs of advertising, sales, and directories, and also associated staff work. Revenue accounting expenses are incurred in the preparation of customer bills and the maintenance of customer accounts. These expenses include stationery and postage and the rentals of business machines for these operations. Like traffic expenses, they generally do not vary in alternative plans being considered in economy studies and therefore can be omitted from the computations.

However, commercial, marketing, and revenue accounting expenses must be included in studies involving changes in these operations, or in special studies used for rate decisions. When such expenses are included in a study, they should be computed specifically for the conditions of that study. People from the responsible departments can be helpful in compiling the necessary data.

Broad averages which apply to the overall operation of a department can be derived from the Monthly Report No. 5 for the Commercial and Marketing Departments, and also for the Revenue Accounting Division of the Accounting Department. Again, a broad average does not represent a specific operation unless the operation is similar to one which makes up the bulk of the department's work.

General Administrative Expenses

General administrative expenses can be summarized from the Monthly Report No. 5 for the Executive Department, Accounting Department (less Revenue Accounting), Treasury Department, and Law Department, as well as other miscellaneous general expenses. This classification may also include insurance, accidents and damages, and operating rents. These expenses have no direct bearing on the method of providing service; and since they are usually common under alternative proposals, they are omitted from comparative studies.

In many cases, certain expenses summarized in reports may include amounts which are cleared to plant accounts as construction costs and are capitalized. Extreme caution must be exercised in economy studies to be certain that the appropriate expense component is treated as expense and that the appropriate capital component is treated as capital. In all cases, it is always important to find out first whether such amounts are truly related to the study plans.

Certain expenses are better treated in studies as separate items. For example, relief and pensions, which are a function of wages, should be loaded on the wage rates that are used. However, license contract fees, which are allocated in accordance with each company's costs for services received, should be treated as lump-sum costs.

Taxes Other than Income Taxes

Taxes of various kinds are levied on the property, operations, or revenues by different levels of public agencies ranging from the local fire or school districts to the federal government. In developing the annual costs of a specific project, the analyst must include all these taxes and in their proper amount. Of course, when the differences in the annual costs of two or more proposals are being compared, only those taxes which are not common to all the plans need to be calculated.

Individual taxes are discussed next; however, a word of caution is needed. There is some danger in compiling composite miscellaneous tax rates to be used in a study. Such rates tend to be accepted for use generally, but they may be misused if some of the taxes do not apply. Except when many repetitive studies are to be made, it is best to list and evaluate the appropriate taxes separately.

AD VALOREM TAXES   As the name implies, *ad valorem* taxes are taxes levied on the value of the plant, with *value* here determined by assessment. Property taxes are generally of this type. A percentage figure is usually developed for a study by relating the actual past taxes of the ad valorem type to the original cost of the plant. This practice is modified for known changes in tax rates, or for changes in the ratio of the assessed value to the original cost. For example, it may happen that older plant being considered for removal will have a lower-than-average assessed valuation. This fact should be recognized in studies in which the removal of plant is compared with continuing plant in service if removal is expected to cause a reduction in the property tax liability.

The expense associated with ad valorem taxes in cost studies will generally be computed and expressed as an annual cost which is a percentage of the first cost. It is advisable to discuss studies in which ad valorem taxes are significant with people in the company who deal with the taxing authorities. In some cases, it may be found that in practice the removal of plant may not result in a reduction in the ad valorem taxes.

GROSS RECEIPTS, OR GROSS REVENUE, TAXES   Gross receipts, or gross revenue, taxes are not directly related to the first cost, but rather to the revenue requirements themselves. They are a tax levied on the revenue and collected by the company. They can usually be omitted from comparative revenue-requirement-type studies, however, because they will increase the revenue requirements for all alternatives by an equal percentage.

MISCELLANEOUS TAXES   Other miscellaneous taxes may often be related either to the first cost or to gross revenues similar to either ad valorem taxes or gross revenue taxes.

SALES TAXES   Sales taxes are levied on part or all of the material sold in some areas, and may apply either to first costs or to expenses. Sales taxes on materials are part of the first cost, and sales taxes on maintenance are part of the expense. In cost studies, these taxes are included in the estimated cost of materials used for construction or maintenance. Occupation and use taxes are similar to sales taxes and are treated in the same manner.

SOCIAL SECURITY AND UNEMPLOYMENT TAXES   Social security and unemployment taxes are treated as a loading on labor costs when first costs are derived. They are also included in maintenance and other operations costs.

Plant-related Nonrecurring Expenses (Noncontinuing Costs)

In general, the expenses of plant operations are continuous—or at least they could occur any time throughout the life of the plant. Nevertheless, there are times when these expenses are one-time expenditures. Examples are costs for the rearrangements of outside plant when a new central office wire center is established; service order work by the local Commercial Department forces to provide new or changed services; sales expenses associated with new services; and plant labor for substituting one type of telephone set for another.

These one-time expenses are paid out of current revenue. If the study requires that such expenses be spread out over a period of time so they can be compared with expenses for other plans, care must be used in selecting the period for computing the equivalent annual cost. This period may coincide with the life of the plant; or it may be a shorter period, determined by an estimate of when such expenses might be incurred.

So-called *cutover* expenses sometimes include capital expenditures as well as operating expense items. Such capitalized cutover expenses should of course be recognized and treated as capital expenditures and not expense.

SUMMARY

The various types of costs used in economy studies fall into two classes: those associated with the investment of capital—*capital costs*—and those associated with physical plant—*operations costs*. Some costs occur at one time only, whereas others are recurring and continue at regular intervals throughout the life of the plant. Operations costs are paid out of current operating revenues. Capital expenditures are paid with investor funds—debt and equity. Because the capital expenditures are converted into future expenses, they have recurring burdens of return, repayment, and income taxes. These recurring capital costs are treated in Chapters 7, 8, and 9.

SUMMARY

7

CAPITAL REPAYMENT AND DEPRECIATION

INTRODUCTION

Chapter 6 defined two different kinds of costs, capital costs and operations costs. It also explained that recurring capital costs are the result of the capitalization of the first cost of long-lived (existing more than one year) plant, and they are therefore composed of three elements: *capital repayment, return,* and *income taxes.* This chapter examines the first of these, capital repayment, and the accountant's provision for it, called *depreciation.* Chapter 8 treats the other two elements of continuing capital costs, return and income taxes.

What is capital repayment, then, and why does it occur? Why capital repayment costs occur is a good starting point, because it also explains what capital repayment is. Capital repayment costs occur because certain physical properties, such as telephone plant, eventually come to the end of their useful lives—often as a natural result of wear and tear. Even if they are not worn out, they may be retired anyway: They are too small; they are replaced by some new technological development; or they no longer meet the customer's requirements. There could be some other "functional" reason why properties must be taken out of service; many times, retirements are caused by a combination of events.

Whatever the reason for the retirement, the original sum of money invested in the property, offset by any salvage, is a cost to the firm which must ultimately be received from the customer. This cost is passed on to the customer as depreciation, and it is called the *capital repayment cost.*

CAPITAL REPAYMENT COSTS

Illustrating Capital Repayment Costs

Although other devices may be used to illustrate capital repayment costs, time-cost diagrams are especially convenient. They are used now to show three of the many ways to account for these costs on a particular item of telephone plant. This item has a total installed first cost of $1,000 at time 0, negligible salvage, and an expected useful life of 4 years.

Diagram A of Figure 7-1 shows the capital repayment cost as $1,000 occurring at the time the item was put into service. Here, the capital repayment cost can be viewed as a loan that comes due immediately.

An expenditure accounted for in this way is said to be *expensed.* If the loan is repaid at time 0, the capital is repaid without any interest expense.

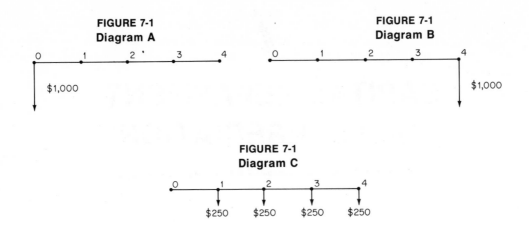

FIGURE 7-1
Diagram A

FIGURE 7-1
Diagram B

FIGURE 7-1
Diagram C

Diagram B of Figure 7-1 shows the capital repayment cost as $1,000 occurring at the end of its expected useful life. Now the capital repayment cost can be viewed as a loan that comes due at retirement.

This way of accounting for capital repayment is called *retirement accounting*. If the loan is repaid at retirement, an interest cost is incurred on the full $1,000 for 4 years.

Diagram C of Figure 7-1 shows the capital repayment cost divided up into four equal parts occurring at the end of each year of expected life. This time, the capital repayment cost can be viewed as a loan that comes due in equal annual installments.

This kind of accounting for capital repayment costs is called *straight-line depreciation*. Capital is repaid as each installment is paid; and interest costs are incurred at the end of each year, but only on the amount left unpaid at the beginning of each year.

Theoretically, any one of these three methods (or, for that matter, countless others) can be used to account for capital repayment costs. In practice, however, telephone plant generally provides service for a number of years before it is retired. As a result, the expenditure for plant is usually translated into a series of charges against revenues for each year of service. For accounting purposes and rate decisions, then, the entire cost of plant is not reflected either when it is bought or when it is retired. Instead, it is reflected as an equal annual expense during the time it produces service—the procedure called *straight-line depreciation*.

Cost of Money

The illustrations demonstrate that the total capital repayment cost—$1,000—remains the same regardless of how the accountants record this expenditure. Nevertheless, the cost of using this capital can be quite different. This difference would be caused by different schedules for repaying the capital.

In economic selection studies, the cost for the use of capital is often expressed as a percentage of the unpaid balance; it is called the *cost of money* or the *cost of capital*. The important point to remember is that the total dollar cost will vary depending on how fast the capital is repaid—even if the percentage rate remains the same.

How Capital Is Repaid and Why

The firm has a continuing stream of costs while it is providing service: costs for current maintenance and supplies, costs for additional plant to provide for growth, and costs for new plant to replace old equipment. If the firm is to remain financially healthy and continue to provide quality service, all these costs—including the costs of plant—must be repaid.

Capital repayment costs may be recovered from increases in revenues, or else by savings generated by the investment. They may also be partially repaid by positive net salvage at the time of retirement. In a continuing business, however, the repayment is not made directly, through a disbursement to the investor. Instead, the money is reinvested in new plant or in other assets. The investors' interests are protected by the transfer of their capital from old plant to new plant in installments as the old plant is used up in service. This recycling of capital not only preserves the integrity of the investors' capital, but it also reduces the need for new capital to support construction. So, even though the firm has no intention of directly returning the capital invested in any specific investment to its investors, capital repayment is vital to the ongoing health of the business and its ability to raise future capital.

Estimates Needed in Computing Capital Repayment Costs

Accountants are generally concerned with past happenings—past expenses and receipts. However, in doing an economy study, the analyst often needs to use estimates of the future in order to determine the capital repayment and other capital costs. These estimates are very important. In fact, it cannot be emphasized often enough that an economy study will be only as accurate as the data—most of them estimates—the study is based upon.

Life of Telephone Plant

One important estimate which is usually needed for an economy study is the life of telephone plant. Since the lives underlying the depreciation rates used on the company books are readily available, the analyst may be tempted to use them in an economy study. However, doing so may introduce a serious error into the study. The following discussion of the development of book depreciation rates shows why.

To develop the lives underlying book depreciation rates, the depreciation engineer studies relatively small building blocks of data and then combines these in several successive steps. This produces what is called an *average service life* for each of the relatively few plant accounts for which book depreciation accruals are calculated.

In this process of combining data into broader and broader groupings, the "average life" becomes less and less representative of specific plant units. For example:

- Data for plant of all ages are combined. This means that the average life of step-by-step switching equipment reflects the relatively long lives of early installations of this equipment, and therefore it would be quite unsuitable for use in a study of a proposed addition to an office which will soon be replaced by electronic equipment.

- Data for diverse plant are combined. The average service life for the Motor Vehicles account includes heavy-construction trucks as well as passenger cars; that of the Buildings account includes repeater huts and office buildings.

Because of the nonhomogeneous nature of the plant it represents, the average service life for an account might change from one depreciation study to another simply because of a shift in the mix of plant—without any change in the life characteristics of individual plant items.

The diagrams in Figure 7-2 illustrate this aggregative process—in Diagram A, schematically; in Diagram B, with illustrative data. The illustrative data show that the average service life for the Furniture and Office Equipment subaccounts would be

FIGURE 7-2
Diagram A: Development of "Average Service Life"
Underlying Depreciation Rates Used in Financial
Statements

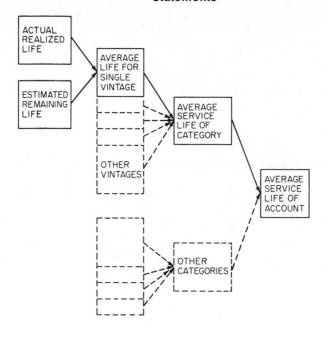

FIGURE 7-2
Diagram B: Illustrative Data as Might Be Found in Typical
Depreciation Rate Study

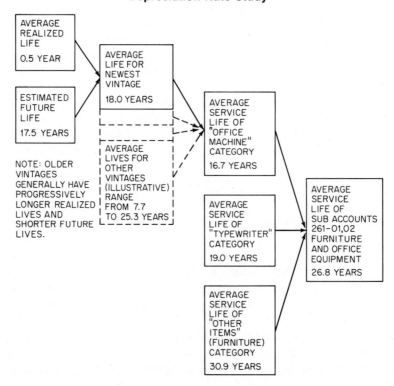

inappropriate to use in a study involving the purchase of electronic calculators, even though calculators are included in this account. The inclusion of furniture and other relatively long-lived plant has significantly influenced the average.

But even if the analyst looked at data for the narrower definition of plant, the depreciation "category," two problem areas remain:

- The average service life of the category reflects plant of all ages, with widely varying lives.

- The category is far from monolithic in composition.

The latter point might well be amplified. The depreciation engineer defines "depreciation categories" of plant in order to subdivide the various accounting classes into more homogeneous groupings. However, it would not be economically feasible to achieve complete homogeneity. So, the category called "Office Machines" in the illustrative study includes not only electronic calculators, but rotary calculators, adding, coin-counting, stencil-cutting, dictating, letter-opening, billing and addressing, and other machines.

Even if the analyst avoided the pitfalls just described and was able to get a life estimate characteristic of a compact, homogeneous grouping of plant, it would be necessary to exercise caution because the plant being studied may well not be subject to "average" conditions.

In some cases, the life of particular plant may be defined by the conditions of the problem. For example, a study plan may call for the installation of additional equipment in an office which is expected to be completely retired in two years. This study plan must of course provide for the repayment in two years of all the capital invested in installation, in engineering, and in any material which is not expected to be reused in another location.

As another example, the lives of certain plant items may depend on the lives of particular components. The life of a specific cable may be limited, say by the life of the pole line, which will be removed because of a highway relocation.

In other situations, the conditions of the problem may not provide any definite indication of the life of the plant. The life estimate would then depend on the judgment of the person making the study, and only by coincidence would this estimated life be the same as the average service life reflected in the group depreciation rate.

However, even though it is not possible to forecast the life of plant exactly, the analyst must at least try to compare the probable relative lives of alternative types of plant.

For example, common-control equipment, such as electronic switching, is more flexible than the electromechanical step-by-step equipment; and it can more easily adapt to existing service offerings, as well as to new and as yet unimagined service offerings. Therefore, in a study comparing the use of the two types of equipment, it would be reasonable to estimate a shorter life for a new installation of step-by-step equipment than for electronic equipment.

Most economy studies deal with single items, each with a single life. In a study dealing with a relatively few items of plant, the usual practice is to estimate the life of each unit. However, if the study involves groups of many items of relatively short life— as, for example, telephone sets—it may be necessary to take into account that the lives of individual items within a group usually vary widely. Recognition of the nonuniformity of life within a group is a complex matter, which is treated separately in Chapter 14.

Gross Salvage

Another estimate that the analyst may need to use in an economy study is that of gross salvage. *Gross salvage* is the dollar amount received from the sale of an asset at the end of its life—either for reuse or for scrap. It might also mean its trade-in value, or its worth in some other location.

Cost of Removal

The analyst must also estimate the *cost of removal*, which refers to the cost of salvaging the material at the end of its life. This cost includes the costs of demolishing, tearing down, removing, or otherwise disposing of plant. The cost of removal is likely to be greater if the plant has to be removed from other working equipment, or if it has to be taken out carefully for reuse instead of being ripped out for scrap.

Net Salvage

The *net salvage* is derived from the two previous estimates. It is equal to the gross salvage less the cost of removal, and is usually thought of as a positive figure. Often, however, the cost of removal will exceed the gross salvage, leaving net salvage a negative figure. An example is the retirement and removal of equipment occupying valuable space in a building. The equipment itself may have little reuse or scrap value but a substantial removal cost. Nevertheless, it is removed in order to make the space it occupies available for another use.

In either case, the net salvage affects the capital repayment costs. When it is positive, net salvage is a partial repayment of capital. When it is negative, it is a capital cost that must also be repaid.

When determining the capital repayment costs for an economy study, the analyst should not be distracted by the accounting transactions—particularly if the assets are retired early or are reused in other locations or projects. The reason is that the analyst and the accountant have different objectives. The accountant's concern is the allocation of receipts and expenditures to particular time periods; it is not valuation. The analyst's concern is valuation, however. The accounting transactions are also quite specialized and complex under the Uniform System of Accounts and usually have little relevance to finding the capital repayment costs for an economy study.

In estimating net salvage, just as in estimating the life of plant, the analyst should use the best information available which is specific to the proposed investment; usually, this is not the accounting average.

The estimate of net salvage could be substantial if reuse is a possibility. However, whether equipment is to be reused does not depend solely on the physical condition of the equipment itself; it also depends on whether there will be a use for this equipment at the end of its first location life. This possibility, in turn, depends on the possibility that a more desirable alternative type of equipment might be developed meanwhile. It also depends upon supply conditions for the same type of equipment at the time of retirement. Such uncertainties indicate the wisdom of pessimism rather than optimism in estimating future possibilities for reuse.

DEPRECIATION

The term *depreciation* has been defined in a great many ways, but the definition that is most important to the analyst is that of the accountants. According to the American Institute of Certified Public Accountants: "Depreciation accounting is a system of accounting which aims to distribute cost or other basic value of tangible capital assets, less salvage (if any), over the estimated useful life of the unit (which may be a group of assets) in a systematic and rational manner. It is a process of allocation, not of valuation." This definition emphasizes that depreciation is a way to designate the capital repayment costs. Essentially, it is the accountant's specialized version of capital repayment; and it is applied, as described earlier, to assets whose lives are greater than one year.

Depreciation Accounting

As pointed out before in this book, accountants maintain the books of the firm and prepare reports which will convey financial information about the firm to managers, stockholders, the financial community and the regulatory bodies. As a result, accountants keep careful records of receipts and expenditures and then assign them to the proper month or year—that is, the "accounting period" to which they apply. Although most of these basic data are actual past receipts and disbursements, some have to be estimates of future receipts and disbursements. Examples of such estimates would include revenues for service given in the current month but which is to be billed and collected in the next month; taxes computed for the current year but which are payable next year; and the anticipated gross salvage and cost of removal for existing plant. All these transactions must be assigned to accounting periods, too, so that revenues and expenses will match well enough in time so that the reported net income of a period will be meaningful.

An illustration earlier in this chapter depicted the capital repayment cost as a dollar amount at the time the investment was made. However, if accountants included this entire expenditure as an expense in the year of the investment, earnings would be distorted for that year and for future years as well. This is why accountants distribute the cost of the investment over the service life of the property, using the procedure known as *depreciation accounting*.

In accounting, the original cost of plant is viewed as a *prepaid expense*, meaning a cost which is to be allocated proportionately to each of the accounting periods during which the plant is used. *Depreciation expense*, then, is the estimated amount of capital consumed during each accounting period. The *depreciation reserve* is the record of how much of the cost of plant currently in service has been recovered.

According to the definition expressed by the American Institute of Certified Public Accountants, the accounting concept is strictly one of cost allocation and not of valuation. In other words, the depreciation expense for an accounting period is not a measure of the decrease in value of the firm's assets. It is recorded as an expense on the income statement, and it is one of the deductions from revenues that is included when earnings are determined. It is also recognized as an allowable deduction when taxable income is computed.

Various techniques are available to the accountants for determining depreciation. Some of these techniques are advantageous from an income-tax standpoint (for example, accelerated depreciation which will be covered in Chapter 9). Various techniques are also available for determining the depreciation to be reported on financial statements. However, as Chapter 3 pointed out, the Uniform System of Accounts has specified that telephone companies must use straight-line depreciation on financial statements. This present chapter is therefore limited to a discussion of the straight-line depreciation technique—which, because it is used for the firm's books, is also called *book depreciation*.

Straight-Line Depreciation

An example earlier in this chapter introduced straight-line depreciation assuming negligible salvage. However, the complete formula, including salvage, is:

$$\frac{100\% - \% \text{ net salvage}}{\text{Average service life}} = \% \text{ depreciation rate}$$

where: Net salvage = gross salvage − cost of removal

Straight-line depreciation allocates the first cost less the net salvage (this difference may be called the *service value*) equally over each year of life. However, when many items are involved, it would not be practical to use the straight-line formula for each item of plant. The Uniform System of Accounts therefore allows the formula to be applied to groups of telephone plant.

To compute depreciation expense for an accounting period, accountants multiply the original cost of an aggregation (account or subaccount) by the depreciation rate for that aggregation. In each accounting period, the depreciation expense is added to the Accumulated Depreciation (reserve) account, as well as being charged to the Depreciation Expense account, as Chapter 3 explained.

If an individual unit lives exactly the average service life, its total accumulated depreciation expense will equal its original cost—provided negligible salvage and a constant average service life are assumed. At retirement, then, the accountant will subtract the original cost from both the Plant account and the Accumulated Depreciation account. This practice effectively cancels out the amount added to the Plant account when the plant item was installed, as well as its own Accumulated Depreciation expense in the Accumulated Depreciation account. Subtracting the same amount from the plant account (an asset) and from the Accumulated Depreciation account (a contra-asset) has left the total assets unchanged. Because the asset side of the balance sheet did not change, the liability side did not change either. Therefore, the process of retiring plant has in no way retired capital.

This point may be clearer with an illustration:

1. Assume the following simplified balance sheet at a point in time:

Plant	$1,000	
Less Accum. Depr.	200	
Total Assets	$ 800	Total Capital $800

2. Now suppose that an expenditure of $100 is made for an item of plant with funds from new capital:

Plant	$1,100	
Less Accum. Depr.	200	
Total Assets	$ 900	Total Capital $900

3. During the life of that item, its cost is recovered through depreciation accruals, which are credited (added) to Accumulated Depreciation. This means that immediately before the retirement of that item, Accumulated Depreciation is larger by $100.

4. First assume that no other new plant has been added:

Plant	$1,100	
Less Accum. Depr.	300	
Total Assets	$ 800	Total Capital $800

   Since no plant additions have been made, Accumulated Depreciation expense has been returned directly to the investors rather than being reinvested in the business. In effect, the capital raised to purchase this plant has been "repaid," returning Total Capital to its level before the purchase of the item.

5. Now consider, instead, a case in which new plant has been added which costs the same as the item which has reached the end of its life, and

depreciation accruals are invested in this new plant rather than being directly repaid to the investor:

Plant	$1,200	
Less Accum. Depr.	300	
Total Assets	$ 900	Total Capital $900

Note that since the new plant has been supported by "internally generated funds," no more new capital was raised.

6. Finally, the retirement of the item of plant which has reached the end of its life occurs, and both the Plant and Accumulated Depreciation accounts are reduced by its cost:

Plant	$1,100	
Less Accum. Depr.	200	
Total Assets	$ 900	Total Capital $900

Note that although plant was retired, neither Total Assets nor Total Capital has been reduced from the level at the beginning of the life of the item. The retirement has not impaired the integrity of the investors' funds. What *has* changed is the plant in which the funds are invested: Old plant has been replaced by newer plant. In a sense, the investors' funds have been "recycled."

The previous example dealt with a single item of plant. This is the context of many accounting texts which typically deal with transactions assuming item accounting. With item accounting, when an asset is retired, any difference between the cost recorded on the Plant account and the amount of accumulated depreciation for the asset (plus net salvage) is explicitly recognized as either a "loss" or a "gain." However, a firm which uses group depreciation accounting does not show gain or loss on the retirement of units, because the assumption is that a unit in the group is fully depreciated when it retires. At that time, the entire cost of that unit is subtracted from the Accumulated Depreciation account, regardless of its age. The age at retirement then becomes part of the mortality data which are reflected in the depreciation rate to be applied to the remaining plant. It is important to recognize that under group depreciation accounting, plant generates depreciation expense as long as it is in service, even if it lives long past the average life underlying the depreciation rate applied to it.

The mechanics for coping with the fact that, in an aggregation of plant items, some will live lives less than average and some will live lives longer than average are inherent in the way the group is defined.

Straight-Line Vintage Group Depreciation

For many years, depreciation studies in the Bell System have been made on what is known as a *straight-line vintage group (SLVG) basis*. With this approach, the finest subdivision of plant is the *vintage group*, which consists of all the plant in a depreciation category added in a single calendar year. The weighted arithmetic average of the lives of all the units in that group is called the *average life*.

As previously explained, the average lives for the many vintage groups in a category are combined into one average service life for the category. In this process, *reciprocal weighting* is used in order to produce depreciation charges for the category which equal the sum of the charges that would be made if each vintage group were depreciated separately. *Reciprocal weighting* basically means that it is the depreciation rates for the categories which are averaged rather than the average lives. Similarly, the average service lives for categories are reciprocally weighted and combined to produce an average service life for an account which will result in depreciation charges equal to those that would be experienced if each category were depreciated separately. This means that the average service lives for accounts and for categories will vary over time as the proportions of component groups change.

However, under the SLVG procedure, unless estimates are revised, the average life for the vintage group remains constant over the group's life span; therefore, depreciation accruals are generated by the vintage group on the basis of a single, constant depreciation rate. Since some items will retire with lives shorter than the average service life, the Accumulated Depreciation account is reduced by more than the depreciation those items have generated. Conversely, items living longer than the average service life will have accrued more than will be subtracted when they retire. However, if the average service life is correctly estimated, the underaccruals for units which live less than the average service life will be exactly offset by the accruals for those units which live longer. The entire cost of the group—but no more than this cost—will have been allocated to operating expenses by the end of the life of the last item of the group.

A survivor curve like that shown in Figure 7-3A is used to determine the average life for each vintage group; the method used is described in Figure 7-3B. Survivor curves are selected after an analysis of historical data and forecasts of future life expectancy. To find the average life, the total dollar-years of existence is divided by the number of dollars. Next, the average lives of all vintage groups are composited into one average service life for the entire plant category (for example, for all poles).

Figure 7-4 assumes that the original $100,000 investment is, in fact, the entire plant category. (There is no need to composite the different vintage group average lives in this case.) It illustrates how the vintage group depreciation rate is calculated and also the yearly accruals, with eventual full recovery of capital.

FIGURE 7-3A
Vintage Group Calculation of Average Life

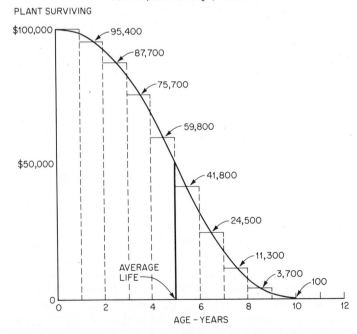

(Most units in the vintage group do not retire at the average life. Some have lives shorter than average; others, longer than average. The typical survivor curve here results from this dispersion of retirements and represents dollars surviving from the initial $100,000 at various ages.)

FIGURE 7-3B
Vintage Group Calculation of Average Life

$$\text{Average life} = \frac{\text{area under curve}}{\text{original amount}}$$

Area under curve = sum of areas of rectangles

Height		Width
100,000	×	1
95,400	×	
87,700	×	
75,700	×	
59,800	×	
41,800	×	
24,500	×	
11,300	×	
3,700	×	
100	×	
Area = 500,000	×	1

$$\text{Average life} = \frac{500,000}{100,000} = 5.0 \text{ years}$$

1. Average life for vintage group is the area under the survivor curve divided by the initial placement.

$$\text{Average years} = \frac{\Sigma(\text{units} \times \text{years})}{\text{total units}}$$

2. The total is conveniently found by dividing the area under the curve into vertical rectangles. Each rectangle is one year in width; the height equals the plant surviving.

FIGURE 7-4
Calculation of Vintage Group Depreciation Rate and Accruals

$$\text{Depreciation rate} = \frac{100\% - \% \text{ average net salvage}^*}{\text{average service life}}$$

$$= \frac{100\%}{5.0 \text{ years}} = 20\% \text{ per year}$$

Year	Plant surviving	Annual rate	Depreciation accruals
1	$100,000	20%	$ 20,000
2	95,400	20%	19,080
3	87,700	20%	17,540
4	75,700	20%	15,140
5	59,800	20%	11,960
6	41,800	20%	8,360
7	24,500	20%	4,900
8	11,300	20%	2,260
9	3,700	20%	740
10	100	20%	20
		Total	$100,000

*Assume zero % average net salvage.

1. Five-year average life equates to 20% annual straight-line vintage group depreciation rate, assuming zero net salvage.
2. Application of 20% rate to plant surviving each year results in the full capital repayment by the time the last plant retires.

Straight-Line Equal-Life Group

The *Equal-Life Group Plan* (ELG) is a refinement of straight-line group depreciation accounting which has a number of advantages. First, ELG makes possible the total allowable lifetime depreciation of any item to be accrued as depreciation expense over its actual life (whether this is longer or shorter than average). In other words, it allocates the full cost of the shorter-lived units to depreciation expense while they are in service, with the longer-lived units bearing only their own costs. This is accomplished by dividing each vintage group into smaller groups, with each group consisting of units which are expected to have the same life. This distribution is based on life tables developed from the mortality data of telephone plant. Although it is not possible (or necessary) to identify the individual units of plant which will have a given life, it is possible to determine statistically the number of units or dollars of plant in each equal-life group—provided the mortality data are adequate.

Figure 7-5 illustrates how equal-life groups are established within a vintage. Figure 7-6 compares the vintage-group depreciation and equal-life-group depreciation.

As Figure 7-6 illustrates, the ELG rate for a vintage of plant changes over time. It is, in effect, a composite of the depreciation rates for the equal-life groups in existence at each point, and it diminishes as short-lived groups with high rates drop out. With ELG, then, the average service life for a class of plant is even less meaningful to the engineering economist, since, in addition to the limitations cited previously, with ELG it is also a function of the average age of plant currently in service.

FIGURE 7-5

**Division of Survivor Curve into Equal-Life Groups for a
Single Vintage of Plant**

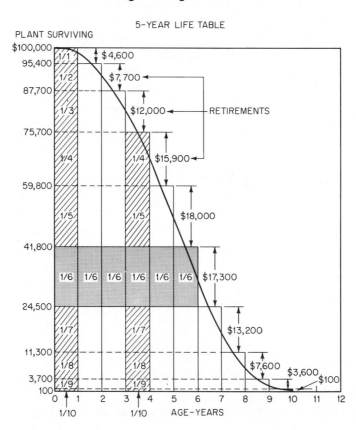

Division of Survivor Curve into Equal-Life Groups

1. The area under the curve can be found by dividing it into horizontal rectangles. The length of each rectangle is defined by an integral value of age at the retirement of some plant; the height represents the amount of plant retiring at that age.

2. Each of these groups of plant having the same value of life expectancy is called an equal-life group (ELG).

3. Under the ELG procedure, depreciation rates are developed such that the investment in each equal-life group is recovered over the life of that group.

4. The depreciation during the first year of service of this vintage group provides for repayment of all the $4,600 retiring at the end of 1 year; plus one-half of the $7,700 with a 2-year life; and so on, down to one-tenth of the amount expected to have a 10-year life.

5. In the fourth year, the depreciation accrual no longer includes anything for the first three equal-life groups, because the capital invested in them was recovered during their service.

6. This calculation is in contrast to SLVG, in which the 20% rate would, in effect, be applied to all these groups.

FIGURE 7-6
Simplified Example; 2.5 Year Average Life

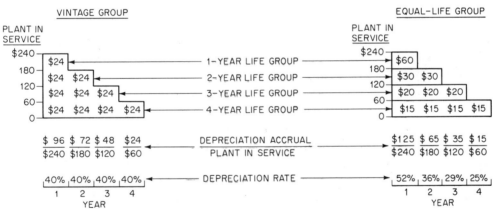

1. Vintage Group:
   A. 4 Units at $60
      1 survives 1 year
      1 survives 2 years
      1 survives 3 years
      1 survives 4 years
      4 units   10 years

$$\frac{10}{4} = 2.5 \text{ years average life}$$

   B. The assumption of zero net salvage gives a 40% depreciation rate.
   C. The 40% depreciation rate applies to plant in service each year:
      applies to $60 surviving only 1 year (accrues $24)
      applies to $60 surviving 4 years (accrues $96)

2. Equal-Life Group
   This method provides for repayment of capital in the group (one unit in this case), with 1-year life over the 2 years. It repays the $60 that will live 2 years at $30 per year; and so on.

7. Individual units of plant are not identified as having specific lives. The procedure is based on statistical expectancy, reflecting years of actuarial analyses of life characteristics.

8. This is straight-line depreciation, *not* accelerated depreciation, even though capital repayment is faster than it is with SLVG. It is straight-line depreciation of the investment in each equal-life group.

Equal-life group depreciation is also known as *unit summation* depreciation, because it will result in the same depreciation expense as the sum of the straight-line depreciation expense for all units with individual life estimates. When compared with the straight-line vintage group, ELG will shift the capital repayment costs for a group towards the early years. With growing plant, this means that the depreciation expense will be greater during the period of growth. If there is sufficient revenue to maintain the same net income as with the vintage group, the higher depreciation accruals resulting from the use of the ELG technique result in a larger proportion of internally generated funds. As a result, less outside financing is needed than would otherwise be required.

HOW THE ANALYST USES DEPRECIATION

Effect of Book Depreciation Methods on Economy Studies

Most economy studies within the scope of this text are not affected by the particular book depreciation method which the accountants use. Normally, the economy study will assume unit depreciation of the individual assets being studied. The use of ELG can then be considered a refinement in the accounting which permits capital repayment costs to be allocated more closely to the charges the economist views as appropriate. However, certain studies must recognize that individual lives in a group will vary about some average. Such studies may be affected by the book depreciation method and are covered in Chapter 14.

Mistaken Concepts of Depreciation

From time to time, decisions may be influenced by thoughts such as this: "We can't retire that equipment because it's not fully depreciated." Or " Let's get rid of that equipment because it's fully depreciated, and we can save that depreciation expense."

Both statements ignore any consideration of whether it is advisable to retire an asset. To the depreciation engineer, no asset is fully depreciated until it is retired; and all assets are fully depreciated when they retire. To the engineering economist, the expenditure for the asset is money spent in the past; and the repayment of that capital, whether it is accomplished yet or not, is an obligation of the firm to its investors which is independent of the actual retirement of the plant. The decision of whether or not to replace plant must be based on a comparison of future expenditures, and it should not be influenced by the extent depreciation accruals have been realized on the existing plant or by any purported "savings in depreciation expense."

However, if one plan calls for the immediate retirement of a large amount of plant, and an alternative plan calls for keeping it in service—perhaps indefinitely—apparently there is a real difference in the company's depreciation expenses in the future, depending on which plan is adopted.

It is true that the actual life of the plant in question would be different under each plan. It is also true that as average depreciation rates for the whole group of property units are changed to reflect all added experience and new views as to future life expectancy of the property, the actual life of the plant realized under the plan adopted would be reflected in the group average.* The important point, however, is that depreciation reflects the cost of plant installed in the past. The amount that was paid for this plant is the obligation to be repaid through depreciation accruals—either accruals already made, or accruals to be made in the future. This obligation is not changed by the timing of the removal of the plant (except as it might affect the amount of net salvage). The most that can be affected is the time or manner in which remaining repayment, if any, is made. This is a matter of concern only if a very large replacement program is undertaken much earlier than had been anticipated. This is not a problem of engineering economy, but rather one of accounting, rate regulation, and possibly of financing. These matters, of course, may have a very important bearing on the decision that is made, but they in no way affect the relative economy of the alternative plans. The depreciation engineers who determine the life of plant on which to base depreciation rates must be alert to this sort of a situation long before it happens; in fact, if they do a first-class job, they will anticipate important changes and innovations. Experience has shown that if competent attention is paid to depreciation study work, the averages will prevail.

Clearing Accounts

There is one final topic which should be discussed because of the misunderstanding it generates. That is the use of clearing accounts for certain depreciation charges.

For some classes of plant, depreciation charges are calculated and credited to the Accumulated Depreciation account; but these charges do not appear as depreciation expenses during the current accounting period. Instead, these credits to the depreciation reserve are offset by debits to other accounts in such a way that the use of the asset is appropriately reflected as a cost of operating the business.

An example will clarify this point. If a truck is used in the construction of a pole line, the depreciation charges on that truck—as well as the cost of its gas, oil, and so on—are considered to be part of the cost of constructing the pole line. So, instead of including the truck's depreciation as a current expense, it is "capitalized" and included in the investment in the Pole Line account. On the other hand, the depreciation charges on a car used by a Marketing person who is servicing accounts is reflected as a current expense—but as "Marketing Expense" rather than as "Depreciation Expense."

The mechanics for making these translations involve the use of the "clearing accounts," which are explained in Chapter 3.

SUMMARY

This chapter shows how capital repayment costs stem from the investments being considered in an economy study; how they can be expressed in various ways; and why they are important to the company and to economy studies. The chapter also explains that depreciation is the accountant's expression of the requirement for capital repayment.

The depreciation procedures required by the Uniform System of Accounts for capital repayment (straight-line group depreciation for the financial reports) can be quite different from capital repayment costs used in economy studies. The analyst must thoroughly understand these differences. Fundamentally, they can be explained by a

---

*In the Bell System, depreciation rate schedules are changed about every three years.

comparison of the objectives of accountants with those of operations managers. The accountants are charged with maintaining the financial records of the firm. They are therefore more concerned with past and current expenditures than with future spending decisions. Operations managers, on the other hand, are charged with the control of future expenditures, and often they have to make decisions among alternatives involving various amounts of future capital. As a result, they are concerned about capital repayment over which they still have control; they are not influenced by capital repayment of past "sunk" costs over which they have no control, and which are common to all of their alternatives. In addition, the operations managers must be careful of accounting averages which do not relate to the specific items they are considering.

The treatment of capital repayment in economy studies can be summed up in this way: The cost for a project must include an amount sufficient to recover the estimated capital investment during the estimated service life of the property being considered. This is true regardless of the depreciation expense which may actually be recorded on the books and in the financial reports. Accounting depreciation rates are established on the basis of past and predicted experience, and usually for broad groups of plant. Economy studies are not based on past experience, but on future estimates of specific plant items. As Chapter 3 explained, operating results should control depreciation accounting; however, the accounting should never be permitted to control future operations.

8

RETURN AND INCOME TAXES

INTRODUCTION

Chapter 6 divided costs into two broad groups, capital costs and operations costs. Chapter 7 emphasized the capital repayment element of capital costs. It also demonstrated that although invested capital can be repaid in various ways over its life, the capital repayment required will always be the same. However, the cost for the use of this capital varies directly with the time it takes to pay it back. An analogous situation for individuals would be the interest costs on the remaining balance of a home mortgage, which depend on the length of the mortgage. For example, more total interest is paid on a 30-year mortgage than on a mortgage lasting 25 years; and more on a 25- than a 20-year mortgage. In economy studies, these interest costs are the return owed the investors for the use of their capital. The sum of the capital repayment and the associated return is called the *capital recovery* cost, or, in some texts, the amortization cost.

Another way to view capital recovery costs is as a set of installment payments required from customers. As the company receives revenues for service, it *could* gradually repay capital to the investors. Furthermore, until the capital is fully repaid, the company would also need to pay a return amount to the investors for the use of their money. Of course, as mentioned previously, a going concern does not pay investors back directly. Instead, it reinvests the capital repayment in new plant on behalf of them.

This chapter will demonstrate that for an assumed interest rate, or rate of return to the investor, the present worth of the capital recovery cost is a constant for any method of capital repayment.

Because the assumed rate of return is the interest cost of an investment, it is often called the *cost of money*. Since the funds for the investment come from two fundamental sources—debt and equity capital—the cost of money is really a composite of the different returns for these two sources. The return on the equity portion of capital is taxable. This means that income taxes have to be paid before the equity investors can realize this portion of the return. As a result, when an investment is partly financed with equity capital, income taxes are a capital cost stemming from the return. Naturally, a return cannot be guaranteed on any investment. However, the return, as well as its taxable component, is a realistic requirement for operating a business successfully. The return and its tax burden must therefore be considered part of the revenue requirement which each investment generates.

Although the term *depreciation* covers the provision for capital repayment both for income taxes and for financial reports, it is important to distinguish between the two purposes. Many different, equivalent techniques can be used to determine repayment and return, and the analyst could apply any of these techniques in an economy study. However, there are few alternatives for determining income tax costs. The Internal Revenue Service has tax regulations prescribing exactly what is allowed for capital repayment when return is computed for tax purposes. These regulations and their applications are quite complex. Chapter 9 covers them in detail.

It is important to understand something else about the problem of determining capital repayment and return. When performing an economy study in order to choose among alternatives, the analyst must be able to estimate the amount and timing of each capital expenditure and each operations cost. The future is of course full of unknowns; but by drawing upon knowledge and experience and the facts available, the analyst has a solid basis for these estimates. However, the problem of estimating the future return component of capital costs, as well as its closely related income tax element, is more difficult. The rest of this chapter discusses these two capital cost items more thoroughly.

RATE OF RETURN

Basic Considerations

Any "going business" must take in enough revenue each year to pay its capital repayment and operations costs. Enough revenue to meet all such costs is just a primary necessity, however; investors will not supply funds to a business unless they expect its revenue to be greater than its expenses. In fact, revenue must exceed expenses by enough of a margin to provide earnings that will give the security holders a competitive return on their investment. The rate of earnings which is enough to attract investment capital to the business, in the amounts and at the times needed, naturally depends on many factors; and finding this rate is another basic financial problem that must be resolved for each business.

Some return is required on the total capital invested in a business. However, the particular return needed to provide adequate compensation depends on two factors. One is the magnitude of the risks involved. Since these risks vary widely from industry to industry and from company to company, there are corresponding variations in the rate of earnings needed to give security holders a satisfactory return on their investment. The investment on which the security holders earn a return is represented on the balance sheet by various debt accounts—and also by the equity accounts, including amounts paid in to the company for common and preferred stock and reinvested earnings.

The common and preferred stock accounts represent the amounts actually paid in by the shareowners; but the reinvested earnings are also part of the shareowners' equity, since they are invested in the business too, just as are any other amounts paid in. In other words, the shareowners are entitled to a return on their entire equity, which includes the reinvested earnings as well as the amounts paid in for common and preferred stock.

Because the holders of debt securities have a legal claim on the assets of the business should bankruptcy occur, and they therefore assume less risk, they require a lower rate of return than do the stockholders.

The second factor that affects the required rate of return is the proportion of debt and equity capital in the company's capital structure. This is commonly expressed as a company's debt ratio:

$$\text{Debt ratio} = \frac{\text{debt capital}}{\text{debt capital} + \text{equity capital}}$$

The reason the capital structure directly affects the return which each class of security holder needs is that every increase in the debt ratio reduces the equity cushion protecting the bondholders; it also increases the risk to the shareowners because a larger portion of the business is legally committed. As the debt ratio increases, both the bondholders and the shareowners require higher rates of return. To a certain extent, there is a trade-off, because the proportion of lower-cost debt capital is increasing. As risk increases, however, financing flexibility decreases; and this decrease, combined with the threat of a reduced bond rating, would eventually increase the overall rate of return required.

Appropriate Rate of Return for an Economy Study

In doing an economy study, the analyst is usually trying to estimate cash flows occurring in the future. Therefore, the rate of return should reflect the cost of raising new money, and it should be reasonably representative of the long-term future average. Fluctuations about the average are not considered because they would prevent the application of simple analysis techniques. For this reason, the rate of return (also called the *cost of money*, or the *cost of capital*) which is appropriate to use in cost studies may differ considerably from the rate of return which may appear to be needed over a short term because of current financial or economic conditions.

The firm must determine the appropriate rate of return by blending past experience with the current and projected future costs of raising new capital. One important consideration is the effect of current and future economic conditions on the problem of maintaining the integrity of the security holder's investment in the business. The firm must also make some assumptions as to what conditions will control rates in the future and in what respects these conditions will differ from past experience.

Even though the rate of return is an important factor in economy studies, it is not possible to predict the right rate to use with a high degree of precision. Nor is precision necessary, since so many assumptions and estimates about the future are made. The usual practice in the Bell System has been to use an expected long-term average rate, rounded to the nearest half or whole percent, which is based on overall Bell System experience rather than on the experience of any of the companies individually.

An example follows of the computations needed for finding the composite cost of future money. The estimates assumed for future ratios and interest costs are given here:

Capital component	Ratio		Cost		
Debt	0.45	×	9.0%	=	4.1%
Preferred stock	0.08	×	9.5%	=	0.8%
Common stock	0.47	×	15.0%	=	7.1%
Composite total	1.00				12.0%

The 12% would be the composite cost of money which a firm with the financial structure illustrated would expect to pay to the composite of its investors.

Present Worth of Capital Recovery Costs

A basic objective of this text is to teach the analyst how to make economic comparisons among alternatives involving various cash flows over time. As stated repeatedly, the concept of the time value of money permits different cash flows to be converted to equivalent amounts and makes direct comparisons possible.

Chapter 7 (pp. 149–150) demonstrated three different ways to repay capital invested in a given project under study. The present chapter pointed out earlier that these three repayment schedules (and any others, for that matter) have an equivalent present worth if return is included. In other words, the present amounts equivalent to their capital recovery costs are equal. Although this fact can be proved mathematically, it can also be convincingly demonstrated in the tables that follow.

As in Chapter 7, the assumption is that a $1,000 investment (the total installed first cost) is made at time 0 with an estimated negligible salvage and a 4-year life.

In the first case, the capital will not be repaid until the end of the 4-year life. (This was actually the second example in Chapter 7.) If a return of 12% is assumed to be paid each year, Diagram A of Figure 8-1 shows this situation from the firm's vantage point as a receiver of cash from the investor:

FIGURE 8-1
Diagram A

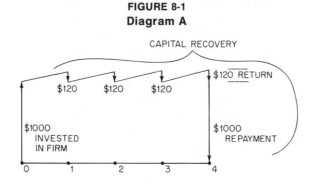

Since the balance goes to zero, the capital recovery is equivalent to the investment. This is also illustrated in Part A of Table 8-1:

TABLE 8-1
Part A

Year (A)	Unpaid balance (B)[1]	Return (0.12 × B) (C)[2]	Capital repayment (D)[2]	Capital recovery costs (C + D) (E)[2]	(P/F, 12%, N) (F)	PW of capital recovery costs (E × F) (G)
1	$1,000	$120	—	$120	0.8929	$ 107
2	1,000	120	—	120	0.7972	96
3	1,000	120	—	120	0.7118	85
4	1,000	120	$1,000	1,120	0.6355	712
					Total =	$1,000

[1]Beginning of year.

[2]End of year.

In the second case, the capital will be repaid on a straight-line basis. (This was the third example in Chapter 7.) Again assuming that the 12% return is paid each year, Diagram B of Figure 8-1 illustrates this situation from the firm's vantage point:

FIGURE 8-1
Diagram B

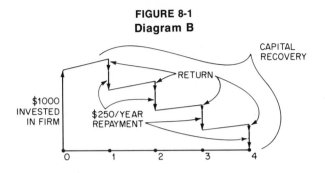

Part B of Table 8-1 below again illustrates that the present worth of this capital recovery is $1,000:

TABLE 8-1
Part B

Year (A)	Unpaid balance (B)[1]	Return (0.12 × B) (C)[2]	Capital repayment (D)[2]	Capital recovery costs (C + D) (E)[2]	(P/F, 12%, N) (F)	PW of capital recovery costs (E × F) (G)
1	$1,000	$120	$250	$370	0.8929	$ 330
2	750	90	250	340	0.7972	271
3	500	60	250	310	0.7118	221
4	250	30	250	280	0.6355	178
					Total =	$1,000

[1] Beginning of year.

[2] End of year.

In the first example in Chapter 7, all the capital was recovered immediately. There is of course no return component; but the total *PW* of the capital recovery costs is still $1,000, because the *PW* factor is 1.0.

It is now possible to generalize and say that, assuming negligible salvage, equation (8.1) holds:

$$PW(\text{capital recovery}) = PW(\text{first cost}) \tag{8.1}$$

This will always be the case regardless of how the capital is repaid. The capital recovery amounts were found by adding return to each capital repayment amount; then discounting each combined amount using the same rate has removed the return. These examples are, in fact, just another demonstration of equivalence examined in Chapter 5.

The example of the $1,000 investment can be extended to include salvage—say a positive net salvage of $200 at the end of its 4-year life. This $200 can be considered a negative capital repayment cost, which therefore reduces the amount of capital recovery that must be derived from the revenues over the life of the investment. The salvage can be expressed as *PW(net salvage)*, and the formula for capital recovery then becomes:

$$PW(\text{capital recovery}) = PW(\text{first cost}) - PW(\text{net salvage}) \tag{8.2}$$

With values substituted, *PW*(capital recovery) becomes:

$$PW(\text{capital recovery}) = 1{,}000(P/F,\ 12\%,\ 0) - 200(P/F,\ 12\%,\ 4)$$
$$= 1{,}000(1.0) - 200(0.6355)$$
$$= \$873$$

The assumption here is that $200 of the original $1,000 investment is being repaid at the end of its life. This leaves $1,000 − $200, or $800, still to be repaid in some manner. Just how it is repaid does not matter when present worth is being considered. The total capital recovery is still $1,000, but the $200 salvage reduces the cost to be passed on to customers.

If the remaining capital is to be repaid on a straight-line basis, the annual capital repayment cost is calculated as follows:

$$\frac{1{,}000 - 200}{4} = \$200 \text{ annual capital repayment cost}$$

Part C of Table 8-1 illustrates:

TABLE 8-1
Part C

Year (A)	Unpaid balance (B)[1]	Return (0.12 × B) (C)[2]	Capital repayment (D)[2]	Capital recovery costs (C + D) (E)[2]	(P/F, 12%, N) (F)	PW of capital recovery costs (E × F) (G)
1	$1,000	$120*	$200	$320	0.8929	$286
2	800	96*	200	296	0.7972	236
3	600	72*	200	272	0.7118	194
4	400	48*	200	248	0.6355	158
		Totals:	$800			$874[3]

[1]Beginning of year.

[2]End of year.

[3]Difference in answer due to rounding.

*An additional column H can be set up, as follows, to show the present worth of the return:

PW of return
(C × F)
(H)
$107
77
51
31
$266

This sum can be converted to an equivalent annual cost sometimes referred to as the *equated cost of money:*

$266(A/P, 12%, 4) = 266(0.32923) = $88

When this figure is added to the straight-line depreciation ($88 + $200), the result is equal to the equivalent annual cost of capital recovery, which is covered in the next section.

The total amount of capital which is recovered also includes the amount obtained from salvage:

$$PW(\text{salvage}) = \$200(P/F,\ 12\%,\ 4)$$
$$= \$200(0.6355) = \$127$$
$$PW(\text{capital recovered}) = PW(\text{capital recovery cost}) + PW(\text{salvage})$$
$$= \$873 + \$127 = \$1{,}000$$

Diagram C of Figure 8-1 shows the balance diagram for the receipt of investors' cash when salvage repays part of the capital:

FIGURE 8-1
Diagram C

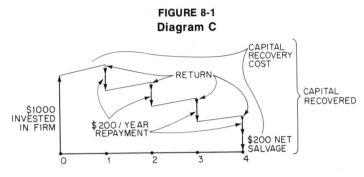

Because the amount of capital which must be recovered is known, and the present worth of any salvage is independent of the repayment method, the present worth of the capital recovery costs is a constant and is independent of the repayment method, too.

Annual Cost of Capital Recovery

In economy studies, it is sometimes convenient to express the capital recovery costs as "levelized" equivalent annual costs (AC) for the life of the investment—meaning the annuity equivalent to the present worth of the capital recovery costs over the life of the plant. Chapter 5 demonstrated how to convert the present worth (PW) to an equivalent annuity.
Therefore:

$$AC(\text{capital recovery}) = PW(\text{capital recovery})(A/P, i\%, N) \qquad (8.3)$$

In the previous example, with the total installed first cost of $1,000, the net salvage was $200 and the life was 4 years. The present worth of capital recovery was found as:

$$PW(\text{capital recovery}) = \$873$$

Then the annual cost of capital recovery is:

$$
\begin{aligned}
AC(\text{capital recovery}) &= (873)(A/P, 12\%, 4) \\
&= (873)(0.3292) \\
&= \$287
\end{aligned}
$$

Most engineering economy texts refer to the factor (A/P) as the *capital recovery factor*, because it is used to convert a present amount into a future annuity which includes both the capital repayment and the return. This annuity is still another capital recovery schedule, which can be demonstrated in Part D of Table 8-1:

TABLE 8-1
Part D

Year (A)	Unpaid balance (B)[1]	Return (0.12 × B) (C)[2]	Capital repayment ($287 − C) (D)[2]	Capital recovery costs (C + D) (E)[2]	(P/F, 12%, N) (F)	PW of capital recovery costs (E × F) (G)
1	$1000	$120	$167	$287	0.8929	$256
2	833	100	187	287	0.7972	229
3	646	78	209	287	0.7118	204
4	437	52	235	287	0.6355	182
		Totals:	$798[3]			$871[3]

[1]Beginning of year.

[2]End of year.

[3]Difference in answer due to rounding.

Except for a small difference that is due to rounding, the totals for capital repayment and the *PW* of capital recovery costs agree with the corresponding figures in Part C of Table 8-1, illustrating capital repayment on a straight-line basis. In the absence of salvage, equal annual payments representing capital recovery costs are analogous to mortgage payments.

It is also possible to find *AC*(capital recovery) over the life of an investment directly from the capital expenditures for the first cost and the net salvage. For example, two relationships have already been illustrated:

$$AC\text{(capital recovery)} = PW\text{(capital recovery)}(A/P, i\%, N) \qquad (8.3)$$
$$PW\text{(capital recovery)} = PW\text{(first cost)} - PW\text{(net salvage)} \qquad (8.2)$$

If capital recovery costs are designated C_{rec} and the first cost is assumed to occur at time 0, equation (8.2) becomes:

$$PW(C_{rec}) = \text{(first cost)}(P/F, i\%, 0) - \text{(net salvage)}(P/F, i\%, N)$$

This relationship can be substituted in equation (8.3) as follows:

$$AC(C_{rec}) = [\text{(first cost)}(P/F, i\%, 0) - \text{(net salvage)}(P/F, i\%, N)](A/P, i\%, N)$$

The equation can be simplified, however, because of the two relationships that follow:

$$\text{(first cost)}(P/F, i\%, 0) = \text{(first cost)}(1)$$
$$(P/F, i\%, N)(A/P, i\%, N) = (A/F, i\%, N)$$

Therefore:

$$AC(C_{rec}) = \text{(first cost)}(A/P, i\%, N) - \text{(net salvage)}(A/F, i\%, N) \qquad (8.4)$$

If equation (8.4) is now applied to the example in which the total installed first cost was \$1,000, the net salvage was \$200, and the life was 4 years, the result is:

$$
\begin{aligned}
AC(C_{rec}) &= \text{(first cost)}(A/P, i\%, N) - \text{(net salvage)}(A/F, i\%, N) \\
&= (1,000)(A/P, 12\%, 4) - (200)(A/F, 12\%, 4) \\
&= (1,000)(0.3292) - (200)(0.2092) \\
&= 329 - 42 \\
&= \$287
\end{aligned}
$$

The equivalent annual capital recovery costs have been found over the life of the investment, and they agree with the result previously found using equation (8.3):

$$AC(C_{rec}) = PW(C_{rec})(A/P, i\%, N) \qquad (8.3)$$

Two alternative formulas can be used to calculate the equivalent costs of capital recovery directly from the first cost and the net salvage. These calculations are illustrated in the section that follows, and they may be of interest to some readers. The relationships explained there are equations (8.5) and (8.6). Although all three formulas for calculating the annual costs of capital recovery are equivalent and result in the same answer, equation (8.4) is preferred. One reason is that it avoids the implication that economy studies are assuming a sinking fund depreciation, which the other two equations do imply. Another is that it is somewhat easier to interpret than these equations.

Calculation of Equations (8.5) and (8.6)

Both equation (8.5) and equation (8.6) are based on the relationship between the capital recovery factor and the sinking fund factor which was first expressed in Chapter 5:

$$(A/P, i\%, N) = (A/F, i\%, N) + i$$

Both equations also start with equation (8.4).

Equation (8.5) is derived as follows:

$$AC(C_{rec}) = (\text{first cost})(A/P, i\%, N) - (\text{net salvage})(A/F, i\%, N) \tag{8.4}$$
$$AC(C_{rec}) = (\text{first cost})[(A/F, i\%, N) + i] - (\text{net salvage})(A/F, i\%, N)$$
$$= (\text{first cost})i + (\text{first cost})(A/F, i\%, N) - (\text{net salvage})(A/F, i\%, N)$$
$$= (\text{first cost})i + (\text{first cost} - \text{net salvage})(A/F, i\%, N) \tag{8.5}$$

The second term represents the equivalent annual cost of capital repayment if capital repayment is made at the end of the investment's life. Although it is sometimes referred to as *annuity or sinking fund depreciation,* it is not depreciation as the accountants define the term. It can be viewed as a sum representing annual amounts paid into a savings account so that the final balance (including accrued interest) will be equal to the first cost minus the net salvage. This amount, along with the net salvage, can be used to repay the original capital at the end of the life of the investment. Because no capital is repaid until then, the return component for each year is equal to the first term—that is, (first cost)i.

Equation (8.6), which is derived from the same sources as equation (8.5), follows:

$$AC(C_{rec}) = (\text{first cost})(A/P, i\%, N) - (\text{net salvage})(A/F, i\%, N) \tag{8.4}$$
$$AC(C_{rec}) = (\text{first cost})(A/P, i\%, N) - (\text{net salvage})[(A/P, i\%, N) - i]$$
$$= (\text{first cost})(A/P, i\%, N) - (\text{net salvage})(A/P, i\%, N)$$
$$+ (\text{net salvage})(i)$$
$$= (\text{first cost} - \text{net salvage})(A/P, i\%, N) + (\text{net salvage})(i) \tag{8.6}$$

Equations (8.5) and (8.6) can now be applied to the example where the total installed first cost was $1,000, the net salvage was $200, and the life was 4 years. First, equation (8.5):

$$AC(C_{rec}) = (\text{first cost})i + (\text{first cost} - \text{net salvage})(A/F, i\%, N)$$
$$= (1000)(0.12) + (1000 - 200)(A/F, 12\%, 4)$$
$$= 120 + 800(0.2092) \tag{8.5}$$
$$= \$287 \text{ (as before)}$$

Now, equation (8.6):

$$AC(C_{rec}) = (\text{first cost} - \text{net salvage})(A/P, i\%, N) + (\text{net salvage})(i)$$
$$AC(C_{rec}) = (1000 - 200)(A/P, 12\%, 4) + 200(0.12)$$
$$= (800)(0.3292) + 24 \tag{8.6}$$
$$= \$287 \text{ (as before)}$$

INCOME TAXES

Basic Considerations

The Introduction to this chapter discussed the capital costs for any investment under study. The statement was made that along with capital repayment and return, income taxes are incurred on the return required for the equity portion of the capital. These income taxes are then a portion of the revenue required to meet the capital costs of an undertaking.

This *revenue requirements* approach to economy studies assumes that in the long run there will be sufficient revenues to cover all legitimate costs, including the income taxes. Sufficient revenues are not guaranteed, of course; but if the business is to remain healthy, those revenues must be realized. The plan with the lowest revenue requirement is the alternative which will permit service to be provided at the lowest cost that will maintain the health of the firm.

Before it is possible to select the alternative which minimizes the revenue requirements, it is necessary to determine the income taxes which would result from the taxable return on any particular investment. Some techniques are needed to make this calculation, and compositing state and federal income tax rates is a good starting point.

Compositing State and Federal Income Tax Rates

The major component of income taxes is the share paid to the federal government. The profits of corporations are taxed progressively. For example, small corporations are taxed at a relatively low rate which changes from time to time. The amount of profit which is taxed at the lower rate is also revised occasionally. Large businesses such as the telephone companies normally exceed these minimums, and for several years they have been taxed at a rate of 48% on incremental profits.

Most of the states also levy income taxes at varying rates. Many state income tax regulations are similar to the federal regulations, which is why it is convenient to composite the rates for federal and state income taxes. State income taxes are deductible in the computation of the federal tax; some states allow the federal income tax to be deducted before the state income tax is computed; and a few states allow a current year's state income taxes to be deducted before the current year's tax is calculated.

Three of the more common possibilities for compositing state and federal income tax rates are illustrated in the section that follows. The examples will be interesting for readers to follow. However, it is well to consider that the problem of compositing taxes can in reality be much more difficult, particularly in multistate companies, which are subject to the conditions in several states. Furthermore, both the tax laws and the tax rates are constantly changing. The analyst should be alert to the impact these changes have on the composite income tax rate (t), and also on the provision for income taxes in the revenue requirements.

The following examples illustrate how to find the composite income tax rate t on $100 under three different sets of conditions. In each case:

- Federal income tax rate = $t_f = 0.48$
- State income tax rate = $t_s = 0.10$

Condition A

Here, both the federal income tax and the state income tax *are not* deductible for computing the state income tax.

Subject to state income tax	$100
State income tax (0.10 × 100)	$10
Subject to federal income tax	$90
Federal income tax (0.48 × 90)	$43.20
Total tax (10 + 43.20)	$53.20
Composite income tax rate (53.20/100)	0.532, or 53.2%

More conveniently, this computation can be restated as a formula:

$$t = t_s + t_f(1 - t_s)$$
$$= 0.10 + 0.48(1 - 0.10) = 0.532 = 53.2\%$$

$$(8.7)$$

Condition B

In this case, the federal income tax is deductible for computing the state income tax, but the state income tax *is not:*

Taxable income	$100
Federal income tax	$x
Subject to state income tax	$100 − x
State income tax [0.10(100 − x)]	$10 − 0.10x
Subject to federal income tax [100 − (10 − 0.10x)]	$90 + 0.10x
Federal income tax (x)	0.48(90 + 0.10x)

Solving for x:

$$x = 0.48(90 + 0.10x)$$
$$x = 43.2 + 0.048x$$
$$x(1 - 0.048) = 43.2$$
$$x = \$45.38$$

State income tax [0.10(100 − 45.38)]	$ 5.46
Total tax (5.46 + 45.38)	$50.84
Composite income tax rate (50.84/100)	0.5084, or 50.84%

This computation can also be restated as a formula:

Taxable income	= 1
Federal income tax	= x
Subject to state income tax	= 1 − x
State income tax	= $t_s(1 - x)$
Subject to federal income tax	= $1 - t_s(1 - x)$
Federal income tax	= $t_f[1 - t_s(1 - x)] = x$

Expand and solve for x:

$$t_f - t_f t_s + t_f t_s x = x$$

$$x = \frac{t_f - t_f t_s}{1 - t_f t_s} = \frac{t_f(1 - t_s)}{1 - t_f t_s}$$

Composite tax = state income tax + federal income tax

$$t = t_s(1 - x) + x$$
$$= t_s - t_s x + x$$
$$= t_s + (1 - t_s)x$$
$$= t_s + \frac{(1 - t_s)t_f(1 - t_s)}{1 - t_f t_s}$$
$$t = t_s + \frac{t_f(1 - t_s)^2}{1 - t_f t_s} \qquad (8.8)$$
$$t = 0.10 + \frac{0.48(1 - 0.10)^2}{1 - 0.48 \times 0.10}$$
$$= 0.5084, \text{ or } 50.84\%$$

Condition C

Now federal income tax *is not* deductible for computing the state income tax, but the current state income tax *is* deductible.

Taxable income $100

State income tax .... $x

Subject to state income tax .... $100 − x

State income tax (x) .... 0.10(100 − x)

Solving for x:

$$x = 0.10(100 - x)$$
$$x = 10 - 0.10x$$
$$x(1 + 0.10) = 10$$
$$x = \$9.09$$

Subject to federal income tax (100 − 9.09) $90.91

Federal income tax (0.48 × 90.91) $43.64

Total tax (9.09 + 43.64) $52.73

Composite income tax rate (52.73/100) .... 0.5273 or 52.73%

This computation can be restated as another formula. The first step is to find an effective state income tax rate which reflects the current state tax as a deductible expense. This rate is represented by the term t_{se}.

$$t_{se} = t_s(1 - t_{se})$$
$$t_{se} = t_s - t_s t_{se}$$
$$t_{se} + t_s t_{se} = t_s$$
$$t_{se}(1 + t_s) = t_s$$
$$t_{se} = \frac{t_s}{1 + t_s}$$

Because federal income tax is not deductible for computing the state income tax, the general equation for Condition A—equation (8.7)—can be used to develop the general equation for Condition C. However, t_{se} must be substituted for t_s.

Therefore:

$$t = t_s + t_f(1 - t_s)$$
$$t = t_{se} + t_f(1 - t_{se})$$

$$(8.7)$$

Since:

$$t_{se} = \frac{t_s}{1 + t_s}$$

$$t = \frac{t_s}{1 + t_s} + t_f\left[1 - \frac{t_s}{1 + t_s}\right]$$

$$t = \frac{t_s}{1 + t_s} + t_f - \frac{t_f t_s}{1 + t_s}$$

$$t = t_f + t_s\frac{1 - t_f}{1 + t_s}$$

$$(8.9)$$

$$t = 0.48 + 0.10\frac{(1 - 0.48)}{(1 + 0.10)}$$

$$t = 0.5273, \text{ or } 52.73\%$$

Computing Income Tax

Because there must be enough revenue to pay all the capital costs over the life of an investment, the revenues required to cover capital costs (RR_c) must include capital repayment, return, and income taxes. In other words:

$$RR_c = C_{\text{rep}} + \text{return} + T \tag{8.10}$$

where C_{rep} = capital repayment = D_b = book depreciation
= straight-line depreciation
Return = return to the investors after income taxes
T = income tax costs

Income taxes are found by multiplying the taxable income by the composite tax rate. Or:

$$T = t(\text{taxable income})$$

Taxable income is the revenue received (RR_c), less the depreciation allowed for tax purposes (D_t), and less the debt interest portion of return which is not taxable.
In other words:

$$T = t(RR_c - D_t - \text{debt interest}) \tag{8.11}$$

Other costs, such as operations costs, would add to the total revenue required but then would be subtracted as tax deductions and would cancel out of equation (8.11).
If equation (8.10) is substituted into equation (8.11):

$$T = t(C_{\text{rep}} + \text{return} + T - D_t - \text{debt interest})$$

Since it is assumed for now that capital repayment is the same for both book and tax purposes, C_{rep} above equals the tax depreciation (D_t). Therefore:

$$T = t(\text{return} + T - \text{debt interest})$$
$$= t(\text{return}) + t(T) - t(\text{debt interest})$$
$$T - t(T) = t(\text{return}) - t(\text{debt interest})$$
$$T(1 - t) = t(\text{return} - \text{debt interest})$$
$$T = \frac{t}{1 - t}(\text{return} - \text{debt interest}) \tag{8.12}$$

When doing many economy studies, it is convenient to relate income taxes directly to the return after taxes. This is achieved by a so-called *income tax factor*, which is designated in this book by the Greek letter ϕ (Phi, pronounced either *fee* or *fie*). To find ϕ, however, the debt interest must be redefined in terms of these parameters:

r = debt ratio
i = overall rate of return $\doteq$ composite cost of money
i_d = debt interest rate

therefore:

$$\text{Total return} = i(\text{capital})$$

and

$$\text{Debt return} = i_d r(\text{capital})$$

The portion of the return which is debt interest, and therefore the amount of debt interest, is as follows:

$$\text{Debt interest portion of return} = \frac{i_d r(\text{capital})}{i(\text{capital})} = \frac{i_d r}{i}$$

$$\text{Debt interest} = \frac{r i_d(\text{return})}{i}$$

If the relationship for debt interest is substituted in equation (8.12):

$$T = \frac{t}{1 - t} (\text{return} - \text{debt interest})$$

$$T = \frac{t}{1 - t} \left[\text{return} - \frac{ri_d(\text{return})}{i} \right] \tag{8.12}$$

$$= \frac{t}{1 - t} (1 - \frac{ri_d}{i}) \text{ return}$$

Now, define the income tax factor (ϕ) as:

$$\text{Income tax factor} = \frac{t}{1 - t} \left(1 - \frac{ri_d}{i} \right) = \phi \tag{8.13}$$

Then:

$$T = \phi \times \text{return} \tag{8.14}$$

It is possible to get a better understanding of the income tax factor, ϕ, by observing its relationship to equation (8.12):

$$\frac{t}{1 - t} (\text{return} - \text{debt interest}) = \text{taxes} \tag{8.12}$$

$$\frac{t}{1 - t} (\text{equity return}) = \text{taxes}$$

$$\underbrace{\frac{t}{1 - t} \left(1 - \frac{ri_d}{i} \right)}_{\phi} \underbrace{\overbrace{\text{return}}^{\text{Equity return}}}_{\text{Total return}} = \text{taxes}$$

At the time an economy study is made, the terms making up ϕ are assumed to be constant during the life of each investment. As a result, ϕ will simplify the calculation of income tax costs. It is important to recall that equation (8.12) will hold only if capital repayment is the same for both book and tax purposes. If this is not the case (for example, when straight-line depreciation is used for book purposes and accelerated depreciation is used for tax purposes), adjustments are necessary. These adjustments are covered in Chapter 9. However, the tax obligation, expressed by the product of ϕ and the return, is a basic obligation, which will permit at least an initial consideration of income tax effects.

Formulas follow for calculating the present worth of income tax costs and the annual cost of income taxes—first, present worth.

Present Worth of Income Tax Costs

Equation (8.14) shows that the income tax is equal to the product of the income tax factor and the return. The present worth of income tax costs is then found from this relationship:

$$PW(T) = PW(\phi \times \text{return})$$

Since ϕ is an assumed constant:

$$PW(T) = \phi PW(\text{return}) \tag{8.15}$$

By definition:

$$C_{rec} = \text{return} + C_{rep}$$

Therefore:

$$PW(C_{rec}) = PW(\text{return}) + PW(C_{rep})$$

and

$$PW(\text{return}) = PW(C_{rec}) - PW(C_{rep})$$

Substituting this statement of $PW(\text{return})$ into equation (8.15):

$$PW(T) = \phi PW(\text{return})$$
$$PW(T) = \phi[PW(C_{rec}) - PW(C_{rep})] \tag{8.15}$$

If straight-line depreciation is assumed for capital repayment (that is, $C_{rep} = D_b$), equation (8.15) becomes:

$$PW(T) = \phi[PW(C_{rec}) - PW(D_b)] \tag{8.16}$$

The expression for the present worth of income tax costs—equation (8.16)—is the present worth of income tax expenditures. It is therefore part of the total present worth of capital costs for the investment.

Two examples follow in which the present worth of income tax is calculated. The first example is for an investment which occurs at time zero; the second, for an investment which is deferred, or which occurs in the future.

For illustrative purposes, the following examples will assume that the income tax factor, ϕ, is based upon the following characteristics of the firm:

r = Debt ratio $\qquad$ = 0.45
i = Overall rate of return, or composite cost of money $\quad$ = 0.12
i_d = Debt interest rate $\qquad$ = 0.09
t = Composite income tax rate $\qquad$ = 0.532
$\qquad$ This is a composite of state and federal income
$\qquad$ taxes. (See Condition A, page 174.)

Therefore:

$$\phi = \frac{t}{1-t}\left(1 - \frac{ri_d}{i}\right)$$

$$\phi = \frac{0.532}{1-0.532}\left[1 - \frac{(0.45) \times (0.09)}{0.12}\right]$$

$$= 0.753$$

Example A. Here, the problem is to find PW of income tax in the following situation:

A No. 5 Crossbar addition is to be put in service with a total installed first cost of $25,000, a life of 12 years, and a net salvage of $1,000.

The PW of the capital recovery costs can be calculated as follows:

$$
\begin{aligned}
PW(C_{rec}) &= PW(\text{first cost}) - PW(\text{net salvage}) \\
&= 25,000(P/F, 12\%, 0) - 1,000(P/F, 12\%, 12) \\
&= 25,000(1) - 1,000(0.2567) \\
&= 25,000 - 257 \\
&= \$24,743
\end{aligned}
$$

Next, the $PW(D_b)$ can be calculated. If straight-line depreciation is assumed for tax purposes, the nonsalvageable part of the investment ($25,000 - $1,000 = $24,000) is to be allocated in equal amounts of $24,000/12, or $2,000 per year. The $2,000 annual amount is converted to a PW by using the (P/A) factor:

$$
\begin{aligned}
PW(D_b) &= 2,000(P/A, 12\%, 12) \\
&= 2,000(6.194) \\
&= \$12,388
\end{aligned}
$$

The PW of income tax is found by substituting these values into the formula of equation (8.16):

$$
\begin{aligned}
PW(T) &= \phi[PW(C_{\text{rec}}) - PW(D_b)] \qquad (8.16)\\
&= 0.753(24{,}743 - 12{,}388)\\
&= 0.753(12{,}355)\\
&= \$9{,}303
\end{aligned}
$$

Example B. If the investment is deferred (that is, does not occur at time 0), the PW calculation is made in the same way but with a change in the PW factors. For example, if the project were to be deferred until the end of year 10 and retired at the end of year 22, the PW of income tax costs would be computed as follows:

$$
\begin{aligned}
PW(C_{\text{rec}}) &= 25{,}000(P/F, 12\%, 10) - 1{,}000(P/F, 12\%, 22)\\
&= 25{,}000(0.3220) - 1{,}000(0.0826)\\
&= \$7{,}967\\
PW(D_b) &= 2{,}000[(P/A, 12\%, 22) - (P/A, 12\%, 10)]\\
&= 2{,}000(7.645 - 5.650)\\
&= 2{,}000(1.995)\\
&= \$3{,}990\\
PW(T) &= 0.753(7{,}967 - 3{,}990)\\
&= \$2{,}995
\end{aligned}
$$

Alternatively, the $PW(T)$ could have been calculated as previously in year 10, and then the present worth found in year 0; that is,

$$
\begin{aligned}
PW(T) &= 9{,}303(P/F, 12\%, 10)\\
&= 9{,}303(0.3220)\\
&= \$2{,}996
\end{aligned}
$$

Annual Cost of Income Taxes

To calculate the present worth of income taxes in the previous section, it was necessary to develop this formula:

$$
PW(T) = \phi[PW(C_{\text{rec}}) - PW(D_b)] \qquad (8.16)
$$

To get the levelized equivalent annual cost of income taxes, another formula can be developed in a similar fashion, giving this relationship:

$$
AC(T) = \phi[AC(C_{\text{rec}}) - AC(C_{\text{rep}})]
$$

If straight-line depreciation is again assumed for tax purposes, C_{rep} may be represented by book depreciation (D_b). This gives:

$$
AC(T) = \phi[AC(C_{\text{rec}}) - AC(D_b)] \qquad (8.17)
$$

Because book depreciation is straight-line depreciation under the Uniform System of Accounts, D_b is already in the form of an annual cost.

An example follows in which the levelized annual income tax cost is calculated. It is important to note that when the investment is deferred (that is, does not occur at time 0), this levelized equivalent annual cost occurs over the life of the investment. It may be left in that form, or it may be converted to a levelized equivalent annual cost over the study period. Which form to use will depend upon the purpose of the study.

Example C. The problem is to find the levelized annual income tax cost in the situation that follows:

A No. 5 Crossbar addition is to be put in service with a total installed first cost of $25,000, a life of 12 years, and a net salvage of $1,000. $\phi = 0.753$.

The annual cost of capital recovery is calculated as follows:

$$
\begin{aligned}
AC(C_{rec}) &= AC(\text{first cost}) - AC(\text{net salvage}) \\
&= 25{,}000(A/P, 12\%, 12) - 1{,}000(A/F, 12\%, 12) \\
&= 25{,}000(0.16144) - 1{,}000(0.04144) \\
&= 4{,}036 - 41 \\
&= \$3{,}995
\end{aligned}
\tag{8.4}
$$

The annual cost of income taxes is found by substituting the $AC(C_{rec})$ and the depreciation amount into equation (8.17). If straight-line depreciation is assumed for tax purposes, the nonsalvageable part of the investment ($25,000 - $1000 = $24,000$) is to be written off in equal amounts of $24,000/12$, or $2,000 per year.

$$
\begin{aligned}
AC(T) &= \phi[AC(C_{rec}) - AC(D_b)] \\
&= 0.753(3{,}995 - 2{,}000) \\
&= 0.753(1{,}995) \\
&= \$1{,}502
\end{aligned}
\tag{8.17}
$$

TOTAL CAPITAL COSTS

As Chapter 6 and also this chapter have pointed out, capital costs consist of three components: capital repayment, return, and income taxes. The sum of the first two components, capital repayment and return, has been defined as *capital recovery*. The present worth of the taxes must also be included in the present worth of the capital costs. The present worth of the capital costs is therefore the sum of the present worth of the capital recovery costs and the present worth of the taxes. The levelized equivalent annual costs are added in the same way as the present worths.

Total capital costs are, then:

$$
PW(\text{cap. cost}) = PW(C_{rec}) + PW(T)
$$

$$
AC(\text{cap. cost}) = AC(C_{rec}) + AC(T)
$$

In Example A, then:

$$
PW(\text{cap. cost}) = 24{,}743 + 9{,}303 = \$34{,}046
$$

In Example B:

$$
PW(\text{cap. cost}) = 7{,}967 + 2{,}995 = \$10{,}962
$$

And in Example C:

$$
AC(\text{cap. cost}) = 3{,}995 + 1{,}502 = \$5{,}497
$$

SUMMARY

The primary objective of this text is to teach the analyst how to make economic comparisons among alternatives that involve different cash flows over time. This chapter has given the analyst some basic tools for meeting that objective. Formulas have been derived to express capital costs in forms that are useful for making the necessary comparisons.

The initial obligation of the firm to its investors is to recover their capital. However, capital recovery includes the repayment of capital with a return. Therefore, for any investment, the firm's revenue requirements must include capital recovery costs that are equivalent to the net capital expenditures:

$$PW(C_{rec}) = PW(\text{first cost}) - PW(\text{net salvage}) \qquad (8.2)$$

This same obligation may also be stated as an equivalent annuity. The levelized equivalent annual costs may be found either from the present worth or from the capital expenditures themselves. Therefore:

$$AC(C_{rec}) = PW(C_{rec})(A/P, i\%, N) \qquad (8.3)$$
$$AC(C_{rec}) = (\text{first cost})(A/P, i\%, N) - (\text{net salvage})(A/F, i\%, N) \qquad (8.4)$$

The expenditure of capital and its eventual repayment and return create taxable income. Taxes must be paid from this sum before the return is realized. At a specified level of return that is required by investors, this is a burden on the equity portion of the return. The income tax factor, ϕ, times the total return will determine the amount of this tax burden, which must also be included in the revenue requirements.

$$\phi = \frac{t}{1-t}\left(1 - \frac{ri_d}{i}\right) \qquad (8.13)$$

Because the capital recovery consists of both repayment and return, the return is found by subtracting the repayment, assumed in this chapter to be book depreciation. The present worth, or the levelized equivalent annual cost of taxes, is then found from the two following equations:

$$PW(T) = \phi[PW(C_{rec}) - PW(D_b)] \qquad (8.16)$$

and

$$AC(T) = \phi[AC(C_{rec}) - AC(D_b)] \qquad (8.17)$$

where   Capital repayment = book depreciation = tax depreciation

These formulas give the basic tax liability, which assumes that the same method of capital repayment is used for both financial reports and income tax purposes. This is straight-line depreciation, which is prescribed by the Uniform System of Accounts for financial reports. Because the tax regulations allow other forms of depreciation, however, further adjustments to the income tax cost formulas are required.

The total capital cost associated with an investment is found by adding the capital recovery costs to the income tax costs:

$$PW(\text{cap. cost}) = PW(C_{rec}) + PW(T)$$
$$AC(\text{cap. cost}) = AC(C_{rec}) + AC(T)$$

9
FEDERAL INCOME TAXES

INTRODUCTION

Chapter 8 identified income taxes as one of the basic capital costs of any investment; and it showed how to find the present worth of the income tax costs for any particular investment, using the formula

$$PW(T) = \phi[PW(C_{rec}) - PW(D_b)] \tag{8.16}$$

The assumption underlying equation (8.16) is that

Book depreciation = tax depreciation

or,

$$(D_b) = (D_t)$$

For many years, this assumption could be made because it was essentially true. However, changes in tax regulations have occurred, which means that it is no longer even a close approximation to the actual situation.

This chapter will describe techniques for reflecting these changed tax regulations when estimating income tax costs in an economic selection study. It will be observed that, in some cases, the analytic technique described may impute a process somewhat different from that used in rate making. This should not be a matter of concern. The economic study attempts to identify the relevant costs associated with a proposal, while the objectives of the rate-making process are much broader.

The changes in tax regulations began with the Internal Revenue Code of 1954, which allowed taxpayers to use an accelerated depreciation method to determine their tax liability, regardless of the method they followed for book purposes. Accelerated methods caused larger tax deductions for depreciation in the early years of the life of a depreciable asset and smaller deductions in the later years, but they did not actually change the total allowance for depreciation over the entire life of the asset. In other words, the total depreciation allowance could not exceed the total cost less the salvage.

However, the Accounting Principles Board and the Securities and Exchange Commission required that when an unregulated business used *accelerated tax depreciation,* it had two options: (1) use accelerated depreciation for both book and tax purposes, or (2) use straight-line depreciation for book purposes and accelerated depreciation for tax purposes and then "normalize" income. Income is normalized by including the tax deferrals generated by accelerated tax depreciation in the total operating taxes. These tax deferrals are also accumulated in a reserve, which will be used up as tax deductions decline in later years. (This normalization procedure is explained in more detail in a later section of this chapter.)

The Bell System could not adopt the first option (that of using accelerated depreciation for both tax and book purposes) because the Uniform System of Accounts requires that straight-line depreciation be used for book purposes. It was also apparent that many regulatory commissions would not accept the second option (that of normalization) for rate-making purposes. These commissions took the view that accelerated tax depreciation produced tax savings, and that these savings should be treated as a reduction in expense and then allowed to flow or pass through to earnings without any adjustments.

This flow-through treatment could bring lower current customer rates, but it would not provide for the ultimate payment of deferred taxes. There is no certainty that the amounts deferred by accelerated tax depreciation can ever be recouped in the future. As a result, the Bell Companies could be in a position where they might be unable to recover their full capital cost. The Companies therefore concluded that they would not elect the accelerated depreciation option granted in the 1954 law, but would instead continue to use straight-line depreciation for both book and tax purposes.

Then, in December 1969, Congress passed the Tax Reform Act of 1969, and the President signed it into law. One of the provisions of this act was to prohibit use of accelerated tax depreciation if the flow-through treatment was required. This provision applied to public utilities which had previously used straight-line tax depreciation or had normalized tax deferrals resulting from differences between book and tax depreciation. Since the use of accelerated tax depreciation now required normalization for accounting and rate making, the Bell Companies adopted accelerated depreciation (subject to regulatory approval) for tax purposes starting in 1970.

The federal government has an obvious interest in controlling the national economy. Such problems as widespread unemployment and inflation are associated with changes in economic conditions. One of the traditional ways to control the economy has always been through changes in the tax law. One good example is the change in the law on the use of accelerated depreciation. Another is a change in the law which gives tax credits on certain investments. The *investment tax credit* is a direct reduction in a corporation's tax obligation. (This topic will be discussed later in the chapter.) Of course, the tax law can change frequently—sometimes from year to year—for both economic and political reasons. As a result, the task of computing income taxes becomes increasingly complex.

The task is further complicated because of differences in interpretation about what part of any given investment should be capitalized. For example, such items as social security taxes and unemployment taxes associated with construction labor are currently capitalized and depreciated for book purposes, but are treated as current expenses for tax purposes. The treatment of salvage varies considerably too, and it also influences income taxes.

Since the Uniform System of Accounts requires Bell System companies to use straight-line depreciation for book purposes, depreciation for book purposes no longer even approximately equals the depreciation for income tax purposes. As a result, the basic income tax obligation given by

$$PW(T) = \phi[PW(C_{\text{rec}}) - PW(D_b)] \qquad (8.16)$$

requires a number of adjustments to reflect all the regulations in effect at the time of the proposed investment.

The objective of this chapter is to enable the analyst to understand, formulate, and use these adjustments. Chart 9-1 at the end of this chapter shows formulas for computing each of these adjustments. When considering income tax effects for a particular study, the analyst must weigh the conditions and decide which adjustments apply. Chart 9-1 should be a helpful reference—not only for use in this text, but also for doing an actual economy study.

ACCELERATED DEPRECIATION

Of all the items mentioned so far requiring an adjustment to the basic income tax formula, accelerated depreciation has the greatest impact. The Internal Revenue Code of 1954, the Tax Reform Act of 1969, and the Revenue Act of 1971 all address the use of accelerated depreciation on new investments for computing income taxes. As already mentioned, greater tax deductions are claimed during the earlier years of an asset's life than during the later years. This contrasts with straight-line depreciation, in which tax deductions are spread out equally over the life of the asset. However, the sum of deductions for depreciation is the same with either straight-line or accelerated methods.

Various forms of accelerated tax depreciation may be allowed, depending on the current income tax regulations. The procedures for making these computations are described next.

SUM-OF-THE-YEARS DIGITS METHOD

One procedure for computing accelerated tax depreciation is called the *sum-of-the-years digits* method (SYD or SOYD). This method applies a declining depreciation rate to the total allowable amount being depreciated. The total amount allowable depends on how the tax regulations treat net salvage, and there are several possible ways. For this example it is treated the same as for book depreciation, where the total amount allowed is the original cost less the net salvage.

Example: Sum-of-the-Years Digits Method

Total installed first cost	$1,000
Less net salvage	− 50
Total allowable depreciation	$ 950

The estimated service life is 4 years. The sum-of-the-years digits is $4 + 3 + 2 + 1 = 10$.

Then,

$$\text{Tax depreciation in the first year} = (950)\frac{4}{10} = \$380$$

$$\text{Tax depreciation in the second year} = (950)\frac{3}{10} = \$285$$

$$\text{Tax depreciation in the third year} = (950)\frac{2}{10} = \$190$$

$$\text{Tax depreciation in the fourth year} = (950)\frac{1}{10} = \$ 95$$

$$\text{Total accumulated depreciation} = \$950$$

Double Declining Balance

Another method for computing accelerated tax depreciation is the *double-declining-balance* method (DDB). In this method, a depreciation rate that is twice the "stripped" (meaning that salvage is excluded) straight-line depreciation rate is applied to the undepreciated balance of the original investment. In other words, a constant depreciation rate is applied to a decreasing balance. The declining-balance rate may also be based on some multiple (other than double) of the stripped straight-line depreciation rate, such as 1.5. However, the double declining balance is the maximum allowed under current regulations, and all the declining-balance examples in this text are based upon it.

A characteristic of this method is that the only way the total accumulated depreciation ever equals the total allowable depreciation is by coincidence. Furthermore, the undepreciated balance never goes to 0. However, the tax laws do allow the firm to make one change from DDB to another depreciation method in order to complete the depreciation. This switch can be made at whatever point would permit the maximum advance of the depreciation deductions.

Example: Double-Declining-Balance Method

Total installed first cost $1,000
Net salvage $50
Estimated service life   4 years

Find double the stripped straight-line depreciation rate.

$$2 \times \frac{100\%}{\text{estimated service life}} = \frac{(2)(100)}{4} = 50\%$$

Tax depreciation in the first year = $1,000(0.50) = $500
New balance = $1,000 − first year's depreciation
= $1,000 − $500 = $500
Tax depreciation in the second year = $500(0.50) = $250
New balance = $500 − $250 = $250
Tax depreciation in the third year = $250(0.50) = $125
New balance = $250 − $125 = $125

At this point, $875 of the tax depreciation ($500 + $250 + $125) has been accumulated. If the declining-balance procedure were continued, the depreciation in the fourth year would be $125(0.50), or $62.50. Since continuing this method would not accumulate the full amount which is allowed, the taxpayer can switch to another depreciation method. In the fourth year, the allowable depreciation is $950 − $875, or $75. In effect, this is a switch to straight-line depreciation for the last year.

Tax depreciation in the fourth year = $950 − $875 = $75

Total accumulated depreciation = $950

Short Life

Probably the simplest method of accelerating depreciation is to accrue the total allowable depreciation over a period shorter than the expected service life. For example, this method might be used when a short tax life is permitted to encourage companies to install pollution-control facilities.

This short-life approach can also be used in combination with other methods. An example would be to use the short-tax-life method with DDB or SYD—as with the Asset Depreciation Range System, which is explained next.

Asset Depreciation Range System

The Revenue Act of 1971 included certain tax-depreciation procedures known as the *Asset Depreciation Range System* (ADR). These procedures are tailored to the needs of firms which use group depreciation accounting, since they allow taxpayers to elect a fixed tax life, which is different from the guideline life, for a group of assets. However, taxpayers are not required to use ADR; and they might not elect to do so because, under certain conditions, ADR might not be as favorable as other procedures. However, the Revenue Act requires that if ADR is elected for the current vintage, it must be applied to all accounts, and it must be used throughout the entire history of that vintage. No commitment is required about subsequent vintages because each vintage is considered separately; and the option to elect, or reject, ADR is exercised each year.

Since the ADR procedures outlined here may change in the future, they should be treated as examples of approaches to accelerated tax depreciation. However, the general principles will probably apply to any future procedures arising from changes in the tax regulations, since the principles consider such items as short tax lives and the various approaches to treating salvage.

In 1962, the Internal Revenue Service specified the current lives (called *guideline class lives*) which industry may use for computing tax depreciation. The Revenue Act of 1971 extended this concept to include the telephone industry. Under this Act, tax depreciation can be more advantageous because tax depreciation can be taken over a life shorter than the guideline life and also because of the treatment of gross salvage and the cost of removal.

ADR TAX LIFE—ACCOUNTING   Table 9-1 shows an example of the asset guideline classes. According to this table, most telephone plant falls into one of four classes. Other plant is in a number of classes, which are common to general business. If the assets under study are not specifically listed in Table 9-1, the Comptroller's Department can provide the guideline-class life.

The ADR System allows tax depreciation to be calculated over a period of 80 to 120% of the guideline life. Normally the most advantageous application is the most rapid write-off possible, since this minimizes current tax payments. Therefore, when computing tax depreciation, the company is inclined to use the lower limit of the range of permissible lives. This is not *always* true; sometimes the selection of a somewhat longer life is advantageous because of other tax considerations. However, for purposes of illustration, this text will assume that the firm selects the lower limit.

Of course, for many reasons, all of these features are subject to change. There is already one important exception: Nonequipment buildings are not eligible for the life-shortening feature. As the regulations change, there may be other exceptions too. In any questionable cases, the analyst should consult the depreciation engineers and/or tax accountants to make certain that the tax effects in any particular study will be reflected properly.

ADR TAX LIFE—ECONOMY STUDIES   The government chose the ADR guideline-class lives based on the average lives of the types of plant within each guideline class. Of course, each depreciation method which the government allows, and the resulting reduction in tax life, is designed to produce some overall effect on the economy. This overall effect is the same as if the government had made some change in the income tax rate itself. Analysts must consider income taxes because they need to determine the appropriate tax burden that a specific investment should bear.

The problem is that the economy studies are examining specific events, whereas the tax regulations were designed to produce an overall effect; the two are not always compatible.

TABLE 9-1
AN EXAMPLE OF ASSET GUIDELINE LIFE CLASSES

Asset guideline class	Description of assets included	Asset depreciation range, years		
		Lower limit	Asset guideline period	Upper limit
48.0	Communication: Includes assets used in the furnishing of point-to-point communication services by wire or radio, whether intended to be received aurally or visually; and radio broadcasting and television			
48.1	Telephone: Includes the assets identified below and which are used in the provision of commercial and contract telephonic services			
48.11	Central office buildings: Includes special-purpose structures intended to house central office equipment and which are classified in Federal Communications Commissions Account No. 212	36	45	54
48.12	Central office equipment: Includes central office switching and related equipment classified in Federal Communications Commission Account No. 221	16	20	24
48.13	Station equipment: Includes such station apparatus and connections as teletypewriters, telephones, booths, and private exchanges as are classified in Federal Communications Commission Accounts Nos. 231, 232, and 234	8	10	12
48.14	Distribution plant: Includes such assets as pole lines, cable, aerial wire and underground conduits as are classified in Federal Communications Commission Account Nos. 241, 242.1, 242.2, 242.3, 242.4, 243, and 244	28	35	42

As this chapter will show, the impact of accelerated depreciation is partly a function of how much more rapid the tax depreciation is than the book depreciation. Chapter 7 emphasized strongly that the engineering economist is concerned with the capital repayment during the economic life. With a fixed-tax-life system like ADR, however, two assets with different economic lives, but in the same guideline class, will have different tax effects. The reason is that even though the two assets have the same overall tax depreciation, the acceleration under a fixed tax life is greater for assets with longer economic lives than for those with shorter economic lives. In effect, a fixed tax life, coupled with varying economic service lives, results in an effective income tax rate which is high for plant with a short economic life and low for plant with a longer economic life.

However, the tax regulations provide for variation in the average economic lives of plant by establishing several guideline classes, and the reduction to a short tax life is in the same proportion for each. As a result, the same effective tax rate is applied to all depreciable plant.

In most economy studies, the appropriate tax burden can be found by using an implied tax life equal to some percentage of the economic life of the plant under study. The total tax burden of the company will then be allocated to specific investments in the same way as the overall effect of the life-shortening feature. It is not a measure of the actual incremental income taxes paid, and it is not supposed to be. The alternative is to use the actual tax life for the guideline class of the specific investment and to use the economic life for capital recovery. This procedure would not produce the actual incremental taxes paid either, but it would encourage the company to invest in long-lived plant rather than shorter-lived plant because of the lower apparent effective tax rate.

Some studies involve the placement of plant for periods considerably shorter than the usual economic life for such plant. In these cases, rigid adherence to the guideline previously cited would probably be inappropriate and the tax life should be some percentage of the usual economic life for such plant. This is a compromise which seeks to avoid the distortion of the atypical situation but at the same time tries to reflect a greater sensitivity to the life characteristics of the specific plant than adherence to the ADR class life would permit.

Consider the following example: A study compares the costs of underground conduit and poles. In the company making the study, the average life of conduit is 65 years; of poles, 25 years. Both fall into the guideline class called "Distribution Plant," which has a guideline period of 35 years.

It may be observed that the use of a 35-year tax life for both types of plant tends to favor the longer-lived conduit. Yet the guideline period is a proxy for the use of many different lives in order to simplify tax calculations. It is based on an average, recognizing the shorter life of poles and longer life of conduit. On this basis, it is recommended that the tax life for poles be a percentage of its 25-year expected life and that a percentage of 65 years be used for the conduit tax life.

A second study involves a temporary pole line with an expected life of 2 years, which defers a major expenditure. Here the use of 2 years as the basis for calculating tax depreciation appears inappropriate and it is recommended that the "usual" life of 25 years be used.

ADR SYSTEM—TAX-DEPRECIATION METHODS   In addition to the life-shortening feature, the ADR System permits several tax-depreciation methods. In general, the most often used methods are:

> *Method 1 (for new property):* Double declining balance (DDB), with a switch to the sum-of-the-years digits (SYD) when SYD allows greater deductions. In the Bell System, this is called the *hybrid* procedure.

Method 2 (for used property acquired outside the Bell System): A 1.5 declining balance (1.5 DB), with a switch to straight-line (SL) when SL allows greater deductions.

Since the interpretation of what constitutes used property is subject to change, the Comptroller's Department should be consulted whenever there is any question. The hybrid procedure will usually be assumed for economy studies because they are usually evaluating future purchases of new material. One exception is in a study dealing with nonequipment buildings. The structural portion of nonequipment buildings is not eligible for the life-shortening feature, and it is also restricted to the 1.5 DB procedure.

ADR SYSTEM—SALVAGE TREATMENT   The ADR System offers one example of various approaches to the treatment of salvage for tax purposes. Under ADR procedures, the two components of net salvage must be considered: gross salvage and the cost of removal. The cost of removal has no effect on the annual tax depreciation, because for tax purposes it is treated as an expense at retirement. However, gross salvage may have an effect. If the gross salvage at retirement is less than 10% of the original cost, it is treated as a taxable gain. If the initial gross salvage estimate is greater than 10%, however, the total allowable tax-depreciation accumulation is reduced by whatever amount the salvage exceeds the 10%. The gross salvage realized at retirement, less the portion which reduces the total allowable tax-depreciation accumulation, is then a taxable gain. Some of these salvage provisions are available independently of the ADR election and may be applied to certain older vintages of plant.

ADR SYSTEM—EXAMPLE   For simplicity, the example on the next page assumes that the full first cost is capitalized for tax purposes. As a result, the first cost is the basis for calculating each annual tax-depreciation amount in this example (except for the last one), regardless of the salvage estimate. However, any gross salvage estimate over 10% determines the size of the final depreciation amount and when it will occur. This final amount, though, will merely be whatever is left of the total tax depreciation allowed.

In addition, at retirement the adjustment for the "cost of the removal less the taxable gain" is treated as if it were a tax-depreciation expense (negative in this example). This is necessary so that the total book depreciation (investment less the net salvage) will equal the total tax depreciation over the life, and the deferred tax reserve will disappear at retirement.

CAPITAL REPAYMENT AND INCOME TAXES

Up to this point, this chapter has discussed the calculation of accelerated tax depreciation using a number of different methods. However, the analyst may have to consider still different methods resulting from future changes in tax regulations. Before continuing to explain other income tax effects, it will be helpful to discuss first the place of tax depreciation in the overall context of capital-repayment costs. The income tax effects depend on the viewpoint taken toward the flow of capital and its repayment. Three viewpoints must be taken into account:

1. Actual capital cash flow (C_{rec})
2. Capital repayment assumed for book purposes (D_b)
3. Capital repayment assumed for tax purposes (D_t)

According to Chapter 7, capital-repayment costs can be shown in a number of different ways. Depreciation (both book and tax) is merely the accountant's allocation of capital repayment for both financial and tax purposes.

CALCULATION OF ANNUAL TAX-DEPRECIATION EXPENSE (D_t)

Year (A)	ADR balance (B − C) (B)	DDB depreciation (B × 2/8) (C)	SYD rate (D)*	SYD depreciation (B × D) (E)	Tax depreciation D_t (F)†	Accumulated tax depreciation (F accumulated) (G)‡
1	$1,000.00	$250.00	8/36	$222.22	$250.00	$250.00
2	750.00	187.50	7/28	187.50	187.50	437.50
3	562.50	140.63	6/21	160.71	160.71	598.21
4	562.50		5/21	133.93	133.93	732.14
5	562.50		4/21	107.14	107.14	839.28
6	562.50		3/21	80.36	10.72	850.00
7	562.50		2/21	53.57		850.00
8	562.50		1/21	26.79		850.00
9						850.00
10					(40.00)¶	810.00

Investment	$1,000
Service life	10 years
ADR tax life	8 years
Gross salvage	$250
Cost of removal	$60
ADR tax-depreciation method	Hybrid (method 1)

*Initially, the denominator *decreases* each year to reflect the SYD *remaining* tax life so it can be compared with DDB. When the switch is made, the denominator becomes constant at SYD of the remaining tax life at the time of the switch.

†Column C is used until column E is greater. Then column E is used until the accumulation (column G) reaches the total allowable: Only $10.72 is shown in year 6, because $80.36 would cause the accumulation to exceed the allowable $850.00.

‡The total allowable tax-depreciation accumulation is equal to the investment less that part of gross salvage which is more than 10% of the investment.

$$\$1,000 - [\$250 - 0.10(\$1,000)] = \$850.00$$

¶Cost of removal less the taxable gain at retirement.

Gross salvage realized	=	$250.00
less gross salvage which is more than 10% [$250 − 0.10($1,000)]	=	−150.00
Taxable gain		$100.00
Cost of removal	=	$ 60.00
less taxable gain		−100.00
Effective retirement adjustment		$−40.00

This last $40 figure is not a tax-depreciation expense, but has exactly the same effect. In the study, it is therefore included as part of the D_t

The simplest way to represent capital-repayment costs would be to show the capital cash flow as it actually occurs (the first of the three viewpoints mentioned). For example,

Total installed first cost	$1,000
Estimated service life	5 years
Gross salvage	$150
Cost of removal	$50
Net salvage	$100

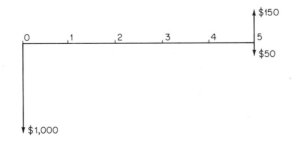

$$\text{Net capital-repayment cost} = \$1,000 - (\$150 - \$50)$$
$$= \$900$$

In Chapter 8, this form of capital repayment was in fact used to find the present worth of capital-recovery costs, using the formula

$$PW(\text{capital recovery}) = PW(\text{first cost}) - PW(\text{net salvage}) \qquad (8.2)$$

or
$$PW(C_{\text{rec}}) = PW(\text{first cost}) - PW(\text{net salvage})$$

Of course, equation (8.2) is an input to the basic income tax formula:

$$PW(T) = \phi[PW(C_{\text{rec}}) - PW(D_b)] \qquad (8.16)$$

Chapter 8 further demonstrated that $PW(C_{\text{rec}})$ is a constant, and so is independent of whatever method the accountant assumes for capital repayment.

When computing income tax costs, however, the accountant's treatment of capital repayment becomes critical because of its impact on taxable income. This treatment has two elements, book depreciation (D_b) and tax depreciation (D_t)—the second and third viewpoints toward capital repayment.

As already mentioned, the Uniform System of Accounts restricts book depreciation to straight-line depreciation. Book depreciation is then treated as an expense when net income is computed for financial reports and regulatory matters. The capital repayment assumed for book purposes would look like this:

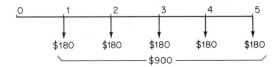

$$\text{Total allowable depreciation} = \$1,000 - (\$150 - \$50)$$
$$= \$900$$

This form of capital repayment is also an input to the basic income tax formula. The expression $PW(C_{\text{rec}}) - PW(D_b)$ measures the return (on a book basis), because $PW(C_{\text{rec}})$ is the sum of the present worth of both capital repayment and return. The expression $\phi[PW(C_{\text{rec}}) - PW(D_b)]$ would then give the present worth of income taxes if book depreciation (D_b) were actually being used to compute those taxes. Of course, it is not, and the expression is no longer true. Adjustments are necessary, using the tax depreciation (D_t).

Tax depreciation, or the capital repayment assumed for tax purposes, is whatever the current tax regulations will allow; therefore, it can be substantially different from book depreciation. If the ADR System is used for this example, the tax-depreciation viewpoint toward capital repayment might appear as follows:

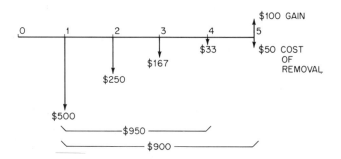

Total allowable depreciation	= $\$1,000 - [\$150 - 0.10(\$1,000)]$	
	= \$	950
Less ($100 - $50)	−	50
Total net tax deductions	\$	900

TABLE 9-2
COMPARISONS OF DEPRECIATION COMPUTED BY DIFFERENT PROCEDURES

Year	SL	SYD	DDB/SL	Short-life SL	ADR Hybrid
Amount of Depreciation					
1	180	300	400	225	500
2	180	240	240	225	250
3	180	180	144	225	167
4	180	120	86	225	33*
5	180	60	30*		(−50)†
Totals	900	900	900	900	900
Computation of the Amounts Above					
1	900/5	(900)5/15	(1,000)2/5	900/4	(1,000)2/4
2	900/5	(900)4/15	(600)2/5	900/4	(500)2/4
3	900/5	(900)3/15	(360)2/5	900/4	(250)2/3
4	900/5	(900)2/15	(216)2/5	900/4	950 − 917*
5	900/5	(900)1/15	900 − 870*		50 − 100†

*Remainder = total allowable depreciation − total accumulated in previous years

†Retirement adjustment = cost of removal − taxable gain

Of course, the actual tax-depreciation amounts can be highly variable, depending on how the current tax regulations are structured. Table 9-2 compares depreciation computed by all the previously described methods and includes straight-line depreciation. For simplicity, except for ADR, the cost of removal and the gross salvage are treated the same as with the book method. Figure 9-1 illustrates the same data graphically. Figure 9-2, which is the accumulation of Figure 9-1, illustrates the progression from no acceleration (SL) to the most acceleration (ADR).

Conceivably, other tax regulations could provide even more acceleration of tax depreciation than the ADR System. Of the methods illustrated in Figure 9-2, though, the ADR System not only repays capital the fastest but also permits a larger accumulation of total depreciation (although, in effect, it is eventually brought back down to the common $900 by the retirement adjustment).

FIGURE 9-1

Annual Amount of Depreciation Computed by Different Procedures

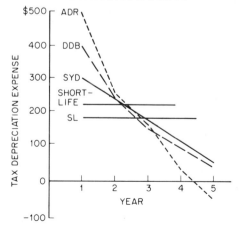

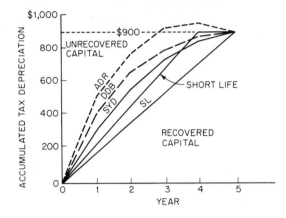

FIGURE 9-2
Accumulated Depreciation Computed by
Different Procedures

Whatever method of tax depreciation is in effect, these three viewpoints toward capital repayment (C_{rec}, D_b, D_t) will be inputs to the income tax costs and will determine the adjustments necessary to the basic formula.

INCOME TAX EFFECTS—*PW* EFFECT OF TAX DEFERRAL

For simplicity, Chapter 8 assumed that D_b was both the book expense and the allowed tax-deductible expense. However, D_t is the actual tax expense; and at given level of revenue there is a difference in income taxes. For any given year, the difference in taxes is $t(D_t - D_b)$. Because the sum of these differences over the life is 0, this is a deferral of taxes rather than a savings. This deferral of income taxes over the life of the investment has a cumulative present-worth effect which may be represented by the expression $PW(t(D_t - D_b))$, or $t[PW(D_t - D_b)]$, or $t[PW(D_t) - PW(D_b)]$. As defined earlier (in Chapter 8), t is equal to the composite income tax rate, which is assumed to be constant over the study period.

It is important to keep in mind that in the early years of an investment's life, D_t is greater than D_b and tax payments are correspondingly reduced. In later years, D_t is smaller than D_b and tax payments are greater than they would be if D_b were still being used. Of course, over the entire life of the investment, the total depreciation is the same because $\Sigma D_t = \Sigma D_b$. However,

$$PW(D_t) > PW(D_b)$$

because D_t is greater than D_b during the early years when the *PW* weighting is heavier. As a result, $PW(D_t) - PW(D_b)$ is positive; and $t[PW(D_t) - PW(D_b)]$ is the present-worth value of the taxes deferred. The quantity $t[PW(D_t) - PW(D_b)]$ is then the present worth of the savings in income tax costs resulting from the use of accelerated tax depreciation.

This present-worth savings in income tax costs is not the full effect of accelerated tax depreciation, however, because the full effect depends on how the difference in taxes is treated. The difference in taxes is a change in the amount of cash which a project has made available for return and repayment. If the difference in taxes is permitted to appear as a difference in income, it is return. If the difference in taxes is prevented from affecting income, it is the same as another source of capital repayment. The current tax law permits accelerated tax depreciation only if the normalization process is used to prevent the difference in taxes from affecting income.

NORMALIZATION

As stated earlier, accelerated depreciation had actually been available for telephone companies to use for some time, but various regulatory bodies required using flow-through accounting with it. Under flow through, income tax differences are passed along to net income (and eventually to customers as a rate change). With flow through, then, the tax difference and the present-worth savings effects are considered to contribute to the return. As a result of the flow-through requirement, the Bell Companies did not initially adopt accelerated depreciation. It was legislation passed by the U.S. Congress that finally changed their position. The Tax Reform Act of 1969 addressed itself to the regulatory treatment of the effects of accelerated tax depreciation. This law stated that if a company was not already using accelerated depreciation with flow through, it could not adopt accelerated tax depreciation, if it was required to use flow through. Instead, the company was required to keep the differences in taxes in certain normalization accounts until the tax-depreciation amounts declined and higher tax payments were required. In other words, the difference in taxes now was viewed as contributing a temporary source of capital repayment rather than return. The new accounts then normalized the books of the company so that there was no change in the company's net income from a regulatory standpoint. The next section explains in more detail how this normalization accounting works.

Accounting

Figure 9-3 on page 196 is an example of a recent AT&T income statement. The total of current and deferred income taxes shown is the tax that would have been incurred had book depreciation been used for tax purposes. However, because accelerated depreciation was used instead, substantial amounts of these taxes were deferred to future periods. Figure 9-3 shows these taxes in the two deferred accounts, marked with arrows. Both current (actually paid) and deferred taxes are included as part of the "Total Operating taxes." Therefore, the use of accelerated tax depreciation has caused no change in "Income before interest deductions" or in "Net income." Also, the depreciation expense shown under "Operating expenses" still reflects the straight-line depreciation prescribed by the Uniform System of Accounts. The tax expense is then in phase with the depreciation expense, and the income statement is therefore "normal."

Each year's deferred income tax is added to the Accumulated Deferred Income Taxes account which is a liability account on the balance sheet shown in Figure 9-4, pages 198–199. This account, also known as the Deferred Tax Reserve, represents a claim of the government on funds that have been temporarily kept in the business in anticipation of paying income taxes. As the tax-depreciation deductions of older investments decline, associated income-tax payments will rise, and their portion of this account will be reduced correspondingly. For any single investment, the cumulative deferred income taxes will at first rise, and then it will decline to 0 over the life of the investment.*

Economy Studies

The present worth of the cumulative deferred income tax has already been shown to be

$$t[PW(D_t) - PW(D_b)]$$

With normalization, this tax deferral is available for investment in plant. As Chapter 4

---

*Accounting procedures have been developed which will achieve this objective even if the tax rate changes during the life of the investment.

FIGURE 9-3
An Example of an AT&T Income Statement

	Thousands of dollars	
	Year 1975	Year 1974
Operating Revenues		
Local service	$14,027,831	$12,812,812
Toll service	13,925,190	12,460,875
Directory advertising and other	1,218,822	1,091,983
Less: Provision for uncollectibles	214,602	191,258
Total operating revenues	28,957,241	26,174,412
Operating Expenses		
Maintenance	5,919,100	5,373,970
Depreciation	4,088,089	3,690,390
Traffic—primarily costs of handling messages	2,132,119	1,996,264
Commercial—primarily costs of local business office operations	1,017,701	917,195
Marketing	1,266,550	1,106,746
Accounting	786,802	728,207
Provision for pensions and other employee benefits	2,363,730	1,852,654
Research and fundamental development	206,279	189,883
Other operating expenses	976,906	860,500
Total operating expenses	18,757,276	16,715,809
Net operating revenues	10,199,965	9,458,603
Operating Taxes		
Federal income:		
Current	129,102	678,407
Deferred ←	1,303,340	1,186,734
Investment tax credits—net	741,646	258,263
State and local income:		
Current	108,330	98,766
Deferred ←	107,825	90,968
Property, gross receipts, payroll-related and other taxes	2,680,819	2,453,537
Total operating taxes	5,071,062	4,766,675
Operating income (carried forward)	$ 5,128,903	$ 4,691,928
Other Income		
Western Electric Company net income	107,308	315,151
Interest charged construction	239,957	240,902
Miscellaneous income and deductions—net	(32,220)	(16,948)
Total other income	315,045	539,105
Income before interest deductions	5,443,948	5,231,033
Interest deductions	2,296,226	2,056,569
NET INCOME	3,147,722	3,174,464

NORMALIZING ACCOUNTS

pointed out, this means that it is a source of internally generated funds during the life of that investment. The use of these tax deferrals to provide capital required in other parts of the business is, in effect, an additional source of capital repayment for the particular investment under study. The present worth of the capital-recovery cost which the customers must bear is then reduced by this present worth of tax deferral.

Chapter 8 defined the present worth of capital-recovery costs as

$$PW(C_{rec}) = PW(\text{first cost}) - PW(\text{net salvage}) \tag{8.2}$$

With normalization, some capital is temporarily recovered through the deferred tax reserve. Therefore, the capital recovery costs for normalization, $PW(C_{rec})_N$, are

$$PW(C_{rec})_N = PW(\text{first cost}) - PW(\text{net salvage}) - t[PW(D_t) - PW(D_b)]$$
$$PW(C_{rec})_N = PW(C_{rec}) - t[PW(D_t) - PW(D_b)]$$

Chapter 8 explained that the total present worth of capital costs was the sum of the present worth of the capital-recovery costs and the present worth of the taxes. Therefore, with normalization the complete expression for the present worth of capital costs becomes

$$PW(\text{capital cost}) = PW(C_{rec})_N + \phi[PW(C_{rec})_N - PW(D_b)]$$

or

$$PW(\text{capital cost}) = PW(C_{rec}) - t[PW(D_t) - PW(D_b)]$$
$$+ \phi\{PW(C_{rec}) - t[PW(D_t) - PW(D_b)] - PW(D_b)\}$$

Because the accelerated depreciation effects are most appropriately considered as entirely an income tax effect, the original definition of capital-recovery costs should be preserved. Therefore,

$$PW(\text{capital costs}) = PW(C_{rec}) + PW(T)$$

Then the present worth of income taxes is

$$PW(T) = PW(\text{capital costs}) - PW(C_{rec})$$
$$= PW(C_{rec}) - t[PW(D_t) - PW(D_b)]$$
$$+ \phi\{PW(C_{rec}) - t[PW(D_t) - PW(D_b)] - PW(D_b)\} - PW(C_{rec})$$
$$= -t[PW(D_t) - PW(D_b)] + \phi\{PW(C_{rec}) - t[PW(D_t) - PW(D_b)] - PW(D_b)\}$$

Rearranging terms yields

$$PW(T) = \phi[PW(C_{rec}) - PW(D_b)] - t(1 + \phi)[PW(D_t) - PW(D_b)] \tag{9.1}$$

The first term is the basic liability from Chapter 8, and the second term represents the effect of the use of accelerated depreciation and normalization of deferred taxes. Equation (9.1) can be converted into the annual cost form simply by substituting the $AC(\ )$ operator for the $PW(\ )$ operator. Referring to Chart 9-1 at the end of this chapter, the expression $-t(1 + \phi)[PW(D_t) - PW(D_b)]$ should be used for computing the income tax effect of accelerated tax depreciation with normalization.

Computation of Capital Costs

The present worth of the capital costs has been shown to be the sum of the present worth of capital recovery, from equation (8.2), and the present worth of the taxes, from equation (9.1), for accelerated depreciation and normalization. Most of the terms in these equations are quite simple to calculate except for the present worth of the accelerated tax depreciation, $PW(D_t)$. This can be difficult to calculate, depending on

FIGURE 9-4
An Example of an AT&T Balance Sheet

| | Thousands of Dollars | |
	December 31, 1975	December 31, 1974
Assets		
Telephone plant—at cost		
In service	$84,618,757	$77,689,387
Under construction	2,881,158	3,335,451
Held for future use	120,865	121,225
	87,620,780	81,146,063
Less: Accumulated depreciation	17,178,862	16,210,674
	70,441,918	64,935,389
Investments		
At equity (F):		
Western Electric Company, Inc.	3,209,232	3,241,924
Other	271,285	246,893
At cost	74,636	87,946
	3,555,153	3,576,763
Current assets		
Cash and temporary cash investments—less drafts outstanding:		
1975, $269,749,000; 1974, $262,817,000	1,123,543	1,121,621
Receivables—less allowance for uncollectibles:		
1975, $76,386,000; 1974, $34,944,000	3,537,664	3,131,733
Material and supplies	450,440	449,828
Prepaid expenses	214,102	155,285
	5,325,749	4,858,467
Deferred charges	833,412	687,607
Total assets	$80,156,232	$74,058,226

the complexity of the tax regulations. To help the analyst make this computation, tables have been prepared which give a present worth of the tax-depreciation factor for several types of tax depreciation. These factors are designated (P/D_t) for the present worth of the tax-depreciation factor and (A/D_t) for its levelized equivalent. They are used in much the same way the time-value factors are used. The product of the factor and the tax base (often the first cost) is the present worth of tax depreciation, or the annuity equivalent to tax depreciation.

$$PW(D_t) = (\text{first cost})(P/D_t)$$
$$AC(D_t) = (\text{first cost})(A/D_t)$$

Tables for these present-worth tax-depreciation factors are included in Appendix C at the end of the book. Levelized tax-depreciation factors may be calculated by multiplying the $PW(D_t)$ factor by the appropriate A/P factor. The example which follows illustrates the use of the sample table in Figure 9-5 (page 200) and the application of the formulas to determine the present worth of capital costs.

	Thousands of Dollars	
	December 31, 1975	*December 31, 1974*
Liabilities and Capital		
Equity		
American Telephone and Telegraph Company		
Preferred shares	$ 3,001,583	$ 3,022,941
(Includes excess of proceeds over stated value)		
Common shares	15,979,281	14,955,272
(Includes excess of proceeds over par value)		
Reinvested earnings	14,787,277	13,816,548
	33,768,141	31,774,761
Minority ownership interest in consolidated subsidiaries	863,784	827,824
	34,631,925	32,602,585
Long- and intermediate-term debt	31,793,326	29,358,326
Current liabilities		
Accounts payable	1,914,082	1,891,308
Taxes accrued	847,646	834,029
Advance billing and customers' deposits	702,912	630,187
Dividends payable	565,266	545,905
Interest accrued	596,802	531,381
	4,626,708	4,432,810
Debt maturing within one year	2,228,782	2,950,124
	6,855,490	7,382,934
Deferred credits		
Accumulated deferred income taxes	4,721,166	3,317,325
Unamortized investment tax credits	2,025,228	1,357,906
Other	129,097	39,150
	6,875,491	4,714,381
Lease commitments		
Total liabilities and capital	$80,156,232	$74,058,226

DEFERRED TAX RESERVE (arrow pointing to Accumulated deferred income taxes)

Example 9-1: PW of Capital Costs with ADR and Normalization. Find the present worth of the capital costs for a proposed addition of toll terminal equipment. Assume the following:

Total installed first cost	$10,000
Gross salvage	$1,500
Cost of removal	$500
Cost of money	12%
Estimated service life (N)	20 years
ϕ	0.67
t	53%
Tax life (20×0.8)	16 years
ADR System (hybrid) with normalization	

FIGURE 9-5
Tax Depreciation Table

```
ASSET DEPRECIATION RANGE                                        ADR LIFE  80% OF SERVICE LIFE
ADR HYBRID                                                             12.0% RATE OF RETURN
DDB WITH SWITCH TO SYD                                                  ANNUAL COMPOUNDING
                                    PRESENT WORTH FACTORS
```

N	0%	10%	15%	20%	30%	40%	50%	60%	70%	80%	90%	100%	*COR ADJ *+CORS X FACTOR	N
3	.87593	.80475	.72602	.64311	.55382	.46454	.37525	.28596	.19668	.10739	.01811		.71178	3
4	.84465	.78110	.71024	.63399	.55427	.47216	.38288	.29359	.20431	.11502	.02573		.63552	4
5	.81732	.76057	.69575	.62457	.54912	.46941	.38969	.30040	.21111	.12183	.03254		.56743	5
6	.79272	.74205	.68128	.61497	.54379	.46751	.38779	.30648	.21719	.12791	.03862		.50663	6
7	.76958	.72435	.66870	.60571	.53729	.46611	.38753	.30781	.22262	.13334	.04405		.45235	7
8	.74796	.70757	.65549	.59640	.53279	.46162	.38810	.30838	.22747	.13818	.04890		.40388	8
9	.72774	.69168	.64353	.58775	.52549	.45844	.38726	.30939	.22967	.14251	.05322		.36061	9
10	.70875	.67656	.63155	.57850	.51957	.45602	.38498	.31060	.23088	.14637	.05709		.32197	10
11	.69005	.66130	.61952	.57001	.51366	.45108	.38330	.31187	.23215	.14982	.06054		.28748	11
12	.67232	.64665	.60795	.56057	.50677	.44700	.38203	.31086	.23342	.15290	.06362		.25668	12
13	.65552	.63260	.59644	.55230	.50083	.44355	.38000	.30988	.23464	.15492	.06637		.22917	13
14	.63956	.61910	.58564	.54322	.49432	.43889	.37704	.30911	.23578	.15606	.06882		.20462	14
15	.62439	.60612	.57468	.53492	.48779	.43430	.37446	.30848	.23683	.15711	.07102		.18270	15
16	.60936	.59304	.56402	.52616	.48177	.43013	.37216	.30795	.23678	.15807	.07297		.16312	16
17	.59504	.58048	.55332	.51773	.47494	.42569	.36964	.30656	.23631	.15894	.07472		.14564	17
18	.58139	.56839	.54318	.50949	.46856	.42077	.36626	.30471	.23590	.15972	.07628		.13004	18
19	.56836	.55675	.53319	.50127	.46252	.41628	.36318	.30302	.23551	.16042	.07768		.11611	19
20	.55590	.54553	.52356	.49349	.45597	.41209	.36035	.30148	.23514	.16103	.07892		.10367	20
21	.54357	.53432	.51390	.48531	.44975	.40702	.35761	.30003	.23479	.16157	.08003		.09256	21
22	.53179	.52353	.50450	.47769	.44360	.40232	.35403	.29851	.23445	.16205	.08102		.08264	22
23	.52051	.51313	.49546	.46987	.43732	.39795	.35070	.29623	.23387	.16246	.08191		.07379	23
24	.50970	.50311	.48657	.46250	.43144	.39324	.34759	.29410	.23268	.16260	.08270		.06588	24
25	.49933	.49344	.47808	.45516	.42545	.38856	.34468	.29210	.23156	.16227	.08340		.05882	25
26	.48909	.48384	.46948	.44789	.41945	.38409	.34120	.29020	.23049	.16194	.08367		.05252	26
27	.47927	.47458	.46126	.44082	.41380	.37967	.33779	.28841	.22947	.16161	.08389		.04689	27
28	.46984	.46565	.45319	.43386	.40793	.37501	.33458	.28621	.22850	.16128	.08407		.04187	28
29	.46077	.45703	.44543	.42719	.40233	.37061	.33155	.28390	.22757	.16096	.08423		.03738	29
30	.45204	.44870	.43786	.42052	.39695	.36645	.32829	.28171	.22669	.16063	.08435		.03338	30
31	.44344	.44046	.43037	.41406	.39134	.36192	.32491	.27962	.22540	.16031	.08445		.02980	31
32	.43517	.43251	.42312	.40761	.38601	.35758	.32171	.27763	.22401	.15999	.08453		.02661	32
33	.42720	.42482	.41606	.40144	.38079	.35346	.31866	.27564	.22269	.15968	.08459		.02376	33
34	.41952	.41739	.40924	.39536	.37557	.34930	.31559	.27329	.22143	.15937	.08463		.02121	34
35	.41210	.41021	.40258	.38945	.37060	.34512	.31233	.27105	.22022	.15906	.08466		.01894	35
36	.40481	.40312	.39604	.38362	.36557	.34108	.30919	.26890	.21906	.15848	.08467		.01691	36
37	.39777	.39626	.38964	.37790	.36064	.33718	.30620	.26685	.21795	.15780	.08467		.01510	37
38	.39098	.38963	.38347	.37239	.35593	.33313	.30330	.26488	.21682	.15715	.08466		.01348	38
39	.38441	.38321	.37745	.36693	.35117	.32925	.30017	.26273	.21538	.15652	.08465		.01204	39
40	.37807	.37699	.37162	.36169	.34658	.32554	.29717	.26052	.21399	.15591	.08462		.01075	40
45	.34881	.34820	.34442	.33687	.32482	.30726	.28291	.25032	.20783	.15316	.08442		.00610	45
50	.32346	.32311	.32043	.31470	.30511	.29054	.26955	.24056	.20139	.15007	.08394		.00346	50
55	.30133	.30113	.29924	.29489	.28725	.27509	.25707	.23127	.19539	.14699	.08292		.00196	55
60	.28188	.28177	.28044	.27713	.27103	.26093	.24534	.22238	.18962	.14413	.08202		.00111	60
65	.26468	.26462	.26367	.26117	.25628	.24788	.23443	.21398	.18388	.14100	.08124		.00063	65
70	.24938	.24934	.24867	.24677	.24287	.23588	.22427	.20606	.17854	.13828	.08055		.00036	70
75	.23569	.23567	.23519	.23375	.23063	.22482	.21481	.19858	.17339	.13531	.07973		.00020	75

The present worth of capital costs equation is

$$PW(\text{capital costs}) = PW(C_{\text{rec}}) + PW(T)$$

To compute $PW(C_{\text{rec}})$,

$$
\begin{aligned}
PW(C_{\text{rec}}) &= PW(\text{first cost}) - PW(\text{net salvage}) \\
&= (10{,}000)(P/F, 12\%, 0) - (1{,}000)(P/F, 12\%, 20) \\
&= (10{,}000)(1) - (1{,}000)(0.1037) \\
&= \$9{,}896
\end{aligned}
$$

Then equation (9.1) can be used to compute $PW(T)$.

$$PW(T) = \phi[PW(C_{\text{rec}}) - PW(D_b)] - t(1 + \phi)[PW(D_t) - PW(D_b)] \qquad (9.1)$$

The first step is to compute $PW(D_b)$, since $PW(C_{\text{rec}})$ is now known.

$$
\begin{aligned}
PW(D_b) &= [(\text{first cost} - \text{net salvage})/\text{service life}](P/A, 12\%, 20) \\
&= [(10{,}000 - 1{,}000)/20](7.469) \\
&= (450)(7.469) \\
&= \$3{,}361
\end{aligned}
$$

Now it is necessary to compute $PW(D_t)$, using the appropriate tax-depreciation table (Figure 9-5).

$$PW(D_t) = (\text{first cost})(P/D_t)$$

The (P/D_t) factor is actually the sum of two components in Figure 9-5. The first component is a function of life and gross salvage; the second component is a function of life and the cost of removal. The first is found at the intersection of the row corresponding to 20 years of estimated service life (N at the extreme left- and right-hand columns), and the gross salvage columns marked 10% and 20% (with interpolation for the 15%).

On the same row (where N is equal to 20 years), the COR ADJ column gives the cost of the removal component for 100% cost of removal. This must be multiplied by the cost of removal expressed as a proportion of the tax basis* to get the second component of the PW of the tax-depreciation factor. These values are marked on Figure 9-5.

Thus,

Present worth of tax-depreciation factor = gross salvage component + COR component

$$(P/D_t) = \overbrace{\frac{0.54553 + 0.52356}{2}} + \overbrace{0.10367 \times 0.05}$$

$$(P/D_t) = 0.53973$$

and

$$PW(D_t) = (10,000)(0.53973)$$
$$= \$5,397$$

Now, values can be substituted into the formula

$$PW(T) = \phi[PW(C_{\text{rec}}) - PW(D_b)] - t(1 + \phi)[PW(D_t) - PW(D_b)]$$
$$= 0.67(9,896 - 3,361) - 0.53(1.67)(5,397 - 3,361)$$
$$= 4,378 - 1,802$$
$$= \$2,576$$

Therefore,

$$PW(\text{capital costs}) = PW(C_{\text{rec}}) + PW(T)$$
$$= 9,896 + 2,576$$
$$= \$12,472$$

Of course, the tax-depreciation table (Figure 9-5) is good only for the assumptions listed at the top. Tables using other assumptions are found in Appendix C at the end of the book.

If it should be desirable to express these present worths as equivalent annual costs (AC)† for the life of the equipment, the conversion is made simple with the (A/P) factor:

$$AC(\text{capital costs}) = PW(\text{capital costs})(A/P, 12\%, 20)$$
$$= (12,472)(0.13388)$$
$$= \$1,670/\text{year}$$

The same result could be found directly by using the annual cost formulas developed in Chapter 8, and then subtracting the expression $t(1 + \phi)[AC(D_t) - AC(D_b)]$ for the adjustment relating to accelerated depreciation with normalization. The appropriate (A/D_t) factor can be calculated by converting the $PW(D_t)$ to an annual equivalent.

ACCELERATED DEPRECIATION WITH FLOW THROUGH

Although flow through of income tax deferrals would cause the loss of accelerated tax depreciation, an understanding of how flow through would affect income taxes will be

---

*The tax basis corresponds to the full investment in this example.
†This is also known as the levelized equivalent annual costs (AC).

helpful in understanding the impact of other tax-affecting conditions covered later in this chapter.

Flow through makes no accounting provision to recognize the taxes deferred when tax depreciation is more rapid than the book depreciation. The difference in taxes paid then flows directly into net income, and thereby causes a reduction in the revenue requirements. Chapter 8 gave the basic equation for the revenue requirements for capital costs as

$$RR_c = C_{rep} + \text{return} + T \tag{8.10}$$

And the equation for taxes was

$$T = t(RR_c - D_t - \text{debt interest}) \tag{8.11}$$

Since capital repayment is considered to be book depreciation (D_b), when equation (8.10) is substituted into (8.11) the result is

$$T = t(D_b + \text{return} + T - D_t - \text{debt interest})$$

With accelerated tax depreciation, the depreciation terms do not cancel, as they did in Chapter 8, because D_b is not equal to D_t. Therefore, the equation for taxes becomes

$$T = t(\text{return} + T - \text{debt interest}) - t(D_t - D_b)$$

Solving for T, as in Chapter 8, yields

$$(1 - t)T = t(\text{return} - \text{debt interest}) - t(D_t - D_b)$$

$$T = \frac{t}{1 - t}(\text{return} - \text{debt interest}) - \frac{t}{1 - t}(D_t - D_b)$$

The quantity $t(D_t - D_b)$, representing the difference in taxes, has reappeared. Only now, with flow through, no reserve is established. The tax difference then becomes an income difference, and ultimately it causes a revenue-requirement difference of $(t/1 - t)(D_t - D_b)$. The first term in the previous tax equation is simply the basic tax liability developed in Chapter 8 [equation (8.12)]; the second term is the result of a different treatment for taxes than for the books. In the present-worth form, then, the equation for present worth of taxes if normalization is not used becomes

$$PW(T) = \phi[PW(C_{rec}) - PW(D_b)] - \frac{t}{1 - t}[PW(D_t) - PW(D_b)] \tag{9.2}$$

This development illustrates that with flow through, the quantity $t[PW(D_t) - PW(D_b)]$ is considered a contribution to return—specifically, equity return. This contrasts with the normalization process, where the same quantity was treated as a source of capital repayment.

Chart 9-1 also shows that equation (9.2) provides for the effect of accelerated tax depreciation with flow through.

INVESTMENT TAX CREDIT

As mentioned before, one way that the federal government controls the economy is through reductions in income taxes for actions which it believes are desirable. These reductions are called *tax credits*. Since investment in new plant benefits the nation's economy, these capital expenditures may be encouraged by tax credits. These credits, called *investment tax credits* or *job development investment credits*, are determined as a specified percentage of eligible construction expenditures, and are a direct reduction in the taxes to be paid.

Investment tax credits have had a varied history in the tax law. They have been offered and withdrawn; the tax credit rate has been varied among industries; the tax credits have been associated with employee stock purchase plans; and various limitations have been imposed.

The Revenue Act of 1971 reinstated an investment tax credit. Its provisions were liberalized in 1975, and again in 1976, and further variations can be expected as economic conditions change. Consequently, the following sections will consider applications of investment tax credit to economy studies with a minimum of reference to specific provisions of the current tax laws. Nevertheless, the general income tax effects discussed here should apply to a wide range of future situations and changes in tax regulations.

The effect of an investment tax credit is a function of many variables. The percentage allowed depends on the tax life chosen for the group the asset is associated with. If the asset is retired before 7 years, however, any credits realized must be partially or wholly given back by a "recapture" provision of the Revenue Act. This means that assets which have an actual service life shorter than their tax life will result in a tax credit effect based on their service life. For most studies, therefore, it is usually satisfactory to assume that the rate is prescribed by whichever is shorter—the tax life or the service life. This rate is applied only to eligible expenditures, and not to the total installed first cost. As the following sections will show, on the tax return certain components of the installed first cost are treated either as an expense or they are not allowed at all as tax deductions. To determine the amount of credit, the investment-tax-credit rate is applied to the *tax basis*. The tax basis is the installed first cost, called the *book basis*, less the tax-expense items and the disallowed costs.

Service-Life Flow Through

The reduction in income taxes from an investment tax credit is actually realized during the tax period in which the expenditure is made. However, to be eligible for the credit, the Bell System elected an accounting option which amortizes the credit over the life of the plant. In other words, instead of having the investment tax credit recognized or flowed through at the time it is realized, its effect is spread over the investment's life. This is called *service-life flow through* (SLFT), and is accomplished by the use of a reserve account similar to the deferred tax reserve, called the *Unamortized Investment-Tax-Credit Reserve*.

Unamortized Investment-Tax-Credit Reserve

The amount of income tax actually paid, labeled "current" on the income statement in Figure 9-3 (page 196) is the tax computed on the tax return; it is reduced by any investment tax credit which is actually realized. The actual credit is placed in the Unamortized Investment-Tax-Credit Reserve. Each year during the life of the investment which created the credit, the reserve is reduced by an amount equal to the credit divided by the life. The net of this transaction (current credits less the amortization of previous credits) is added to the taxes paid, plus the taxes deferred, in order to determine the total operating taxes. Here is how it works:

    Taxes payable without credit
    − Tax credits
    ────────────────
    Taxes paid   (current)
    + Taxes deferred by accelerated depreciation (deferred)
    + Tax credits   ⎫
    − Amortized tax credits   ⎬ (net tax credit)
    ────────────────   ⎭
    Total operating taxes

Because the current tax credits are first subtracted and then added back in, they are canceled out of total operating taxes. Total operating taxes are then

Taxes payable without credits
+ Taxes deferred
− Amortized credits
Total operating taxes

Without this transaction, net income would be increased by the amount of the credit at the time it is actually realized. Instead, net income is increased in small amounts each year over the life of the plant which caused the credit. The amount of tax credit left in the reserve at any time is then available for investment in plant until it is completely amortized. Because this reserve represents funds belonging to the investors, it is shown as a liability on the balance sheet, along with the deferred tax reserve. (See Figure 9-6.)

FIGURE 9-6
An Example of an AT &T Balance Sheet

	Thousands of Dollars	
	December 31, 1975	December 31, 1974
Assets		
Telephone plant—at cost		
In service	$84,618,757	$77,689,387
Under construction	2,881,158	3,335,451
Held for future use	120,865	121,225
	87,620,780	81,146,063
Less: Accumulated depreciation	17,178,862	16,210,674
	70,441,918	64,935,389
Investments		
At equity:		
Western Electric Company, Inc.	3,209,232	3,241,924
Other	271,285	246,893
At cost	74,636	87,946
	3,555,153	3,576,763
Current assets		
Cash and temporary cash investments—less drafts outstanding: 1975, $269,749,000; 1974, $262,817,000	1,123,543	1,121,621
Receivables—less allowance for uncollectibles: 1975, $76,386,000; 1974, $34,944,000	3,537,664	3,131,733
Material and supplies	450,440	449,828
Prepaid expenses	214,102	155,285
	5,325,749	4,858,467
Deferred charges	833,412	687,607
Total Assets	$80,156,232	$74,058,226

Income Tax Effects of Investment Tax Credit
with Service-Life Flow Through

Unlike the effects of accelerated depreciation, where the deferred taxes were considered a temporary source of capital repayment under normalization but additional return under flow through, the investment tax credit affects both return and capital repayment. When it is flowed through, it is return; until it is flowed through, it is a temporary source of capital repayment.

An earlier discussion showed that in the absence of tax credits, the taxes paid are

$$T = t \text{ (return} + T - \text{debt interest)}$$

With a tax credit (TC), the taxes paid are

$$T = t(\text{return} + T - \text{debt interest}) - TC$$

	Thousands of Dollars	
	December 31, 1975	*December 31, 1974*
Liabilities and Capital		
Equity		
American Telephone and Telegraph Company		
Preferred shares	$ 3,001,583	$ 3,002,941
(Includes excess of proceeds over stated value)		
Common shares	15,979,281	14,955,272
(Includes excess of proceeds over par value)		
Reinvested earnings	14,787,277	13,816,548
	33,768,141	31,774,761
Minority ownership interest in consolidated subsidiaries	863,784	827,824
	34,631,925	32,602,585
Long- and intermediate-term debt	31,793,326	29,358,326
Current liabilities		
Accounts payable	1,914,082	1,891,308
Taxes accrued	847,646	834,029
Advance billing and customers' deposits	702,912	630,187
Dividends payable	565,266	545,905
Interest accrued	596,802	531,381
	4,626,708	4,432,810
Debt maturing within one year	2,228,782	2,950,124
	6,855,490	7,382,934
Deferred credits **INVESTMENT TAX**		
Accumulated deferred income taxes **CREDIT RESERVE**	4,721,166	3,317,325
Unamortized investment tax credits	2,025,228	1,357,906
Other	129,097	39,150
	6,875,491	4,714,381
Lease commitments		
Total liabilities and capital	$80,156,232	$74,058,226

However, because of the service-life flow through of tax credits (neglecting the effect of accelerated depreciation for simplicity), the operating taxes are

$$T = t(\text{return} + T - \text{debt interest}) - TC + TC - ATC$$

The *ATC* here stands for *Amortized Tax Credit*, which is the tax credit divided by the life each year.

Solving this equation for T,

$$(1 - t)T = t(\text{return} - \text{debt interest}) - ATC$$

$$T = \frac{t}{1 - t}(\text{return} - \text{debt interest}) - \frac{ATC}{1 - t}$$

The first portion of this expression is again the basic tax liability from Chapter 8. In the present-worth form, the equation is then

$$PW(T) = \phi[PW(C_{\text{rec}}) - PW(D_b)] - \frac{PW(ATC)}{1 - t}$$

However, that is not the complete effect, because during the life of the investment the Investment Tax Credit Reserve is a temporary source of capital repayment. Recall from the normalization development that the quantity $PW(C_{\text{rec}})$ in the preceding equation should represent the capital-recovery cost. The Investment Tax Credit Reserve is the original tax credit, less the amortization of the credit, or

$$\text{Investment Tax Credit Reserve in year } n = TC - \sum_{j=1}^{n} ATC_n$$

This quantity is the initial tax credit and reduces to zero by the end of the life. Therefore, it provides an intermediate source of capital repayment. Since there is a positive balance in the reserve during the life, the present worth of that reserve is not zero, and the present worth of the capital-recovery cost with service-life flow through $[PW(C_{\text{rec}})_{\text{SLFT}}]$ is

$$PW(C_{\text{rec}})_{\text{SLFT}} = PW(C_{\text{rec}}) - [PW(TC) - PW(ATC)]$$

where

$$PW(C_{\text{rec}}) = PW(\text{first cost}) - PW(\text{net salvage}) \text{ and } [PW(TC)$$
$$- PW(ATC)] = PW(TC \text{ reserve})$$

Like the normalization development, the reserve affects both parts of the present worth of capital costs. Therefore, the combined effect of investment tax credit and service-life flow through on the present worth of capital costs is

$$PW(\text{capital cost}) = PW(C_{\text{rec}})_{\text{SLFT}} + \phi[PW(C_{\text{rec}})_{\text{SLFT}} - PW(D_b)] - \frac{PW(ATC)}{1 - t}$$

$$PW(\text{capital cost}) = PW(C_{\text{rec}}) - [PW(TC) - PW(ATC)]$$

$$+ \phi\{PW(C_{\text{rec}}) - [PW(TC) - PW(ATC)] - PW(D_b)\} - \frac{PW(ATC)}{1 - t}$$

To keep the effects of the investment tax credit entirely within the tax component of capital costs, the $PW(T)$ is then

$$PW(T) = PW(\text{capital cost}) - PW(C_{\text{rec}})$$

Then the combined present worth of tax effects is

$$PW(T) = -[PW(TC) - PW(ATC)]$$

$$+ \phi\{PW(C_{rec}) - [PW(TC) - PW(ATC)] - PW(D_b)\} - \frac{PW(ATC)}{1 - t}$$

Rearranging terms yields

$$PW(T) = \phi[PW(C_{rec}) - PW(D_b)] - (1 + \phi)[PW(TC) - PW(ATC)] - \frac{PW(ATC)}{1 - t} \qquad (9.3)$$

The last two terms represent the effects of the investment tax credit and service-life flow through. As with equations (9.1) and (9.2), the first term in equation (9.3) is the basic tax liability. It is important to keep in mind that the effects of accelerated depreciation and of investment tax credit are independent of each other. Therefore, as Chart 9-1 shows, the complete expression for taxes is actually a combination of several elements. The basic tax liability is first followed by an adjustment from equation (9.1) or equation (9.2) for accelerated depreciation; and, finally, the adjustment for investment tax credit from equation (9.3). Again, equation (9.3) can be converted into the annual cost form if the $AC(\ )$ operators are substituted for the $PW(\ )$ operators.

In equation (9.3) the notation $PW(TC)$ represents the present worth of the actual tax credit. The tax reduction will be realized some time after the investment is made. For most study situations, it can be assumed that the entire credit occurs at the end of the year in which the investment is made. This means that the tax credit and the first amortization amount of that credit coincide in time; also, no capital repayment from the credit is available until the reduced tax payment is made, which is assumed to occur at that same time. The present worth of the tax credit is then

$$PW(TC) = TC(P/F, i\%, 1)$$

This assumption is consistent with the other assumptions about capital costs, but it means that it must be considered in the $AC(TC)$ operation. The equivalent annual cost of a tax credit occurring in 1 year is then

$$AC(TC) = TC(P/F, i\%, 1)(A/P, i\%, N)$$

This is consistent with the present-worth analysis.

Example 9-2: Accelerated Depreciation and Investment-Tax-Credit Effects. Find the present worth of the capital costs for the investment in Example 9-1, including the effects of an investment tax credit.

Recall the data of Example 9-1,

Total installed first cost	$10,000
Gross salvage	$1,500
Cost of removal	$500
Cost of money	12%
Estimated service life (N)	20 years
ϕ	0.67
t	53%
Tax life	16 years
ADR System (hybrid) with normalization	
Also assume 10% investment tax credit with service-life flow through ($10,000 × 0.10)	$1,000

In Example 9-1, the *PW* of the capital costs of this same example was found without a consideration of the investment tax credit. The results were

$$PW(\text{capital costs}) = PW(C_{\text{rec}}) + PW(T)$$
$$= \$9{,}896 + \$2{,}576$$
$$= \$12{,}472$$

Although investment tax credit must also be considered now, the present worth of capital recovery will remain unchanged. However, the present worth of income tax costs must be adjusted for the investment-tax-credit effects, in addition to the effects of ADR with normalization.

The $PW(T)$ in Example 9-1 was found from equation (9.1). To include the effects of investment tax credit, equation (9.3) must be combined with equation (9.1). The $PW(T)$ is, therefore,

$$PW(T) = \phi\,[PW(C_{\text{rec}}) - PW(D_b)] - t(1 + \phi)[PW(D_t) - PW(D_b)]$$
$$- (1 + \phi)\,[PW(TC) - PW(ATC)] - \frac{PW(ATC)}{1 - t}$$

Most of the components were evaluated in Example 9.1. Now, the $PW(TC)$ and the $PW(ATC)$ components must be found in order to determine the effect of the investment tax credit.

$$PW(ATC) = (TC/N)(P/A,\ 12\%,\ 20)$$
$$= (1{,}000/20)(7.469)$$
$$= \$373$$

$$PW(TC) = (TC)(P/F,\ 12\%,\ 1)$$
$$= (1{,}000)(0.8929)$$
$$= \$893$$

NOTE: The assumption here is that the investment tax credit is actually realized 1 year after the investment is made.

The $PW(T)$ is then found by combining the results from Example 9-1 with the above investment-tax-credit effects. From Example 9-1,

$$\phi[PW(C_{\text{rec}}) - PW(D_b)] - t(1 + \phi)[PW(D_t) - PW(D_b)] = \$2{,}576$$

and from the tax credit calculations above,

$$-(1 + \phi)[PW(TC) - PW(ATC)] = -1.67(893 - 373) = \ -868$$
$$-\frac{PW(ATC)}{1 - t} = -\frac{373}{1 - 0.53} \qquad\qquad = \ \underline{-794}$$
$$PW(T) = \$\ \ 914$$

Finally, the total present worth of capital costs is:

$$PW(\text{capital costs}) = PW(C_{\text{rec}}) + PW(T)$$
$$= \$9{,}896 + \$914 = \$10{,}810$$

The levelized equivalent annual cost is:

$$AC(\text{capital costs}) = PW(\text{capital costs})(A/P,\ 12\%,\ 20)$$
$$= (\$10{,}810)(0.13388)$$
$$= \$1{,}447\ \text{per year}$$

Immediate Flow Through

Although, the investment tax credit is realized at the time the investment is made, an accounting treatment is used which spreads the impact over the life of the plant. (This procedure is designed to match cost and revenues appropriately, because earnings arise from the use of plant facilities and not from their purchase.)

The tax law prohibits the immediate flow through of investment tax credits by the Bell System. However, the assumption of immediate flow through for an economy study introduces only a small error; yet it simplifies the math greatly and is therefore a useful approximation.

If immediate flow through is assumed, the tax equation becomes

$$T = t(\text{return} + T - \text{debt interest}) - TC$$

Solving for T,

$$(1 - t)T = t(\text{return} - \text{debt interest}) - TC$$

$$T = \frac{t}{1 - t}(\text{return} - \text{debt interest}) - \frac{TC}{1 - t}$$

Therefore, the present-worth form is

$$PW(T) = \phi[PW(C_{\text{rec}}) - PW(D_b)] - \frac{PW(TC)}{1 - t} \tag{9.4}$$

In Example 9-2, this approximation would have found that the effect of investment tax credit was

$$\frac{PW(TC)}{1 - t} = \frac{\$893}{1 - 0.53} = \$1,900$$

$$PW(T) = \$2,576 - \$1,900 = \$676$$

$$PW(\text{capital cost}) = \$9,896 + \$676 = \$10,572$$

The present worth of the capital costs is only 2% less than the more accurate result ($10,810) in Example 9-2. A lower rate for investment tax credit would result in even less error.

This effect of an investment tax credit with immediate flow through is also shown in Chart 9-1 on page 228.

INCOME TAX EFFECTS—FLOW DIAGRAM

This chapter has presented formulas to represent various effects on income taxes, assuming either normalization or flow through. These effects were related to accelerated depreciation and the investment tax credit. Since the math tends to become abstract, it would be helpful to visualize what is happening with a flow diagram.

Figure 9-8 is an expansion of a section of the cash flow diagram used in Chapter 6. (That diagram is reproduced here as Figure 9-7). This section of the diagram represents the capital-cost portion of the revenue requirements. The items flowing to the left are kept in the company to help fund the construction program. Besides book depreciation and reinvested earnings, deferred income taxes and net investment tax credits are shown too. Deferred income taxes are indicated as a function of the difference between D_t and D_b, or $t(D_t - D_b)$. They reduce the net income in the normalization process, and become internally generated funds contributing to the construction program.

Investment tax credit appears as a direct reduction in income taxes before it too crosses over to the left to help finance the construction program. A small increase in net income is shown as portions of the investment tax credit are flowed through in the amortization process. The sum of taxes paid, taxes deferred, and the net of investment tax credits is carried on the income statement as total operating taxes.

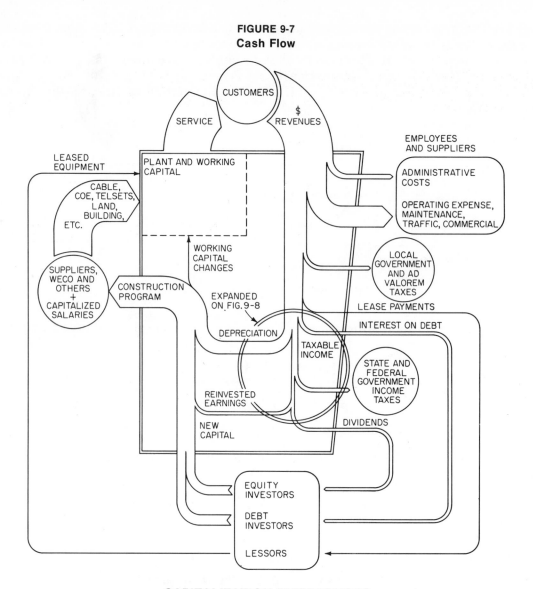

FIGURE 9-7
Cash Flow

CAPITALIZATION DIFFERENCES

The tax effects discussed so far in this chapter are based on the assumption that the entire first cost of plant is capitalized and depreciated over the life. This assumption is correct for the company's financial records; however, for tax depreciation, certain components of the first cost are not considered as depreciable investments.

This kind of difference affects the investment tax credit, too, because the amount of investment tax credit is determined by the amount of the first cost which the tax regulations consider to be capital. This point was not emphasized during the discussion of investment tax credit, because most studies can assume that the entire first cost is capitalized. However, there can be large differences between the amounts capitalized in labor-intensive accounts, such as station connections, and in investments with long construction periods. In studies involving such plant, the tax effects resulting from these differences should be recognized.

First-Cost Components Expensed for Income Tax Purposes

The total installed first cost of an investment may include a substantial amount of operating company installation and engineering costs. These costs consist largely of wages and salaries of the operating-company personnel involved. They include appropriate prorations of the following company expenditures:

1. Amounts paid into the pension trust fund which provide for payment of service pensions

2. Amounts paid into a second, separate pension trust fund which provides for payment of mandatory death benefits

3. Amounts paid for other employee benefits, including the following:
   a. Group life insurance plan
   b. Basic medical expense plan
   c. Extraordinary medical expense plan
   d. Special medical expense plan
   e. Sickness disability benefits
   f. Disability pensions
   g. Accident disability expenses

4. Taxes which include:
   a. State and federal unemployment taxes which are paid only by the telephone companies.
   b. That portion of the Federal Insurance Contribution Act taxes (Social Security) which the companies pay.

FIGURE 9-8
Expansion of Cash Flow

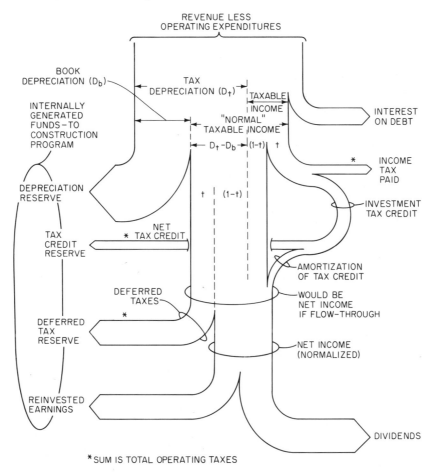

*SUM IS TOTAL OPERATING TAXES

These "loadings" are included in the plant's first cost to the extent that engineering and craft time is charged to asset-type accounts when plant is engineered and installed. As a result, they are capitalized on the books of the company and depreciated over the investment's life as required by the Uniform System of Accounts. However, the tax law allows these loadings as tax-deductible expenses in the year in which they occur. (For a study, it will be accurate enough to assume that these costs are tax deductions coincident with the first tax-depreciation amount.) In effect, part of the first cost is therefore being written off as a tax deduction early in the service life instead of being spread out over the entire period.

However, the tax effects of advancing these tax deductions (relative to their depreciation on the books) are flowed through to net income. At the same time, the tax effects resulting from the use of accelerated depreciation on the major part of the investment are normalized. The easiest way to visualize these income tax effects is to consider each investment in two parts. Equation (9.1) gives the tax effects of the part which is depreciated both on the books and for the tax return. Equation (9.2) gives the tax effects on the remainder of the investment. The total tax effect is the sum of the two parts, plus any investment-tax-credit effect from equation (9.3).

The portion of the installed first cost which is allowed to be depreciated for tax purposes is called the *tax basis*. It is equal to the total installed first cost, the *book basis,* less the tax-deductible items and any disallowed costs. To determine the deferred taxes to be normalized, a portion of the actual book depreciation is used. This is known as *book-tax depreciation* (D_{bt}). It is the amount of book depreciation which would occur if the tax basis were actually the investment. Therefore,

$$PW(D_{bt}) = \frac{\text{tax basis} - \text{net salvage}}{N} (P/A, i\%, N)$$

where N is the expected service life. If the portion of first cost which is considered tax expense is designated E_t, then

$$\text{Tax basis} = \text{first cost} - E_t$$

The tax depreciation is found as illustrated earlier in this chapter, but the tax basis, instead of the first cost, is used as the investment. The tables in Appendix C at the end of the book can still be used to find $PW(D_t)$, since

$$PW(D_t) = (\text{tax basis})(P/D_t)$$
or
$$AC(D_t) = (\text{tax basis})(A/D_t)$$

The $PW(T)$ for the tax-basis portion of the investment is found from equation (9.1), ignoring the investment-tax-credit effects.

$$PW(T) = \phi[PW(C_{\text{rec}}) - PW(D_b)] - t(1 + \phi)[PW(D_t) - PW(D_b)]$$

$$PW(T \text{ on tax basis}) = \phi[PW(C_{\text{rec}} \text{ of tax basis}) - PW(D_{bt})] - t(1 + \phi)[PW(D_t) - PW(D_{bt})]$$

For the other portion of the investment, the $PW(E_t)$ is the tax deduction, and has the same effect as tax depreciation (D_t). The difference between the book depreciation (D_b) and the book-tax depreciation (D_{bt}) is the portion of book depreciation which can be attributed to the E_t. In other words,

$$PW(D_b \text{ of } E_t) = PW(D_b) - PW(D_{bt})$$

The remaining tax effects are then found from equation (9.2).

$$PW(T) = \phi[PW(C_{\text{rec}}) - PW(D_b)] - \frac{t}{1 - t}[PW(D_t) - PW(D_b)]$$

$$PW(T \text{ on } E_t) = \phi\{PW(C_{\text{rec}} \text{ of } E_t) - [PW(D_b) - PW(D_{bt})]\}$$

$$- \frac{t}{1 - t}\{PW(E_t) - [PW(D_b) - PW(D_{bt})]\} \quad (9.2)$$

The total present worth of taxes is then the sum of these two parts.

$$PW(T) = \phi[PW(C_{rec}) - PW(D_b)] - t(1 + \phi)[PW(D_t) - PW(D_{bt})]$$

$$- \frac{t}{1-t}\{PW(E_t) - [PW(D_b) - PW(D_{bt})]\} \quad (9.5)$$

Chart 9-1 shows equation (9.5) for tax effects from both normalization and flow-through when the book basis and the tax basis are not equal. It is the most general of all the equations for accelerated depreciation. It can also be used for other effects which are covered in the remainder of this chapter.

Example 9-3 illustrates the application of equation (9.5). For simplicity, this example neglects the effects of investment tax credit.

Example 9-3. Find the present worth of the capital costs for an investment with the following characteristics.

Total installed first cost (includes $30,000 of costs which will be expensed for income-tax purposes)	$130,000
Gross salvage	$5,000
Cost of removal	$3,000
Estimated service life	20 years
Cost of money	12%
t	48%
ϕ	0.6115
Tax life	16 years
ADR System (hybrid) with normalization	

$$
\begin{aligned}
PW(C_{rec}) &= PW(\text{first cost}) - PW(\text{net salvage}) \quad (8.2)\\
&= 130{,}000(P/F, 12\%, 0) - 2{,}000(P/F, 12\%, 20)\\
&= 130{,}000(1) - 2{,}000\,(0.1037)\\
&= \$130{,}000 - 207\\
&= \$129{,}793
\end{aligned}
$$

$$PW(T) = \phi[PW(C_{rec}) - PW(D_b)] - t(1 + \phi)[PW(D_t) - PW(D_{bt})]$$

$$- \frac{t}{1-t}\{PW(E_t) - [PW(D_b) - PW(D_{bt})]\} \quad (9.5)$$

$$
\begin{aligned}
PW(D_b) &= \frac{130{,}000 - 2{,}000}{20}(P/A, 12\%, 20)\\
&= 6{,}400(7.4694)\\
&= \$47{,}804
\end{aligned}
$$

$$
\begin{aligned}
PW(D_{bt}) &= \frac{\text{tax basis} - \text{net salvage}}{20}(P/A, 12\%, 20)\\
&= \frac{130{,}000 - 30{,}000 - 2{,}000}{20}(7.4694)\\
&= \frac{100{,}000 - 2{,}000}{20}(7.4694)\\
&= \$36{,}600
\end{aligned}
$$

$$PW(D_t) = (\text{tax basis}) (P/D_t)$$

$$\% \text{ gross salvage} = \frac{\text{gross salvage}}{\text{tax basis}} = \frac{5,000}{100,000}$$

$$= 0.05$$

$$= 5\%$$

$$\% \text{ cost of removal} = \frac{\text{cost of removal}}{\text{tax basis}} = \frac{3,000}{100,000}$$

$$= 0.03$$

$$= 3\%$$

$$(P/D_t) = \frac{0.55590 + 0.54553}{2} + (0.03)(0.10367)^*$$

$$= 0.55383$$

$$PW(D_t) = \$100,000(0.55383)$$

$$= \$55,383$$

$$PW(E_t) = E_t(P/F, 12\%, 1)$$

$$= \$30,000(0.8929)$$

$$= \$26,787$$

$$PW\,(T) = 0.6115(\$129,793 - \$47,804) - 0.48\,(1 + 0.6115)(\$55,383 - \$36,600)$$

$$- \frac{0.48}{1 - 0.48}\,[\$26,787 - (\$47,804 - \$36,600)]$$

$$= \$50,136 - \$14,529 - \$14,384$$

$$= \$21,223$$

$$PW(\text{capital cost}) = PW(C_{\text{rec}}) + PW(T)$$

$$= \$129,793 + \$21,223$$

$$= \$151,016$$

First-Cost Components Not Recognized for Income Tax Purposes

Interest during Construction

Because many undertakings often require a lengthy construction interval, the business must get and spend some of the funds necessary for acquiring an asset well before it can actually be placed into service. Accounting handles this situation by first recording such expenditures in Account 100.2, Plant Under Construction; then, it transfers the expenditure to Account 100.1, Plant in Service, when the plant is actually put in service. However, investors require a return on their capital whether it is invested in plant in service or not. As a result, there is a return obligation on these funds during the construction period. Of course, the accountants are concerned with the appropriate allocation of the costs of business. There is a difference of opinion concerning what is appropriate. Many utilities have been directed to allocate the return obligation on investments in plant not yet in service to the time the plant is actually in service. *Interest during construction* (IDC) is the means that accountants use to recognize and allocate this return obligation. The term is somewhat of a misnomer, however. It is

---

*From Figure 9-5.

not really interest; it is a recognition of a return requirement. Other references may call it by a different name. *Allowance for funds used during construction* (AFDC) is a popular term and is more descriptive.

The IDC is analogous to the increase in the mortgage balance made by a mortgage holder for the interest owed in case of a missed or delayed payment. During the time the funds are in the Plant Under Construction account (100.2), a return obligation is recognized and accrued; this return is then included in the amount finally transferred to Plant in Service (100.1). The result is that the amount customers eventually pay for the service includes the costs incurred because investor's funds were tied up during the construction of plant.

Since Plant Under Construction is a nondepreciable account, the capital-repayment portion of the capital cost is deferred on these investments until their cost is transferred to Account 100.1. In many telephone companies, Plant Under Construction is not included in the regulator's rate base; therefore, there are no earnings and no income taxes incurred until the transfer is made. In other words, all capital costs, repayment, return, and income taxes are deferred until the plant is actually in service. Some utilities do include plant under construction in their rate base, thus negating the requirement for interest during construction, by actually having a return requirement during the construction period.

In most companies, interest during construction serves another purpose. Although funds used for plant under construction can be excluded from the regulator's earnings base, the stock and/or bonds which provided those funds cannot be excluded from the capital on which investors expect a return. The IDC provides for the eventual return on this capital; but unless IDC is also included in income, the income statement earnings on capital are understated. For this reason, the Uniform System of Accounts requires that IDC be added to the after-tax income so that the firm's condition can be expressed to its investors more accurately.

The IDC is the accounting and regulatory way of recognizing the time value of money. Because engineering economy studies explicitly recognize the time value of money, they should exclude any IDC amounts and work with the actual cash expenditures.

Unfortunately, when the installed first cost of some asset is taken from accounting data, it often includes a component for IDC because the cost is recorded as of the time of the transfer to Account 100.1, Plant in Service. Little or no information may be available about when the capital funds were committed, or how much was committed. In the absence of such information, the economy study is forced to consider IDC. In this case, present-worth computations result which are not related to the time the capital was committed, but to a later time when the plant is placed in service. Furthermore, since the IDC rate is normally less than the cost of capital used for the economy study, and because IDC is not usually compounded, the use of the accounting data will underestimate the actual obligation to investors.

The IDC is included in the installed first cost (book basis), and it is capitalized and depreciated for book purposes. It is not recognized as a tax expense, and may not be capitalized for tax purposes. It is not depreciable for tax purposes, and it is not included in the tax basis when the allowance for investment tax credit is computed. Therefore, equation (9.5) can be used to recognize the effects by not including IDC in the tax basis. Suppose, for example, that the $130,000 of the installed first cost in Example 9-3 had included $5,000 of IDC besides the $30,000 of capitalized tax expense.

The present worth of the capital-recovery costs would not change; the accountants have simply done part of the work already, by including IDC. The first cost they give is their approximation of the single amount equivalent to the actual expenditures at the IDC rate.

$$PW(C_{\text{rec}}) = \$129,793$$

The present worth of book depreciation is not changed either. That is,

$$PW(D_b) = \$47,804$$

And the present worth of the tax-deductible, but capitalized, expenditures is not changed.

$$PW(E_t) = \$26,787$$

However, the tax depreciation and the book-tax depreciation do change because the tax basis is changed.

$$\text{Tax basis} = \text{book basis} - E_t - \text{IDC}$$
$$= \$130,000 - \$30,000 - \$5,000$$
$$= \$95,000$$

$$PW(D_t) = (\text{tax basis})\,(P/D_t)$$

$$\% \text{ gross salvage} = \frac{\text{gross salvage}}{\text{tax basis}} = \frac{\$5,000}{\$95,000}$$
$$= 0.0526$$
$$= 5.26\%$$

$$\% \text{ cost of removal} = \frac{\text{cost of removal}}{\text{tax basis}} = \frac{\$3,000}{\$95,000}$$
$$= 0.0316$$
$$= 3.16\%$$

$$(P/D_t) = 0.55590 - (0.55590 - 0.54553)\frac{0.0526}{0.10} + 0.0316(0.10367)^*$$
$$= 0.55372$$

$$PW(D_t) = \$95,000(0.55372)$$
$$= \$52,603$$

$$PW(D_{bt}) = \frac{\text{tax basis} - \text{net salvage}}{20}(P/A, 12\%, 20)$$
$$= \frac{95,000 - 2,000}{20}(7.4694)$$
$$= \$34,733$$

The present worth of taxes is then found using equation (9.5).

$$PW(T) = \phi[PW(C_{\text{rec}}) - PW(D_b)] - t(1 + \phi)\,[PW(D_t) - PW(D_{bt})]$$
$$- \frac{t}{1 - t}\{PW(E_t) - [PW(D_b) - PW(D_{bt})]\} \quad (9.5)$$

$$PW(T) = 0.6115(\$129,793 - \$47,804) - 0.48(1 + 0.6115)\,[\$52,603 - \$34,733]$$
$$- \frac{0.48}{1 - 0.48}\,[\$26,787 - (\$47,804 - \$34,733)]$$
$$= \$50,136 - \$13,823 - \$12,661$$
$$= \$23,652$$

$$PW(\text{capital cost}) = \$129,793 + \$23,652$$
$$= \$153,445$$

*From Figure 9-5, page 200.

Example 9-3 found that the $PW(T)$ was \$21,223 without recognizing the effect of IDC. The reason for the increase in $PW(T)$ resulting from IDC is the need to recover this cost without the shelter of a tax deduction.

Western Electric Company Tax Deferral

GENERAL   Under current federal income tax regulations, affiliated companies which are owned 80% or more by other companies within the group may file a single, consolidated income tax return. Except for the Southern New England Company and Cincinnati Bell, all the Bell System Companies qualify (including Western Electric and its subsidiaries). As a result, most double taxation is eliminated, since subsidiary companies are not taxed separately from the parent AT&T. Consequently, the profits on Western's sales of plant to the consolidated companies are not taxable in the year of the sale. Instead, they are excluded from the investment base used by the consolidated companies in computing the depreciation allowable for tax purposes.

The overall result is that income taxes on Western's profits for a specific item of plant are deferred over the plant's life through reduced tax-depreciation deductions. These reductions are equal to the original profit before taxes. The actual tax amount involved is the income tax rate times the Western pretax profit. This amount is what is called the *Western Electric Company tax deferral*. Instead of paying this tax directly at the time of sale, Western passes it on (through AT&T) to the purchasing company. The company then uses it to reduce the plant's investment on the books. The reason for this transaction is that "profit" is not created by intracorporate transfers. "Profit" is created only when revenues are received, and this occurs over the life of the plant; therefore, the tax should be payable over the life.

Here is an example. Assume an operating company buys telephone equipment for \$1,000, including a \$100 Western profit before taxes. The equipment is put into service with a total installed first cost of \$1,400. This total also goes on the books of the company.

At the time of the sale, there is a tax liability on the Western profit. Assuming a federal income tax rate of 48%, it would equal \$100 × 0.48, or \$48. Because Western and the operating company are using a consolidated income tax return, the \$48 tax is not paid to the federal government at the time of sale. Instead, it is passed on through AT&T to the operating company as a \$48 reduction in the \$1,400 recorded on the company's books. Thus, \$1,400 − \$48, or \$1,352, would be depreciable for book purposes instead of \$1,400 (assuming 0 net salvage). For tax purposes, however, the depreciable amount would be \$1,400 − \$100, or \$1,300. Over the life of the equipment, this would increase the federal income taxes paid by \$100 × 0.48, or \$48. So instead of paying \$48 tax in the form of a higher purchase price, the operating company defers it to future years through reduced depreciation deductions for tax purposes.

The Western Electric Company reports the amount of tax and profit on its sales to the operating companies several times a year, so that the accounting entries can be made. Because the tax basis of property is reduced by the pretax profit (profit plus tax), the investment tax credit and tax depreciation are also reduced. The total effect of the Western Electric Company tax deferral is then a complex series of partially offsetting events.

ECONOMY STUDIES   The Western Electric Company tax deferral will result in a small reduction in the capital costs associated with any particular investment. This reduction is partially offset by the reduction in investment tax credit and tax depreciation. When the analyst is studying alternatives involving only Western equipment, this Western tax deferral can be ignored because it rarely causes significant differences between plans. But if the analyst is comparing Western equipment with equipment of an outside supplier, the significance increases somewhat (though it is still small), since the Western tax deferral of course does not apply to the outside supplier's equipment.

If it is necessary to determine the effect of this tax deferral, equation (9.5) and the adjustment for investment tax credit from equation (9.3) can be used. In equation (9.5), however, the $PW(C_{rec})$ is found as if the first cost were reduced by the tax deferral.

However, if the PW(capital costs) has already been determined with the Western Electric deferred tax neglected, the effect can be approximated with the following adjustment.

$$\text{Reduction in } PW(\text{capital cost}) = (1 + \phi)(\text{tax deferral}) \left[1 - (P/D_t) - \frac{TC \text{ rate}}{t_w} \right]$$

where (P/D_t) = present worth of tax-depreciation factor
$\quad TC$ rate = investment tax credit rate
$\quad t_w$ = Western Electric tax rate

The adjustment is derived by subtracting the entire expression for capital costs with the tax deferral from the expression without it. Several terms then combine and cancel; and several terms almost cancel, leading to the approximation.

COMPREHENSIVE EXAMPLE

The following example illustrates the computation of PW(capital costs) with all of the income tax considerations in this chapter included. Assume the purchase of plant with the following estimates.

Example 9-4

Western Electric price	$100,000
includes 3% WECO tax deferral	
Telco charges	50,000
includes 20% loading for social	
security, relief, and pension costs	
Shipping costs	10,000
Sales taxes	5,000
Interest during construction	5,000
Total installed first cost	$170,000

Expected service life	20 years
Gross salvage	$0
Cost of removal	$5,000
Tax Life	16 years
ADR System (hybrid) with normalization	
Investment tax credit rate	10%
Cost of money	12%
Debt interest rate	9%
Debt ratio	45%
Composite income-tax rate	48%

$$\text{Book basis} = \text{first cost} - \text{WECO tax deferral}$$

$$\text{WECO tax deferral} = \$100,000 \times 0.03 = \$3,000$$

$$\text{Book basis} = \$170,000 - \$3,000 = \$167,000$$

$$PW(C_{\text{rec}}) = PW\,(\text{book basis}) - PW\,(\text{net salvage})$$

$$\text{Net salvage} = \text{gross salvage} - \text{cost of removal}$$

$$= 0 - \$5,000 = -\$5,000$$

$$PW(C_{\text{rec}}) = \$167,000(P/F,\,12\%,\,0) - (-\$5,000)\,(P/F,\,12\%,\,20)$$

$$= \$167,000(1) + \$5,000(0.1037)$$

$$= \$167,519$$

$$PW(T) = \phi[PW(C_{\text{rec}}) - PW(D_b)] - t(1 + \phi)\,[PW(D_t) - PW(D_{bt})] - \frac{t}{1-t}\{PW(E_t)$$

$$- [PW(D_b) - PW(D_{bt})]\} - (1 + \phi)[PW(TC) - PW(ATC)] - \frac{1}{1-t}\,PW(ATC)$$

$$PW(D_b) = \frac{\text{book basis} - \text{net salvage}}{N}(P/A,\,i\%,\,N)$$

$$= \frac{\$167,000 - (-\$5,000)}{20}(P/A,\,12\%,\,20)$$

$$= 8,600(7.469)$$

$$= \$64,233$$

$$PW(D_t) = (\text{tax basis})(P/D_t)$$

$$\text{Tax basis} = \text{first cost} - E_t - \text{IDC} - \text{WECO pretax profit}$$

$$E_t = 50,000 \times 0.20$$

$$= \$10,000$$

$$(\text{WECO pretax profit})t = \text{WECO tax deferral}$$

$$\text{WECO pretax profit} = \frac{\$3,000}{0.48} = \$6,250$$

$$\text{Tax basis} = \$170,000 - \$10,000 - \$5,000 - \$6,250$$

$$= \$148,750$$

$$\%\text{ gross salvage} = \frac{\text{gross salvage}}{\text{tax basis}}$$

$$= \frac{0}{\$148,750}$$

$$= 0\%$$

$$\%\text{ cost of removal} = \frac{\text{cost of removal}}{\text{tax basis}}$$

$$= \frac{\$5,000}{\$148,750} = 0.0336$$

$$= 3.36\%$$

$$(P/D_t) = 0.55590 + 0.10367(0.0336)$$

$$= 0.55938$$

$$PW(D_t) = \$148,750(0.55938)$$

$$= \$83,208$$

$$PW(D_{bt}) = \frac{\text{tax basis} - \text{net salvage}}{N}(P/A,\,i\%,\,N)$$

$$= \frac{\$148,750 - (\$-5,000)}{20}(P/A,\,12\%,\,20)$$

$$= \$7,687.50(7.469)$$
$$= \$57,418$$

$$PW(E_t) = \$10,000(P/F, 12\%, 1)$$
$$= \$10,000(0.8929)$$
$$= \$8,929$$

$$PW(TC) = (\text{tax basis})(TC \text{ rate})(P/F, 12\%, 1)$$
$$= \$148,750(0.10)(0.8929)$$
$$= \$13,282$$

$$PW(ATC) = \frac{(\text{tax basis})(TC \text{ rate})}{N}(P/A, i\%, N)$$

$$= \frac{\$148,750(0.10)}{20}(P/A, 12\%, 20)$$

$$= \$743.75(7.469)$$
$$= \$5,555$$

$$\phi = \frac{t}{1-t}\left[1 - \frac{ri_d}{i}\right]$$

$$= \frac{0.48}{1-0.48}\left[1 - \frac{0.45 \times 0.09}{0.12}\right]$$

$$= 0.6115$$

$$PW(T) = 0.6115(\$167,519 - \$64,233) - 0.48(1 + 0.6115)(\$83,208 - \$57,418)$$

$$- \frac{0.48}{1-0.48}[\$8,929 - (\$64,233 - \$57,418)]$$

$$- (1 + 0.6115)(\$13,282 - \$5,555) - \frac{1}{1-0.48}(\$5,555)$$

$$= \$63,159 - \$19,949 - \$1,951 - \$12,452 - \$10,683$$

$$= \$18,124$$

$$PW(\text{capital cost}) = PW(C_{\text{rec}}) + PW(T)$$
$$= \$167,519 + \$18,124$$
$$= \$185,643$$

FLOW DIAGRAM—CAPITALIZATION DIFFERENCES

Figure 9-8, an expanded flow diagram, reflected the simplifying assumption that the book basis and the tax basis were equal. The more general diagram, Figure 9-9, illustrates the same portion of the flow, but with the addition of the effect of differences between the book and tax bases.

Figure 9-9 is, in effect, a graphical representation of the complex tax equation used in the comprehensive example. Although not evident on the diagram, it does include the effects of WECO tax deferral and interest during construction. These effects are included because both they and the tax-deductible expenses (E_t) are reflected in the difference between book depreciation and book-tax depreciation.

APPROXIMATIONS

The calculations in the previous section are quite lengthy. Without the aid of a computer, however, it is often accurate enough to make some approximating assumptions

FIGURE 9-9
Capitalization Differences: Expansion of Cash Flow

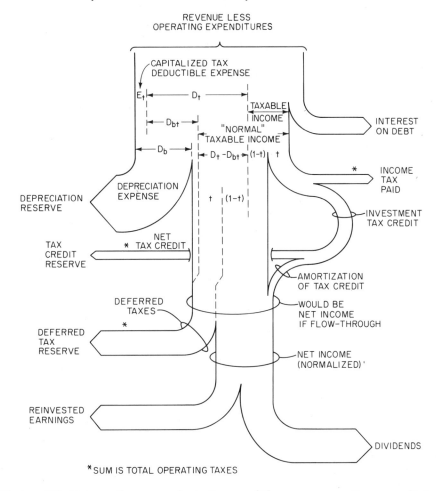

*SUM IS TOTAL OPERATING TAXES

which will simplify the math somewhat. Some of these assumptions are listed, more or less in the order of their impact on the results, from the least to the most.

- Ignore the effects of the WECO tax deferrals.

- Ignore the effects of interest during construction.

- Assume 100% normalization of tax deferrals.

- Assume immediate flow through of tax credits.

- Assume book basis equals tax basis.

- Ignore investment tax credit.

- Ignore accelerated tax depreciation.

If all of these assumptions are made, the problem is reduced to the basic income tax equation in Chapter 8, which is probably too crude to be very useful. However, by making only the first four assumptions, the approximation is quite close and the math is considerably simplified. This is because when 100% normalization is assumed, the $t/(1 - t)$ term in equation (9.5) is replaced by $t(1 + \emptyset)$, and the equation is simplified.

The general equation is

$$PW(T) = \phi[PW(C_{rec}) - PW(D_b)] - t(1 + \phi)[PW(D_t) - PW(D_{bt})]$$

$$- \frac{t}{1 - t}\{PW(E_t) - [PW(D_b) - PW(D_{bt})]\} \quad (9.5)$$

When $t/(1 - t)$ is replaced with $t(1 + \emptyset)$ to assume normalization of the tax effects of E_t, the result is

$$PW(T) = \phi[PW(C_{\text{rec}}) - PW(D_b)] - t(1 + \phi)[PW(D_t) + PW(E_t) - PW(D_b)] \qquad (9.6)$$

The effect of immediate flow through of investment tax credit is then found from the last term in equation (9.4).

$$PW(T) = \phi[PW(C_{\text{rec}}) - PW(D_b)] - \frac{PW(TC)}{1 - t} \qquad (9.4)$$

If the effects of WECO tax deferrals and IDC are ignored, then the complete expression for taxes under the four simplifying assumptions is

$$PW(T) = \phi[PW(C_{\text{rec}}) - PW(D_b)] - t(1 + \phi)[PW(D_t) + PW(E_t) - PW(D_b)] - \frac{PW(TC)}{1 - t}$$

Using this expression for taxes in the comprehensive example, Example 9-4 has the following result.

$$
\begin{aligned}
PW(C_{\text{rec}}) &= PW(\text{first cost}) - PW(\text{net salvage}) \\
&= \$170,000(P/F, 12\%, 0) - (-\$5,000)(P/F, 12\%, 20) \\
&= \$170,000(1) + \$5,000(0.1037) \\
&= \$170,519
\end{aligned}
$$

$$
\begin{aligned}
PW(D_b) &= \frac{\text{first cost} - \text{net salvage}}{N}(P/A, i\% \ N) \\
&= \frac{170,000 - (-5,000)}{20}(P/A, 12\%, 20) \\
&= 8,750(7.469) \\
&= \$65,354
\end{aligned}
$$

$$PW(D_t) = (\text{tax basis})(P/D_t)$$

Book basis = first cost = \$170,000

$$
\begin{aligned}
\text{Tax basis} &= \text{book basis} - E_t \\
&= 170,000 - 10,000 \\
&= \$160,000
\end{aligned}
$$

$$
\begin{aligned}
\% \text{ cost of removal} &= \frac{\text{cost of removal}}{\text{tax basis}} \\
&= \frac{5,000}{160,000} = 0.03125 \\
&= 3.125\%
\end{aligned}
$$

$$
\begin{aligned}
(P/D_t) &= 0.55590 + 0.10367(0.03125) \\
&= 0.55914
\end{aligned}
$$

$$
\begin{aligned}
PW(D_t) &= 160,000(0.55914) \\
&= \$89,462
\end{aligned}
$$

$$
\begin{aligned}
PW(E_t) &= 10,000(P/F, 12\%, 1) \\
&= 10,000(0.8929) \\
&= \$8,929
\end{aligned}
$$

$$
\begin{aligned}
PW(TC) &= (\text{tax basis})(TC \text{ rate})(P/F, 12\%, 1) \\
&= 160,000(0.10)(0.8929) \\
&= \$14,286
\end{aligned}
$$

$$PW(T) = 0.6115(170{,}519 - 65{,}354) - 0.48(1 + 0.6115)(89{,}462 + 8{,}929 - 65{,}354)$$

$$- \frac{14{,}286}{1 - 0.48}$$

$$= 64{,}308 - 25{,}555 - 27{,}473$$

$$= \$11{,}280$$

$$PW(\text{capital cost}) = PW(C_{\text{rec}}) + PW(T)$$
$$= 170{,}519 + 11{,}280$$
$$= \$181{,}799$$

The answer with these simplifying assumptions is $3,844 less than the more precise answer of $185,643, for an error of approximately 2%. A 2% error is often tolerable, considering that the errors in estimating values may be significantly greater.

OTHER TAX-RELATED EFFECTS

State Income Taxes

This chapter has assumed that a composite income tax rate can be used to determine the total income taxes. Many state governments do impose income taxes on corporations using regulations which are sufficiently similar to the federal regulations that total taxes can be found using a composite state and federal tax rate. However, certain states may have significant differences which may need to be recognized. In these states, the federal income tax should be found from the appropriate equation in this chapter, using a tax rate (t'_f) which is equal to the federal rate reduced to allow for the deduction of state taxes. This rate is dependent on the state regulations and can be found from the equations for the composite tax rate (t) in Chapter 8. In each case, it is the federal tax divided by the taxable income.

$$t'_f = \frac{\text{federal tax}}{\text{taxable income}}$$

The state taxes should be found using the appropriate equation, where the state tax rate is the composite rate less the rate used in the federal equation, or

$$t'_s = t - t'_f$$

Three conditions were illustrated in Chapter 8.

Condition A. State and federal taxes are not deductible for state taxes.

$$\text{Taxable income} = \$100$$
$$\text{Federal income tax} = \$43.20$$

$$t = 0.532$$
$$t'_f = \frac{\$43.20}{\$100} = 0.432$$
$$t'_s = 0.532 - 0.432 = 0.10$$

Condition B. Federal tax is deductible for state taxes.

$$\text{Taxable income} = \$100$$
$$\text{Federal income tax} = \$45.38$$

$$t = 0.5084$$

$$t_f' = \frac{\$45.38}{\$100} = 0.4538$$

$$t_s' = 0.5084 - 0.4538 = 0.0546$$

Condition C. State tax is deductible for state taxes.

$$\text{Taxable income} = \$100$$
$$\text{Federal income tax} = \$43.64$$

$$t = 0.5273$$

$$t_f' = \frac{\$43.64}{\$100} = 0.4364$$

$$t_s' = 0.5273 - 0.4364 = 0.0909$$

The sum of the federal income tax and the state income tax is the total income tax. Because the tax rates are different, the income tax factor (ϕ) will be different in the different equations. The company tax accountants will supply the applicable state tax regulations.

Income tax regulations in some states permit or require deferred income taxes to be administered on a flow-through basis instead of normalized as are deferred federal income taxes. Although this can be precisely reflected with the development of additional methods, such detail is unnecessary because the effect is negligible for economic comparison purposes.

Capital Gain Taxes on Land

This chapter has been discussing income taxes associated with return required on the capital invested in plant. The gain realized when property is sold for more than its original cost may be subject to a different tax treatment under the "capital gain tax" provisions of the tax regulations.

Because telephone plant is not usually acquired in anticipation of an increase in value, capital gains are seldom a consideration in an economy study. Realization of capital gains in the Bell System is very rare; and when it does occur, it generally involves the sale of land. This discussion is therefore oriented towards the sale of land. Analysts doing studies which anticipate an appreciation in value in other classes of plant should consult the depreciation engineers and tax accountants to determine how the gain should be treated.

The expression for the present worth of capital recovery has been shown to be

$$PW(C_{\text{rec}}) = PW(\text{first cost}) - PW(\text{salvage})$$

The present worth of the basic income taxes has been shown to be

$$PW(T) = \phi[PW(C_{\text{rec}}) - PW(D_b)]$$

With land, the various adjustments required because book and tax depreciation are unequal do not apply because land is a unique form of asset which is not depreciable. Land is expected to keep its value during its life in the firm, which means that all the capital repayment is expected from its eventual sale. The expression for taxes therefore simplifies to

$$PW(T) = \phi[PW(C_{\text{rec}})]$$

where the future sale price or salvage is exactly equal to the first cost. Adding the taxes and the capital recovery gives

$$PW(\text{capital cost}) = (1 + \phi) \, PW(C_{\text{rec}})$$

or $\quad PW(\text{capital cost}) = (1 + \phi) \, [\, PW(\text{first cost}) - PW \, (\text{first cost as salvage})]$

If the future sale price exceeds the first cost, there is a capital gain and perhaps a tax on that gain. If a gain is anticipated, the easiest way to include its effect is to first determine the capital costs as if future salvage were equal to the first cost, then adjust for the gain effects. With a capital-gain tax rate of t', the amount of the gain remaining after payment of the tax is

$$(1 - t')(\text{gain})$$

This flows through to net income, and similar to flow through, the result is a reduction in the capital costs equal to

$$\frac{1 - t'}{1 - t} \, (\text{gain})$$

Therefore, the present worth of the capital costs is

$PW(\text{capital cost})$

$$= (1 + \phi) \, [\, PW(\text{first cost}) - PW(\text{first cost as salvage})] - \frac{1 - t'}{1 - t} \, PW(\text{gain})$$

Here, the gain is the sale price less the first cost.

The analyst should consult the company tax accountants for the appropriate value of t'. This value should reflect the fact that capital gains are often offset by capital losses in the tax computation, in which case the effective value of t' is 0.

Effective Income Tax Rate

Most of the effects on basic income taxes are reflected in the corporate annual report as an effective federal income tax rate. Income taxes are actually computed as a percentage of taxable income. Taxable income does not appear on the income statement because of the different approaches the IRS and the financial community take in determining profit. The effective income tax rate is the rate which, if it is applied to the income-statement pretax income, would have resulted in the same taxes as those appearing in the financial report. Net income is the amount left after income taxes have been deducted from pretax income. Therefore,

$$\text{Pretax income} = \text{net income} + \text{income taxes}$$

$$(\text{Net income} + \text{income taxes}) \, (\text{effective income-tax rate}) = \text{income taxes}$$

$$\text{Effective income tax rate} = \frac{\text{income taxes}}{\text{net income} + \text{income taxes}}$$

For example, to find the effective income tax rate of AT&T, federal income taxes can be divided by the sum of federal income taxes, net income, and minority ownership interest in net income. The last term applies only to AT&T, and not to the affiliated companies, and is usually found and described in the notes to the financial statements contained in the Annual Report.

The calculation of the effective income tax rate, with figures in thousands of dollars drawn from the 1975 AT&T Annual Report, income statement, is as follows:

$$\frac{2{,}174{,}088^{a}}{2{,}174{,}088^{a} + 3{,}147{,}772^{b} + 77{,}027^{c}} = 0.403, \text{ or } 40.3\%$$

NOTES:

a. Total federal income taxes. (See Figure 9-3, page 196.)

b. Net income. (See Figure 9-3.)

c. Minority ownership interest in net income. This figure is found in the notes to the financial statements. It is a deduction from the Miscellaneous Income and Deductions line shown in Figure 9-3, and is added back in for the sake of the computation.

As explained in the notes to the financial statements, several factors determine the difference between the effective and the nominal (48% at that time) federal income tax rates. These include all of the effects discussed in this chapter. Although the effective income tax rate is a useful tool for corporate-wide analysis work, it should not be used for economy studies. The important element in economy studies is any difference in tax effects, and this can not be measured by use of an effective rate that is the result of overall operations.

SUMMARY

Chapter 8 introduced, and this chapter reviewed, the basic income tax formula:

$$PW(T) = \phi[PW(C_{rec}) - PW(D_b)] \tag{8.16}$$

Both Chapter 8 and this chapter have emphasized that the formula can be used only if the same method of capital repayment is assumed for both book and income tax purposes. In the past, the methods were essentially the same and employed straight-line depreciation. Since 1969, however, the Bell companies have been adopting advantageous forms of capital repayment for tax purposes. Since straight-line depreciation has continued to be used for book and regulatory purposes, the assumption of equal book and tax depreciation can no longer be considered even approximately accurate. A series of adjustments to the basic formula is therefore necessary. The objective of this chapter has been to explain, formulate, and apply these adjustments. They are now summarized.

The use of accelerated tax depreciation does not save any total taxes, but it does defer some into later years. These deferred taxes are included in the total operating taxes shown on the income statement to avoid overstating income by the amount deferred. This treatment is called *normalization,* which means that the taxes deferred are actually considered a source of capital repayment. While the taxes are not actually reduced, the present worth of the taxes is reduced with accelerated tax depreciation.

Equation (9.1) gives the present worth of the income tax costs, assuming accelerated depreciation for tax purposes with normalization of deferred taxes.

$$PW(T) = \phi[PW(C_{rec}) - PW(D_b)] - t(1 + \phi)[PW(D_t) - PW(D_b)] \tag{9.1}$$

Accelerated tax depreciation can take on many forms. This chapter explains most of them—including sum-of-the-years digits (SYD), double declining balance (DDB), short life, and combinations of these, such as the Asset Depreciation Range System (ADR). No doubt other forms of tax depreciation will develop in the future as tax regulations change. However, equation (9.1) will continue to hold if tax deferrals are normalized.

If tax deferrals are not normalized, they flow through to net income, and equation (9.2) must be used rather than equation (9.1).

$$PW(T) = \phi[PW(C_{rec}) - PW(D_b)] - \frac{t}{1 - t}[PW(D_t) - PW(D_b)] \tag{9.2}$$

The investment tax credit (*TC*) was also discussed in this chapter. This is a reduction in taxes, rather than a deferral, which was the case for accelerated depreciation. The amount of investment tax credit available because of a specific investment is a complex matter, dependent on the most current tax regulations.

However, if the amount of the investment tax credit is known for a particular investment, the adjustment in the present worth of income tax costs is often substantial. The adjustment expressed by equation (9.3) can be attached to either (9.1) or equation (9.2), depending on the accelerated tax-depreciation treatment.

$$TC \text{ effect} = -(1 + \phi)\,[PW(TC) - PW(ATC)] - \frac{PW(ATC)}{1 - t} \qquad (9.3)$$

where ATC = amortized tax credit = TC/N
      N = estimated service life

Equation (9.3) assumes that the investment tax credit is amortized over the service life. This is because only a portion of this credit is released each year to reduce income taxes and increase earnings (service-life flow through). In effect, flow through of the credit is deferred over the service life.

An often useful approximation for economy studies is to assume the immediate flow through of the investment tax credit. Then the adjustment to the present worth of income tax requirements is expressed by equation (9.4) rather than equation (9.3).

$$TC \text{ effect} = -\frac{PW(TC)}{1 - t} \qquad (9.4)$$

When an item of plant is put in service, the total installed first cost is capitalized on the books of the company and is depreciated on a straight-line basis for the financial reports. However, a portion of that first cost is often actually expensed for tax purposes during the year the investment is made. This portion includes such prorated cost items as pensions, employee insurance, social security, and unemployment taxes. This means that the basis for depreciation on the books may be different from the basis for tax depreciation. Assuming accelerated depreciation with normalization for tax purposes, along with the expensing of a portion of the first cost for tax purposes, the complete general formula for the present worth of taxes is given as equation (9.5).

$$PW(T) = \phi[PW(C_{\text{rec}}) - PW(D_b)] - t(1 + \phi)\,[PW(D_t)$$
$$- PW(D_{bt})] - \frac{t}{1 - t}\{PW(E_t) - [PW(D_b) - PW(D_{bt})]\} \qquad (9.5)$$

where E_t = portion of the total installed first cost capitalized on the books but expensed
        for income-tax purposes
      N = estimated service life
      D_{bt} = book depreciation on a tax basis

Several other factors also may affect the income tax costs.

Interest during construction (IDC) is the accountant's way of recognizing the return obligation on capital invested in plant not yet in service, and allocating that return requirement to the time the plant is actually in service. Although IDC is not normally a consideration in economy studies, it may be the only way to approximate unknown information on the timing of expenditures. With the correct interpretation of the amounts, equation (9.5) can also be used to handle IDC. The Western Electric Company tax deferral associated with a particular investment in plant, purchased from Western, results in a deferral of the tax obligation incurred on Western's profit on the item. This tax deferral applies only to associated companies that are included in the consolidated income statement. The advantages of this deferral are passed on to the associated company, producing a small reduction in the capital costs. However, the adjustment is usually small enough to be ignored in a comparative cost study, especially if the alternatives involve only Western equipment. Because the tax deferral is not available on plant purchased from outside suppliers, it is possible that the effect might be significant if equipment manufactured outside the Bell System is being

CHART 9-1
PRESENT WORTH OF INCOME TAX AND MAJOR ADJUSTMENTS

Basic tax liability	Accelerated tax depreciation		Investment tax credit
$D_b = D_t$　　　　(8.16)	tax basis　　=	Normalization:　　　　(9.1) $-t(1 + \phi)[PW(D_t) - PW(D_b)]$	Service-life flow through:　　　　(9.3)
		Flow through:　　　　(9.2) $-\dfrac{t}{1 - t}[PW(D_t) - PW(D_b)]$	$-(1 + \phi)[PW(TC) - PW(ATC)]$ $-\dfrac{PW(ATC)}{1 - t}$
	Book basis		
$\phi[PW(C_{rec}) - PW(D_b)]$	tax basis　　≠	Normalization and flow through (general case):　　　(9.5) $-t(1 + \phi)[PW(D_t) - PW(D_{bt})]$ $-\dfrac{t}{1 - t}\{PW(E_t) - [PW(D_b)$ 　　　　　　　　$- PW(D_{bt})]\}$	Immediate flow through:(9.4) $-\dfrac{PW(TC)}{1 - t}$
	Book basis	Normalize all (approximation):　　　　(9.6) $-t(1 + \phi)[PW(D_t) + PW(E_t) -$ 　　　　　　　　$PW(D_b)]$	

compared with Western equipment. An approximate formula was provided for this purpose, or equation (9.5) can also be used.

Which of the many income tax formulas applies in a particular study depends on the assumptions made in the study and on the current or anticipated tax regulations. Chart 9-1 illustrates the most significant formulas presented in this chapter and is intended to help in the selection of the appropriate one. At this writing, the Bell System Companies normalize taxes deferred because of accelerated tax depreciation, and they use service-life flow through for the investment tax credit. Studies involving investment in classes of plant with little or no Telco labor, such as motor vehicles, can probably safely assume that the book basis equals the tax basis. However, certain classes of plant, such as station connections, have a very high Telco labor element. This means that the expenditures capitalized for the books but expensed for the tax return will have a large effect on the capital costs.

The complete income tax expression for a study will consist of (1) the basic income tax liability, (2) one of the four expressions for accelerated tax-depreciation effects, and (3) one of the two expressions for investment-tax-credit effects. The appropriate sections of this chapter describe how to include the effects of further adjustments for interest during construction, Western Electric tax deferrals, or capital gains.

10
INFLATION AND THE
COST OF MONEY

INTRODUCTION

Earlier chapters have discussed separately operations costs, capital costs, the time value of money, and the concept of equivalence. All of these concepts are combined in the discipline known as *engineering economy*. The purpose of this chapter is to examine the effect of inflation on costs, and to show how this consideration is combined with the others in an economy study. The chapter will show first that inflation is reflected in one cost, the cost of money; therefore, it must be recognized in other costs as well.

Three techniques for handling inflation in economy studies are described: the explicit calculation of future costs, the use of an arithmetic shortcut (the convenience rate) to find the present worth of future inflated costs, and the systematic elimination of inflation from all costs.

The problem of forecasting future cost levels is also discussed, as well as the development of separate cost-trend rates for individual cost elements which will be consistent with the composite inflation rate reflected in the cost of money.

DETERMINING THE COST OF MONEY

A firm's *cost of money* is the return which investors require for the use of their money. Thus it is also the price that the firm must pay for one of the resources it uses. However, estimating the future cost of money with precision is very difficult, if not impossible, because so many outside factors must be considered. These factors include the investors' other investment opportunities and their perception of the future economy.

In the Bell System, corporate financial officers estimate the anticipated cost of money rate after consulting with members of the investment community, economists, and other knowledgeable financial advisors. Because this estimate is strongly influenced by current and anticipated economic conditions, it will change from time to time as both the economy and forecasts of the economy change. A major economic consideration which is reflected in the estimated cost of money is the anticipated future rise in general price levels—that is, inflation.

INFLATION AND THE COST OF MONEY

The return which investors require has three fundamental components:

1. A payment for the use of funds which is large enough to compensate the investor for postponing the consumption of resources

2. An additional amount to reflect the riskiness of the investment

3. An increment to offset the effects of inflation

Perhaps the best way to understand the inflation component is to view investors as consumers who are willing to postpone present consumption in order to gain additional consumer goods in the future. This gain in goods is their real return, or profit; money is merely the trading medium which they use. If the price of the goods is increasing because of inflation, investors must earn a rate of return equal to the rate of change in price—the rate of inflation—just to stay even, ignoring, for simplicity, any income taxes investors must pay on their return.

Assume that a prospective investor has $1,000 which could be used now to buy 100 units of some consumer goods costing $10 each. However, this investor may be willing to invest the $1,000 and forego buying these goods for 5 years. If the price is increasing at a 5% rate, it would go up as follows.

Time	Price
Now	$10.00
1 year	10.50
2 years	11.03
3 years	11.58
4 years	12.16
5 years	12.76

In 5 years, then, this investor will need $12.76 × 100, or $1,276, to buy exactly the same 100 units. If the investor gets back only $1,276 in 5 years though, there has been no real gain, because in 5 years the $1,276 will not buy any more goods than the $1,000 now.

In general, however, investors are not satisfied to just stay even. They expect a reward for foregoing consumption—a payment for the use of their funds. In other words, investors look for some absolute gain. As an example, assume, that our investor wants to acquire a 7% gain in goods each year. The following table shows the cash that must be realized from the investment.

Time	Quantity		Price		Cost
Now	100	×	$10.00	=	$1,000
1 year	107	×	10.50	=	1,124
2 years	114	×	11.03	=	1,257
3 years	123	×	.11.58	=	1,424
4 years	131	×	12.16	=	1,593
5 years	140	×	12.76	=	1,786

If the price of the goods were not increasing, the firm would need to return only $1,400 to the investor, that is,

$$\$10 \times 140 = \$1,400$$

This is just a 7% rate of return. To achieve the same result when the rate of inflation is 5%, the investor must have $1,786. The reason is that a portion of this sum is needed to compensate for the lower purchasing power of those future dollars.

If this $1,786 is related to the initial investment of $1,000, the apparent rate of return is somewhat greater than 12%.

$$\$1,000(F/P, i\%, 5) = \$1,786$$
$$i = 12.35\%$$

At least, this is the rate of return which investors would appear to be earning according to the published reports of the company. (These reports do not distinguish between dollars of different purchasing power, so that this rate of return is diluted by purchasing power.) Furthermore, this is also the return that investors would demand to assure the additional compensation needed for protection against inflation. This return represents the cost of money which is used in economy studies. To avoid ambiguity, in this chapter it will be called the *nominal* rate of return, whereas the inflation-free rate (7% in the example) will be called the *real* cost of money. The term *real* here means that this rate measures the return in units of constant purchasing power, with the effect of inflation removed.

Measurement of Inflation

The United States government, the Bell System, and others measure the historical rates of price change in portions of the economy for a variety of purposes. They use many different systems of indexes to make these measurements. An index is developed by sampling the segment of the economy that the index is designed to measure. The sample is often referred to as a *market basket* of goods representative of that segment of the economy. An index is a set of ratios, usually expressed as percentages but without the percent sign. The index of the cost of a market basket of goods is the ratio of the cost of the goods in one year divided by the cost of the same goods in some base year.

The federal government, through the Department of Commerce (Bureau of Economic Analysis) and the Department of Labor (Bureau of Labor Statistics), compiles the Consumer Price Index (CPI), the Wholesale Price Index (WPI), the Implicit Price Index for the Gross National Product (IPI-GNP), and many others. Of these indexes, the CPI is the most commonly used measure of prices in the United States. It is designed to show the effect of retail price changes on a selected fixed standard of living, which is represented by certain selected consumer goods and services. The WPI measures the impact of inflation at the wholesale level, for both consumer goods and industrial products. The prices for personal services are not included in the WPI, however. The IPI-GNP is a measure of the effect of general price-level changes on the Gross National Product. The GNP is the total market value of all final goods and services—that is, end products—produced in the economy during some given period of time. Several of the Bell Companies compute indexes for the market basket of products they buy, and the AT&T Company provides a composite of Bell System Telephone Plant Indexes (BSTPI). The BSTPI, and also the various similar Bell Companies' indexes, are one source of information on how the prices of capitalized expenditures have behaved in the past. These indexes are made up of component indexes which reflect the historical cost trends for practically every kind of plant which may be studied. Furthermore, their data bases contain enough information so that indexes can be constructed reflecting the cost trends in some noncapitalized expenditures.

Since all indexes are historical, their primary purpose is for making comparisons of today's values with those of the past. For example, the government uses its indexes to estimate whether wage earners have actually improved their position with their wage increases, or to estimate the real growth in the economy. The Bell Companies use the BSTPI in rate proceedings, construction budget analyses, property tax matters, sale of telephone plant, cost of service studies, and many other applications.

In the absence of specific knowledge about future costs, historical costs are a basis for projecting future cost trends. Historical cost trends can be measured by examining the applicable price index. The *cost-trend rate* implied by an index is the rate of change in that index. For example, if an index at one point in time was 120, and 126 exactly 1 year later, a cost-trend rate of 5% is appropriate for that year, since

$$\frac{126 - 120}{120} = \frac{6}{120} = 0.05 = 5\%$$

Generally, cost-trend rates for a single year are not very useful for measuring price inflation because they are so volatile. A better measure is the average cost-trend rate for the index over some longer period. This is the annually compounded rate of growth which occurred in the index during that period. Consider the following sample index.

Date	Index
1/1/77	126
1/1/76	120
1/1/75	110
1/1/74	107
1/1/73	108
1/1/72	104

The cost-trend rate (I) for the last 3 years is:

$$107(1 + I_3)^3 = 126$$

$$(1 + I_3)^3 = \frac{126}{107} = 1.1776$$

$$1 + I_3 = \sqrt[3]{1.1776}$$

$$1 + I_3 = 1.056$$

$$I_3 = 0.056 = 5.6\% \text{ per year}$$

For the last 5 years, it is

$$104(1 + I_5)^5 = 126$$

$$I_5 = 3.9\% \text{ per year}$$

One of the two rates could be used to represent the historical cost-trend rate in the index, depending upon the application.

Relationship between Return and Inflation

Theoretically, investors anticipate future inflation in the economy, and reflect it in the return they require. If the yield on corporate debt over recent years is plotted against a broad measure of inflation in the economy, there is a noticeably strong correlation between them. As an example, Figure 10-1 compares the corporate AAA bond yield with the change in the Implicit Price Index for the Gross National Product (IPI-GNP). The points on the IPI-GNP curve were found by using the previous 3-year-average percentage change as the inflation that investors expect for the following year. Of course, the relationship between inflation and the investors' required return is not this simple. However, Figure 10-1 does demonstrate the correlation between the debt interest rates and the inflation which investors anticipate. But equity investors, too, want to protect the value of their investment from inflationary erosion. As more dramatic evidence of this investor sensitivity to the effects of inflation, consider the very high interest rates which occur in countries with rampant inflation.

FIGURE 10-1
Inflation versus Debt Interest

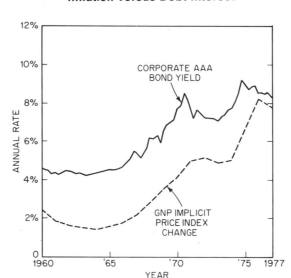

Relevance to Economy Studies

From the early 1920s until the early 1960s, the Bell System recommended a cost of money of 7% for economy studies. This was during a period which had little inflation overall, and improvements in productivity largely offset what little inflation there was. Figure 10-2 shows the annual percentage change in the Consumer Price Index for the period 1960 through 1976. This measurement of inflation can then be compared with the cost-of-money changes during this same period. In 1963, the recommended Bell System cost of money was increased to 8%; in 1969, to 8.5%. As inflation became even more pervasive in the early 1970s, the cost of money was raised to 9.5%, and then to 10.5%. In 1974, it was increased to 12%, as short-term inflation rates became greater than 10% and productivity increases simply could not keep up. There is substantial support for the view that these cost-of-money increases largely represented a reaction by investors to the steadily increasing rate of inflation in the U.S. economy.

This means that one element of cost in an economy study—the cost of money—has already been adjusted by investors to include the impact of inflation. As a result, when this nominal cost of money is used as a discount rate in an economy study, it may have an undesired effect on other study costs. This effect is best illustrated by what happens at the time of a change in the cost of money. If investors anticipate future inflation increases, the cost of money will probably increase too unless other offsetting forces are operating; and this new, higher cost of money will be used to discount future costs. Unless the study cost estimates are changed as well, the higher discount rate will make these future costs appear less important than they were before the increase; yet, the reason for the increase in the cost of money was to recognize that future costs are expected to increase. The point is that if a cost of money which includes inflationary effects is used, the estimates of all other future costs (such as first costs, maintenance costs, and ad valorem taxes) must also include the cost trends caused by inflation. Otherwise, the results will be misleading.

The effect of using the nominal cost of money in a present-worth analysis, but ignoring the changing price level of the materials themselves, can be illustrated simply. Assume that this investment is being considered: $1,000 investment now, 20-year life, zero net salvage, and $100 recurring operations costs.

FIGURE 10-2
CPI, Annual Percentage Change

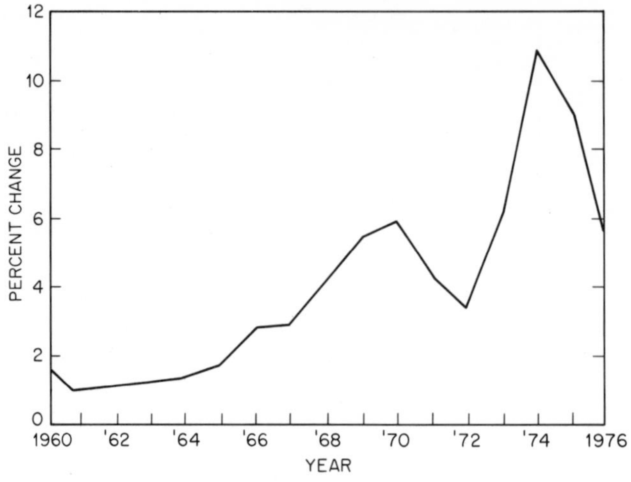

Ignoring income taxes, the present worth of the capital costs for this investment is $1,000. The total present worth of the expenditures (*PWE*) is the capital costs plus the present worth of the operations costs or

$$PWE = \$1,000 + \$100(P/A, i\%, 20)$$

This *PWE* expression can be evaluated for different values of the cost of money, as here.

i, %	PWE
2	$2,635
4	2,359
6	2,147
8	1,982
10	1,851
12	1,747

These *PWE* results demonstrate that increasing the cost of money reduces the economic effect of any future expenditures. In other words, an increase in the cost of money places less emphasis on future costs. Therefore, when the recommended cost of money was changed from 7 to 12% over a period of years, the influence of first costs on present-worth results increased relative to the influence of future expenditures.

Now consider the effects of recognizing inflation in the operations costs of this example. Assume that the operations costs have been estimated to increase by 6% per year over the 10-year period. This annual cost can be expressed as

$$\$100(1 + I)^n \quad \text{or} \quad \$100(1 + 0.06)^n$$

where n is the year the cost will be incurred.

Note the analogy between inflation and compound interest. The quantity $(1 + I)^n$ is simply the $(F/P, I\%, n)$ factor, or the future worth of the present amount factor. An interest rate is the rate of growth of the invested money, and the cost-trend rate is the rate of growth of a price index.

The total *PWE*, again ignoring taxes, is then

$$PWE = \$1,000 + \sum_{n=1}^{20} \$100(F/P, 6\%, n)(P/F, i\%, n)$$

This expression can also be evaluated at various rates for the cost of money:

	PWE	
i,%	**With 6% inflation**	**Without inflation**
2	$4,070	$2,635
4	3,458	2,359
6	3,000	2,147
8	2,653	1,982
10	2,387	1,851
12	2,179	1,747

These *PWE* results demonstrate that when inflation is reflected in the study costs, the discounting effect of the cost of money is partially offset. In other words, discounting costs and inflating them are counteracting procedures. A clear example of this interaction is the *PWE* calculated when the discount rate equaled the 6% inflation rate. When the inflated operations costs were discounted, the *PWE* was $2,000; the total *PWE* was $3,000, including the $1,000 initial expenditure. The $2,000 is the sum of the noninflated $100 yearly expense over the 20-year period. Although the costs were inflated, the discount rate effectively removed the inflation. When the cost of money is not equal to the inflation rate used, this offsetting effect is not as apparent. Nevertheless, it is important to understand this relationship between the discount rate and the inflation rate, or rates, used in a study.

Figure 10-3 illustrates this offsetting effect. Assume that an item currently costs $1, and that this price is not expected to change over the next 20 years. A curve can be drawn by finding the present worth at 12% of each annual future $1 expenditure in this cost series. Now suppose the price is estimated to increase by 4% per year. The curve showing the present worth of the inflated price for each year can then be compared with

FIGURE 10-3
Impact of Inflation on Present-Worth Results

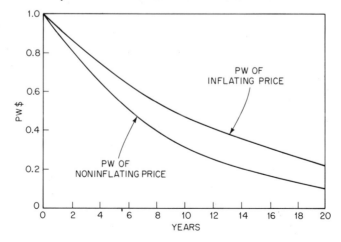

the present-worth curve with no inflation assumed. The difference between them is significant. It might perhaps be better appreciated if the scale of the chart were changed to millions of dollars. For example, assume that plant is to be purchased at the end of year 20 which currently costs $1 million but that this price will increase by 4% per year. This present-worth difference between the inflated and the noninflated costs will be approximately $123,500.

INFLATIONARY EFFECTS ON COSTS AND REVENUES

Some particular rate of inflation is inherent in any specific cost of money; it is therefore important that estimates of future cost levels used in an economy study be consistent with the rate of inflation included in the cost of money. These estimates of future cost levels must also recognize that the various costs are influenced by inflationary effects in varying degrees. Therefore, consistent cost-trend rates should be developed by specific cost category—for example, for each plant account and for labor expense. Consistency is achieved when the composite of all the individual cost-trend rates agrees with the inflation component of the cost of money. Of course, an individual cost-trend rate can be higher or lower than this composite rate and still be consistent with the cost of money. The methods of forecasting future costs will be discussed later in this chapter.

Some study costs are fixed and should not be adjusted to reflect future inflation. An example is a lease, because it commits the lessee to fixed payments for a specified period of time. Another example is depreciation, because capital repayment allowances under the FCC and income tax regulations do not recognize changes in the purchasing power of the dollar. Therefore, both book and tax depreciation associated with an asset are limited to the cost of the original asset less any salvage amount.

This chapter has emphasized the need to reflect future price increases in the study costs. For some study applications, it may be desirable to reflect the estimated revenues associated with an investment alternative. These revenues may also be increased over the study period because of an expected expansion in the customer usage of a service, or because of anticipated price increases. However, it is necessary to be cautious about estimating increases in revenues due to price increases because this action implies that future rate increases will be granted to the company, which is not all that certain.

RECOGNIZING INFLATION IN STUDIES

The three principal approaches to recognizing inflation in cost studies are

1. An explicit definition of future costs
2. The use of a "convenience rate" as an arithmetic shortcut
3. A systematic elimination of inflation, using "real" costs

Study results from all three methods can be equivalent provided they are applied correctly. The method selected will generally depend on computational advantages it may have and also on how well it is understood.

Explicit Definition of Future Costs

The technique which is usually simplest to understand is to explicitly define future costs that reflect relevant cost trends. For example, if an item of plant costs $1,000 today and this cost is expected to increase at 7% per year, the proposed addition of this item 1 year from now should be shown in a study as costing

$$\$1,000 \times (1.07)^1 = \$1,070$$

Similarly, the purchase of this item 10 years from now would mean an expenditure of

$$\$1,000 \times (1.07)^{10} = \$1,967$$

Estimated future expenditures can be converted to present worths, or equivalent annual amounts, by using time-value factors based on the firm's recommended cost-of-money rate (the nominal rate).

The Convenience Rate

It is not difficult to understand how to calculate a future expenditure based on an estimated cost-trend rate, and to then find the present worth of that expenditure by using the cost-of-money rate. However, the calculations can be tedious, since each expenditure must be treated individually.

For example, if a maintenance expense is currently $100 but is expected to increase at a rate of 8% per year for 20 years, the separate present-worth calculations for each expenditure would be

Year	Maintenance expense	PW factor at 12%	PW of maintenance expense
1	$108.00	0.8929	$ 96.43
2	116.64	0.7972	92.99
3	125.97	0.7118	89.67
.	.	.	.
.	.	.	.
.	.	.	.
18	399.60	0.1300	51.95
19	431.57	0.1161	50.11
20	466.10	0.1037	48.33
Total *PW*			$1,395.40

Fortunately, there is a simpler way. Since these costs are assumed to increase at some constant compound rate, they are actually a geometric progression. For this type of annual inflating cost, it is helpful to take advantage of an arithmetic shortcut and use a convenience rate which represents the net effect of the two operations: the restatement of costs for inflation, and also for present worth. Recall from Chapter 5 that the convenience rate is

$$i_g = \frac{i - I}{1 + I}$$

where i = nominal cost of money
I = constant rate of change (cost-trend rate)

This formula was derived by assuming a nonuniform series which changes each period by a constant rate, or percentage. Of course, it is important to realize that inflation need not be projected as a constant percentage increase in costs; discrete price increases, or a linear progression, may well be appropriate for some studies. However, if the convenience rate is used, it implies an assumption of geometrically increasing costs.

For the maintenance expense example, the convenience rate is

$$i_g = \frac{0.12 - 0.08}{1.08} = 3.7037\%$$

The present worth of the maintenance expense series can now be determined. That is,

$$PW = \$100(P/A, 3.7037\%, 20)$$
$$= \$1,395.40$$

The convenience rate was purposely used with four significant decimals to show that the resulting answer exactly equals that found by the explicit technique. For most study purposes, however, it is probably satisfactory to round off to the nearest integer percentage.

The convenience rate is useful, when coupled with the explicit technique, as an arithmetic shortcut for reflecting inflation in recurring annual costs such as maintenance expense. However, the convenience-rate technique is not always suitable because it is very easily misused. An incorrect application of the convenience rate can lead to erroneous study results. As Chapter 5 pointed out, the convenience rate is intended only as a mathematical shortcut to finding the *present worth* of a series which is inflating geometrically. It should not be construed as a discount rate. Figure 5-6 (page 104) illustrated how an inflating series can be represented by equivalent present, future, and levelized annual amounts. The convenience rate was used only to find the present worth of the inflating series. The cost of money had to be used to determine the equivalent future and levelized annual amounts.

Unfortunately, because the convenience rate can be used to find the present worth of the geometric series by using the $(P/A, i_g, N)$ factor, it seems to be a natural mistake to use it in the other time-value functions as well. For example, for the inflating maintenance expense, the levelized equivalent annual cost using i_g is

$$AC = PW(A/P, i_g, 20)$$
$$= \$1,395.40(A/P, 3.7037\%, 20)$$
$$= \$100$$

This $100 is the maintenance cost at the start of year 1—that is, the cost in today's dollars. It is the base of the inflating-cost series. For this simple example, it is obvious what this AC represents. However, when the present worth used is the sum of the present worths for capital recovery, taxes, and operating costs, the analyst may not realize that the AC amount is too low. In other words, when the convenience rate is used in the (A/P) factor, the result is the AC that would be required if there were no price inflation. To reflect inflation in the equivalent annual cost, the cost-of-money rate must be used, and not the convenience rate. The AC in nominal dollars is then

$$AC = PW(A/P, 12\%, 20)$$
$$= \$1,395.40(0.1339)$$
$$= \$186.84$$

Systematic Exclusion of Inflation

Earlier, this chapter pointed out that the cost of money already includes a component to protect the investor against inflation, which means that it is necessary to recognize inflation in the cost of other resources as well. It is sometimes proposed that inflation be removed from the cost of money—that the "real" cost of money be used in studies. In that case, inflation in other costs could just be ignored. However, the "real" cost of money cannot be determined exactly. Financial analysts, mathematicians, statisticians, and economists have analyzed the relationship between inflation in the economy and the performance of the stock and bond markets over the last two decades. While no absolute conclusions can be drawn, they have demonstrated that investors have required a range of "real" return during this period.

To use this method of analysis, costs must first be stated in terms of a base year; then an estimated real cost of money can be used to determine the present worth of the constant-dollar expenditures. For example, assume that the real cost of money is 7%. The present worth of the inflating maintenance expenditures in the previous example is then

$$PW = \$100(P/A, 7\%, 20)$$
$$= \$100(10.594)$$
$$= \$1,059.40$$

The present-worth result using the convenience rate was $1,395.40. This difference in answers occurs because the use of the real cost of money ascribes the same rate of inflation to all cost components. In this case, the implied inflation rate is about 5%. In the example of the convenience rate, the maintenance costs were estimated to increase at 8% per year.

This real-cost method is valid and useful if all costs inflate at the same rate, and this rate is identical to that portion of the nominal cost of money which represents the investor's anticipation of inflation. However, more than one cost-trend rate might well apply in a study, and it would be only coincidental that any of them equaled the inflation rate embedded in the cost of money. In fact, as was discussed earlier, some study costs will not increase in an inflationary environment. One example was a set of fixed lease payments. As a result, special modifications are necessary to compensate for the effect of the first big adjustment—that is, of stating "real" costs. These later modifications introduce an arithmetic complexity that rapidly erases the appeal of the method.

Furthermore, this method requires that all expenditures be translated into dollars of constant purchasing power. Although this is an effective device if it is used carefully, it can become conceptually difficult. For example, if this technique were used, the analyst might have to explain that an expenditure shown in the study as $1,000 will really mean a payment of $1,800. The difference is that one cost is in terms of study-date dollars, and the other is in terms of dollars of the year the cost will occur. At the same time, it might be noted that an amount of $200 shown for income taxes is in terms of future dollars because an inflation adjustment does not apply here!

For most applications, the real-cost-of-money approach will not be appropriate because it can be difficult to understand, and also because in practice it may offer no computational advantage. Example 10-1 illustrates the use of the explicit and the convenience-rate techniques.

Example 10-1. Assume a service requirement that may be satisfied by the installation of either a buried cable or an aerial cable. The details of these alternatives are given here.

1. Buried cable ($500 yearly maintenance expense)
   $20,000 initial investment
   20-year life
   $0 gross salvage
   $0 cost of removal

2. Aerial cable ($400 yearly maintenance expense for each investment)
   $10,000 initial investment $10,000 investment in 4 years
   20-year life   16-year life
   $1,000 gross salvage $1,000 gross salvage
   $400 cost of removal $400 cost of removal

All study costs are in terms of today's prices and must be adjusted for future anticipated price increases. The cost-trend rate for new cable is estimated as 7%, and is also considered to reflect the increasing gross salvage of cable. But since the removal expense is expected to be dependent upon the future labor costs, a labor-trend rate of 8% will be used to develop an estimate of this future cost.

The higher maintenance costs for the aerial cable are due to more exposure to adverse environmental conditions. For both alternatives, the maintenance expense is expected to increase by 8% per year. In both alternatives, the capital investments are eligible for the 10% investment tax credit and accelerated tax depreciation.

The long-term financial parameters which apply to this study are

$$i = 12\% \text{ cost of money}$$
$$i_d = 9\% \text{ cost of debt}$$
$$r = 45\% \text{ debt ratio}$$
$$t = 48\% \text{ tax rate}$$

Therefore,

$$\phi = \frac{t}{1-t}\left(1 - r\,\frac{i_d}{i}\right) = 0.6115$$

Both the explicit and the convenience-rate techniques can be used to analyze these study alternatives.

Explicit Analysis

In an explicit analysis, all study costs must be stated in terms of dollars as of the time they will occur. The restated study costs for both cables are now given.

For buried cable,

$$\text{Capital expenditure} = \$20{,}000$$
$$\text{Net salvage} = 0$$

Year	Maintenance Expense
1	$ 540.00
2	583.20
3	629.86
.	.
.	.
.	.
18	1,998.01
19	2,157.85
20	2,330.48
PW at 12%	$6,976.98

For aerial cable,

$$\text{Capital expenditure No. 1} = \$10{,}000$$
$$\text{Capital expenditure No. 2} = \$10{,}000(F/P, 7\%, 4)$$
$$= \$13{,}108$$

The gross salvage and cost of removal at the end of 20 years is expected to be the same for both aerial cable investments.

$$\text{Gross salvage} = \$1{,}000(F/P, 7\%, 20) = \$3{,}870$$
$$\text{Cost of removal} = \$400(F/P, 8\%, 20) = \$1{,}864$$

$$\text{Net salvage} \quad \begin{aligned} &= \$3,870 - \$1,864 \\ &= \$2,006 \end{aligned}$$

Year	Maintenance Expense	
	No. 1	No. 2
1	$ 432.00	
2	466.56	
3	503.88	
4	544.20	
5	587.73	$ 587.73
.	.	.
.	.	.
.	.	.
18	1,598.41	1,598.41
19	1,726.28	1,726.28
20	1,864.38	1,864.38
PW at 12%	$5,581.59	$4,119.43

Now that all costs are restated to reflect the effect of inflation, it is possible to find the present worth of these expenditures (*PWE*). The *PWE* will be determined by the sum of the present worths for capital recovery, income taxes, and maintenance expenses.

BURIED CABLE ANALYSIS   The present worth of the capital-recovery cost is

$$\begin{aligned} PW(C_{\text{rec}}) &= PW(\text{first cost}) - PW(\text{salvage}) \\ &= \$20,000 \end{aligned}$$

The present worth of income tax is

$$PW(T) = \phi[PW(C_{\text{rec}}) - PW(D_b)] - t(1 + \phi)PW(D_t - D_b)$$
$$- \frac{1}{1 - t} PW(ATC) - (1 + \phi)[PW(TC) - PW(ATC)]$$

The present worth of book depreciation is

$$\begin{aligned} PW(D_b) &= \frac{\$20,000 - 0}{20}(P/A, 12\%, 20) \\ &= \$7,469 \end{aligned}$$

The investment in buried cable is a current expenditure and the future net salvage is zero. As discussed earlier in this chapter, the recovery of capital is limited to the original first cost less any net salvage amount.

The present worth of tax depreciation is

$$PW(D_t) = (P/D_t)(\text{first cost})$$

From the tax-depreciation table in Chapter 9 (page 200) the (P/D_t) factor for a 20-year service life, 0 gross salvage, and 0 cost of removal is 0.55590.

$$\begin{aligned} PW(D_t) &= 0.55590(\$20,000) \\ &= \$11,118 \end{aligned}$$

The investment tax credit is

$$\begin{aligned} TC &= \text{tax-credit rate} \times \text{first cost} \\ &= 0.10(\$20,000) \\ &= \$2,000 \end{aligned}$$

And the present worth of this investment tax credit is

$$PW(TC) = TC(P/F, 12\%, 1)$$
$$= \$2,000(0.8929)$$
$$= \$1,786$$

The present worth of the amortized tax credit is

$$PW(ATC) = \frac{TC}{\text{life}}(P/A, 12\%, 20)$$
$$= \frac{\$2,000}{20}(7.469)$$
$$= \$747$$

Now, the present worth of income tax is

$$PW(T) = 0.6115(\$20,000 - \$7,469)$$
$$- 0.48(1 + 0.6115)(\$11,118 - \$7,469)$$
$$- \frac{1}{1 - 0.48}(\$747) - (1 + 0.6115)(\$1,786 - \$747)$$
$$= \$1,729$$

The *PWE* for the buried cable alternative is

$$PWE(\text{buried cable}) = PW(C_{\text{rec}}) + PW(T) + PW(M)$$
$$= \$20,000 + \$1,729 + \$6,977$$
$$= \$28,706$$

AERIAL CABLE ANALYSIS   The first capital expenditure for aerial cable occurs at the beginning of the study period. The capital-recovery costs for this expenditure are

$$PW(C_{\text{rec}}) = \$10,000 - \$2,006(P/F, 12\%, 20)$$
$$= \$9,792$$

The *PW* of book depreciation is

$$PW(D_b) = \frac{\$10,000 - \$2,006}{20}(P/A, 12\%, 20)$$
$$= \$2,986$$

The allowances for both book and tax depreciation are a function of the historical original cost of the plant, its salvage, and its life. Both the original cost, the salvage, and the life of the plant will be influenced by inflation. The effect of inflation on historical costs will be discussed later in this chapter.

The income tax obligation of this investment is determined as follows:

$$\% \text{ gross salvage} = \frac{\$3,870}{\$10,000} = 38.70\%$$

$$\% \text{ cost of removal} = \frac{\$1,864}{\$10,000} = 18.64\%$$

Using the tax-depreciation table, the present worth of tax depreciation is

$$(P/D_t) = GS \text{ component} + COR \text{ component}$$
$$= 0.49349 - (0.49349 - 0.45597)\frac{8.7}{10.0} + 0.10367(0.1864)$$
$$= 0.49349 - 0.03264 + 0.01932$$
$$= 0.48017$$

$$PW(D_t) = (P/D_t)(\text{first cost})$$
$$= 0.48017(\$10,000)$$
$$= \$4,802$$

The tax-credit present-worth effects are

$$TC = (TC \text{ rate})(\text{first cost})$$
$$= 0.10(\$10,000)$$
$$= \$1,000$$
$$PW(TC) = TC(P/F, 12\%, 1)$$
$$= \$1,000(0.8929)$$
$$= \$893$$
$$PW(ATC) = \frac{TC}{\text{life}}(P/A, 12\%, 20)$$
$$= \frac{\$1,000}{20}(7.469)$$
$$= \$373$$

The present worth of income tax is

$$PW(T) = 0.6115\,(\$9,792 - \$2,986) - 0.48(1 + 0.6115)(\$4,802 - \$2,986)$$
$$- \frac{1}{1 - 0.48}(\$373) - (1 + 0.6115)(\$893 - \$373) = \$1,199$$

The *PWE* for the first aerial cable investment can now be found by combining the intermediate results.

$$PWE(\text{aerial cable No. 1}) = PW(C_{\text{rec}}) + PW(T) + PW(M)$$
$$= \$9,792 + \$1,199 + \$5,582$$
$$= \$16,573$$

The above analysis must be repeated for the second capital expenditure to determine the total *PWE* for aerial cable.

$$PW(C_{\text{rec}}) = \$13,108(P/F, 12\%, 4) - \$2,006(P/F, 12\%, 20)$$
$$= \$8,122$$
$$PW(D_b) = \frac{\$13,108 - \$2,006}{16}(P/A, 12\%, 16)\,(P/F, 12\%, 4)$$
$$= \$3,075$$
$$\% \text{ gross salvage} = \frac{\$3,870}{\$13,108} = 29.52\%$$
$$\% \text{ cost of removal} = \frac{\$1,864}{\$13,108} = 14.22\%$$

From the tax-depreciation table, for a 16-year service life, the present worth of tax depreciation is

$$(P/D_t) = GS \text{ component} + COR \text{ component}$$
$$(P/D_t) = 0.56402 - (0.56402 - 0.52616)\frac{9.52}{10.00}$$
$$+ 0.16312(0.1422)$$
$$= 0.56402 - 0.03604 + 0.02320$$
$$= 0.55118$$

$$PW(D_t) = (P/D_t)(\text{first cost})(P/F, 12\%, 4)$$
$$= 0.55118(\$13,108)(0.6355)$$
$$= \$4,591$$

The tax-credit present-worth effects are

$$TC = (TC \text{ rate})(\text{first cost})$$
$$= 0.10(\$13,108)$$
$$= \$1,311$$
$$PW(TC) = TC(P/F, 12\%, 1)(P/F, 12\%, 4)$$
$$= \$1,311(0.5674)$$
$$= \$744$$
$$PW(ATC) = \frac{TC}{\text{life}}(P/A, 12\%, 16)(P/F, 12\%, 4)$$
$$= \frac{\$1,311}{16}(6.974)(0.6355)$$
$$= \$363$$

The present worth of income tax is then

$$PW(T) = 0.6115(\$8,122 - \$3,075) - 0.48(1 + 0.6115)(\$4,591 - \$3,075)$$
$$- \frac{1}{1 - 0.48}(\$363) - (1 + 0.6115)(\$744 - \$363)$$
$$= \$601$$

The *PWE* for the second aerial cable investment can now be determined.

$$PWE \text{ (aerial cable No. 2)} = PW(C_{\text{rec}}) + PW(T) + PW(M)$$
$$= \$8,122 + \$601 + \$4,119$$
$$= \$12,842$$

The total aerial cable *PWE* can be found by combining the intermediate results.

$$PWE(\text{aerial cable}) = PWE(\text{aerial cable No. 1}) + PWE(\text{aerial cable No. 2})$$
$$= \$16,573 + \$12,842$$
$$= \$29,415$$

The difference between the buried cable and the aerial cable *PWEs* is:

$$PWE(\text{buried cable}) - PWE(\text{aerial cable}) = \$28,706 - \$29,415$$
$$= -\$709$$

The two alternatives are essentially equivalent, although the aerial cable is slightly more expensive.

Convenience-Rate Analysis

The convenience-rate concept will produce the same results. The first step is to calculate the convenience rate that applies to each study cost.

$$\text{Capital-expenditure rate} = \frac{12 - 7}{1.07} = 4.67\%$$

$$\text{Maintenance-expense rate} = \frac{12 - 8}{1.08} = 3.70\%$$

To use the convenience-rate technique, study costs must be in terms of today's cost levels. The example data can be used in the analysis without any adjustment.

BURIED CABLE ANALYSIS   The investment in buried cable is a current expenditure, and the future net salvage is zero. The only cost for this alternative, which must be adjusted to reflect inflation, is the maintenance expense. So the required calculations for $PW(C_{rec})$, $PW(D_b)$, and $PW(T)$ are identical to the corresponding calculations in the explicit analysis. For this simple alternative, the only use of the convenience rate is in calculating the present worth for the 20 years of inflating maintenance expense.

$$PW(M) = 500(P/A, 3.70\%, 20)$$
$$= \$6,979$$

The individual present-worth values resulting from the convenience-rate analysis should be compared with the corresponding results of explicit analysis. The slight difference in answers is due to the rounding of the convenience rate to two significant decimals and the rounding of the intermediate results to the nearest whole dollar amount.

Combining the $PW(C_{rec})$, $PW(T)$, and $PW(M)$ results produces the PWE for the buried cable alternative.

$$PWE = \$20,000 + \$1,729 + \$6,979$$
$$= \$28,708$$

These calculations illustrate how the convenience rate can be used together with the explicit-analysis technique.

AERIAL CABLE ANALYSIS   The first aerial cable expenditure occurs at the start of the study period, which means that the first cost is not subject to inflation. However, the gross salvage and cost of removal which are stated in terms of today's cost levels must be adjusted for anticipated inflation. Gross salvage has been estimated to increase at the same rate as the price of new cable; thus, the cable convenience rate of 4.67% can be used to reflect the net effect of inflation and the discount rate on this salvage amount. The present worth of the cost of removal expense can be determined by using the maintenance (labor) convenience rate (3.70%). The capital-recovery costs for this investment are

$$PW(C_{rec}) = PW(\text{first cost}) - PW(\text{net salvage})$$
$$= PW(\text{first cost}) - [PW(\text{gross salvage}) - PW(\text{cost of removal})]$$
$$= \$10,000 - [\$1,000(P/F, 4.67\%, 20) - \$400(P/F, 3.70\%, 20)]$$
$$= \$9,792$$

The use of the convenience rate allows the gross salvage and cost of removal to be stated in terms of today's cost levels.

As discussed earlier, the recovery of capital through depreciation expense is limited to the original first cost less any net salvage amount. The tax depreciation allowed by the tax regulations is also limited by the original first cost. These depreciation amounts will have a declining purchasing power in an inflationary economy. To find the present worth of a nominal dollar series, the nominal cost of money must be used as the discount rate. So the calculations for the present worth of depreciation expense and the present worth of income tax are exactly like those using the explicit-analysis technique. This observation tends to support the recommendation that the convenience-rate technique be reserved for special applications where it simplifies the arithmetic, such as finding the present worth of an inflating series of annual expenses. The present worth of income tax is then

$$PW(T) = 0.6115(\$9,792 - \$2,986) - 0.48(1 + 0.6115)(\$4,802 - \$2,986)$$

$$-\frac{1}{1 - 0.48}(\$373) - (1 + 0.6115)(\$893 - \$373)$$

$$= \$1,199$$

The present worth of the maintenance expense for the aerial cable is

$$PW(M) = \$400(P/A, 3.70\%, 20)$$
$$= \$5,583$$

The *PWE* for the first aerial cable investment can now be found by adding the $PW(C_{rec})$, $PW(T)$, and $PW(M)$ values.

$$PWE(\text{aerial cable No. 1}) = \$9,792 + \$1,199 + \$5,583$$
$$= \$16,574$$

Now, the present worth of expenditures for the delayed aerial cable installation can also be calculated:

$$PW(C_{rec}) = \$10,000(P/F, 4.67\%, 4) - [\$1,000(P/F, 4.67\%, 20) - \$400(P/F, 3.70\%, 20)]$$
$$= \$8,122$$

Again, the calculations for the present worth of book depreciation and the present worth of income taxes are the same as those for the explicit analysis. Therefore, the present worth of income taxes is

$$PW(T) = 0.6115(\$8,122 - \$3,075) - 0.48(1 + 0.6115)(\$4,591 - \$3,075)$$

$$-\frac{1}{1 - 0.48}(\$363) - (1 + 0.6115)(\$744 - \$363)$$

$$= \$601$$

The maintenance expense will begin 4 years into the study period, and it will be incurred for 16 years instead of 20 years. The present worth of this inflating maintenance expense is then

$$PW(M) = \$400(P/A, 3.7\%, 16)(P/F, 3.7\%, 4)$$
$$= \$4,121$$

The *PWE* for the second aerial cable installation is

$$PWE(\text{aerial cable No. 2}) = \$8,122 + \$601 + \$4,121$$
$$= \$12,844$$

By combining the two present-worth results, the total present worth of expenditures for the aerial cable alternative can be determined.

$$PWE(\text{aerial cable}) = PWE(\text{aerial cable No. 1}) + PWE(\text{aerial cable No. 2})$$
$$= \$16,574 + \$12,844$$
$$= \$29,418$$

The difference between the *PWE* for the buried cable and that for the aerial cable is

$$PWE(\text{buried cable}) - PWE(\text{aerial cable}) = \$28,708 - \$28,418$$
$$= -\$710$$

This comparative example has demonstrated two methods of reflecting inflation in study costs. The convenience-rate present-worth results are essentially the same as the values obtained by using the explicit-analysis technique. As indicated earlier, the small difference between answers occurs because the convenience rates were rounded to two significant decimals and the intermediate results rounded to the nearest whole dollar. The explicit specification of future expenditures, which reflect relevant cost trends, is generally simplest to understand and apply in studies. However, the convenience-rate technique is a valuable tool to find the present worth of an inflating-cost series, such as maintenance expense. This technique may also be useful in special situations, e.g., converting existing economy-study computer programs to allow reflection of inflation in study costs.

FORECASTING FUTURE COSTS

Economy studies are an analysis of events which are expected to occur in the future. When future prices or costs must be forecast, it becomes evident that economy studies are more an art than a science. The astronomer can forecast the future location of planets and other celestial bodies with great accuracy because there are precise mathematical rules describing their movement. There are rules, certainly, to describe the future economy, but they are complicated by subjective attitudes and factors which cannot be reduced to mathematical relationships. As a result, the most difficult part of including future cost trends in an economy study is to forecast what those trends may be.

The AT&T forecast of a composite cost of money is a first step towards estimating future cost trends. Since this estimate of the cost of money reflects some amount to protect investors against the inflationary erosion of their funds, it therefore implies a forecast of the inflation rate which investors anticipate in the things they spend their money on. However, the investors' market basket of goods is not the same as the telephone company's market basket of goods. Some of the same things are of course in both market baskets, such as furniture and tools, but they are needed in different proportions. The composite inflation which the company anticipates is therefore not necessarily the same as the inflation which investors anticipate. Nevertheless, because the same economic forces are operating, inflation should affect both market baskets in a similar way.

The investors' view of inflation is probably best represented by some macroeconomic price index like the CPI, the WPI, or the IPI-GNP. Figure 10-4 compares the yearly percentage rate of change of the CPI, the IPI-GNP, and the BSTPI. Note that even though the rates of change fluctuate year by year, there is a correlation in their movements. It is not unreasonable, therefore, to assume that, over some future period, the composite BSTPI rate of change will be approximately equal to the investors' anticipated rate of inflation.

FIGURE 10-4
Rates of Change

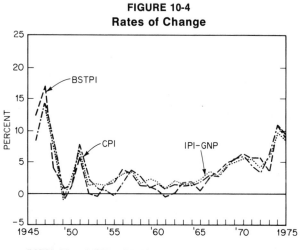

BSTPI: BELL SYSTEM TELEPHONE PLANT INDEX

IPI-GNP: IMPLICIT PRICE DEFLATOR, GROSS NATIONAL PRODUCT
        —TOTAL SECTOR

CPI: CONSUMER PRICE INDEX

Historical Cost-Trend Rates

The BSTPI consists of component indexes which reflect the historical cost trends for telephone plant. These historical cost trends can serve as a basis for allocating the inflation anticipated by investors across the plant accounts.

Figure 10-1 demonstrated that investors compensate for anticipated future inflation by requiring a higher return. Their estimate of future inflation is highly influenced by the recent history of inflation in the economy. It is appropriate, then, in developing historical cost trends from the BSTPI data, to use cost data for the recent past, e.g., the last 5 years.

A detailed discussion of forecasting techniques is beyond the scope of this text. However, a simple method of determining cost-trend rates from an index was illustrated earlier in this chapter. Various index forecasts are also available from both government and business sources and may be useful in developing cost-trend rates. For most applications, however, an analyst will not have to calculate these rates but will utilize the rates developed for use by the total company. In any case, the cost-trend rates used in the study must be consistent with the inflation component of the cost of money.

Consistent Cost-Trend Rates

Typically, the purpose of an economy study is to consider buying a limited number of things—perhaps central-office equipment and labor. The market basket of goods for an economy study is a small sample, and future price increases for this sample will not necessarily equal the overall inflationary trend. In fact, economy studies may often compare alternatives with different small market baskets, and different cost-trend rates. The cost-trend rates for these small samples are components of the overall inflation rate, and an estimate of them must be a reasonable component of the composite rate.

The necessity for reasonable and consistent estimates cannot be overemphasized, because study decisions can be extremely sensitive to the cost-trend rates used. However, once an estimate of cost-trend rates is developed which is consistent with the inflation component of the cost-of-money rate, the data may generally be used without modifications for future changes in economic conditions. This is true because the difference between the cost-of-money and inflation rates will tend to remain fairly constant. This will be discussed further in the following section.

LONG-TERM EFFECT

Some analysts have expressed concern that the assumption of a constant annual rate of inflation produces inordinately high future cost levels. Others point to economic forecasts that indicate lower rates of inflation. They have therefore proposed lowering the cost-trend rates at various points in the study period.

Because the *actual* cost-trend rates will be consistent with the *actual* cost of money, a significant change in the cost-trend rates implies a change in the composite cost-of-money rates. If the cost-trend rates are changed to maintain consistency, the cost-of-money rate should also be adjusted. If comparable changes are made in both the cost-of-money rate and the cost-trend rates, the present worth of future costs will be very close to the values found, without changes in either rate. This point is illustrated by the table on p. 249.

A change of the same amount in both the cost-of-money and the inflation rates produces very little difference in the study results. Therefore, the most significant

Case	Cost at year 0	Composite inflation rate, %	Cost at year 10	Cost-of-money rate, %	PW at year 0
1	$1,000	4	$1,480	10	$571
2	1,000	6	1,791	12	577
3	1,000	8	2,159	14	582

factor is the difference between the rates rather than the absolute level of the cost-of-money or the cost-trend rates.

Usually, it is best not to modify the assumed cost-trend rates during the study period, to avoid what appears to be unduly high cost levels, or because the analyst feels that general economic conditions will change during the study period. Intuitively objectionable cost levels in the future might seem more acceptable if present-day cost levels were related to those which existed in the past.

To use any inflation projection, the composite inflation projection must be consistent with the inflation projection implied in the cost-of-money estimate. Because these projections may have different sources, they may not be consistent, and one or the other must be modified. Since the recommended cost-of-money rate has been determined as the rate suitable for use in economy studies, the inflation projection should be modified to be compatible.

Sensitivity Testing

Since the accuracy, reasonableness, and consistency of projected inflation rates is rather uncertain, it is necessary to examine decisions to see how sensitive they are to the rates selected. Under a given cost of money, ranges of reasonable cost-trend rates should be defined and the study results examined at various rates within those ranges. However, sensitivity testing should always remain plausible. For example, the change in cost of labor has a large effect on the composite rate. If a 12% cost of money was determined in an atmosphere of, say, 8% anticipated labor cost increase, then it would be unreasonable to use the 12% cost of money over a range of from 0 to 14% cost trends for labor. If the cost trend for labor were 0%, the cost of money would be considerably less than 12%; but if the labor cost trend were 14%, it would be considerably higher. Since the change in the cost of labor is such a large component of the composite inflation, its range of reasonableness is probably quite narrow to be consistent with the cost of money. The inflation rates of other, less significant components could vary over a wider range, yet still be reasonable.

Estimating Current Costs

To make the best possible estimate of the capital and operations costs associated with a future purchase of telephone plant, the analyst needs to estimate the costs that would be incurred if the purchase were made now. Recall that both the explicit and convenience-rate techniques required that study costs be known in terms of current, or today's, dollars. These current costs served as the base from which the future costs were determined, i.e., the base of the inflating amount or series.

For many years, the Bell System has used an estimating procedure which expresses study parameters and costs in terms of a percentage-of-first-cost factor. Examples of study items expressed in this fashion are gross salvage, cost of removal, maintenance expense, and ad valorem taxes. The total capital costs for an investment can also be represented by an annual percentage of first cost. This estimating procedure tends to encourage the reliance upon average cost data. The gross salvage, cost of removal, maintenance expense, and ad valorem tax factors are obtained by dividing the *total* cost, or value, by the original *total* cost of the associated plant. The capital-cost percentage can be determined for any service life, gross salvage, and cost of removal, but all too often the average service life and the average net salvage are used to obtain this cost factor. Even if an average cost is considered appropriate for a current study cost, the reflection of inflation in future costs is complicated because some amount of inflation is already included in several of these cost factors.

This unintentional reflection of inflation occurs because the current expenditures and the historical costs for the associated plant are out of phase. The current expenditures for cost of removal, maintenance, and ad valorem taxes—used in deriving the percent-of-first-cost factors—represent current levels of cost. However, the associated plant investment represents equipment installed in previous years at different, and generally lower, cost levels. So if the numerator of the percent-of-first-cost fraction is based upon current cost levels but the denominator represents earlier, lower cost levels, the resulting cost factor is overstated. It is overstated relative to the result that would be obtained if the current cost were related to the associated plant installed at current cost levels. But another viewpoint is that this cost factor simply has some amount of inflation built into it, since it reflects the fact that expenses associated with plant will increase with inflation while the plant investment itself, once made, stays the same.

Chapter 6 suggested several sources of cost information that the analyst might use in developing current-dollar cost estimates. However, the reader was cautioned that average costs were not appropriate for evaluating individual investments. The cost data that an analyst uses to estimate current costs should be specific to the study situation. The economic life, rather than the account average service life, plus associated estimates for gross salvage and cost of removal should be used to determine the capital costs. Cost estimates for expenses such as maintenance and ad valorem taxes should be based upon recent cost data rather than historical cost averages.

It is usually preferably to express these costs in terms of current dollars. However, if costs must be expressed in terms of annual percentages of the first cost, these factors should be developed on a unit cost basis rather than on an account-average cost basis. For example, an annual percentage factor for maintenance can be obtained by dividing the current maintenance cost for an item of plant by the current first cost for that plant.

SUMMARY

Investors include an element in their return requirements to protect them against anticipated price inflation. Because the firm uses this cost of money in its economy studies, it must also include a similar and consistent anticipation of trends in the other study costs. It is possible to do studies either by explicitly calculating future expenditures or by combining this calculation with the discounting operation. This latter approach, with the use of convenience rates, is a valuable tool for finding the present worth of an inflating-cost series, but the explicit specification of future expenditures is generally the simplest technique to apply and understand. For either type of analysis, a firm must somehow estimate the effect of inflation on future costs in order to make reasonable decisions between alternatives.

The analysis of cost trends may be an important help in making the best possible estimates of future costs. Past cost trends are measured by the use of various price indexes. The various Bell Company indexes—the Bell System Telephone Plant Indexes (BSTPI)—or the several United States government indexes may be useful in economy study work.

Since the forecasts of the inflationary trends are quite uncertain at best, it is necessary to test study results for their sensitivity to ranges of values. Various projections of the several indexes can be made, or are available from other sources. Although these projections can be useful, for economy study work they must be reasonable and consistent with the inflationary element implied in the cost-of-money estimate. If not, they must be modified or adjusted.

To utilize the techniques for reflecting inflation in economy studies, an estimate is required of the costs that would be incurred if the contemplated purchase were made now. This current-cost estimate should be based upon the specifics of the investment being considered. Cost estimating procedures based upon historical average costs do not accurately estimate current costs and complicate the reflection of inflation in future costs.

11

IDENTIFYING ALTERNATIVES

INTRODUCTION

The purpose of each chapter in the book so far has been to give the reader the basic tools for doing an economy study: some understanding of accounting, finance, financial mathematics, and costs (including the return on capital and income taxes). Now the reader is ready for the topics discussed in the rest of the book, which describe the actual procedures for doing a study.

This chapter describes the first step: identifying alternative ways to solve a problem which requires the expenditure and receipt of money. Next, the text describes techniques which are used in comparing alternatives. Two chapters follow, on the characteristics of retired equipment and their replacements. Then, after a consideration of the special topic of leasing versus owning, this fundamental treatment of the economy study ends with a chapter on techniques for evaluating and presenting the results of an economy study.

A GENERAL DISCUSSION OF ECONOMY STUDIES

Why Are They Called "Engineering" Economy Studies?

Many groups and individuals besides engineers do economy studies of the type covered in this book. How is it, then, that this branch of economics has acquired the label "engineering"? The answer lies in the type of work which engineers do and in their frequent need for a time-value cost study. For example, engineers are continually asked to develop devices which will function well and economically over long periods of time. The type of analysis presented in this book gives them a way to measure both present and future costs, and to then determine how economic their creations will be compared with other ways of accomplishing the same goal.

Although the training which an engineer receives is an excellent background, an engineering degree is certainly not required for doing economy studies. All managers who have need for this kind of analysis should learn to do an economy study themselves, or else ask an experienced analyst to do a study for them. However, managers who choose to have others do the studies for them should at least acquire a working knowledge of the techniques so they will have a good understanding of the results.

Why and When Are Engineering Economy Studies Made?

One of the long-standing objectives of the Bell System is to provide good service at the lowest overall cost that is consistent with earning a reasonable rate of return on the investment. The economy study is a way of finding out whether proposed plans contribute toward that objective. Other methods, such as an itemized listing of advantages and disadvantages, offer few, if any, meaningful measures of what each proposal contributes.

Because of the failings of these other methods, economy studies should be performed whenever a decision is required which will influence the cost of doing business. The studies need not always be formal, but they should always be sufficient to provide the answers to these three questions:

- Why *do* it?
- Why do it *this way*?
- Why do it *now*?

Experience will indicate when to do a formal rather than an informal study. However, even for an informal study, it is important to follow the principles of good economy studies, because there is no other way to examine the real differences among alternatives.

WHY DEVELOP ALTERNATIVES?

When confronted with a problem to be solved, or an objective to be reached, most managers do consider the available alternatives—even in simple matters. For instance, managers who need to transmit information to others normally pick the means of transmittal after considering whether to use a telephone call, a personal visit, or a written message of some type. The alternatives developed for use in economy studies are normally far more complex and, at least initially, are likely to be more numerous than in the example of transmitting information. However, the reason for considering alternatives is the same.

Alternatives are developed and compared in order to increase the likelihood of solving a problem, or of achieving an objective, in the most efficient manner possible. Usually, any solution in a set of solutions will solve a problem; but to stop with the first method that comes to mind leaves the maximization of results almost entirely to chance. In other words, the use of this narrow approach would make it virtually impossible to say whether or not the solution of a problem has contributed at all toward the overall goal of providing good service at the lowest cost that is consistent with earning a reasonable rate of return on the investment.

Perhaps the two most common reasons people give for not considering more alternatives are (1) there was not enough time; and (2) the job is "always done that way." There are certainly times when a snap decision is required because of a deadline. However, managers should try to avoid operating in the shadows of deadlines, so they can control the business instead of letting the business control them. When time is limited and a number of decisions must be made, snap decisions should be reserved for problems posing the least amount of threat to the business. Whatever time is available should then be spent on examining the various alternatives for solving problems which are critical to the company's health.

The second reason, that of precedence for doing a job in a particular way, is decidedly weak. There is no more logic in continuing to blindly repeat actions of the past, just because they have become routine, than there is in change for the sake of change. The history of the human race has been one of technological development; and if something has been done one way for a long period of time, there is almost certainly a different way of doing it. It cannot be assumed that the methods of the past can never be improved. However, it is absolutely necessary to consider each alternative separately and carefully, because *different*—or even *newer*—does not necessarily mean *better*.

HOW TO DEVELOP ALTERNATIVES

Developing alternatives is an art requiring an inquisitive and innovative nature. Still, the following four conditions must be satisfied before it is possible to construct any reasonable set of alternatives.

- The present situation must be known.
- The future objective must be defined in some manner.
- The composition of the interval between the present and future situations must be known or assumed.
- The constraints on the problem, and the possible solutions, must also be known or assumed.

The order in which these conditions will be met depends upon the particular problem. Sometimes, a planner will be asked to develop a view of the future from a known present position. At other times, the desired future objective will be given, and the planner has to figure out what the starting point is. At still other times, the question might be: "How will this new technology affect the company?" In each case, a different condition is known and the others must be found. The need to satisfy these four conditions is discussed in the following sections.

Understanding the Present Situation

An understanding of the present situation is essential for developing a reasonable set of alternatives. The quip about not being able to get there from here certainly applies if the location of "here" is not known. Someone who is lost cannot look at a map and plot alternative routes (or, for that matter, one route) back home before finding some kind of landmark or location sign which will help establish the present location. Similarly, a telephone company planner cannot suggest alternative ways of continuing service for a business subscriber with a PBX before knowing how much of the machine capacity is now being used.

Understanding the Future Objective

Closely associated with the present situation is the condition of the future. Until the future objective has been defined in some way, the planner will have no way of knowing when an alternative has satisfied the problem. For example, usually there is no need to plan for a PBX which will serve 1,000 stations if the customer's ultimate workforce will never require more than 125 phones. Nor will an 80-station unit continue to be satisfactory for that customer after some point in time. Even if the problem being worked on warrants nothing more than a projection of possible courses for the future conduct of the business, no alternative which examines possibilities through 5 years, say, will be sufficient if what was needed was a 10-year solution. However, before the planner had become aware that "future" in this case meant 10 years, there would have been no way to know that the 5-year plan was unsatisfactory.

Understanding the Composition of the Interval in Between

A knowledge of where things stand at the moment, and also where the alternatives are supposed to end up, is necessary but not sufficient. The composition of the interval between the present and the future situations must be known or assumed. Going back to the example of the person who was lost and trying to find the way home, it would do no good to plot an alternative route which goes over a bridge if the bridge had been washed out. Neither does it pay to make plans for an equipment installation in 6 months' time when the technology required for such an installation has not yet been developed and is not even being considered for development within the foreseeable future.

When assumptions must be made, it may be wise to develop contingency plans and to determine how sensitive the results are to variations in those assumptions. (For more discussion on sensitivity analysis, see Chapter 16.)

Understanding the Constraints on the Problem

Knowing, or assuming, the constraints on the problem and the possible solutions defines how far to go in developing alternatives. In the example of the lost traveler, it would be foolish to look for ways home that required swimming a river if the traveler could not swim. Likewise, it would do no good to develop a set of alternatives requiring capital expenditures of between $1 and $2 million when the absolute maximum allowable expenditure is $500 thousand.

Specifying the Features of Each Alternative

Satisfying these four conditions will not automatically give the planner all the alternate solutions to a problem. However, it will do two very important things: (1) It will satisfy the question of whether there really is a problem; and (2) it will provide the basic knowledge needed before the analyst can apply the creative thinking required to finish the job.

As each alternate plan is developed, it should be stated clearly in writing. Such statements should include the basic assumptions and the details of all items which vary from plan to plan. All the plans should be comparable in terms of what results are achieved and the length of time these results can be sustained. If one plan calls for something which will inherently provide more than the competing plans, the extra savings or revenues should be credited accordingly.

HOW MANY ALTERNATIVES ARE NEEDED?

Once the process of developing alternatives has begun, the next logical question is: "How many is enough?" The answer is: "Enough alternatives have been developed when the set includes all the viable ones." No alternative which has a chance of being the best one should be left out.

This does not necessarily mean that there must be a large number of alternatives. In fact, it is not outside the realm of possibility that there would be one and only one way of doing something, and that this plan must be carried out. However, this situation is highly improbable. What usually happens is that an opportunity arises in which the expenditure of some amount of money can result in the receipt or savings of money. Although it might be possible to take advantage of this opportunity in only one way, rejecting the opportunity is also a viable alternative. In other words, the present operation could continue as if the opportunity had never arisen. As an example of this kind of situation, assume that a machine is developed that could provide a unique new telephone service for which subscribers could be charged. If marketing tests show that subscribers are interested, the service can be offered—provided the particular machine is bought and installed. However, if the tests show that the subscribers are not interested, the telephone company has the alternative of rejecting the opportunity to buy the machine, and of continuing with present services as if the new one had never been developed.

Usually, however, many alternatives might be viable. Each of them should be considered to make sure that a poor decision is not made by default. In other words, no action should be taken that is just somewhat better than what has been done previously, when another action which would be far more beneficial is bypassed simply because no one gave it any thought. Consideration of all viable alternatives might be regarded as a highly time-consuming task. The following example illustrates that a logical approach will simplify the job.

The dispatcher for the XYZ Trucking Company finds that during a certain 2-hour period, three trucks and three drivers are idle. (One truck is a small delivery van; the second is an 18-foot stake bed; and the third is a tractor-trailer, or "semi.") Wanting to keep as much as possible of the equipment productive, the dispatcher accepts a delivery of 1,000 pounds, which normally would have been rejected, for a fee of $10. The delivery is to be made to a place 30 miles (round trip) away by city streets, 45 miles by toll road (a 50-cent toll), and 50 miles by a free limited-access road. The drivers are all equally qualified to drive any of the vehicles. Because they have been with the company varying lengths of time, they make $5, $6, and $7 an hour, respectively; and they are paid whether they are active or idle. The following table shows different costs per mile driven by the vehicles (gas, oil, and wear and tear).

Road	Van	Truck Stake	Semi
City	12¢	16¢	20¢
Other	10¢	14¢	18¢

The dispatcher knows that the owner wants each delivery to be made in the most economic manner possible, and therefore begins to look at all 27 alternatives. These are shown in the decision tree in Figure 11-1.

A decision tree is a graphic illustration showing how various segments of a decision are related. For instance, Figure 11-1 shows that a decision about a delivery can be broken down into the three smaller decisions of which truck, which driver, and which road. Choosing any one of the trucks still leaves a choice of one of three drivers; and when the driver has been selected, one of three roads must be picked. Since the three decisions in this problem are independent, the order in which they appear (truck, driver, or road) could be changed. However, if the choice of truck depended on which driver was selected, the driver would have to be selected before the truck. In other words, the dependent choices would have to follow the independent one.

Before going through every branch of the decision tree and finding the cost for each, the dispatcher considers some facts. All three trucks are idle, and each can handle the weight of the delivery in one trip; but the van is less expensive to operate in any kind of driving situation than either of the other two. Therefore, the van can automatically be selected over the other two, and 18 of the 27 alternatives drop out. Furthermore, since the drivers are to be paid during the 2-hour period whether they drive or not, the decision of which driver to send is not an economic one; any driver can go. This eliminates 6 more of the alternatives, leaving only the three routes to be examined. The final decision is then easily and quickly made by calculating the cost of each of the 3 alternatives:

City route: 30 miles × 12¢ per mile = $3.60

Toll road: 45 miles × 10¢ per mile + 50¢ toll = $5.00

Limited-access road: 50 miles × 10¢ per mile = $5.00

FIGURE 11-1
Decision Tree for Delivery Problem

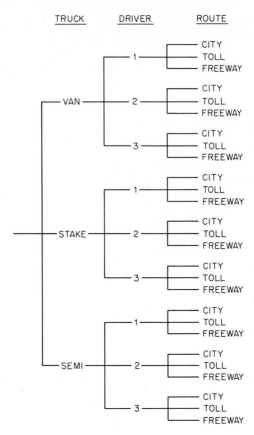

So, the dispatcher assigns any driver, and gives instructions to take the van and deliver the 1,000 pounds using the city route.

In practice, the dispatcher probably would not have made this kind of analysis in such a simple situation. However, this example was carefully constructed to show that all the alternatives can be considered, and, usually, in a reasonable amount of time. It is recommended that the planner use this approach to eliminate as many alternatives as possible, and then document the reasons why they were eliminated before proceeding to compare the remaining alternatives. A great deal of time and work can be saved in this way.

In the preceding example, the solution became trivial because the decisions were independent. However, an analyst should be careful not to eliminate alternatives before investigating the situation thoroughly, to make certain that the decisions are truly independent and separate. For instance, if the load had been larger and had required multiple trips, the vehicles could not have been eliminated so easily, and the number of alternatives requiring calculations would have increased. Or, if it were illegal to drive the semi over city streets, the decisions about which vehicle and which route to take would not be independent. In both of these examples, the decisions are not separate; as a result, the number of alternatives to consider will increase.

QUANTIFYING THE ALTERNATIVES

Before alternatives can be compared, they must be converted to dollars and cents. Unless this is done, the advantages and disadvantages can be listed but not compared, because there is no way to measure them.

In any comparison, it is the differences which are important. Those things which are common can be ignored since they will occur under any alternative. In an economy study, these differences are called *incremental* costs; the similarities, *common* costs. Although this distinction may seem clear, there are times when it becomes difficult to say which costs are common and which are not.

Common Costs

Assume the problem is to decide between buying one of two cars. Both cars have the same tires and accessories, but different bodies and engines. Car A costs $4,500 to buy and $750 per year to operate; Car B costs $5,000 to buy and $700 per year to operate. The following table shows a detailed breakdown of these costs.

	Car A	Car B
First expenditures		
Body and engine	$3,500	$4,000
Tires and accessories	1,000	1,000
	$4,500	$5,000
Operations costs		
Insurance	$ 100	$ 150
Gasoline	500	400
Upkeep	150	150
	$ 750	$ 700

At least three approaches can be taken to the determination of common costs. The first approach is to argue that two different automobiles are being considered. Even though they have certain similarities, nothing is really common about them. The costs that should be used for the comparisons are therefore $4,500, $5,000, $750, and $700. The comment that the tires and accessories are the same would be disputed with the statement that no two tires or two radios or two of anything else are ever the same; they will wear and operate differently, even though they are the same model. A second approach, at the other extreme, is to argue that a dollar is the same as any other dollar when it is spent at the same point in time. The costs that should then be considered are therefore the incremental costs—or $0, $500, $50, and $0, as the next table shows.

	Car A	Car B
First cost	$4,500	$5,000
Common costs	4,500	4,500
Incremental capital expenditure	$ 0	500
Operations costs	$ 750	$ 700
Common costs	700	700
Incremental operations costs	$ 50	$ 0

Finally, one of the many middle-ground approaches would be to agree that the capital costs are $3,500 and $4,000. These can be found simply by removing the $1,000 common cost for tires and accessories. The operations costs are $100 and $50, and these could be found by considering the $100 for insurance, $400 for gasoline, and $150 for upkeep—all as common costs.

This example shows that the treatment of common costs does not really affect the incremental costs, because the same differences will still be there. However, as Chapter 12 will point out, the treatment of common costs does affect the measurement of the significance of these differences.

The final decision about common costs depends on the purpose for making a cost estimate. As an example, if the car were going to be used in a commercial venture, it might be appropriate to include all costs in order to see how much total revenue would be required to provide the desired return. On the other hand, if someone were interested only in deciding which of the cars would be most economical to own, any of the three approaches for removing common costs would be satisfactory.

Sunk Cost

A special type of common cost which can normally be disregarded is the *sunk cost*, or the original cost of existing assets. In the earlier example about buying a car, the fact that the prospective buyer may have just spent $2,500 to build a garage would be disregarded. Whether or not the car is bought, the garage has been built and must be paid for. The cost of the garage is common to each alternative, and is a sunk cost.

Retirement of Assets

On the other hand, there is a need to consider receipts or disbursements that will occur because existing assets will be retired at some point in the future. The salvage that can be realized from these assets, the costs of operating them, the taxes that are paid on them, and the costs for replacing them—all of these are relevant costs. For instance, suppose the purchase of a new car required a decision about whether to keep an old one for convenience, or else to sell it for $1,000. Of course, this convenience will cost at least as much as the cost to operate the vehicle. One might think that because the car had already been paid for, there would certainly be no additional capital cost for keeping it. However, in addition to the operating costs, there is a capital cost, which possibly could exceed $1,000. This would be the case if the owner kept the car without trying to sell it again, until there was no longer any market for it, and, as a result, had to pay to have it hauled away. In other words, the owner not only passed up $1,000, but paid to have the car removed. That money could have been invested, and it would have brought a return. Or it could have been used to avoid the interest which has to be paid on borrowed money. Under these conditions, the convenience of the car would cause a capital cost of more than $1,000. Similarly, if the owner had not held the car quite so long, but had sold it later for $300, the capital cost (not including the income tax considerations) would have been over $700, that is, $1,000 − 300(P/F, i\%, N)$.

Whenever there is a choice of either retiring an existing asset or keeping it, as in the preceding example, its salvage value at the time it could have been sold should be recognized as a capital cost in the plan which calls for keeping it. An equally satisfactory way of handling the situation would be to credit the plan, or plans, that would retire the asset with a capital savings equal to the salvage value. However, at no time should one plan be charged and the alternative credited for the same sum. That would be double-counting. Chapter 13 will discuss replacement study considerations.

Using Spare Capacity

Another problem in pricing alternatives is that very often one plan makes use of *spare capacity* in existing plant, whereas a competing plan does not. For example, Plan 1 may use spare channels in an existing coaxial route; Plan 2 may establish a microwave system. Or Plan A may place a cable, such as a trunk cable, between points X and Y using a spare duct in a conduit structure, whereas Plan B may bury an equivalent cable over an alternate route between the same two points.

The question in each case is this: How much should be charged to the first plan for using spare capacity? The answer is:

1. The immediate cost of making the spare capacity serviceable, including any necessary repairs

2. Any increase in the cost of operations for keeping the plant in service over the costs of idle status

3. The cost of advancing the time when new capacity must be added because the spare has been used

For example, Plan A places a cable in a spare duct of an existing conduit run. Assume that this duct is one of three spares available now. Further assume that ducts cannot be made available through retirements of old cables. For item 1, assume that no repairs are required, and that the cost of rodding the duct will be charged to the first cost of the cable. For item 2, assume that maintenance costs will not increase. Even under these conditions, the answer is not yet clear. Item 3, the cost of advancing future conduit relief, suggests these two questions.

- When will the conduit be reinforced if Plan A *is not* carried out?

- When will the conduit be reinforced if Plan A *is* carried out?

The exchange plan indicated that a cable will be added every 3 years for an indefinite period. If the spare duct is not used now, the next cable addition is scheduled in year 3.

Without Plan A	*With Plan A*
Year 3: Spare duct No. 1 to be used	Now: Spare duct No. 1 to be used
Year 6: Spare duct No. 2 to be used	Year 3: Spare duct No. 2 to be used
Year 9: Spare duct No. 3 to be used	Year 6: Spare duct No. 3 to be used
Year 12: Conduit to be reinforced	Year 9: Conduit to be reinforced

Plan A therefore advances the reinforcement of the duct run by 3 years—from year 12 to year 9. It should therefore be charged with the present worth of the costs associated with that advancement.

The Cost Assumptions

Any economy study of the future always requires estimates of future costs for all of the items appearing in it. Chapter 6 has already discussed these costs, as well as the sources for basic information. Chapter 6 also included words of caution about assumptions made from these data. For instance, the use and misuse of average costs was treated, and it was also pointed out that direct savings or costs often are not accompanied by proportionate changes in indirect values. However, even though it is difficult to estimate future costs, it is also unavoidable. No valid study can be put together without estimates of costs by which to measure the alternatives.

The cost assumptions which an analyst makes for an economy study should be tailor-made to fit the exact problem at hand. Generally, these assumptions will result from the best experience and judgment of those involved in the project. Since these assumptions are just that—assumptions—they will continually be subjected to review as the project moves from its early planning and study phases through the proposal, implementation, and monitoring stages. It is therefore a good practice to keep accurate records of each data source and the resulting assumptions.

COMPARABLE ALTERNATIVES

A very important matter in all economy studies is to be sure that the alternatives being compared are truly comparable. This means that all alternatives must provide essentially the same service and over the same period of time. For example, two alternatives would not be comparable if one provided service for 5 years and the other for 10 years. In order to make a valid comparison here, it would be necessary to assume that something is done to provide service for an additional 5 years in the one plan, or that service under the second plan would be terminated at the end of 5 years.

Coterminated Plant or Repeated Plant

The time period over which each alternative provides service is often determined by the economic life of the plant used. Often it is possible to design alternatives whose various elements have economic lives with a coincident termination point. Sometimes, alternatives can have a common terminating point because of some common factor other than economic life. Studies of sets of alternatives which have a common termination date are said to be *coterminated,* and the assumption of a common termination date is known as the *coterminated plant assumption.*

Many studies do not have a common factor which determines the period; nor do the economic lives of the various elements coincide. This mismatch in timing must be recognized in order to get valid results. The best way to handle the problem would be for the analyst to have infallible foresight, and to be able to forecast the events which would occur under any alternative as far into the future as necessary. But since few are blessed with infallible foresight, the analyst needs to make an assumption which will make the alternatives comparable. The assumption often made is that when existing plant is retired, it will be replaced by plant which is essentially the same as the retired plant, both in operation and in cost. This assumption is known as the *repeated plant assumption,* which is a very powerful tool, as will be shown here. If the repeated plant assumption is made, there is some point in time when the annual equivalent costs will either be common to each alternative and can be ignored; or else they will have a constant difference which can be evaluated.

The following example illustrates the contrast between the coterminated plant assumption and the repeated plant assumption.

Example. Suppose the firm has an old truck which was bought several years ago for $6,000, and is now wearing out. The firm estimates that it has two alternatives:

> *Alternative A:* Sell the old truck now for $1,800, and buy a new truck for $7,500. Upkeep and taxes for the new truck will be $800 each year during its life.

Alternative B: Keep the old truck for 2 more years, with upkeep and taxes of $950 in the first year and $990 in the second. At the end of the second year, sell the old truck for $1,500 and buy a new one for $7,500. At that time the upkeep and taxes for the new truck will also be $800 per year.

The $6,000 cost of the old truck (whether it has been paid for or not) is a sunk cost common to both alternatives and it is irrelevant to the decision about buying a new vehicle. However, this does not mean that there are no capital costs associated with keeping the old truck. As long as there is some salvage value now which is not obtained, there will be some capital cost. This capital cost will be the result of the foregone opportunity to realize salvage currently, which is offset by the realization of any future salvage. In both of the following sets of time-cost diagrams, the salvage value of the existing truck has been shown as a charge to the plan that keeps the truck. However, it could just as well have been shown as a credit to the plan which retires the truck now.

The decision between the two alternatives depends, in part, on the appropriate assumptions about the life of the new truck.

Coterminated Plant Assumption

As an example of the coterminated plant assumption, suppose the truck being considered is a rather special truck, which will be needed for only 8 more years. If the truck is bought now, it is estimated to have a salvage value of $2,000 in 8 years. If it is bought 2 years from now, it is estimated to have a salvage value of $2,300 upon retirement. Figure 11-2A illustrates this situation.

These two alternatives are directly comparable because they provide essentially the same service over the same period of time—the two plans are coterminated.

FIGURE 11-2A
The Coterminated Plant
Assumption

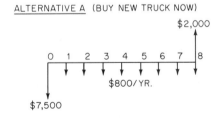

ALTERNATIVE A (BUY NEW TRUCK NOW)

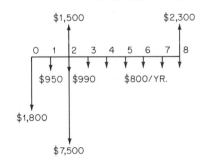

ALTERNATIVE B (KEEP OLD TRUCK 2 YEARS;
THEN BUY NEW TRUCK)

FIGURE 11-2B
The Repeated Plant Assumption

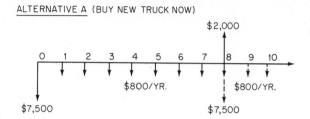

ALTERNATIVE A (BUY NEW TRUCK NOW)

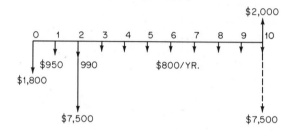

ALTERNATIVE B (KEEP OLD TRUCK 2 YEARS;
THEN BUY NEW TRUCK)

Repeated Plant Assumption

As an example of the repeated plant assumption, suppose now that the truck is used in the normal business operation. The need for a truck of a similar type is expected to continue indefinitely. In this case, a truck would be expected to have about the same useful life whether it is bought now or 2 years from now. If this type of truck is expected to have an economic life of 8 years and a salvage value of $2,000, the plans are not coterminated, as illustrated in Figure 11-2B.

Figure 11-2B, without the expenditures depicted by dotted lines, shows that these two alternatives do not provide service over the same period of time. In fact, even if replacements are assumed (dotted lines), the two alternatives will never coterminate as long as the replacements have an equal economic life. However, suppose replacements can be assumed to have costs equal to the costs of the original trucks. Then, after year 2, the equivalent annual costs in each year of an economic life cycle will be the same in either alternative. In that case, it is not important that the replacement cycles are not occurring coincidently, and the common equivalent annual costs can be neglected. Therefore, by converting the costs into annual equivalent costs, the plans are made comparable provided the study period is any length longer than 2 years.

In this example, the cost estimates did not change with time as they would have with inflation. If inflation is a factor, the fact that replacement cycles are out of phase may prevent the assumption of common future costs. In fact, the assumption of inflation contradicts an important part of the repeated plant assumption; and the problem of comparing non-coterminated alternatives becomes more complex, as Chapter 13 illustrates.

INTANGIBLES

As discussed earlier in the chapter, an economy study is a tool to use in decision making. However, to maximize the effectiveness of this tool, the analyst must convert as many details of the various alternatives as possible into dollars and cents, because only in this form are they directly comparable. Nevertheless, some very important factors defy accurate conversion. These are called *intangibles*, or *irreducibles*. An example might be a plan which provides a more attractive building at a higher cost than an alternative plan. The problem is, how much is beauty worth? The answer is that beauty holds a different value for different people. Yet no study is complete until it has accounted for all differences among alternatives.

It is important for the analyst to identify and evaluate any intangible factors in a study. When the study later becomes part of a proposal, the decision makers can consider those intangibles and decide whether such factors as moral responsibility, goodwill, or beauty justify the higher cost. Of course, the judgment will be subjective, and is likely to vary from study to study, and from decision maker to decision maker. Even though the analyst's primary responsibility is to evaluate tangible factors, it is necessary to evaluate intangible factors too, because often they influence the final decision.

SUMMARY

The principles of engineering economy should be applied whenever decisions are required which affect costs. Even when the decision appears obvious, the situation will benefit from an engineering economy interpretation—if not a complete study. It is very important to consider all viable alternatives, because a study can select the best alternative only from those alternatives which were evaluated. Developing the alternatives requires inquisitiveness and innovation. However, knowledge of the following four conditions will help to provide the basic framework:

1. Knowing the present situation.
2. Defining the future objective situation.
3. Knowing the composition of the interval between the present and the future situations.
4. Knowing the constraints on the problem.

Alternatives which are going to be compared must really be comparable. They must provide essentially the same service over the same period of time. Studies in which all plant is retired by a certain common date (the coterminated plant assumption) satisfy the time period criterion. Studies in which all plant does not retire by a coincident date may be able to make use of the repeated plant assumption to make the time periods comparable. The repeated plant assumption says that when plant is retired, it is replaced with similar plant, having similar costs, so that at some point in the future the equivalent annual costs of all alternatives become common and can be excluded. The repeated plant assumption relies upon an assumption that costs will not change with time. As a result, its versatility and power are eroded when inflation does cause changes in the costs. Since the repeated plant assumption is just a way of allocating the costs of alternatives to make them comparable, some other means of allocating costs must be used to achieve this goal if inflation is a factor.

Complete identification of each alternative requires that all possible items be quantified in terms of money. Engineering economy concepts are applied only to those factors which can be expressed in monetary terms. Although intangibles cannot be quantified in the economy study, they must be identified, too. The ultimate decision maker can then weigh the significance of the intangibles, together with the economic considerations, to make the final decision.

12

ECONOMY STUDY TECHNIQUES

INTRODUCTION

Once the alternative plans for an economy study have been developed, the next logical step in the decision-making process is to compare them. If the plans have been developed objectively, all of the viable alternatives will have been identified and supported by cost data that is as complete and as reliable as possible. Several techniques can be used to compare plans and to select the most economical one. However, if the alternatives are not identified objectively, or if the cost data are not complete and reliable, the best possible application of any study technique will be worthless.

Sometimes, it is possible to make a meaningful and valid comparison of alternatives directly from raw cost data. Such direct comparisons are possible if there are few costs, if the costs occur at the same time, and if the differences between costs are always in the same direction. For instance, suppose a company must choose between machine A and machine B. Each would satisfy the project requirements as well as the other; and they would have identical service lives with equal salvage values at the end of that service life. However, machine A may cost $10,000 to buy and $3,000 a year to operate, whereas machine B costs only $7,500 to buy and $2,000 a year to operate. Machine B is the easy choice from the viewpoint of economy, since it is obviously less expensive in every respect.

Yet, it would take just a simple reversal of the operations costs to create a significantly harder problem. With annual operations costs of $2,000 for machine A and $3,000 for machine B, it becomes necessary to apply the concepts of the time value of money and equivalence which were developed in Chapter 5. When these concepts are applied, at least three additional pieces of information are required which were not needed before.

First, it is now necessary to know the expected service lives of the machines. In the original problem, it was enough to know only that the machines would have unspecified, but equal, service lives. This was enough because B would always be more economical than A so long as A cost more to buy and more to operate in each year of their common life. However, as soon as B becomes more expensive to operate, the lives become critical because they determine how long it will take A to overcome the disadvantage of the $2,500 higher purchase cost, given the advantage of the $1,000 lower yearly operating cost. Obviously, a $1,000 savings for only 1 year will not adequately compensate for the $2,500; but a $1,000 savings for 10 years will, unless income taxes and the cost of money are very high.

The income taxes and cost of money, then, are the other two pieces of information now required to complete the comparison. Again, as long as A was more expensive to buy and to operate than B, it was unimportant to know what the taxes and cost of money were because they are directly proportional to the capital spent. However, as soon as A becomes operationally cheaper, the taxes and cost of money affect the length of time required to compensate for the higher purchase price. There is another very fundamental reason why it is necessary now to know the cost of money. As earlier chapters pointed out, the return required on capital is actually the cost of money. As a result, it is a very basic value in analyses which apply either a *present-worth technique* or a *rate-of-return* technique. Both of these techniques will be discussed in great detail in this chapter. For the moment, though, it is enough to know that the cost of money is primarily the discount rate in present-worth analyses, and is the dividing line between accepting or rejecting an alternative in rate-of-return analyses.

Incremental Cost Differences

Most of the time, economy study techniques are applied to problems like the machine A versus machine B problem. These techniques develop values representing (on an equivalent time basis) the various costs and revenues of competing, *mutually exclusive* alternatives. A set of alternatives is said to be mutually exclusive when choosing one of them rules out the choice of any of the others. Regardless of how many alternatives are in the set, the choice of one of them depends on the incremental differences in the long-term costs and revenues for each.

Sometimes, one of the alternatives will be a *null* alternative—meaning that no costs or revenues need to be estimated for it. For example, continuing the present method of operation might be a null alternative for which costs are not expected to change. If a new method of operation is proposed, its changes in costs are the incremental costs which are compared with the null. Examples 12-1A and 12-1B demonstrate situations in which there are only two alternatives. In Example 12-1A, one alternative is a null alternative, so that costs must be developed only for the other alternative. In Example 12-1B, however, costs must be developed for both alternatives.

Example 12-1A. Suppose a machine shop owner has an opportunity to buy a machine which will allow the addition of a new product line. This machine will cost $5,000. The receipts, less taxes and operations costs, are expected to result in a net cash increase of $2,000 each year for N years. If this machine is the only one which will produce the specific product, the decision is really between two alternatives: (1) buy the machine; or (2) do not buy the machine. If the yearly $2,000 net cash is enough at the owner's cost of money to cover the $5,000 capital expenditure, the decision is simple: The machine should be bought. If the $2,000 is not enough, the machine should not be bought.

Example 12-1B. Now suppose another machine shop owner already owns a machine which is an integral part of a complex process. This machine is worn out, though, and a replacement is needed. Two machines, Y and Z, which will do the job, are on the market. It would cost $15,000 to buy Y, and the annual operating costs would run $3,000. On the other hand, Z would cost $20,000; but it is fully automatic and would cost only $1,000 a year to operate. Assuming that the shop's revenues would be unaffected by the choice of machine, and neglecting the effects of income taxes for simplicity, the following time-line diagrams for these machines would represent the situation.

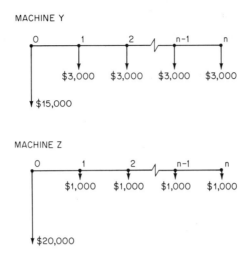

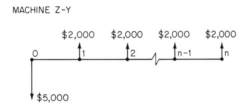

The owner must have one of the machines; the question is, which one? In order to make the decision, the owner looks at the incremental cost differences, which appear in the following time-cost diagram, labeled MACHINE Z-Y. This diagram is found by subtracting the diagram for Y from the diagram for Z.

Notice here that if the common revenue had been included for Y and Z, it would have disappeared in the incremental Z - Y cash flow anyway. Notice, too, that this incremental cash flow is the same as that confronting the owner in Example 12-1A. This owner's decision should be to buy Z if the $2,000 in saved operating expenses covers the additional $5,000 capital expenditure at the owner's cost of money. If it does not, the owner should buy Y.

The incremental cost analysis of Example 12-1B is sometimes called the *with and without* principle, because the problem is considered from the viewpoint of buying or not buying Z. Not buying Z, or "without" it, would require a capital expenditure of $15,000. Buying Z, or "with" it, requires $20,000. Similarly, the operating costs would be $3,000 "without" Z and $1,000 "with" it. The conclusion is that Z costs $5,000 (since a minimum of $15,000 will be spent anyway), and it contributes $2,000 in operating savings each year (when compared with the operating costs of the alternative). The decision to buy Z or not (buy Y) is based upon determining if, at the owner's cost of money, the $2,000 incremental annual benefits are enough to cover the incremental first cost of $5,000.

Three Basic Techniques

This chapter discusses three basic cost comparison techniques which have been found to be effective and consistent with each other: (1) rate of return (*ROR*), (2) present worth of expenditures (*PWE*), and (3) present worth of annual costs (*PWAC*).

When applied correctly, each of the three techniques will lead to the same choice from a set of mutually exclusive alternatives. This point is important to keep in mind because the "numbers" in the input and output are different, depending on which technique is used. This is because each method represents a different viewpoint of the firm. This difference in viewpoint is demonstrated in the following diagram of a project.

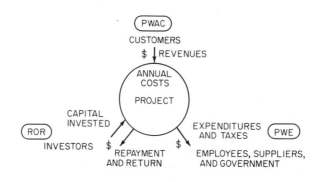

PROJECT DIAGRAM

The rate-of-return technique (ROR) is developed around the viewpoint that each accepted project should represent the most efficient use of money possible. Money is said to be used efficiently in a project if as much as possible is available to be returned in the form of repayment and return to the investor for a given amount invested. This of course is the end goal of any investor—to get back more than was put in. For regulated businesses like the Bell System, this viewpoint is fully compatible with the natural desire of the customer—low bills. Since the regulatory process limits the overall rate of return, any investment which is more efficient than that limit will benefit the customers by reducing the revenues required for the business as a whole.

The viewpoint of the present-worth-of-expenditures technique (PWE) is that the firm should minimize its expenditures for plant, operations costs, and taxes without sacrificing good telephone service. Customers find this concept agreeable too, because in the long run the revenues collected from them must be equivalent to the costs of doing business. If expenditures are minimized, then the need for revenues is also a minimum.

The last of the three viewponts, the present worth of annual costs (PWAC), is almost the same as PWE. The goal of either concept is to minimize expenditures. The difference is, though, that with PWAC the capital costs, and sometimes the operations costs, are converted into levelized equivalent annual costs before the present worth is calculated. Over the total life of an investment, PWAC will be equal to the PWE. For some period different from the life, PWAC is an implied allocation of capital costs. To handle the same situation, the PWE technique requires an explicit allocation of capital costs, and this may not be the same as the PWAC allocation.

The use of each of these three techniques, as well as some of the problems which can arise if they are not applied correctly, are explained in detail in the three main sections of this chapter.

RATE-OF-RETURN TECHNIQUE

The first technique, rate of return, is a measure of how attractive an alternative is from the viewpoint of investors. Keep in mind that this technique is developed around the concept of making the most efficient use of money. Investors in the firm provide the business with capital in anticipation of a future return. The firm must then assume the role of the investor in specific projects. One way for the firm to decide how to invest the funds entrusted to it is to make sure that each increment of those funds is invested at an efficiency equal to, or greater than, the rate its investors would want. Rate-of-return analyses therefore treat an incremental cash flow as a time-value-of-money problem. As Chapter 5 explained, any time-value-of-money problem has four elements:

1. Some amount, or series, of money

2. An equivalent amount, or series, of money

3. An interest rate

4. The number of periods of time

However, rate-of-return analyses usually combine the two amounts, or series, of money into one series, called the *net cash flow*. The amount to be invested is either known or estimated, and the receipts, or cost savings, less the expenditures for other costs and taxes are estimated too. The result is the net cash flowing from the firm into the project, and back to the firm from the project. It is important to recognize that the net cash flow actually combines two equivalent amounts or series of money. The rate of return is the interest rate at which they are equivalent.

The two principal elements of net cash flow are the investment and its recovery. Of course, in complex situations, it may be difficult to tell which is which; but because the two are equivalent, they have the same present worths, except that they have opposite signs. Therefore, the usual definition of the rate of return is that interest rate which will cause the present worth of the net cash flow to be zero:

$$0 = \sum_{n=0}^{N} NCF_n(P/F, r\%, n) = \sum_{n=0}^{N} \frac{NCF_n}{(1 + r)^n} \tag{12.1}$$

where n = each year of the period
$\quad N$ = number of years
$\quad NCF_n$ = net cash flow in year n
$\quad r$ = rate of return (interest rate)

Given this definition, the rate of return is called an *internal rate of return* (*IROR*). It is called *internal* because the rate is based solely upon the cash flow, and there are no external influences.

Rate of return can also be defined as that interest rate which will cause the final balance of the net cash flow to equal 0. This definition is based upon the balance diagrams discussed in Chapter 5 and the future worth of the net cash flow. Because the present worths of the two elements of the net cash flow are equal and of opposite sign, their future worths must be equal too. This definition provides some additional insight about what the rate of return actually represents. Recall from Chapter 5 that the balance diagram for the deposit of funds in a bank, followed by withdrawals, resulted in a zero final balance when the diagram was constructed at the rate of interest paid by the bank. This means that the rate of return is the interest rate paid by the bank. For projects in which a business invests, the rate of return is the interest rate which the project is paying on funds which are deposited or invested in it.

The usual way to find the internal rate of return (*IROR*) is to solve equation (12.1) by trial and error. First, the *IROR* is estimated, and the equation is evaluated using this estimate. If the estimate is not equal to the solution rate, the equation will have some non-zero value which is called the *net present value* (*NPV*). This *NPV* is important because it shows whether the solution—the *IROR*—is actually greater or less than the estimate. If the *NPV* is positive, the *IROR* is greater than the estimate; if the *NPV* is negative, the *IROR* is less than the estimate.

Example 12-2 illustrates how equation (12.1) is applied for a project which has a null alternative.

Example 12-2. Suppose equipment can be bought for $1,000 which would have combined operations and income tax costs of $200, $225, and $260 during the 3 years of its proposed life. The revenues resulting from the operation of that equipment for these years would be $600, $650, and $725, respectively. At the end of the final year of operation, the equipment can be removed and junked for a net salvage of $50. The *IROR* of this proposed investment compared with doing nothing (i.e., not buying the equipment) would be calculated using formula (12.1). The cash-flow diagram for this project is

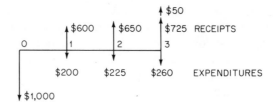

Subtracting expenditures from receipts gives the net cash flow (*NCF*):

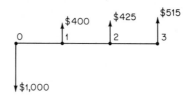

These values can now be substituted into equation (12.1):

$$0 = \sum_{n=0}^{N} \frac{NCF_n}{(1 + r)^n}$$

$$0 = NPV = \frac{-1,000}{(1 + r)^0} + \frac{400}{(1 + r)^1} + \frac{425}{(1 + r)^2} + \frac{515}{(1 + r)^3}$$

The next step is to select a trial rate (r), and to begin a trial-and-error solution. Experience will help to determine a suitable starting rate.

Table 12-1 illustrates the trial-and-error solution for *IROR* in Example 12-2, and finds that the *IROR* is 15.45%. For most purposes, however, it would have been satisfactory to interpolate between the first two trials, or to graphically solve for the *IROR* by plotting the *NPV* versus the *IROR*.

The graphic solution for Example 12-2 is:

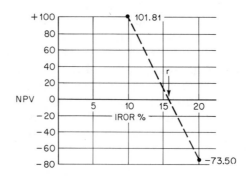

TABLE 12-1
INTERNAL RATE OF RETURN
(Trial-and-Error Solution for Example 12-2)

Equation to be solved for *IROR* (r): $\dfrac{-1,000}{(1+r)^0} + \dfrac{400}{(1+r)^1} + \dfrac{425}{(1+r)^2} + \dfrac{515}{(1+r)^3} = 0$

Assumption	Value of terms				Net present value	Next step suggested by the net present value
	$\dfrac{-1,000}{(1+r)^0}$	$\dfrac{400}{(1+r)^1}$	$\dfrac{425}{(1+r)^2}$	$\dfrac{515}{(1+r)^3}$		
r = 10%*	$-1,000	$363.64	$351.24	$386.93	$101.81	Raise value of r to reduce NPV toward zero.
r = 20%	−1,000	333.33	295.14	298.03	−73.50	Reduce value of r to about 15% since 10% and 20% produce similar NPVs but with different signs.
r = 15%	−1,000	347.83	321.36	338.62	7.81	Raise value of r to reduce NPV toward zero.
r = 16%	−1,000	344.83	315.84	329.94	−9.39	Reduce value of r to about 15.5%
r = 15.5%	−1,000	346.32	318.58	334.24	−0.86	Reduce value of r still more.
r = 15.4%	−1,000	346.62	319.14	335.11	0.87	Raise value of r to about 15.45%.
r = 15.45%	−1,000	346.47	318.86	334.68	0.01	STOP. Equation is essentially solved. IROR = 15.45%.

*The starting assumption is somewhat arbitrary. The cost of money is often a good starting point, but experience will provide the best guide.

Either a graphic or an interpolated solution is an approximation which is usually satisfactory and will save the many trials needed to gain a precision that is probably unwarranted by the original cost estimates.

The balance diagram for this net cash flow can now be constructed to demonstrate the future-worth definition for the rate of return. At 15.45%, it is as follows:

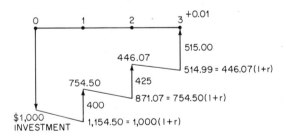

The balance diagram illustrates that this cash flow is analogous to the cash flow of a depositor who places $1,000 in a bank and then makes three withdrawals. If the bank (the project) pays 15.45% interest, the account balance is exactly zero at the end.

In this very simple case, the decision is relatively simple. This project is beneficial if the cost of money, or the investor's required rate of return, is less than the 15.5% rate of return. It is not beneficial if the cost of money is greater. However, this decision views the project in isolation. If the business can get $1,000 with a cost of money less than 15.5%, this purchase should be made. If there are other places to spend the $1,000, this rate of return by itself is not sufficient evidence to justify this project, and to eliminate another project with a lower rate of return. As Example 12-3 will illustrate, the *IROR* determines only if the project itself is beneficial, but it is not a measure of that benefit at the firm's cost of money.

Incremental Internal Rate of Return

Example 12-2 illustrated the computation of an *IROR* on a project compared with a null alternative. In that illustration, it was possible to estimate the future revenue for the project. In most practical situations, though, it is not possible to identify specific revenues for specific alternatives. Besides, even if it is possible to identify all of the costs and revenues for specific alternatives, it is necessary to use the incremental *IROR* to compare alternatives because *IROR* considers each alternative in isolation. Therefore, if *IROR* is used as a decision-making tool, it must be based on the incremental differences between alternatives.

It is very important to keep in mind that the *IROR* is the critical cost of money which one project could tolerate and still break even. There is, of course, some cost of money at which the firm will operate. To compare alternatives, it is necessary to determine the relative operating level. Example 12-3 illustrates this, and in a simpler way, the following analogy to rate of return also illustrates it.

Peanuts—An Analogy to Internal Rate of Return

Suppose you are going into the peanut business. You buy the bagged nuts and hire a vendor to sell them for you, using your old vending cart. Peanuts cost you 3 cents per bag wholesale, and they sell for 25 cents per bag. This leaves 22 cents per bag to pay the wages of the vendor; your profit is whatever is left over.

How do you want to pay your vendor? Let's say you have settled on one of two ways:

- $200 per week salary
- $100 per week plus 10 cents per bag commission

If 22 cents per bag are available to pay the salary of $200 per week, the vendor must sell

$$\frac{200}{0.22} = 909 \text{ bags/week}$$

to earn just enough to pay the salary. At 909 bags/week, you break even; if the vendor sells more, you make money.

On commission the vendor makes

$$\text{Wages} = \$100 + \$0.10 \text{ (sales)}$$

Sales to break even are then

$$\text{Sales} = \frac{100 + 0.10 \text{ (sales)}}{0.22}$$

$$0.22 \text{ (sales)} = 100 + 0.10 \text{ (sales)}$$

$$0.12 \text{ (sales)} = 100$$

$$\text{Sales} = \frac{100}{0.12} = 833 \text{ bags/week}$$

If the vendor is on commission and sells more than 833 bags/week, you make money. Now you know that any peanut vendors worth their salt (not included in the costs) can sell an average of 1,200 bags of peanuts per week. So you decide that the commission is a better way to pay, because it has a lower break-even sales level; this means you are getting the profits on more bags.

That may in fact be the correct conclusion, but further analysis is needed. If your vendor sells 1,200 bags of peanuts a week, you take in

$$1,200 \times \$0.25 = \$300$$

and you paid, for the peanuts,

$$1,200 \times \$0.03 = \$36$$

This leaves

$$\$300 - \$36 = \$264$$

to pay the vendor and for your profit. If the vendor is on salary, your profit for the week is

$$\$264 - \$200 = \$64$$

On commission, your profit for the week is

$$\$264 - (\$100 + 1,200 \times \$0.10) = \$44$$

Obviously, if the sales are the same 1,200 bags under either plan, you are much better off paying the vendor a flat salary. This is also evident from the following sketch.

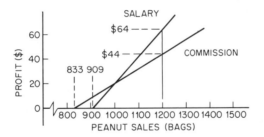

Another way to determine this relationship is with an incremental analysis. This would produce the sales level at which the wages of the vendor were the same whether on a flat salary or on commission:

$$\$200 = \$100 + 0.10 \text{ (sales)}$$

$$\text{Sales} = \frac{100}{0.10} = 1,000$$

Now you would know that below 1,000 sales the commission pays the vendor less than the salary, which means you would make more. Above 1,000 sales, the commission is more than the salary, and you make less. Of course, the intent of the commission plan is to encourage the vendor to sell more. But how much more is the critical factor, not the break-even levels.

The internal rate of return on each alternative provides the critical cost of money for breaking even, for each alternative. However, to make a decision between two alternatives, it is necessary to know the relative advantage of one alternative over the other at the expected operating level. The incremental *IROR* provides this information because it determines the intersection—the critical cost of money at which the alternatives switch. Neither *IROR* or incremental *IROR* measures the profitability at the actual operating level, however.

Example 12-3. For simplicity, Example 12-2 excluded the computations for income taxes. Example 12-3, however, illustrates incremental *IROR* including the income tax computations for a firm with 100% equity financing to demonstrate these tax effects without the complexities of deductible interest.

Suppose a company must buy either equipment A or equipment B in order to perform a necessary task. Equipment A would cost $4,000 to buy and install, with operations costs of $1,300 per year for a 3-year life. It would have a gross salvage of $400 and would cost $300 to remove. On the other hand, equipment B would cost $5,000 to install and have operations costs of $700 per year during the 3 years. The gross salvage of B would be $500, with a $300 cost of removal. Suppose further that the firm can identify $4,000 per year of revenues which it will receive for the task either of these machines performs.

The following tables illustrate how the firm could find the net cash flow for each alternative if the income tax rate is 48% and straight-line depreciation is used for both tax and book purposes.

Alternative A

	0	*1*	*2*	*3*
			Year	
1. Revenue		$4,000	$4,000	$4,000
2. Operating expenses (excluding depreciation)		1,300	1,300	1,300
3. Depreciation (4,000 − 100)/3		1,300	1,300	1,300
4. Net operating revenue (1 − 2 − 3)		1,400	1,400	1,400
5. Income tax (4 × 0.48)		672	672	672
6. Net income (4 − 5)		728	728	728
7. After-tax cash flow (1 − 2 − 5)		2,028	2,028	2,028
8. Capital/salvage	4,000			−100
9. Net cash flow (7 − 8)	−4,000	2,028	2,028	2,128

Alternative B

	0	*1*	*2*	*3*
			Year	
1. Revenue		$4,000	$4,000	$4,000
2. Operating expenses (excluding depreciation)		700	700	700
3. Depreciation (5,000 − 200)/3		1,600	1,600	1,600
4. Net operating revenue (1 − 2 − 3)		1,700	1,700	1,700
5. Income tax (4 × 0.48)		816	816	816
6. Net income (4 − 5)		884	884	884
7. After-tax cash flow (1 − 2 − 5)		2,484	2,484	2,484
8. Capital/salvage	5,000			−200
9. Net cash flow (7 − 8)	−5,000	2,484	2,484	2,684

The *IROR* for each alternative can now be found by trial-and-error solution of equation (12.1) for line 9, net cash flow.

$$\text{For } IROR_A: 0 = \frac{-4{,}000}{(1 + r_A)^0} + \frac{2{,}028}{(1 + r_A)^1} + \frac{2{,}028}{(1 + r_A)^2} + \frac{2{,}128}{(1 + r_A)^3}$$

$$IROR_A = 25.2\%$$

$$\text{For } IROR_B: 0 = \frac{-5{,}000}{(1 + r_B)^0} + \frac{2{,}484}{(1 + r_B)^1} + \frac{2{,}484}{(1 + r_B)^2} + \frac{2{,}684}{(1 + r_B)^3}$$

$$IROR_B = 24.3\%$$

The fact that $IROR_A$ (25.2%) is higher than $IROR_B$ (24.3%) does not mean that Alternative A is better than Alternative B. The $IROR_A$ of 25.2% indicates only that Alternative A is good if the firm's cost of money is less than 25.2%. The $IROR_B$ of 24.3% indicates only that Alternative B is good if the firm's cost of money is less than 24.3%. Those quantities do not indicate anything about the value of one alternative over the other at the firm's actual cost of money. It is necessary now to apply the with-and-without principle and an incremental analysis to decide between these alternatives. Because Alternative B requires an incremental investment of $1,000 more than A, the incremental analysis should be done on Alternative B – A—that is, the incremental cash flow is found by subtracting the cash-flow table for A from the table for B.

Incremental Alternative B – A

	B – A	Year 0	1	2	3
1. Revenue	(4,000 – 4,000)				
2. Operating expenses (excluding depreciation)	(700 – 1,300)		$-600	$-600	$-600
3. Depreciation	(1,600 – 1,300)		300	300	300
4. Net operating revenue (1 – 2 – 3)			300	300	300
5. Income tax (4 × 0.48)			144	144	144
6. Net income (4 – 5)			156	156	156
7. After-tax cash flow (1 – 2 – 5)			456	456	456
8. Capital/salvage	$\frac{(5{,}000 - 4{,}000)}{[-200 - (-100)]}$	1,000			-100
9. Net cash flow (7 – 8)		-1,000	456	456	556

The incremental-cash-flow table can be constructed even though some of the values in the table of each alternative are not known. The only values that must be known are the following differences: revenues, line 1; operations costs, line 2; depreciation, line 3; and capital expenditures and salvage, line 8. With these values lines 4, 5, 6, 7, and 9 can be computed for the incremental table even though some values in the A and B tables are not known.

The incremental $IROR_{B-A}$ can now be found by solving equation (12.1), using the incremental net cash flow on line 9 of the B – A table.

$$\text{For } IROR_{B-A}: 0 = \frac{-1{,}000}{(1 + r_{B-A})^0} + \frac{456}{(1 + r_{B-A})^1} + \frac{456}{(1 + r_{B-A})^2} + \frac{556}{(1 + r_{B-A})^3}$$

$$IROR_{B-A} = 21.1\%$$

This incremental $IROR$ of 21.1% indicates that the $1,000 incremental cost of B over A is an attractive investment if the cost of money is less than 21.1%. The firm should therefore decide in favor of B if its cost of money is less than 21.1%, but in favor of A if its cost of money is greater than 21.1%.

Figure 12-1 illustrates this analysis graphically. The trial-and-error solutions of the net cash flows for A and for B result in a *NPV* for each estimate of the *IROR*. These *NPV* amounts are shown in Table 12-2 and are the data points plotted in Figure 12-1. The trial-and-error solution of the incremental net cash flow (B − A) produces the differences in *NPV* between these two curves.

The plot of *NPV* versus the cost of money shows that the real indicator of economic merit of one alternative over the other is the *NPV* at the firm's cost of money. If the slope of the two lines were the same, the point at which they intersect the abscissa (the cost of money), which is the *IROR* for each, could be used to determine the most economic alternative. However, since the two curves do not have the same slope, they intersect each other. The incremental *IROR* determines the point at which the two curves intersect each other, and this point can be used as an indicator. To the left of that intersection, representing a lower cost of money, Alternative B is the choice because it has a greater *NPV* than Alternative A. To the right of that intersection, representing a higher cost of money, Alternative A is the choice because it has a higher *NPV*.

It was possible to draw Figure 12-1 in this example because revenues were known. A different revenue level, or the absence of any revenue, would shift the curves up or down, causing the *IROR* of each to change substantially. However, a different revenue level would not change the relationship of the curves; nor would it affect the incremen-

FIGURE 12-1
Comparison of Alternatives A and B in Example 12-3

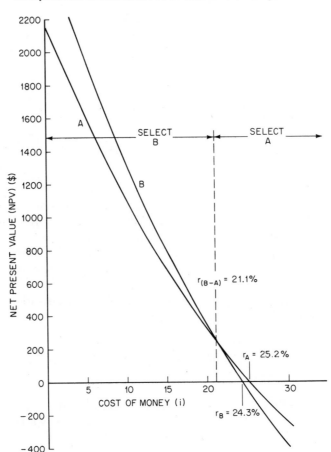

TABLE 12-2
NET PRESENT VALUE DATA FOR FIGURE 12-1

	NCF_n, $			
	n = 0	n = 1	n = 2	n = 3
A	−4,000	2,028	2,028	2,128
B	−5,000	2,484	2,484	2,684
B − A	−1,000	456	456	556

	NPV_i, $						
	i = 0%	i = 5%	i = 10%	i = 15%	i = 20%	i = 25%	i = 30%
A*	2,184	1,609	1,118	696	330	10	−271
B*	2,652	1,937	1,328	803	348	−49	−398
B − A	468	328	209	107	18	−59	−126

*Data rows plotted.

tal *IROR*. A different revenue level, then, would not influence the decision between alternatives.

Decision Rule for *IROR*

Once a valid incremental *IROR* has been computed, it can be used effectively to decide between two mutually exclusive alternatives. The rule follows:

> When considering the *IROR* derived from the incremental cash flows produced by Alternative X minus Alternative Y:
>
> (a) Reject Alternative X if r is less than i.
>
> (b) Accept Alternative X if r is greater than i.
>
> where $r = IROR_{X-Y}$ and i = composite cost of money

If the decision rule is applied to Example 12-2, the recommendation would be to buy the equipment provided the composite cost of money is 15.5% or less. At a cost of money higher than 15.5%, the revenues are not enough to make the project economically worthwhile and the equipment should not be bought.

If the decision rule is applied to Example 12-3, the recommendation would be to buy equipment B provided the cost of money is 21.1% or less. If the cost of money is greater than 21.1%, the savings of B are not enough to support the extra purchase price, and equipment A should be bought instead.

Efficiency, Not Profitability

When managers apply the *IROR*, it is most important that they do not misinterpret their results and believe they have determined the profitability of an alternative or a project. Equipment B compared with A of Example 12-3 had a reasonably high incremental *IROR*. However, Alternatives A and B could easily represent the purchase of something like pollution-control equipment, which would not increase company revenues. In this case, neither alternative would be profitable since the only positive cash flow associated with them would be the gross salvage.

Instead of profitability, the incremental *IROR* measures the economic efficiency of an incremental cash flow. In other words, if pollution-control equipment were being bought in Example 12-3, it would be economically more efficient to buy the higher-priced type B equipment in order to save on operations costs, provided the composite cost of money is 21.1% or less. If the cost of money is higher than 21.1%, it would be more efficient economically to buy A (since one or the other has to be bought) because the operations savings from B would not justify the additional capital costs of B.

Multiple Alternatives

When the *IROR* technique is used to evaluate more than two mutually exclusive alternatives, the problem becomes more complex, and care is needed to ensure that all necessary comparisons are made. Example 12-4 illustrates the computations necessary when three alternatives are being considered.

Example 12-4. Suppose in addition to equipment types A and B, equipment type C can be bought at an installed cost of $5,500. This unit would have operations costs of $450 each year for the 3 years. The gross salvage for type C would be $500, and the cost of removal would be $400. Assuming the same $4,000 of revenue per year with equipment C, the *IROR* for C can be found from the following cash-flow table:

Alternative C

	Year			
	0	*1*	*2*	*3*
1. Revenue		$4,000	$4,000	$4,000
2. Operating expenses (excluding depreciation)		450	450	450
3. Depreciation (5,500 − 100)/3		1,800	1,800	1,800
4. Net operating revenue (1 − 2 − 3)		1,750	1,750	1,750
5. Income tax (4 × 0.48)		840	840	840
6. Net income (4 − 5)		910	910	910
7. After-tax cash flow (1 − 2 − 5)		2,710	2,710	2,710
8. Capital/salvage	5,500			−100
9. Net cash flow (7 − 8)	−5,500	2,710	2,710	2,810

The *IROR* for Alternative C can now be found just as for A and B.

$$\text{For } IROR_C: 0 = \frac{-5,500}{(1 + r_C)^0} + \frac{2,710}{(1 + r_C)^1} + \frac{2,710}{(1 + r_C)^2} + \frac{2,810}{(1 + r_C)^3}$$

$$IROR_C = 23.1\%$$

Again, if the cost of money is less than 23.1%, Alternative C is satisfactory. However, to find if C is better than A or B at the actual cost of money for this firm, it is necessary to find the *IROR* on C over A and also the *IROR* on C over B.

Incremental Alternative C−A

	Year			
	0	*1*	*2*	*3*
1. Revenue				
2. Operating expenses (excluding depreciation)		$−850	$−850	$−850
3. Depreciation		500	500	500
4. Net operating revenue (1 − 2 − 3)		350	350	350
5. Income tax (4 × 0.48)		168	168	168
6. Net income (4 − 5)		182	182	182
7. After-tax cash flow (1 − 2 − 5)		682	682	682
8. Capital/salvage	1,500			
9. Net cash flow (7 − 8)	−1,500	682	682	682

Because the *NCF* amounts after year 0 are equal, the *IROR* for C − A does not require a trial-and-error solution.

$$\text{For } IROR_{C-A}: 0 = -1,500 + 682(P/A, r_{C-A}\%, 3)$$

$$(P/A, r_{C-A}\%, 3) = \frac{1,500}{682} = 2.199$$

Interpolation in the tables gives

$$IROR_{C-A} = 17.4\%$$

More accurately, on a programmed calculator

$$IROR_{C-A} = 17.3\%$$

However, the picture is not complete without a consideration of the $IROR$ for C − B.

Incremental Alternative C−B

		Year		
	0	1	2	3
1. Revenue				
2. Operating expenses (excluding depreciation)		$-250	$-250	$-250
3. Depreciation		200	200	200
4. Net operating revenue (1 − 2 −3)		50	50	50
5. Income tax (4 × 0.48)		24	24	24
6. Net income (4 − 5)		26	26	26
7. After-tax cash flow (1 − 2 − 5)		226	226	226
8. Capital/salvage	500			100
9. Net cash flow (7 − 8)	−500	226	226	126

The trial-and-error solution is necessary for $IROR_{C-B}$.

$$\text{For } IROR_{C-B}: 0 = \frac{-500}{(1 + r_{C-B})^0} + \frac{226}{(1 + r_{C-B})^1} + \frac{226}{(1 + r_{C-B})^2} + \frac{126}{(1 + r_{C-B})^3}$$

$$IROR_{C-B} = 8.4\%$$

The $IROR_{B-A}$ was previously found to be 21.1%.

The diagrams at the top of the next page may make it easier to visualize how the three incremental rates can be assembled to make a decision among the three alternatives.

The diagrams show that if the decision rule is applied sequentially to the incremental $IROR$ for each pair of alternatives, the results are:

If the firm's cost of money is less than 8.4%, choose C.

If the firm's cost of money is between 8.4% and 21.1%, choose B.

If the firm's cost of money is greater than 21.1%, choose A.

The only way the $IROR$ for C − A of 17.3% would be a factor in this example is if Alternative B did not exist. Figure 12-2 illustrates this fact and is also a graphic comparison of all three alternatives. Figure 12-2 is the same as Figure 12-1 except that the NPV versus cost-of-money curve was added for Alternative C. The additional data are included in Table 12-3 on page 284. Figure 12-2 shows that the complex application of the decision rule is required to determine which of the three alternatives has the greatest NPV in each range of possible costs of money.

The firm making this decision must know its cost of money—or, at least, the range in which its cost of money lies. A different firm, with a different cost of money, could make exactly the same study yet reach a different decision.

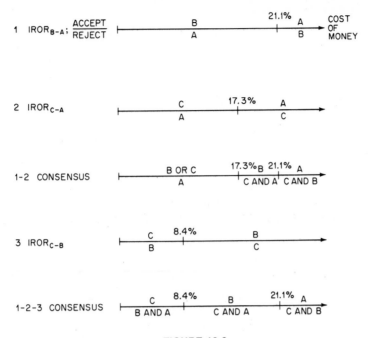

FIGURE 12-2
Comparison of Alternatives A, B, and C in Example 12-4

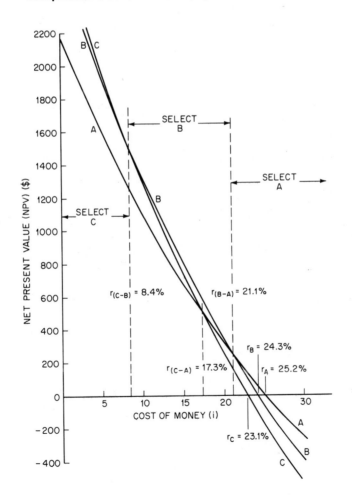

TABLE 12-3
NET PRESENT VALUE DATA FOR FIGURE 12-2

	NCF_n, $\$$			
	$n = 0$	$n = 1$	$n = 2$	$n = 3$
A	−4,000	2,028	2,028	2,128
B	−5,000	2,484	2,484	2,684
B − A	−1,000	456	456	556
C	−5,500	2,710	2,710	2,810
C − A	−1,500	682	682	682
C − B	−500	226	226	126

	NPV_i, $\$$						
	$i = 0\%$	$i = 5\%$	$i = 10\%$	$i = 15\%$	$i = 20\%$	$i = 25\%$	$i = 30\%$
A*	2,184	1,609	1,118	696	330	10	−271
B*	2,652	1,937	1,328	803	348	−49	−398
B − A	468	328	209	107	18	−59	−126
C*	2,730	1,966	1,315	753	266	−159	−533
C − A	546	357	197	57	−64	−169	−262
C − B	78	29	−13	−50	−82	−110	−155

*Data rows plotted.

Number of Comparisons

When the *IROR* is used to analyze multiple alternatives, the computations can become quite complex. The problem is that as the number of alternatives increases, the number of comparisons which must be made also increases, but at a faster pace. The following table shows the number of comparisons which must be made in order to reach a decision.

Alternatives	Comparisons
2	1
3	3
4	6
5	10
.	.
.	.
.	.
n	$\dfrac{n(n-1)}{2}$

If there were very many alternatives, the number of necessary comparisons would be quite large. For instance, 10 alternatives would require 45 comparisons. Fortunately, there are not many times when the number of alternatives is actually that high. Still, there are studies in which the number is high enough to make the use of this technique awkward.

Further Income Tax Considerations

Up to now, all the effects of income taxes have not been included in the *IROR* analysis, so that the concepts of *IROR* could be presented without the complexities introduced in Chapter 9. Clearly, the actual effects of income tax are more complicated than anything in the examples so far. However, once the taxes have been found, their total effect is included in the net cash flow, and the principles for applying *IROR* remain the same. The point is, however, that when tax effects are considered, the computation of the incremental net cash flow is significantly more complex, since the cash-flow table has many more entries.

These five are the major factors which must be recognized to include all relevant income tax effects.

1. Composite debt and equity capital

2. Accelerated tax depreciation

3. Normalization

4. Capitalized, tax-deductible expenses

5. Investment tax credit

The fact that investments are made with composite debt and equity capital means that one of the tax-deductible items, interest on debt, is proportional to the capital which has not yet been recovered each year during the life. As a result, the cash-flow table must include a computation of that remaining capital balance. There is a certain amount of controversy among economists as to just how this remaining capital balance should be found. This controversy has led to the creation of several competing economic models of the firm. Appendix 12A at the end of the chapter provides a brief description of the more commonly used other models. The model used in this text will be evident from the cash-flow table below (Table 12-4).

The factor of accelerated tax depreciation requires that both it and the book depreciation have entries in the cash-flow table. This is because it causes a difference between the operating revenue and the taxable income.

Normalization of deferred taxes, the third factor, means that the deferred taxes must be included in the cash-flow table because these amounts are a source of internally generated funds and therefore influence the remaining capital balance.

Expenses which are capitalized on the books but expensed for taxes also cause differences between operating revenue and taxable income, and they influence the normalization of deferred taxes.

Finally, the investment tax credit reduces the tax payment, and the amortization of this credit affects the remaining capital balance. The investment tax credit therefore requires two entries on the cash-flow table—one for the credit and one for the amortization of the credit.

Example 12-5 illustrates the use of a cash-flow table (see Table 12-4) to determine the net cash flow recognizing the above tax effects. The table is constructed for a single alternative. However, if the various entries were differences between alternatives instead, the result would be an incremental net cash flow.

TABLE 12-4
NET CASH FLOW INCLUDING INCOME TAX COMPUTATIONS

			Year			
	0	1	2	3	4	5
1. Revenue		$3,000	$3,000	$3,000	$3,000	$3,000
2. Operating expense (excluding depreciation)		600	600	600	600	600
3. Depreciation						
3a. Tax depreciation $\}$ ($\Sigma16$)		2,250	1,125	750	25	−150
3b. Capitalized tax expense		1,000				
3c. Book depreciation ($\Sigma16 \div$ life)		1,000	1,000	1,000	1,000	1,000
3d. Book-tax depreciation [($\Sigma16 - 3b) \div$ life]		800	800	800	800	800
4. Net operating revenue (1 − 2 − 3c)		1,400	1,400	1,400	1,400	1,400
5. "Normal" tax liability (1 − 2 − 3b − 3d − 13)t		181	696	718	737	748
6. Taxable income (1 − 2 − 3a − 3b − 13)		−1,073	1,126	1,546	2,310	2,509
7. Tax computed (6 × t)		−515	540	742	1,109	1,204
8. Investment tax credit (TC rate × tax basis)		150				
9. Amortized TC (8 ÷ life)		30	30	30	30	30
10. Taxes paid (7 − 8)		−665	540	742	1109	1204
11. Taxes deferred (5 − 7)		696	156	−24	−372	−456
12. Operating income (4 − 5 + 9)*		1,249	734	712	693	682
13. Interest on debt (18_{n-1} × debt ratio × i_d)		223	149	104	65	41
14. Net income (12 − 13)		1,026	585	608	628	641
15. After-tax cash flow (1 − 2 − 10)		3,065	1,860	1,658	1,291	1,196
16. New capital/salvage	5,500					−500
17. Net cash flow (15 − 16)	−5,500	3,065	1,860	1,658	1,291	1,696
18. Remaining capital balance (18_{n-1} + 12 − 17)	5,500	3,684	2,558	1,612	1,014	0

*Or [4 − (10 + 11 + 8 − 9)] on the income statement.

Debt ratio = 45%, i_d = 9%, t = 48%

Example 12-5. Find the rate of return for a 5-year project which has the following characteristics:

Investment	$5,500
Life	5 years
Gross salvage	$800
Cost of removal	$300
ADR tax depreciation for:	80% of life
Amount of investment capitalized for books, but expensed for taxes	$1,000
Investment-tax-credit rate (reduced for short life)	⅓ of 10%
Maintenance and operations costs	$600 per year
Normalization of deferred taxes	
Revenue	$3,000 per year
Financing	
Debt ratio	45%
Debt interest rate	9%
Income tax rate	48%

The explanation of some entries on Table 12-4 follows.

Depreciation (line 3)

Tax Depreciation (line 3*a*):

$$\text{ADR for 4 } (0.80 \times 5) \text{ years on tax basis}$$
$$\text{DDB rate} = 2/4 = 0.5$$
$$\text{Tax basis} = \text{investment} - \text{capitalized tax expense}$$
$$= 5,500 - 1,000 = \$4,500$$
$$\text{Allowable tax depreciation} = \text{tax basis} - \text{gross salvage in excess of 10\% of tax basis}$$
$$= 4,500 - (800 - 0.10 \times 4,500) = \$4,150$$

Year (A)	Base (B)	DDB B × 0.5 (C)	SYD rate (D)	SYD B × D (E)	Salvage; taxable gain − COR (F)	Tax depreciation (C or E) − F (G)
1	$4,500	$2,250	4/10	$1,800		$2,250
2	2,250	1,125	3/6	1,125		1,125
3	1,125	563	2/3	750		750
4	1,125		1/3	375		25*
5					(450 − 300)	−150

*Reaches limit of allowable tax depreciation.

Capitalized expenditures, expensed for taxes (line 3*b*):

$$\text{Capitalized tax expense} = \$1,000$$

Book depreciation (line 3c):

$$\text{Book depreciation} = \frac{\text{book basis} - \text{net salvage}}{\text{life}}$$

$$\text{Book basis} = \text{investment} = \$5,500$$

$$\text{Book depreciation} = \frac{5,500 - 500}{5}$$

$$= \$1,000 \text{ per year}$$

Book-tax depreciation (line 3d):

$$\text{Book-tax depreciation} = \frac{\text{book basis} - \text{capitalized tax expense} - \text{net salvage}}{\text{life}}$$

$$= \frac{5,500 - 1,000 - 500}{5}$$

$$= \$800 \text{ per year}$$

Investment tax credit (line 8):

$$\text{Investment tax credit} = TC \text{ rate} \times \text{tax basis}$$
$$= 1/3 \times 0.10 \times 4,500$$
$$= \$150$$

Amortization of TC (line 9):

$$\text{Amortization of } TC = \frac{TC}{\text{life}}$$

$$= \frac{150}{5}$$

$$= \$30 \text{ per year}$$

All other entries are directly from the original information, or are found from the line relationships shown in Table 12-4.

Equation (12.1) can now be used to solve line 17 by trial and error.

$$\text{For } IROR: 0 = -5,500 + \frac{3,065}{(1 + r)^1} + \frac{1,860}{(1 + r)^2} + \frac{1,658}{(1 + r)^3} + \frac{1,291}{(1 + r)^4} + \frac{1,696}{(1 + r)^5}$$

$$IROR = 25.8\%$$

If this project were one of a set of mutually exclusive alternatives, it would of course be necessary to find the incremental cash flows between the alternatives, as in Example 12-4. In that case, it would not be necessary to fill in all the entries for each alternative. To complete an incremental cash-flow table like Table 12-4, it would be necessary to know just the differences between the following lines for each pair of alternatives.

Line 1 Revenues (if there are any differences)

Line 2 Operations expenses (excluding depreciation)

Line 3 Depreciation (3a, 3b, 3c, 3d)

Line 8 Investment tax credit

Line 16 New capital/salvage

The remaining lines can then be computed according to the directions on the table. The computation of the incremental net cash flow is quite complex. However, once it is found, the incremental rate of return is used exactly the same as illustrated in Example 12-4.

Effect of Common Costs

The most frequently claimed advantage of the internal rate of return is that it is insensitive to the treatment of common costs. Because each value in an incremental-cash-flow table is a difference, either including or excluding a common amount does not affect the outcome, since

$$A - B = (A + X) - (B + X)$$

The reason for this apparent advantage is that the *IROR* does not measure the amount of economic merit one alternative has over another. It is simply a way to find the critical cost of money at which the decision would swing from one to the other. This point was illustrated in Figure 12-1, and again in Figure 12-2. It is important to keep in mind that the *IROR* does not reflect the size of the *NPV* at the firm's cost of money—it reflects only the points of intersection. As a result, common costs are not a factor because the *IROR* does not attempt to resolve the issue of how much better one alternative is than another.

Capital Budgeting

What the *IROR* analysis does do is provide information over a wide range of possible costs of money, rather than just at a single point (as in the present-worth techniques). This makes it a valuable analysis tool, particularly for capital budgeting. A firm with only a limited amount of money to spend must pick and choose from among its various projects within the limit of the available capital. This limit exists because the firm is unwilling to pay the higher cost of obtaining additional capital funds. This limit is then a result of the relationship between the price of a commodity (money) and its *supply*.

The *IROR* analysis provides data like that in Figure 12-2, which is fairly typical. If the cost of money is low, Alternative C is the choice, spending $5,500. If the cost of money is high, Alternative A, spending $4,000, is the choice; and in the mid-range, Alternative B wins, spending $5,000. As the possible cost of money goes up, the amount of economically justified capital goes down. The *IROR* analysis therefore shows the relationship between the price of the commodity (money) and the *demand* for it. The capital budgeting problem is to match these two relationships. Careful interpretation of *IROR* data can help managers do this.

Application of *IROR* to Selecting Alternatives

In practice, the strict interpretation of *IROR* data presented up to now is often modified. In simple examples like Example 12-4, the incremental *IROR* may also be interpreted as the rate of return of an incremental investment, rather than as the critical point at which the merit of alternatives switches. With this slightly different interpretation, the *IROR* is then used to select the most economic alternative at a known cost of money. The various alternatives are arranged in increasing order of first cost, or net cash flow at time 0. Each increment of capital must be economically justified. The most economic alternative is the one for which the incremental *IROR* over each lower economically justified first-cost alternative exceeds the known cost of money or the required rate of return. Example 12-4A illustrates how the alternative from Example 12-4 would be arranged.

Example 12-4A

Alternative	First cost
A	$4,000
B	5,000
C	5,500

$$IROR_{B-A} = 21.1\% \qquad IROR_{C-B} = 8.4\% \qquad IROR_{C-A} = 17.3\%$$

At a known cost of money of 10%, Alternative B is justified over A because the rate of return of the incremental investment in B over A is greater than 10%. Alternative C is eliminated because its incremental $IROR$ over B is less than 10%, even though its incremental $IROR$ over A is greater than 10%. Alternative B remains the economic choice.

Two problems occur when the $IROR$ is used to determine the most economic alternative at a particular cost of money. The first problem is that the decision-making process illustrated in Example 12-4A implies a single answer for the incremental $IROR$ for each pair of alternatives. Unfortunately, the traditional definition of the $IROR$ may result in multiple answers, which complicates the application. This occurs because the incremental $IROR$ determines the point, or points, at which two NPV versus cost-of-money curves intersect. The multiple answers occur when the curves have multiple intersections. This does not present any difficulty if the $IROR$ information is used to represent the change in decision with a change in the cost of money, as in Figure 12-2. But multiple roots do not satisfy the interpretation of $IROR$ as the rate of return of an incremental investment.

The second problem is actually the result of an easily made error in finding an $IROR$. If the incremental cash flow is reversed—that is, if equation (12.1) is solved for $IROR_{B-A}$ even though A spends more at time 0 than B—the answer is the same as it would be for $IROR_{A-B}$. This can cause a misinterpretation of the answer. It happens because the reversal has effectively multiplied both sides of equation (12.1) by -1, which does not change the answer.

Multiple roots may occur when the yearly incremental cash flows change signs more than once. The following time-cost diagram illustrates this situation.

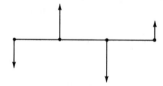

The balance diagram for such a cash flow may not conform to an interpretation of $IROR$ as the rate of return of an incremental *investment*. If there are multiple roots, the balance diagram would look like this:

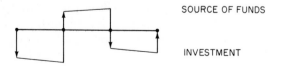

This kind of balance diagram means that the cash flow is not just an internal investment, but is a combined source of funds and investment.

The problem of the misinterpretation of roots occurs because the balance diagram for the reversed cash flow may be entirely a source of funds. Then the traditional definition for *IROR* determines the cost of money of the incremental source, and not a rate of return.

Because of these two related problems, the *IROR* based on the traditional definition may not be a suitable tool for selecting the most economic of mutually exclusive alternatives at a specified cost of money. Appendix 12B at the end of this chapter discusses these problems further, and develops a *modified rate of return (MROR)*, which is suitable for evaluating alternatives at a particular cost of money. However, the *MROR* does not replace *IROR* as an indicator of the relationship between two curves which show *NPV* versus a range of cost-of-money when these curves have several intersections. All *MROR* does is indicate the accept-reject relationship of the two curves at a specified cost of money.

Rate of return is so complex and so easily misused and misinterpreted that its use should be reserved for experts. It would be attractive to have such an apparently simple index of goodness. Unfortunately, rate of return is not simple, nor is it necessary for economic decision making at a known cost of money. Either the *PWE* or the *PWAC* technique can be used to indicate the economic decision at a known cost of money; and using either of these techniques will avoid many of the problems associated with rate-of-return analysis.

PRESENT-WORTH-OF-EXPENDITURES TECHNIQUE

The present-worth-of-expenditures (*PWE*) technique is a measure of how attractive an alternative is from the viewpoint of how much money the firm must spend in order to undertake each alternative. These funds come from two major sources: (1) the investment, or reinvestment, of investors' funds; and (2) the revenues of customers. Because investors provide funds only in anticipation of its repayment plus a suitable return, the firm incurs that obligation when it accepts money from its investors. If the business is to remain healthy, this obligation must be met in the long run from revenues. Because a portion of the investors' return is considered taxable income, the income tax on that portion is part of the total obligation which must also come from revenues. Certain of the firm's expenditures do not require investors' funds, because they are considered current expense and should be covered by current revenue. In this way, all the expenditures of the firm ultimately become a requirement for revenue from the customers.

The discussion of the internal rate of return pointed out that at the investors' required rate of return, or the cost of money, the present worth of their net cash flow, over the life of the investment, must be zero. If the firm meets its obligation, then the present worth of the expenditures must be equal to the present worth of revenues required from the customers. This was illustrated in the project diagram on page 270. This diagram must be in balance at the rate of return to the investors. This means that, in the long run, revenues must be equivalent to expenditures at the cost of money, even though they may not be equivalent at any time during the life. Because the present worth of the expenditures is equivalent to the present worth of the revenue required to satisfy investors, the *PWE* technique is also known as the present worth of revenue requirements (*PWRR*).

By finding the *PWE* of each alternative, and selecting the one which has the least *PWE*, the firm will choose the alternative which is able to satisfy its investors while still providing service to its customers at the least possible cost. The *PWE* technique does not require any estimate of revenues; however, if a difference in revenues is anticipated because of a difference in service, then these anticipated revenues must be subtracted from the *PWE* of the alternative which generates them to maintain comparability. Revenues may also enter the decision to accept or reject an incremental investment compared with a null alternative. Since the *PWE* of a null alternative is zero, an alternative project must have a negative *PWE* in order to be economical. The only ways to achieve a negative *PWE* are either through cost savings or additional revenues.

The *PWE* of an alternative is the sum of the present worth of the capital expenditures less the present worth of any net salvage (positive net salvage is the same as a negative expenditure), plus the present worth of the income taxes and the present worth of any operations costs or savings. The first two elements were combined in Chapter 8, and are the present worth of capital recovery, $PW(C_{rec})$. The present worth of the capital costs, PW(capital costs), is then the $PW(C_{rec})$ plus the present worth of income taxes, $PW(T)$, from Chapters 8 and 9.

The *PWE* technique is illustrated in Example 12-6, using the same alternatives and the same firm (with 100% equity) as in Examples 12-3 and 12-4.

Example 12-6. From Examples 12-3 and 12-4, we get the following table.

| | **Alternatives** | | |
	A	**B**	**C**
First cost	$4,000	$5,000	$5,500
Gross salvage	400	500	500
Cost of removal	300	300	400
Operations costs/year	1,300	700	450

Service life, 3 years

Book depreciation = tax depreciation = straight line

Income-tax rate = 48%

Financing, 100% equity

Assuming a cost of money of 10%, the *PWE* for each alternative would be found as follows.

$$\phi = \frac{t}{1-t}\left(1 - \frac{ri_d}{i}\right) = \frac{0.48}{1-0.48}(1-0) = 0.9231$$

For $PWE(A)$,

$PW(C_{rec})$	$= PW$(first cost) $- PW$(net salvage)	
$PW(C_{rec})$	$= 4,000\ (P/F, 10\%, 0) - 100(P/F, 10\%, 3)$	
	$= 4,000\ (1.0) - 100(0.7513)$	$= \$3,925$
$PW(T)$	$= \phi[PW(C_{rec}) - PW(D_b)]$	
	$= 0.9231\left[3,925 - \dfrac{4,000-100}{3}(P/A, 10\%, 3)\right]$	
	$= 0.9231[3,925 - 1,300(2.4869)]$	$= \$\ \ \ 638$
PW(operations costs)	$= 1,300(P/A, 10\%, 3)$	
	$= 1,300(2.4869)$	$= \$3,233$
	$PWE(A)$	$= \$7,796$

For $PWE(B)$,

$PW(C_{rec})$
$= 5,000 - 200(P/F, 10\%, 3)$
$= 5,000 - 200(0.7513)$ $= \$4,850$

$PW(T)$
$= 0.9231\left[4,850 - \dfrac{5,000 - 200}{3}(P/A, 10\%, 3)\right]$
$= 0.9231[4,850 - 1,600(2.4869)]$ $= \$\ \ \ 804$

$PW(\text{operations costs}) = 700(P/A, 10\%, 3)$
$= 700(2.4869)$ $= \$1,741$

$PWE(B)$ $= \$7,395$

For $PWE(C)$,

$PW(C_{rec})$
$= 5,500 - 100(P/F, 10\%, 3)$
$= 5,500 - 100(0.7513)$ $= \$5,425$

$PW(T)$
$= 0.9231\left[5,425 - \dfrac{5,500 - 100}{3}(P/A, 10\%, 3)\right]$
$= 0.9231[5,425 - 1,800(2.4869)]$ $= \$\ \ \ 876$

$PW(\text{operations costs}) = 450(P/A, 10\%, 3)$
$= 450(2.4869)$ $= \$1,119$

$PWE(C)$ $= \$7,420$

Therefore,

$$PWE(A) = \$7,796$$
$$PWE(B) = \$7,395$$
$$PWE(C) = \$7,420$$

The alternative with the least PWE is B, but it is only slightly better than C. This is reflected by an examination of the differences and the percentage differences in PWE between the alternatives.

$$PWE(A - B) = 7,796 - 7,395 = \$401$$

Thus, A is more than B by 5.4%.

$$\frac{401}{7,395} = 0.054 = 5.4\%$$

$$PWE(C - B) = 7,420 - 7,395 = \$25$$

And, C is more then B by 0.3%.

$$\frac{25}{7,395} = 0.003 = 0.3\%$$

Examples 12-3 and 12-4 included \$4,000 of revenue each year which is common to all alternatives. If that common revenue is included in the PWE analysis, it is subtracted from each alternative as if each alternative were to be compared with a null alternative.

$$PW(\text{revenue}) = 4,000(P/A, 10\%, 3)$$
$$= 4,000(2.4869) = \$9,948$$

Alternative	PWE	PW(revenue)	Net PWE	Difference From B	Difference % of B
A	$7,796	− $9,948 =	−$2,152	$401	15.7
B	7,395	− 9,948 =	−2,553		
C	7,420	− 9,948 =	−2,528	25	1.0

Including the common revenue has not changed the *PWE* differences themselves, but it has significantly changed the percentage of those differences. If the common revenue had been equal to the *PWE* of B, or $7,395, the percentage differences would both have been infinity (difference divided by 0). In summary, even though the difference in *PWE* is not affected by the treatment of common costs or revenues, the significance of those differences can be quite sensitive to this treatment.

Effect of the Treatment of Common Costs

The fact that the treatment of common costs does not influence the *PWE* differential, but does affect the significance of the differences, can cause problems in interpreting study results. Example 12-7 illustrates this problem of interpretation.

Example 12-7. Suppose a study has been made on three alternatives:

- Alternative A represents a continuation of present practices
- Alternative B represents a thorough revamping of present practices.
- Alternative C represents some refinements of Alternative B.

The *PWE* results in the study are

- Alternative A: $5,000,000
- Alternative B: $3,000,000
- Alternative C: $2,950,000

Assume that $2,750,000 of the *PWE* are common between B and C, but not with A. One manager's interpretation of these results might proceed as follows:

1. Alternative B results in 40% less *PWE* than A.

$$\frac{5,000,000 - 3,000,000}{5,000,000} = 0.40 = 40\%$$

2. Alternative C results in 1.7% less expense, or more savings, than B.

$$\frac{3,000,000 - 2,950,000}{3,000,000} = 0.017 = 1.7\%$$

3. Alternative C does not provide a significant savings over B, since errors in the study data could easily exceed 1.7%. Therefore, a very few minor intangible factors can cause Alternative B to be selected.

The problem is that another manager might offer this analysis.

1. Alternative B results in 40% less expense than A.

2. Alternative C results in 20% less expense than B, since $2,750,000 of the *PWE* values for the two alternatives is common and should be removed.

$$\frac{3,000,000 - 2,750,000 - (2,950,000 - 2,750,000)}{3,000,000 - 2,750,000} = 0.20 = 20\%$$

3. Alternative C is therefore significantly better than B, which was significantly better than A. Any minor intangible factors favoring B are not important, and Alternative C should be adopted.

Even if both managers agree on which savings percentage was significant, there would still be a problem in defining which costs are common. Chapter 11 pointed out this difficulty, but at the time, it said that it did not matter, because the *PWE* differential would be the same regardless of which common values were removed. That is true: There will be a $50,000 *PWE* advantage for Alternative C over B regardless of which common values might be removed.

However, the basic problem which Example 12-7 reveals is that unlike *IROR*, the *PWE* technique does measure the economic merit of each alternative at a specific cost of money. The treatment of common costs shifts the base, thus influencing the percentage differences. The two techniques, *PWE* and *IROR*, can then be used to complement one another. If common costs are unknown, or for some reason cannot be excluded, the incremental *IROR* may help to determine if differences in *PWE* are significant.

Relationship Between *PWE* and *ROR*

The *PWE* and *ROR* techniques can be used to complement each other because, when correctly applied, they are equivalent and will not contradict each other's result. In Example 12-6, at a 10% cost of money, the *PWE* indication was Alternative B. In Example 12-4, the 10% cost of money was within the range where the *IROR* indication was also Alternative B. Furthermore, Figure 12-2 shows that at 10% cost of money, Alternative B is only slightly better than Alternative C, just as in the *PWE* results.

This similarity in results does not occur merely because the two approaches are similar. It occurs because the two techniques are mathematically equivalent at the firm's cost of money. This is illustrated in the following explanation. The net *PWE* for Example 12-6 was found as

$$\text{Net } PWE = PW(C_{\text{rec}}) + \frac{t}{1-t}[PW(C_{\text{rec}}) - PW(D_b)]$$
$$+ PW(\text{operations cost}) - PW(\text{revenue}) \quad (12.2)$$

The net present value for Example 12-3, at a 10% cost of money, could have been found from a similar formula, which restates the cash-flow tables.

$$NPV = PW(\text{revenue}) - PW(\text{operations cost}) - PW(\text{taxes})$$
$$- PW(\text{first cost}) + PW(\text{salvage}) \quad (12.3)$$

When the last two terms in equation (12.3) are combined, the result is $-PW(C_{\text{rec}})$.
This is because

$$PW(\text{taxes}) = t[PW(\text{revenue}) - PW(\text{operations cost}) - PW(D_b)]$$

The taxes in equation (12.3) are

$$PW(\text{taxes}) = t[PW(\text{revenue}) - PW(\text{operating cost}) - PW(D_b)]$$

Therefore, equation (12.3) can be rewritten as

$$NPV = -PW(C_{\text{rec}}) + (1 - t)[PW(\text{revenue}) - PW(\text{operations cost})] - t\, PW(D_b) \quad (12.4)$$

If Equation (12.4) is divided by $(1 - t)$ and multiplied by -1, the result is

$$\frac{-NPV}{(1-t)} = \frac{PW(C_{\text{rec}})}{(1-t)} - [PW(\text{revenue}) - PW(\text{operations costs})] - \frac{t}{1-t}PW(D_b)$$

Then, since

$$1 + \frac{t}{1-t} = \frac{1}{1-t}$$

Rearranging terms and substitution gives

$$\frac{-NPV}{(1-t)} = \left(1 + \frac{t}{1-t}\right) PW(C_{\text{rec}}) - \frac{t}{1-t} PW(D_b) + PW(\text{operations costs}) - PW(\text{revenue})$$

$$\frac{-NPV}{(1-t)} = PW(C_{\text{rec}}) + \frac{t}{1-t}[PW(C_{\text{rec}}) - PW(D_b)] + PW(\text{operations costs}) - PW(\text{revenue})$$

The right-hand side is the expression shown in equation (12.2) for net *PWE*. Therefore,

$$\frac{-NPV}{(1-t)} = \text{net } PWE \tag{12.5}$$

The relationship in equation (12.5) is very important, because it means that the *ROR* technique and the *PWE* technique are equivalent. The *ROR* technique determines which alternative has the greatest *NPV* in the range in which the firm's cost of money lies, whereas *PWE* finds the least *PWE* at the firm's cost of money. Because, the *NPV* divided by a negative constant $-(1 - t)$ equals the net *PWE*, the economic indications are identical.

The *NPV* data of Table 12-3 can be converted into *PWE* data.

Alternative	NPV, 10% (Table 12-3)	$\dfrac{-NPV}{1 - 0.48}$	Net PWE (Example 12-6)
A	$1,118	$-2,150	$-2,152
B	1,328	-2,554	-2,553
C	1,315	-2,529	-2,528

The relationship of equation (12.5) also holds for firms which are financed with composite capital, and have the very complex income tax effects illustrated in Example 12-5 and the cash-flow table, Table 12-4. However, the derivation of the relationship with all income tax effects is quite complex. It can be demonstrated by the computation of the *PWE* for the investment in Example 12-5. Example 12-8 finds the *PWE* for this same investment, with the same tax effects as in Example 12-5, if the composite cost of money is 12%. This example requires the use of the full tax equations developed in Chapter 9.

Example 12-8. Find the net *PWE* for an investment which has the following characteristics.

From Example 12-5,

Investment	$5,500
Life	5 years
Gross salvage	$800
Cost of removal	$300
ADR tax depreciation for:	80% of life

Amount of investment capitalized for books, but expensed for taxes $1,000

Investment tax credit ⅓ of 10%

Maintenance and operations costs $600 per year

Normalization of deferred taxes

Revenue $3,000 per year

Financing:

Composite cost of money	12%
Debt interest rate	9%
Debt ratio	45%
Income tax rate	48%

$$PW(C_{rec}) = PW(\text{first cost}) - PW(\text{net salvage})$$

$$= 5{,}500(P/F, 12\%, 0) - 500(P/F, 12\%, 5)$$

$$= 5{,}500(1.0) - 500(0.5674) \qquad = \$5{,}216$$

$$PW(T) = \phi\,[PW(C_{rec}) - PW(D_b)] - t(1 + \phi)\,[PW(D_t) - PW(D_{bt})]$$

$$- \frac{t}{1-t}\left\{PW(E_t) - [PW(D_b) - PW(D_{bt})]\right\} - \frac{1}{1-t}\,PW(ATC)$$

$$- (1 + \phi)\,[PW(TC) - PW(ATC)]$$

$$PW(D_b) = \frac{5{,}500 - 500}{5}(P/A, 12\%, 5)$$

$$= 1{,}000(3.6048) \qquad = \$3{,}605$$

$$PW(D_{bt}) = \frac{5{,}500 - 1{,}000 - 500}{5}(P/A, 12\%, 5)$$

$$= 800(3.6048) \qquad = \$2{,}884$$

$$PW(D_t) = \sum_{n=1}^{5}(\text{line } 3a \text{ of Table 12-4})(P/F, 12\%, n)$$

$$= (\text{calculation omitted}) \qquad = \$3{,}370$$

$$PW(E_t) = 1{,}000(P/F, 12\%, 1)$$

$$= 1{,}000(0.8929) \qquad = \$\ \ 893$$

$$PW(TC) = (5{,}500 - 1{,}000)(0.10 \times 0.333)(P/F, 12\%, 1)$$

$$= 150(0.8929) \qquad = \$\ \ 134$$

$$PW(ATC) = \frac{150}{5}(P/A, 12\%, 5)$$

$$= 30(3.6048) \qquad = \$\ \ 108$$

$$\phi = \frac{t}{1-t}\left(1 - \frac{ri_d}{i}\right)$$

$$= \frac{0.48}{1 - 0.48}\left[1 - \frac{(0.45)(0.09)}{0.12}\right] = 0.6115$$

$$PW(T) = 0.6115(5{,}216 - 3{,}605) - 0.48(1.6115)(3{,}370 - 2{,}884)$$

$$- \frac{0.48}{1 - 0.48}[893 - (3{,}605 - 2{,}884)] - \frac{1}{1 - 0.48}(108) - 1.6115(134 - 108)$$

$$= 985 - 376 - 159 - 208 - 42 \qquad = \$\ \ 200$$

$$PW\text{(operations cost)} = 600(P/A, 12\%, 5)$$
$$= 600(3.6048) \qquad\qquad\qquad = \$ \quad 2{,}163$$
$$PWE = PW(C_{\text{rec}}) + PW(T) + PW\text{(operations cost)}$$
$$= 5{,}216 + 200 + 2{,}163 \qquad\qquad = \$ \quad 7{,}579$$
$$PW\text{(revenue)} = 3{,}000(P/A, 12\%, 5)$$
$$= 3{,}000(3.6048) \qquad\qquad\qquad = \$ \quad 10{,}814$$
$$\text{Net } PWE = PWE - PW\text{(revenue)}$$
$$= 7{,}579 - 10{,}814 \qquad\qquad\qquad = \$ \; -3{,}235$$

Net present value is found from line 17 on Table 12-4

$$NPV = \sum_{n=0}^{5} (\text{line 17 of Table 12-4})(P/F, 12\%, n)$$

$$= (\text{calculations omitted}) = \$1{,}682$$

$$\frac{-NPV}{1 - t} = \frac{-1{,}682}{1 - 0.48} = -\$3{,}235 = \text{net } PWE$$

This result demonstrates that the NPV at the cost of money, divided by $-(1 - t)$ is equal to the net PWE at the same cost of money, even with composite financing and complex income tax effects.

Decision Rule for *PWE*

As Example 12-6 illustrated, the decision rule for PWE data is not restricted to pairs of alternatives. The generalized decision rule for the PWE technique is

> Of any number of mutually exclusive alternatives for doing a specified job; pick, as the most economic alternative, the one with the lowest PWE at the cost of money.

The result of this decision rule has already been demonstrated to be identical with the result of the $IROR$ decision rule if the critical cost of money is the same. This is because the serial application of the $IROR$ decision rule is the same as an application of a net present value decision rule, which could be stated as follows:

> Of any number of mutually exclusive alternatives for doing a specified job, pick, as the most economic alternative, the one with the greatest NPV at the cost of money.

Because PWE equals $-NPV/(1 - t)$, these two decision rules will reach the same conclusion—but with an important difference from the $IROR$ decision rule. The use of the PWE decision rule requires only one calculation for PWE for each alternative. Because the $IROR$ decision rule requires $n(n - 1)/2$ calculations for n alternatives, the PWE technique results in a savings in time, although this may be negated somewhat by the use of computers. More importantly, though, PWE is a greatly simplified application, which is less subject to errors of interpretation.

The use of the PWE technique and its decision rule avoids, but does not solve, the multiple-answer problem which occurred with the $IROR$ technique. The existence of multiple roots for the $IROR$ indicates that the economic advantage switches back and forth between two alternatives at various costs of money; but there is no ambiguity at a single specific cost of money. By restricting the analysis to a single cost of money, the PWE technique is not affected by what may occur at another cost of money. The relationship which causes the multiple answers is then ignored. The problem would appear if the *PWE* technique were used to test the economic decision between two alternatives for sensitivity to the cost of money. If there is more than one root for *IROR*, the *PWE* results at several costs of money will indicate that the decision switches.

Application to Economy Studies

The *PWE* technique is most applicable to those studies which can assume that all plant in all alternative plans is retired by a common date. Many studies can be arranged to make this coterminated plant assumption. The following example illustrates the use of the *PWE* technique in the usual way, with revenues not known; and it includes many of the income tax effects of Chapter 9. Example 12-9 also illustrates a convenient assumption which simplifies the income tax calculation. In this example, although the investment tax credit is determined from the tax basis, both tax depreciation and book depreciation are determined from the book basis. As a result, the slight difference between them, owing to the capitalization of tax-deductible expenditures, is neglected. This approximation is usually satisfactory for investments in plant which do not include a significant amount of company loadings on labor costs.

Example 12-9. A company has the following financial characteristics.

Composite cost of capital	12%
Debt interest rate	9%
Debt ratio	45%
Income tax data	
Composite rate	50%
Uses ADR life	80% of service life
Investment tax credit	4% of qualified investment
	(no life limitation)

This company is faced with an economic decision between two alternatives which have the anticipated expenditures shown in Table 12-5 on the next page.

Alternative I

PW(capital recovery)

$$PW(C_{\text{rec}}) = PW(\text{first cost}) - PW(\text{net salvage})$$

(A) $10,000(P/F, 12\%, 0) - 1,000(P/F, 12\%, 10)$
   $= 10,000(1.000) - 1,000(0.3220)$ $= \$9,678$

(B) $6,500(P/F, 12\%, 5) - 500(P/F, 12\% \ 10)$
   $= 6,500(0.5674) - 500(0.3220)$ $= \$3,527$

PW(income taxes)

$$PW(T) = \phi[PW(C_{\text{rec}}) - PW(D_b)] - t(1 + \phi)[PW(D_t) - PW(D_b)]$$

$$- \frac{PW(ATC)}{1 - t} - (1 + \phi)[PW(TC) - PW(ATC)]$$

$$\phi = \frac{t}{1 - t}\left(1 - \frac{ri_d}{i}\right)$$

$$= \frac{0.5}{1 - 0.5}\left[1 - \frac{(0.45)(0.09)}{0.12}\right]$$

$$= 0.6625$$

TABLE 12-5

ANTICIPATED EXPENDITURES FOR EXAMPLE 12-9

Item	Time point	Cost
Alternative I		
Installed cost		
Unit A (70% qualifies *TC*)	0	$10,000
Unit B (80% qualifies *TC*)	5	6,500
Gross salvage		
Unit A	10	2,000
Unit B	10	800
Cost of removal		
Unit A	10	1,000
Unit B	10	300
Rearrangements (expense)		
Unit A	0	1,000
Unit B	5	300
Ad valorem taxes		
Unit A	1–5	400
Unit A + Unit B	6–10	660
Unit operation and maintenance	1	1,600
	2	1,680
	3	1,760
	4	1,850
	5	1,940
	6	2,840
	7	2,980
	8	3,130
	9	3,290
	10	3,450
Alternative II		
Installed cost		
Unit C (90% qualifies *TC*)	0	$ 8,000
Unit D (90% qualifies *TC*)	3	9,800
Gross salvage		
Unit C	10	750
Unit D	10	750
Cost of removal		
Unit C	10	750
Unit D	10	800
Rearrangements (expense)		
Unit C		0
Unit D	3	600
Ad valorem taxes		
Unit C	1–3	320
Unit C + Unit D	4–10	710

TABLE 12-5

ANTICIPATED EXPENDITURES FOR EXAMPLE 12-9 (*Continued*)

Item	Time point	Cost
Unit operation and maintenance	1	1,300
	2	1,360
	3	1,430
	4	3,000
	5	3,160
	6	3,320
	7	3,480
	8	3,660
	9	3,840
	10	4,040

(A) $\quad PW(D_b) = \dfrac{10,000 - 1,000}{10} (P/A,\ 12\%,\ 10)(P/F,\ 12\%,\ 0)$

$\qquad\qquad\quad = 900(5.650)(1.000)$

$\qquad\qquad\quad = \$5,085$

$\quad\ PW(D_t) = (\text{first cost})(P/D_t)\ (P/F,\ 12\%,\ 0)$

$\quad\ (P/D_t) = \text{factor for 10-year life, 20\% gross salvage, and cost of removal adjustment}$

$\qquad\qquad\quad = 0.63155 + 0.1(0.32197)$

$\qquad\qquad\quad = 0.66375$

$\ PW(D_t) \quad = 10,000(0.66375)(1.000)$

$\qquad\qquad\quad = \$6,638$

$\ PW(TC) \quad = 10,000(\text{qualified }\%)(TC\ \text{rate})(P/F,\ 12\%,\ 1)$

$\qquad\qquad\quad = 10,000(0.7)(0.04)(0.8929)$

$\qquad\qquad\quad = \$250$

$PW(ATC) = ATC(P/A,\ 12\%,\ 10)(P/F,\ 12\%,\ 0)$

$\ ATC \qquad = \dfrac{TC}{L} = \dfrac{280}{10} = \$28$

$PW(ATC) = 28(5.650)(1.000)$

$\qquad\qquad\quad = \$158$

$\quad\ PW(T) = 0.6625(9,678 - 5,085) - 0.5(1.6625)(6,638 - 5,085)$

$\qquad\qquad\quad - \dfrac{158}{1 - 0.5} - 1.6625(250 - 158) = \$1,283$

(B) $\quad PW(D_b) = \dfrac{6,500 - 500}{5} (P/A,\ 12\%,\ 5)(P/F,\ 12\%,\ 5)$

$\qquad\qquad\quad = 1,200(3.605)(0.5674)$

$\qquad\qquad\quad = \$2,455$

$\quad\ PW(D_t) = 6,500[0.74561 + 0.046(0.56743)]\ (P/F,\ 12\%,\ 5)$

$\qquad\qquad\quad = 5,016(0.5674)$

$\qquad\qquad\quad = \$2,846$

$$PW(TC) = 6{,}500(0.8)(0.04)(P/F,\ 12\%,\ 6)$$
$$= 208(0.5066) = \$105$$

$$PW(ATC) = \frac{208}{5}\,(P/A,\ 12\%,\ 5)(P/F,\ 12\%,\ 5)$$
$$= 41.6(3.605)(0.5674) = \$85$$

$$PW(T) = 0.6625(3{,}527 - 2{,}455) - 0.5(1.6625)(2{,}846 - 2{,}455)$$
$$- \frac{85}{1 - 0.5} - 1.6625(105 - 85) = \$182$$

PW(rearrangements, ad valorem taxes, maintenance, and other operations costs)

(A)	$1{,}000(P/F,\ 12\%,\ 0)$	$= 1{,}000(1.000)$	$= \$\ 1{,}000$
	$400(P/A,\ 12\%,\ 5)$	$= 400(3.605)$	$=\ \ \ 1{,}442$
(B)	$300(P/F,\ 12\%,\ 5)$	$= 300(0.5674)$	$=\ \ \ \ \ \ 170$
(A & B)	$660(P/A,\ 12\%,\ 5)(P/F,\ 12\%,\ 5)$	$= 660(3.605)(0.5674)=$	$1{,}350$
	$1{,}600(P/F,\ 12\%,\ 1)$	$= 1{,}600(0.8929)$	$=\ \ \ 1{,}429$
	$1{,}680(P/F,\ 12\%,\ 2\ )$	$= 1{,}680(0.7972)$	$=\ \ \ 1{,}339$
	$1{,}760(P/F,\ 12\%,\ 3\ )$	$= 1{,}760(0.7118)$	$=\ \ \ 1{,}253$
	$1{,}850(P/F,\ 12\%,\ 4\ )$	$= 1{,}850(0.6355)$	$=\ \ \ 1{,}176$
	$1{,}940(P/F,\ 12\%,\ 5\ )$	$= 1{,}940(0.5674)$	$=\ \ \ 1{,}101$
	$2{,}840(P/F,\ 12\%,\ 6)$	$= 2{,}840(0.5066)$	$=\ \ \ 1{,}439$
	$2{,}980(P/F,\ 12\%,\ 7)$	$= 2{,}980(0.4523)$	$=\ \ \ 1{,}348$
	$3{,}130(P/F,\ 12\%,\ 8)$	$= 3{,}130(0.4039)$	$=\ \ \ 1{,}264$
	$3{,}290(P/F,\ 12\%,\ 9)$	$= 3{,}290(0.3606)$	$=\ \ \ 1{,}186$
	$3{,}450(P/F,\ 12\%,\ 10)$	$= 3{,}450(0.3220)$	$=\ \ \ 1{,}111$
	Total present worth		$= \$16{,}608$

Present worth of expenditures for Alternative I

Capital recovery $(A + B)$ + income taxes $(A + B)$
　　　　　　　　+ rearrangements and operations costs $(A + B)$
$$(9{,}678 + 3{,}527) + (1{,}283 + 182) + (16{,}608) = \$31{,}278$$

Alternative II

PW(capital recovery)

(C) $8{,}000(P/F,\ 12\%,\ 0) - 0(P/F,\ 12\%,\ 10)$
$= 8{,}000(1.000)$ 　　　　　　　　　$= \$8{,}000$

(D) $9{,}800(P/F,\ 12\%,\ 3) - (-50)(P/F,\ 12\%,\ 10)$
$= 9{,}800(0.7118) + 50(0.3220)$ 　　$= \$6{,}992$

PW(income taxes)

(C) $PW(D_b) = \dfrac{8{,}000 - 0}{10}\,(P/A,\ 12\%,\ 10)(P/F,\ 12\%,\ 0)$
$$= 800(5.650)(1.000) = \$4{,}520$$

$$PW(D_t) = 8{,}000\,[0.67857 + 0.094(0.32197)]\,(P/F,\ 12\%,\ 0) = \$5{,}671$$

$$PW(TC) = 8{,}000(0.9)(0.04)(P/F,\ 12\%,\ 1)$$
$$= 288(0.8929) = \$257$$

$$PW(ATC) = \frac{288}{10}(P/A, 12\%, 10)(P/F, 12\%, 0)$$

$$= 29(5.650)(1.000) = \$164$$

$$PW(T) = 0.6625(8,000 - 4,520) - 0.5(1.6625)(5,671 - 4,520)$$

$$- \frac{164}{1 - 0.5} - (1.6625)(257 - 164) = \$866$$

$$(D)\ PW(D_b) = \frac{9,800 - (-50)}{7}(P/A, 12\%, 7)(P/F, 12\%, 3)$$

$$= 1,407(4.564)(0.7118) = \$4,571$$

$$PW(D_t) = 9,800\ [0.73497 + 0.082(0.45235)](P/F, 12\%, 3)$$

$$= 7,566(0.7118) = \$5,385$$

$$PW(TC) = 9,800(0.9)(0.04)(P/F, 12\%, 4)$$

$$= 353(0.6355) = \$224$$

$$PW(ATC) = \frac{353}{7}(P/A, 12\%, 7)(P/F, 12\%, 3)$$

$$= 50(4.564)(0.7118) = \$162$$

$$PW(T) = 0.6625(6,992 - 4,571) - 0.5(1.6625)(5,385 - 4,571)$$

$$- \frac{162}{1 - 0.5} - 1.6625(224 - 162) = \$500$$

PW(rearrangements, ad valorem taxes, maintenance, and other operations costs)

(D)	$600(P/F, 12\%, 3)$	$= 600(0.7118)$	$= \$$	427
(C)	$320(P/A, 12\%, 3)(P/F, 12\%, 0)$	$= 320(2.402)(1.000)$	$=$	769
(C & D)	$710(P/A, 12\%, 7)(P/F, 12\%, 3)$	$= 710(4.564)(0.7118)$	$=$	2,307
	$1,300(P/F, 12\%, 1)$	$= 1,300(0.8929)$	$=$	1,161
	$1,360(P/F, 12\%, 2)$	$= 1,360(0.7972)$	$=$	1,084
	$1,430(P/F, 12\%, 3)$	$= 1,430(0.7118)$	$=$	1,018
	$3,000(P/F, 12\%, 4)$	$= 3,000(0.6355)$	$=$	1,906
	$3,160(P/F, 12\%, 5)$	$= 3,160(0.5674)$	$=$	1,793
	$3,320(P/F, 12\%, 6)$	$= 3,320(0.5066)$	$=$	1,682
	$3,480(P/F, 12\%, 7)$	$= 3,480(0.4523)$	$=$	1,574
	$3,660(P/F, 12\%, 8)$	$= 3,660(0.4039)$	$=$	1,478
	$3,840(P/F, 12\%, 9)$	$= 3,840(0.3606)$	$=$	1,385
	$4,040(P/F, 12\%, 10)$	$= 4,040(0.3220)$	$=$	1,301
	Total present worth		$=$	$\$17,885$

Present worth of expenditures for Alternative II

Capital recovery $(C + D)$ + income taxes $(C + D)$

$$+ \text{ rearrangements and operations costs } (C + D)$$

$$(8,000 + 6,992) + (866 + 500) + (17,885) = \$34,243$$

$$PWE \text{ for Alternative I} = \$31,278$$

$$PWE \text{ difference in favor of Alternative I} = \$\ 2,965$$

Alternative I would be adopted if the company decision were to be based solely on economy.

Shortcut to Finding the Present Worth of a Series

Example 12-9 illustrated the usual way of finding the present worth of a nonuniform series which is neither a geometric nor an arithmetic progression. This is a bothersome procedure if done manually on a calculator, because it is necessary to look up each factor, write it down, multiply by the appropriate amount, and then add up the results. There are many opportunities for error, and the final result is somewhat inaccurate because of the continued use of rounded factors.

On a manual calculator, a better way to proceed is to begin with the last amount, find its present worth 1 year earlier, add it to the next-to-last amount, find that sum's 1-year present worth, and so on through the entire series. To do this, it is necessary to enter 1 plus the interest rate $(1 + i)$ as a constant; enter the last amount; divide by the constant; add the next-to-last amount; divide by the constant; and add, divide, through the series—with the last operation being to divide by the constant. This procedure works because the present-worth factor for 1 year, $(P/F, i\%, 1)$, is $1/(1 + i)$. It takes longer to explain it than to do it. The following illustrates the operation applied to the maintenance cost in Example 12-9, Alternative I.

Operation	Amount	Display
(Enter Constant)	(1.12)	(1.12)
1. Enter (last amount)	3,450	$ 3,450
Divide by	constant	3,080
2. Add	3,290	6,370
Divide by	constant	5,688
3. Add	3,130	8,818
Divide by	constant	7,873
4. Add	2,980	10,853
Divide by	constant	9,690
5. Add	2,840	12,530
Divide by	constant	11,188
6. Add	1,940	13,128
Divide by	constant	11,721
7. Add	1,850	13,571
Divide by	constant	12,117
8. Add	1,760	13,877
Divide by	constant	12,390
9. Add	1,680	14,070
Divide by	constant	12,563
10. Add (first amount)	1,600	14,163
Divide by	constant	$12,645 (Answer, *PW*)

Finding the present worth this way is quicker because it is not necessary to look up any factors; it is more accurate because the effective factors used include all the significant digits up to the capacity of the calculator.

Application to Non-coterminated Studies

Even though the *PWE* technique is most useful for coterminated studies, it can be applied to non-coterminated studies too. In that case, however, some explicit method is needed for allocating the *PWE* of one or more alternatives. There are several ways to make such allocations.

In any *PWE* study, the study period should be at least as long as the life of the earliest retiring plant. Then, the simplest way to allocate the present worth of the capital costs of any plant which lives longer than the study period is to charge any plan having such plant with only the levelized equivalent annual costs (*AC*) of this plant for the number of years it lives within the study period. The present worth of that portion of the *AC* is then an allocation of the total *PWE* for that plant.

For instance, if the second installations in the 10-year study in Example 12-9 (plant B in Alternative I, and plant D in Alternative II) had been expected to live 10 years (like the original plant), the present worth of 5 years of *AC* would be included in Alternative I for plant B. Similarly, the present worth of 7 years of *AC* for plant D would be included in Alternative II. This allocated *PWE* is found by first finding the *PWE* over the expected life, multiplying by the (*A/P*) factor over the expected life, and then multiplying by the (*P/A*) factor over the life in the study period.

$$
\begin{aligned}
\text{Allocated } PW(\text{capital cost}) = {} & \text{total life } PW(\text{capital cost}) \\
& \times (A/P, i\%, \text{total life}) \\
& \times (P/A, i\%, \text{life during study period})
\end{aligned}
$$

With this allocation, the *PWE* study will then be identical with the *PWAC* study, which is covered in the following section. If inflation in the first cost of plant is a significant factor in the study, this allocation may not be satisfactory because it fails to recognize that the value of having plant with life remaining is enhanced by not having to buy new plant at an inflated price.

Another allocation of *PWE* can be made by explicitly selecting some future value to be used in the study, like salvage, which reflects the value of the remaining life of plant after the study period. There are a number of ways to make this artificial estimate of salvage. Chapter 13 has an example which illustrates one way of estimating this salvage value and recognizing the effect of price changes in the first costs.

PRESENT-WORTH-OF-ANNUAL-COSTS TECHNIQUE

The *present-worth-of-annual-costs* (*PWAC*) technique is essentially the same as the *PWE* technique except capital costs are converted into levelized equivalent annual costs (*AC*) before their present worth is found. Operations costs should be treated as they are expected to occur. If operations costs are expected to be equal each year, then they can be combined with the levelized equivalent annual capital costs. In many instances, the results of a *PWAC* analysis will be exactly the same as those for a *PWE* analysis because the present worth of an annuity which is equivalent to an expenditure is the expenditure itself. Therefore, if the present worth of the *AC* is found over a period equal to the life of the plant, the *PWE* equals the *PWAC*.

The decision rule for *PWAC* studies is therefore almost identical with that for *PWE* studies.

> Of any number of mutually exclusive alternatives for doing a specified job: pick, as the most economic alternative, the one with the lowest *PWAC* at the cost of money.

The *PWAC* technique has been very popular for two reasons. The first is that it allows a grouping of equipment into specific categories; average lives, salvage values, maintenance costs, operations costs, and ad valorem tax rates can then be assigned to each category. Of course, combining the capital costs and operations costs into a single rate assumes that operations costs are equal each year. With this assumption, it is a relatively simple matter to calculate the annual cost rates as a percentage of installed costs for the different categories.

An economy study using these all-inclusive annual costs can be put together very quickly, as Example 12-10 will demonstrate. The second reason for the popularity of *PWAC* is that working with annual costs can simplify the treatment of non-coincident equipment placements and retirements.

Because the *PWAC* technique first converts expenditures into a levelized amount and then finds the present worth, these two operations can be done separately. If the present worth is found over some period not equal to the life of the plant, the *PWAC* technique is then just an allocation of costs. Many studies are not coterminated. That is, some plant is expected to live longer in one plan than in another. The *PWAC* technique is most advantageous in such studies. Unfortunately, because inflation and its effects must be included in studies, the power and simplicity of the *PWAC* technique are strained severely.

Applying the *PWAC* technique to non-coterminated studies requires that two time periods within the total study period be defined:

1. The *planning period,* which is the period from time 0 to the time of the last placement

2. The *complementary period,* which is the period from the time of the last placement to the time of the last retirement

The planning period, then, is that span of years in which plant additions, removals, and changes are designed in alternative ways to meet growth forecasts or other service requirements. The planning period is restricted to that number of years ahead for which reasonable estimates of the future can be made.

The complementary period is the span of years beyond the end of the planning period for which annual costs generated during the planning period are assumed to continue under any alternative. It makes use of the repeated plant assumption, and depends on the life estimates used in developing the annual costs and the longest-surviving plant under these estimates. Strictly speaking, the annual costs for the alternatives are not comparable until all the annual costs under all the plans are included. In practice, however, if the longest-surviving plant is relatively minor, the complementary period may be closed at the last retirement of significance. Any costs which occur after the close of the complementary period are neglected or are considered common.

A time-cost diagram of a comparison involving non-coincident placements and retirements might look something like the following:

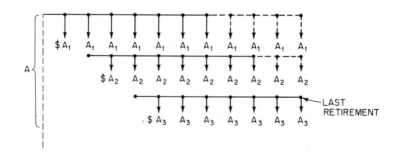

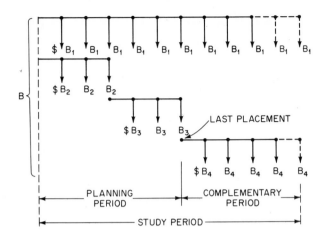

Example 12-10. A company with a 12% cost of capital is trying to decide between equipment A and equipment B. Equipment A would be bought as a single unit for $10,000, and is expected to have a useful life of 15 years. At the end of the 15-year period, it would have a gross salvage value of $600; expectations are that it will cost about $500 to remove. A lifetime maintenance contract is available for $300 per year, and other operations costs on the fully automatic unit are expected to be about $50 a year. Equipment B, on the other hand, could be bought as a smaller initial unit; it would then be supplemented at time points 5 and 10 with additional units, which, like the original, would cost $4,000 installed. Like A, each equipment B unit is expected to last 15 years. A contract is available from the supplier which will not only guarantee a unit purchase price of $4,000 for each of the three units, but also provide maintenance for $200 per unit per year over the life of the equipment. These units are also fully automatic, and operations costs are estimated at $25 per unit per year. The gross-salvage estimate for each unit of equipment B is $250, and the cost for removal is expected to be $200. The company's financial structure is such that it has a ϕ of 0.6625. Finally, both alternatives would qualify for a 7% investment tax credit on 95% of either unit, and the company uses an ADR life equal to 80% of the service life.

The annual costs as a percentage of the installed cost are as follows.

Equipment A:

Capital-recovery costs:

$$AC(C_{\text{rec}}) = 100\%(A/P,\ 12\%,\ 15) - \frac{600 - 500}{10{,}000}(100\%)(A/F,\ 12\%,\ 15)$$

$$= 100\%(0.14682) - 1\%(0.02682) = 14.66\%$$

Income taxes:

$$AC(T) = \phi[AC(C_{\text{rec}}) - AC(D_b)] - t(1 + \phi)[AC(D_t) - AC(D_b)]$$
$$- \frac{ATC}{1 - t} - (1 + \phi)[AC(TC) - ATC]$$

$$AC(D_b) = 99\% \div 15 = 6.60\%$$

$$AC\ (D_t) = (\text{first cost})(P/D_t)(A/P,\ 12\%,\ 15)$$

$$(P/D_t) = \text{factor for 15-year life, 6\% gross salvage, and cost of removal adjustment}$$

$$= 0.61343 + (0.18270)(0.05) = 0.62257$$

$$AC(D_t) = 100\%\ (0.62257)(0.14682) = 9.14\%$$

$$AC(TC) = 100\%(\text{qualified }\%)(\text{ITC rate})(A/P, 12\%, 15)^*$$
$$= 100\%(0.95)(0.07)(0.14682)$$
$$= 0.98\%$$

$$ATC = \frac{TC}{N} = \frac{6.65\%}{15} = 0.44\%$$

$$AC(T) = 0.6625(14.66\% - 6.60\%) - 0.5(1.6625)(9.14\% - 6.60\%)$$
$$- \frac{0.44\%}{1 - 0.5} - (1.6625)(0.98\% - 0.44\%) = 1.45\%$$

Ad valorem taxes, maintenance, and other operations costs:

Ad valorem taxes	=	2.00%
Maintenance		
$\frac{300}{10,000}(100\%)$	=	3.00
Other operations costs		
$\frac{50}{10,000}(100\%)$	=	0.50
Total	=	5.50%

Equipment A's annual cost rate = $AC(C_{rec}) + AC(T) + AC$ (operations costs)
$$= 14.66\% + 1.45\% + 5.50\% = 21.61\%$$

Equipment B:
Capital-recovery costs:

$$AC(C_{rec}) = 100\%(A/P, 12\%, 15) - \frac{250 - 200}{4,000}(100\%)(A/F, 12\%, 15)$$

$$= 100\%(0.14682) - 1.25\%(0.02682) = 14.65\%$$

Income taxes:

$AC(D_b)$	= 98.75%/15 = 6.58%
$AC(D_t)$	= first cost $(P/D_t)(A/P, 12\%, 15)$
(P/D_t)	= factor for 15-year life, 6.25% gross salvage, and cost of removal adjustment = 0.61297 + (0.18270)(0.05) = 0.62211
$AC(D_t)$	= 100% (0.62211)(0.14682) = 9.13%
$AC(TC)$	= 0.98% ⎫ (same as equipment A calculation)
ATC	= 0.44% ⎭
$AC(T)$	= 0.6625(14.65% − 6.58%) − 0.5(1.6625)(9.13% − 6.58%)

$$- \frac{0.44\%}{1 - 0.5} - (1.6625)(0.98\% - 0.44\%) = 1.45\%$$

Ad valorem taxes, maintenance, and other operations costs:

Ad valorem	=	2.00%
Maintenance		
$\frac{200}{4,000}(100\%)$	=	5.00%

---

*Assumes tax credit occurs at time 0; it must be multiplied by (P/F, 12%, 1) if credit is assumed to be coincident with the first annual cost.

Other operations costs

$$\frac{25}{4,000}(100\%) \qquad = \underline{\quad 0.62\%\quad}$$

Total $\qquad\qquad = 7.62\%$

Equipment B's annual cost rate $= AC(C_{rec}) + AC(T) + AC \text{ (operations costs)}$
$$= 14.65\% + 1.45\% + 7.62\% = 23.72\%$$

The annual costs are as follows:

Equipment A: $\$10,000(0.2161) = \$2,161.00$
Equipment B: $\$4,000(0.2372) = \$948.80$

The time-cost diagram is as shown.

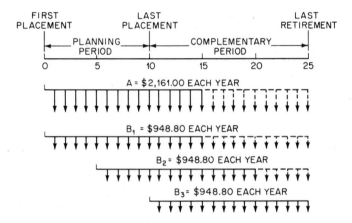

NOTE: The dotted lines in the time-cost diagram represent the assumption of replacing the equipment represented by the solid lines at the cost paid for those original units.

The final step in the analysis is to find the present worth of these levelized equivalent annual costs.

Equipment A:

$$\$2,161.00(P/A, 12\%, 25) = 2,161.00(7.843) = \$16,948.72$$
$$PWAC \text{ of Equipment A} = \$16,948.72$$

Equipment B:

$$\$948.80(P/A, 12\%, 25) = 948.80(7.843) \qquad\qquad = \$7,441.44$$
$$\$948.80(P/A, 12\%, 25 - 5) = 948.80(7.843 - 3.605) = 4,021.01$$
$$\$948.80(P/A, 12\%, 25 - 10) = 948.80(7.843 - 5.650) = \underline{2,080.72}$$
$$PWAC \text{ of Equipment B} \qquad\qquad\qquad\qquad\quad = \$13,543.17$$
$$PWAC \text{ difference in favor of B} \qquad\qquad\quad = \$3,405.55$$

It may be desirable to express these costs in terms of an AC over the 25 years. The annual cost for equipment A is already in the form of an AC, but the costs for B must be converted.

$$AC(B) \qquad\qquad\qquad = \$13,543.17(A/P, 12\%, 25)$$
$$= 13,543.17(0.1275)$$
$$= \$1,726.77$$
$$AC(A) \qquad\qquad\qquad\qquad = \$2,161.00$$

$$AC \text{ difference in favor of B} = \$434.23$$

As mentioned earlier, the *PWAC* technique is popular because the average annual cost rates can be used for specific equipment categories, and also because non-coincident retirements can be treated simply by extending an annual cost throughout the complementary period. However, a few serious pitfalls are associated with the *PWAC* technique that make the *PWE* technique preferable—even though both techniques are equivalent when done properly.

One of these pitfalls is to assume nonrepresentative "average" costs. For instance, in a typical annual cost rate, an average maintenance cost would be represented as a percentage of the installed first cost. If this figure were 5% and it was applied to a first cost of $1,000, the maintenance cost would be $50 per year. A series of annual maintenance costs that also averages $50 per year is: $5, $10, $15, $20, $25, $35, $50, $80, $110, $150. At a discount rate of 10%, the *PWAC* for the $50 average annual cost is $307.20, whereas the *PWE* for the series is $240.48. The *PWAC* value for the average is more than 25% higher than the *PWE* for a series that represents the natural tendency of increasing maintenance costs with aging equipment. Furthermore, the average maintenance cost will not reflect the fact that the current first cost may be higher than the former first cost because of maintenance-saving features now built into the product. The use of such averages, then, may ignore exactly what the study is supposed to evaluate.

A second pitfall is the inclination to use average annual cost rates for a specific equipment item with an expected life and salvage value which is different from the category average. For example, the annual cost rate for a class of equipment with the following characteristic would be 15.95%.

Average life	40 years
Gross salvage	10% of first cost
Cost of removal	6% of first cost
Cost of capital	12%
ϕ	0.6625
ADR life	80% of service life
ITC rate	4%
Eligible portion of investment	95%

If a specific unit of this class of equipment were to be installed for 10 years with the expectation of 26% gross salvage as the only other change, the proper annual cost rate would be 18.53%. The misuse of the class average would therefore produce an error of almost 15%.

One final example of a pitfall of the *PWAC* technique is the tendency of some to regard these annual costs as real costs rather than as equivalents. Popular misconceptions include the ideas that the yearly cost of doing business can be measured by the levelized equivalent annual costs, and that annual costs can be saved by removing the equipment which are prompting these costs. In reality, these costs are the annual equivalents of expenditures which in many cases have already been made; therefore, they cannot be used to measure current costs of doing business. As for being able to save annual costs by equipment removals, the only costs which can be saved are those involving maintenance, operations, and ad valorem taxes. The capital costs for repayment, return, and income taxes cannot be saved.

SUMMARY

This chapter describes techniques for making economy studies involving mutually exclusive alternatives for accomplishing a specific job. Although it is sometimes possible to make a valid comparison of alternatives without the aid of a formal study, this is usually not the case. This chapter has presented three study techniques which can be used effectively to select the most economic alternative of a set of mutually exclusive alternatives. The three are internal rate of return, present worth of expenditures, and present worth of annual costs. When they are applied correctly, they are consistent with each other for deciding among mutually exclusive alternatives.

The internal rate of return ($IROR$) is the interest rate which causes the present worth of the net cash flow for a project to be zero. However, it is strictly a measure of the efficiency of an incremental cash flow. It is not a measure of profitability. To use the internal rate of return to decide among several mutually exclusive alternatives, it is necessary to find the incremental rate of return between all the pairs of alternatives. Because of this, the number of comparisons can grow to an awkward level.

There are other problems too. The more serious of these is that the internal-rate-of-return analyses may be complicated by the existence of multiple answers, and these can easily be misinterpreted. Overall, the internal rate of return is a very complex technique, and should be interpreted only by experts. The major advantage of this technique is that when common costs can not be excluded, it may provide a way to determine if differences among the alternatives are significant.

The use of the other two techniques will avoid many of the problems with the $IROR$ technique and will provide entirely satisfactory economic answers. The present worth of expenditures (PWE) technique determines the present worth at a given discount rate of all expenditures for each project. Because the present worth of the revenue received from the customer must be equivalent to the present worth of the firm's expenditures, the alternative with the least PWE will be the most economic. The difference in PWE is a measure of the size of the potential savings which can be attributed to one alternative when it is compared against others in the same mutually exclusive set. Although the treatment of common costs affects the PWE value of individual alternatives, it does not affect the PWE differential between alternatives which have common costs. Although the differential is unchanged, the significance of that difference may be obscured by the treatment of common costs. The major advantage of the PWE technique is that it avoids the complex incremental analysis required by the $IROR$ technique. This is because the incremental analysis is implicit in the PWE comparison.

The present-worth-of-annual-costs ($PWAC$) technique is usually equivalent to PWE, except that it converts all expenditures to annualized costs. Like the PWE, the alternative with the least $PWAC$ will be the most economic.

By first converting expenditures into levelized equivalent costs, the $PWAC$ technique permits costs to be allocated to study periods different from the total life of plant. The $PWAC$ technique is therefore most applicable to studies involving non-coterminated plans and using the repeated-plant assumption. When the effects of inflation are included in studies, however, the $PWAC$ technique is less likely to be applicable. Furthermore, the $PWAC$ technique seems to encourage the use of average annual cost percentages, which seldom apply to a particular study; and $PWAC$ also shares the problems with PWE about interpreting the significance of the differences.

The Discount Rate

The "time value of money" concept is fundamental to engineering economy. This concept involves translating sums of money into equivalent amounts at different points in time based on some "discount rate" which reflects the cost of money.

Within this text, the appropriate rate for use in this discounting process has been indicated to be the firm's composite cost of money. However, there is not universal agreement on how the discount rate should be defined, and the reader may observe different approaches being used or recommended by other texts.

While it would be inconsistent with the role of this book to examine each of these approaches in detail, it might be well to provide a brief description of some that the reader may encounter.

PRETAX COST OF MONEY

In this approach, the income taxes associated with return are included in the discount rate along with the composite cost of money to reflect the intimate association of the two.

EQUITY COST OF MONEY

Some advocate using the cost of equity capital as the discount rate on the rationale that the residual effect of changes in cash flow impact on the stockholder. When the equity rate is used as the discount rate, interest on debt is viewed as an expense apart from the cost-of-money calculations.

TAX-ADJUSTED COST OF MONEY

This discount rate is less than the composite cost of money by a factor representing the tax effect of debt interest. The argument for use of this rate is that the effective cost of capital to the firm is reduced by the tax advantage of the interest deduction. (In the other methods, the debt deduction is recognized but as part of the computation of tax expense rather than in the cost of money).

SUMMARY

Considerable controversy has been generated relative to which discount rate is "correct." The pretax rate is marred by difficulty in reflecting tax-depreciation write-offs. However, compelling arguments can be marshalled for either the equity rate or the tax-adjusted rate as competitors to the composite cost of capital.

Essentially each of these various definitions constitutes a model of the firm representing a view of the way in which changes in cash flow impact the business. Arguments for any of the approaches are based on certain subtle assumptions about the economic dynamics of the firm—assumptions which are not clearly "right" or "wrong."

This text has recommended continuation of the Bell System practice of using the composite cost of capital as the discount rate in economic selection studies. In addition to being broadly accepted and familiar to Bell System analysts, this model portrays very closely the financial and regulatory environment in which a public utility operates.

Rate of Return

The chapter proper pointed out that internal rate of return *(IROR)* was a complex subject, and there are problems in applying it to decisions at a specified cost of money. This appendix discusses some of those problems, and also explains a solution using a modified rate of return *(MROR)*.

PROBLEMS WITH *IROR*

Two of the more critical problems resulting from the use of *IROR* to select the economic alternative at a known cost of money are:

- The possibility of multiple answers
- The misinterpretation of results

Engineering economy authorities often dismiss the possibility of multiple answers because the cash flows of normal real-life projects seldom produce them. That is true. But when *IROR* analysis is applied properly for choosing among several mutually exclusive projects, or for capital budgeting, it is necessary to find the *incremental* differences between cash flows and the incremental internal rate of return. An incremental cash flow, which is just the difference between two simple cash flows, can quite often have multiple answers.

The problem of misinterpreting results usually arises when the analyst computing the incremental cash flow reverses the order of the alternatives. For example, if A is known to be more economic than B, the incremental rate of return on the A-B cash flow is of course expected to be greater than the decision rate. But, in doing a study, it is not certain that A is more economic than B. If the incremental rate of return is found on the B-A cash flow, it will be the same as the incremental rate of return on A-B. If the rate of return is then the sole criterion for deciding between the two alternatives, an unwary analyst could easily choose B—even though A is actually more economic.

The two problems are directly related. They exist primarily because the traditional definition of *IROR* does not recognize the signs (+ or −) of the balances represented by the cash flows. It therefore does not consider whether the cash flow is an investment or a source of funds, or a combination of the two. A simplified interpretation of rate of return—which considers the higher, the better—is not valid. If the cash flow represents a source of funds, the indicated interest rate will be the cost paid for money; in that case, the lowest-cost source is best. If the cash flow is a combination of both an investment and a source of funds, the criterion is unclear. The incremental (A-B) and (B-A) cash flows will produce the same rate of return even though one is an investment and the other a source of funds, because the equation for one is the same as the equation for the other; the only difference is that both sides have been multiplied by −1. The multiple answers occur because the cash flow is a combination of an investment and a source of funds.

Most writers on internal rate of return warn users to watch out for cash flows with more than one change in sign. For example,

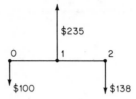

Such cash flows may have multiple answers because they can represent both an investment and a source of funds. The balance diagram for such a cash flow will demonstrate this quickly. The final cash flow is negative. Therefore, the balance before this must have been positive. Since the first amount was negative, the balance must have changed. The balance diagram for the cash flow in the time-line diagram must look like this:

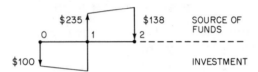

The traditional definition of internal rate of return will result in two solutions:

$$0 = \frac{-100}{(1 + r)^0} + \frac{235}{(1 + r)^1} - \frac{138}{(1 + r)^2}$$
$$r = 15\% \text{ or } 20\%$$

Neither of these two solutions is "correct," in the sense of indicating a rate of return on an incremental investment. The cash flow is simply not a "normal" investment. It is an investment (negative balance) during the first year, but a source of funds (positive balance) during the second. Two balance diagrams can be drawn, each of which goes to zero.

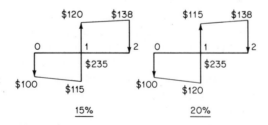

The usual interpretation of the traditional definition for *IROR*, which presumes that high rates are good, assumes that the cash flow is an investment and that the *NPV* of that cash flow will be positive at rates less than the rate of return. Figure 12-2 illustrated that the incremental rates of return must be used to find the ranges of cost of money in which one of several alternatives may have the greatest *NPV*. The *NPV* of the cash flow at the expected cost of money is the real indicator of the economic merit. In other words, the internal rate of return works only if it is consistent with the *NPV*.

The *NPV* of this example cash flow is plotted in the following diagram.

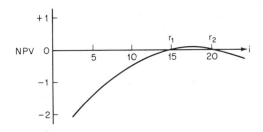

In this example, this cash flow is favorable only if the cost of money lies between 15 and 20%. If the cost of money is 10%, the project should not be selected. If this diagram represents an incremental rate of return on the difference between two projects (A − B), project A should be rejected in favor of B at the same 10% cost of money. If all signs in the cash flow were reversed (meaning incremental B − A), the roots would be the same; but now the *NPV* versus *i* diagram would be reversed. At a 10% cost of money, then, Alternative B is still the choice.

As this chapter indicated, finding the internal rate of return was the same as finding the interest rate that a bank must be paying on a deposit (investment). A cash flow resulting in a balance diagram with both positive and negative balances means that the bank has allowed the withdrawal of more than the amount deposited. In other words, the cash flow is in part external to the bank. As a result, the *internal* rate of return, as a rate of return on an incremental investment, just does not exist.

The *NPV* at a known cost of money does not suffer from the same problem because the *NPV* presumes that money costs the known rate during any period when a cash flow is a source of funds. If the rate of return during the period when the cash flow is an investment is greater than the cost of money, the *NPV* will be positive. Because the *NPV* is the real indicator of economic merit, the solution to the problem is to find an index, like rate of return, which is consistent with the *NPV* concept. This index is then the rate of return during the time the cash flow represents an investment, provided the cost of money during any time the cash flow is a source of funds is the known cost of money.

MODIFIED RATE OF RETURN *(MROR)*

The *modified rate of return (MROR)* is a way to avoid some of the problems of *IROR* while still keeping an index with characteristics similar to the *IROR*. The *MROR* cannot be defined in a simple mathematical equation like the *IROR*, because it evolves from a serial operation for computing each sequential balance. Each balance must be tested to determine the status of the cash flow as an investment or source of funds so that the proper rate can be applied. In the simple example used so far, the operation is quite simple. The *MROR* (m) is defined as the rate of return during the time the cash flow is an investment if the rate paid during any time the cash flow is a source of funds is the known cost of money (i). In the example, the balance diagram would be as follows:

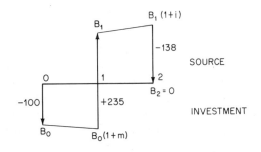

This computation is easier to do by proceeding backwards. The last cash flow was $138, and it paid off a source of funds. If funds are expected to cost 10%, then B_1 must have been:

$$B_1 = \frac{B_2 + 138}{1 + i} = \frac{138}{1.10} = \$125.45$$

The investment, B_0, must have grown at some rate (m) such that the following holds.

$$B_0(1 + m) + 235 = B_1$$
$$-100(1 + m) + 235 = 125.45$$
$$-100(1 + m) = -109.55$$
$$m = 0.0955 = 9.55\%$$

Since the $MROR$ of 9.55% is less than the 10% cost of money, the indication is to reject. This decision is consistent with the NPV indication at 10%, illustrated on page 317. If this cash flow is tested at a higher cost of money ($i = 17\%$), the calculation is as follows:

$$B_0(1 + m) + 235 = \frac{138}{1 + 0.17}$$
$$-100(1 + m) + 235 = \frac{138}{1.17} = 117.95$$
$$-100(1 + m) = -117.05$$
$$m = 0.1705 = 17.05\%$$

The $MROR$, 17.05%, is slightly greater than the cost of money (now 17%), which again is consistent with the NPV indication to accept between 15 and 20%, illustrated on page 317.

If the cash flows are reversed, the balance diagram reverses, and the $MROR$ will change.

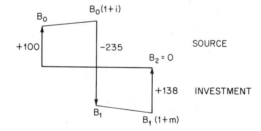

This time it is easier to go forward. At 10% cost of money,

$$B_0(1 + i) - 235 = B_1$$
$$100(1.10) - 235 = -125 = B_1$$
$$B_1(1 + m) + 138 = B_2 = 0$$
$$-125(1 + m) = -138$$
$$1 + m = \frac{138}{125} = 1.104$$
$$m = 0.104 = 10.4\%$$

Again, the $MROR$ is consistent with the NPV indication for the reversed cash flow, because the $MROR$ is greater than the cost of money.

This example has been kept very simple in order to illustrate the concept of the $MROR$. However, $MROR$ can be found for extremely complex cash flows, with multiple changes in sign and balance. The illustration of a solution of a complex cash flow for $MROR$ is tedious and unnecessary, however. Like $IROR$, the $MROR$ must usually be

FIGURE 12B-1
Flow Chart

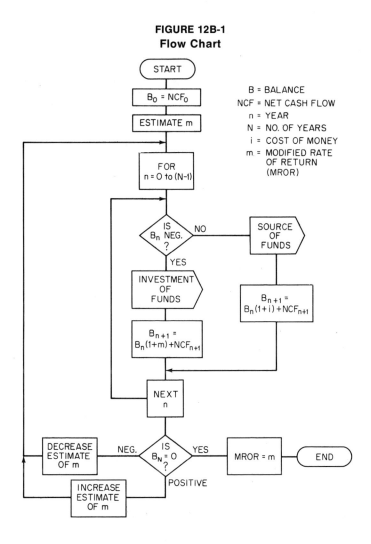

found through a trial-and-error process. This type of solution is obviously best done by computer. Therefore, the sequential procedure by which *MROR* can be found is illustrated by the flow chart in Figure 12B-1. In some simple cases, the *MROR* can be found by a single expression based on the interpretation of the cash flow as in the example in this appendix. However, this is not generally true.

A Single Solution

It is mathematically impossible to have more than one answer for *MROR* at a known cost of money. This is because *MROR* treats the *IROR* problem of multiple answers in the same way as *NPV*, *PWE*, or *PWAC* techniques. As with those techniques, however, the problem is not really solved; it is merely ignored because the analysis is restricted to the relationship of alternatives at a specified cost of money. At another cost of money, the *IROR* values are unchanged, but *NPV*, *PWE*, *PWAC*, and perhaps *MROR* do change. For most cash flows where there is only a single *IROR* answer, the *MROR* will equal the *IROR* at all costs of money. But if there is a possibility of multiple answers for *IROR*, the *MROR* will change with a change in the cost of money. If the cost of money equals any *IROR* answer, the *MROR* will equal that *IROR* answer. Figure 12B-2 illustrates this relationship for the simple example used in this appendix.

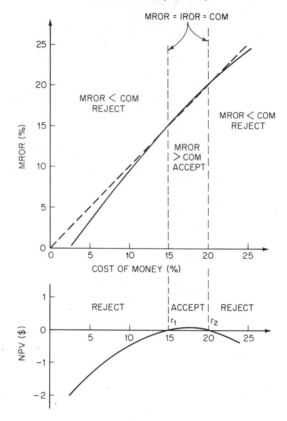

FIGURE 12B-2
Modified Rate of Return (MROR) at Various
Costs of Money (COM)

The dashed line in the upper graph is the line for *MROR* equal to the cost of money. The value of *MROR* versus the cost of money fluctuates around this line in phase with the *NPV* versus cost-of-money curve. As a result, the single *MROR* value is consistent with the *NPV* decision at any cost of money.

Problems with *MROR*

Although the *MROR* helps to solve some of the problems with *IROR*, it is not without problems itself. Just as with *IROR*, if there are n alternatives, then $n(n - 1)/2$ incremental *MROR* values must be found. It is a lot of work to compute *IROR*, but the result is data over the entire range of the cost of money. It takes just as much effort to compute *MROR*. However, if there are any multiple-solution *IROR* situations, the *MROR* data are valid only at the specific cost of money. This means that if *MROR* is used to avoid a multiple-root solution to *IROR*, the cost-of-money versus demand characteristic of *IROR* data is lost entirely.

Another problem with *MROR* is that even though an incremental *IROR* can be found, an *MROR* may not exist; or else it can be infinitely positive or negative. This occurs for one major reason. An incremental cash flow which is a pure source of funds does not have an *MROR* because there is no period of investment during which a rate of return can be found. Cash flows which are investments for only a very short period of time may also have a very distorted *MROR*. This problem is two-faced. On the one hand, it is the reason the *MROR* avoids the possible misinterpretation of an *IROR* root on reversed alternatives. On the other hand, it may not provide an answer when one is required.

Computer programs which encounter this problem can be arranged to reverse the cash flow so that a valid solution is found. *However,* if this happens, the resulting answer must be marked to indicate that the evaluation must also reverse because the *MROR* is now reflecting the *cost* of the incremental source of funds rather than a rate of return on an incremental investment.

The mere substitution of *MROR* for *IROR* does not make the rate of return a theoretically correct index for ranking dissimilar projects for capital budgeting. For example, the fact that an incremental $1 million has an *MROR* of 25% when invested in a TSPS conversion does not indicate that the TSPS is a better place for the money than an ESS-Crossbar replacement project with a 20% incremental *MROR*. If the firm's cost of money is 10%, either of these two alternatives is satisfactory. With *IROR* or *MROR* information, the better one can be determined only by an incremental analysis of the two incremental cash flows. In a practical sense, then, it is probably not feasible to conduct this kind of incremental analysis on the many projects competing for a limited amount of capital. For one thing, the $1 million may have to be committed to one of the projects before anyone even realizes that there may be a better opportunity.

The capital rationing job is a very difficult undertaking. The *MROR* helps by not compounding this difficulty with multiple answers. The *MROR* information, when combined with other economic data from *PWE* or *PWAC* analyses, plus other noneconomic factors, may help managers who are responsible for capital budgeting make their decisions.

Summary of Advantages of *MROR*

The advantages of MROR can be summarized as follows.

1. If *MROR* exists, its interpretation will be consistent with *NPV, PWE,* or *PWAC* analysis.

2. If the cash flow is a pure investment, *MROR* will equal *IROR* because *MROR* is then simply a future-worth definition of *IROR*.

3. The *MROR* can have only a single value for a known cost of capital.

4. The decision rules for *IROR* also apply to *MROR*, but are valid only at the firm's specified cost of money.

13

RETIREMENT, REPLACEMENT, AND OTHER ECONOMY STUDY CONSIDERATIONS

INTRODUCTION

In the past, the telephone industry has been characterized by continual technological change. It must be anticipated that the future, too, holds numerous technological changes. As a result, economy studies must often be done to decide when and if some evolving technology should replace a technology which may be obsolete. These types of studies are somewhat different from a study which does not consider existing plant, because of the need to establish the value, any future life, and the costs associated with plant which was obtained in the past. This chapter discusses the concepts of sunk costs, retirement, salvage, and reuse, and the special considerations involving the replacement of property owned by more than one company (jointly owned). While these concepts are covered here in the context of a replacement analysis, they may also be useful in other studies where the lives of alternative items of plant do not coincide in the future. With those types of studies, too, there is a need to establish the value and costs associated with plant which will live longer than alternative plant.

Several terms are used throughout this chapter, and it will be helpful to review their definitions as applied in the field of engineering economy.

1. *Retirement:* The removal of an existing asset from service.

2. *Replacement:* The substitution of another asset to perform a service which has been performed by an asset that is to be retired.

3. *Obsolescence:* The gradual process of becoming obsolete.

4. *Salvage:* The value received for an asset upon its retirement.

5. *Defender:* An existing asset being considered for retirement.*

6. *Challenger:* A unit being considered as a replacement for an existing asset (the defender).*

As suggested in Chapter 7, there are many reasons why a replacement might be considered before a defender is completely worn out (e.g., new technology, lack of

---

*The words *challenger* and *defender,* as applied to replacement analysis, were popularized by George Terborgh in his extensive writing on engineering economy. In this text, the definitions of *defender* and *challenger* may be extended to include the entire alternative of which each is a part.

defender capacity, or new requirements of customers). Once the decision is made to investigate a replacement, however, the question is: "Which is more economical—the installation and operation of the challenger, or the retention and maintenance of the defender?" This chapter describes how to answer that question. A good starting point for such a study is the life considerations which are common to all studies.

STUDY-LIFE ASSUMPTIONS

When lives are assigned to the various components of an alternative, it is important that they represent the *service* or *economic* life, that is, the period when the components fulfill the specific service being investigated. The service or economic life should be used because that is when the actual costs of the service are experienced. In estimating the lives of alternatives, probably the most common error is to use the time period over which the accountants allocate or depreciate the cost of an asset. While it is important for the accountants to try to match the expenses of the business with its revenues, the period over which they allocate the cost of an asset does not necessarily match the intended service life of that asset. This is particularly true under the group plan of depreciation, in which averages are used for broad equipment categories. Of course, it is possible that this period and the economic life will be the same, but only by coincidence; assuming they are the same is probably invalid.

In a replacement study the most important thing to realize is that the life already lived by the defender is common to its retention or its retirement. As a result, the past life of the defender is irrelevant to the study decision. The important part of the defender's life is any possible future life. Therefore, in a replacement study the account averages are even less significant than in other studies, because the result of the replacement study will affect the account averages. In effect, the reason for the replacement study is to determine the future life of the defender. Estimating reasonable and specific economic future lives for the defender or the challenger is not easy because it requires some assessment of the future. However, this is no different from other studies which require judgment of the future. Just as in other studies, the most important consideration in the life estimate is the best judgment of the analyst and others familiar with the plant.

Normally, the beginning point of the study is the earliest practical date that the challenger could be implemented. When the end point of the study is dictated by the conditions of the problem or by the lives of other items (as suggested in Chapter 7), there is no difficulty in finding a reasonable study life. However, when the end point is indefinite, the length of the study period will depend upon the study maker's judgment. Some of the more important considerations are: (1) the expectation of further changes in technology; (2) the expectation of changes in future prices; and (3) any known planning which is not a part of the project being studied but will influence it. For example, if major technological improvements have occurred on an ever-shorter interval for the kind of asset being studied, the economic life should reasonably be estimated as shorter than the one for the asset of the technological era immediately preceding that of the challenger. Future prices also have a large bearing on asset lives. Expectations of changing salvage values in the future will influence asset lives, and so will expectations of higher costs. The prospect of increasing labor costs will tend to shorten the lives of assets associated with labor-intensive operations. Finally, other plans, such as those which require the eventual elimination of certain types of operations, would suggest a shorter life for the assets required. Because study lives are so heavily influenced by the judgment of the study maker, it is also a good practice to test decisions for their sensitivity to the life assumptions made in the study.

COST CONSIDERATIONS

The relevant costs to be considered in a replacement study are similar to those considered in other studies. However, there are often questions raised because of the existence of a defender in the replacement study. Any discussion of costs for replacement studies should therefore begin with the costs of the defender; and perhaps the best starting point is sunk costs.

Sunk and Common Costs

Sunk costs were defined in Chapter 11 as the costs of the existing assets. In other words, they are the expenditures for the defender up to the point at which any challenger becomes available. Take a one-car taxicab company as an example. If this company were considering replacing a 2-year-old cab with a new vehicle, the sunk costs would be the purchase price of the old cab; any return paid to investors for funds used to purchase the cab; the income taxes paid because of the cab; and any operations costs such as gas, oil, repairs, and ad valorem taxes during those 2 years.

There is no need to consider the sunk costs in a replacement study, though, because past receipts and disbursements and the method of accounting for them are common and therefore irrelevant. The original costs of the old asset, the depreciation that has accrued to it or remains to be accrued, and the net book cost are all things of the past; whether they have been done wisely or foolishly, correctly or incorrectly, is immaterial.

However, any depreciation expense remaining to be accrued seems to cause particular problems. Here, it is important to recognize that the depreciation expense is not the cost; it is, rather, an allocation of a past expenditure. Once the expenditure which created the depreciation has occurred that capital has to be recovered, with return and income taxes; and this need is independent of any present or future decision. No capital costs associated with a past expenditure are germane to the retirement decision. Costs associated with the existence of the asset, or operations costs, such as ad valorem taxes and maintenance, to be incurred in the future, are germane, however, and must be considered.

Sunk costs are irrelevant, then, because they are common between those plans which would keep the assets that caused the costs and those plans which would require those assets to be removed. For instance, if the cab had been financed by a bank loan, the amount of the loan still outstanding would not be relevant in an economy study investigating the replacement of the cab, because the loan is an obligation that is common to either keeping the cab or replacing it. The time-cost diagrams of Figure 13-1 illustrate this condition.

All of the costs to the left of the dotted line (time point −2 through the left side of time point 0) in the defender plan are identical with those to the left of the dotted line in the challenger plan. They are therefore common and can be omitted from the comparison without affecting the outcome.

Another important aspect of defender-associated costs is the question of which costs are eliminated when replacement occurs and which ones are not. Again, it is helpful to look at the cab company example. Suppose the company has a radio dispatcher. The cost of employing this dispatcher could be allocated to the operation of the cab. However, the cost of this dispatcher would not be eliminated with the replacement of the cab and therefore should not be claimed as a savings due to replacement. On the other hand, costs such as gasoline used by the cab would properly be claimed in a replacement study. The key question is, "Will this cost really be eliminated by replacement?" If the answer is "yes," then the cost should be claimed as a savings. If the answer is "no," the cost should be left out of the study.

Salvage

Figure 13-1 shows the net salvage received when the defender is sold or junked as a receipt in the challenger plan. This is a satisfactory approach for many studies, and is in fact what actually would happen if the challenger were selected. However, as Chapter 12 pointed out, it is the differences between plans which are important to the economy study. If the net salvage is shown instead as a cost in the defender plan, as in Figure 13-2, the same difference between the two plans will result. Examination of Figure 13-1 also illustrates that the elimination of the sunk or past costs as common will not affect this difference.

As Figure 13-2 shows, the capital costs of keeping the defender in service can be considered the same as the cost of purchasing the remaining life of the defender with the salvage not obtained (salvage foregone), offset by salvage that ultimately *would* be received if it is kept in service. Salvage can be treated in either one of these two ways: (1) a receipt in the plan which retires the plant (Figure 13-1); (2) a cost in the plan that keeps the plant (Figure 13-2). The second way is often preferable, and for two reasons.

- The effect of income taxes is simpler to determine.

- The cost of keeping can more easily be compared with the cost of several different replacing plans.

This treatment of salvage can be represented mathematically. Assume that Plan A has been proposed which would retire a defender now, with a salvage value of S_0. Plan B would keep the defender for n years, with a salvage of S_n and operations costs of O_b. Under Plan A, the cost of the replacing plant can be determined separately. The other operations of the firm are common to A or B, but will be included here so that the effect of the salvage on A or B can be measured. Those other common operations can be represented by some capital investment of an amount C, living exactly n years, with a salvage of S_c and operations costs of O_c.

FIGURE 13-1
Effect of Sunk Costs

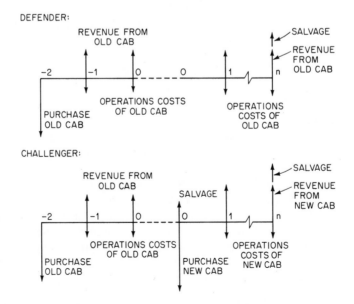

FIGURE 13-2
Effect of Salvage Foregone

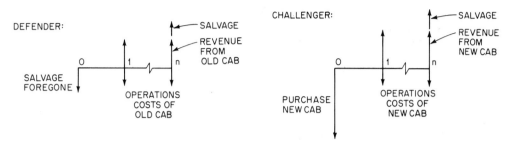

Under Plan A, the capital required to buy the common plant for C is reduced by S_0, and the present worth of the net expenditures is

$$PW(C_{rec})_{A + C} = (C - S_0) - S_c(P/F, i\%, n)$$

The capital which must be repaid from book depreciation is changed by the receipt of S_0, so that

$$PW(D_b)_{A + C} = \frac{(C - S_0) - S_c}{n}(P/A, i\%, n)$$

Although the tax depreciation on the common plant is unaffected by the salvage associated with the plant considered for retirement, S_0 results in an increase in tax liability just as a reduction in tax depreciation would.* Therefore,

$$PW(D_t)_{A + C} = PW(D_t)_C - S_0$$

where $PW(D_t)_C$ is the tax depreciation on the common plant in C.

The present worth of income taxes can now be calculated using equation (9.1), page 197.

$$PW(T)_{A + C} = \phi\left[(C - S_0) - S_c(P/F, i\%, n) - \frac{(C - S_0) - S_c}{n}(P/A, i\%, n)\right]$$
$$-t(1 + \phi)\left[PW(D_t)_C - S_0 - \frac{(C - S_0) - S_c}{n}(P/A, i\%, n)\right]$$

Finally,

$$PW(\text{operating cost})_{A + C} = O_c(P/A, i\%, n)$$

Under Plan B, the capital recovered from future salvage is increased by S_n, and the present worth of the net expenditures is

$$PW(C_{rec})_{B + C} \qquad = C - (S_c + S_n)(P/F, i\%, n)$$

$$PW(D_b)_{B + C} \qquad = \frac{C - (S_c + S_n)}{n}(P/A, i\%, n)$$

$$PW(D_t)_{B + C} \qquad = PW(D_t)_C - S_n(P/F, i\%, n)\dagger$$

---

*Receipt of gross salvage affects tax deductions in either of two ways: (1) it may eliminate a like amount of tax depreciation, in which case it has the effect of a negative tax deduction; (2) it may increase the taxable gain, in which case it also looks like a negative tax deduction. Since cost of removal is tax deductible, the combined effect, net salvage foregone S_0, is the net tax deduction.

†This treatment of S_n is similar to the treatment of S_0.

$$PW(T)_{B+c} = \phi\left[C - (S_c + S_n)(P/F, i\%, n) - \frac{C-(S_c+S_n)}{n}(P/A, i\%, n)\right]$$

$$-t(1+\phi)\left[PW(D_t)_c - S_n(P/F, i\%, n) - \frac{C-(S_c+S_n)}{n}(P/A, i\% \ n)\right]$$

$PW(\text{operating cost})_{B+c} = (O_c + O_b)(P/A, i\%, n)$

Because Plan B keeps the salvageable plant, subtracting the cost of Plan A from the cost of Plan B, that is, (B + C) − (A + C), eliminates the costs of the common plant and gives the net capital cost of the retained plant.

$$PW(C_{rec})_{B-A} = S_0 - S_n(P/F, i\%, n) \tag{13.1}$$

$$PW(T)_{B-A} = \phi\left[S_0 - S_n(P/F, i\%, n) - \frac{S_0 - S_n}{n}(P/A, i\%, n)\right] \tag{13.2}$$

$$-t(1+\phi)\left[S_0 - S_n(P/F, i\%, n) - \frac{S_0 - S_n}{n}(P/A, i\%, n)\right]$$

$PW(\text{operating cost})_{B-A} = O_b(P/A, i\%, n)$

The results of Plan B minus Plan A are exactly what they would have been if the retaining plan "bought" the existing asset for its salvage value now. The incremental impact on the tax liability is the same as if the salvage foregone were a tax-deductible expense and the ultimate salvage a taxable receipt. The quantity $S_0 - S_n(P/F, i\%, n)$ therefore serves the dual role of representing both $PW(C_{rec})$ and $PW(D_t)$ for the existing and retained asset. In other words, the tax equation above is simply equation (9.1), or

$$PW(T) = \phi[PW(C_{rec}) - PW(D_b)] - t(1+\phi)[PW(D_t) - PW(D_b)] \tag{9.2}$$

where $PW(D_t) = PW(C_{rec}) = S_0 - S_n(P/F, i\%, n)$

$$PW(D_b) = \frac{S_0 - S_n}{n}(P/A, i\%, n)$$

Reuse

Assets that are replaced are not always junked or sold to someone else; they are sometimes kept and reused for other purposes. In that case, the replaced (to be reused) asset may have a value higher than if junked. There is a certain amount of intuitive appeal to the concept that an asset which can be reused has more value than one that is to be junked or sold to someone else. The parable in Example 13-1 illustrates exactly how this conclusion is reached.

Example 13-1. Ed Owner says this isn't his year! He has two taxi-cab companies and both are giving him headaches. One of Ed's companies serves a city known for oppressive summer weather. Ed finds that he is losing revenues because the car he operates in the city does not have air conditioning. Ed figures he has to replace this cab with an air-conditioned car—even though his present car is in mint condition!

To add to his troubles, Ed finds that he also has to replace the car he uses in his second company. This company serves a shore community and air conditioning is not a problem; but the car he uses there is approaching total collapse. It makes so much noise that the town's dogs howl as it approaches.

Right now, Ed is bent over a desk trying to reassure himself that buying an air-conditioned cab for the city operation will pay off. One factor Ed knows he has to recognize is that he will get some value out of freeing up the existing car. One measure of the value of the existing car is the amount it could be sold for. Ed is sure he could get $1,500 for it and wonders if this is what he should credit the replacement proposal with.

On the other hand, Ed thinks, why not solve both problems at once? "If I buy the air-conditioned cab for the city, replacing the existing car, why the existing car can then be used at the shore and I can avoid spending $4,000 for a new car there." On this basis Ed thinks—and correctly—that the value of the existing car in the air-conditioning study is not its resale value but rather $4,000 . . . the cost that would be avoided by reusing the car.

Ed Owner's analysis is an oversimplification which assumes that the operations costs, revenues, and life of the almost new cab are the same as those of the avoided new cab for the shore. If the costs, revenues, and life were not the same, the reuse value would be the present worth of the savings realized because the reuse postpones the need to buy the new cab.

A similar question in the telephone industry might be whether to buy an electronic switching system (ESS) to replace a step-by-step (SXS) switching unit. The SXS equipment could be reused for SXS growth elsewhere, and in that way relieve the company of the need to buy new SXS equipment. A reuse value higher than salvage will tend to encourage the introduction of the new technology. But, if this reusable equipment is not properly considered in subsequent studies, the reuse value may not be valid. This is illustrated in a continuation of the parable of Ed Owner's taxi-cab companies.

Example 13-1 (continued). Ed's brother, Ned, manages the shore company for Ed. Ned has been trying to decide what make of car to buy to replace the old jalopy. When Ed called to ask what the cost of a new cab would be, Ned gave him the average price of the cars he was considering. Ed said that was the real value of the slightly used car that he was going to have Ned reuse at the shore. Ned is confused. One of the two cars he had in mind costs less than Ed's value on the used one. In fact, reusing the city car means they have to drive it down and paint the shore rate schedule on it. The price of a new one includes delivery and sign painting. Ned thinks that Ed is nuts, that he might just as well buy a new car for the shore.

So, Ned calls Ed: "I can't afford to reuse your car, Ed, a new one is cheaper". Ed thinks, "Boy, just when I got things figured out, up jumps the devil, again."

"Ned, you're confusing the problem *I* was trying to solve with the question *you* want to answer. Look, the dealer will only give me $1,500 for this practically new car on a trade-in. If I do that, we've got to hit Uncle Red for another $2,500, or so, to get either one of the cars you like. Ned, the cost of your reusing the city cab is the trade-in we don't get plus the cost of getting it down there and repainted. That's got to be less than either one of the two cars you're considering.

"When I say the value of this car is $4,000, that means I have *assumed* we're going to reuse it in deciding about an air-conditioned car. You're looking at *whether* we should reuse it or not. Those are two different questions."

Of course Ed and Ned can see the whole picture in their small operation. In a large business it may be very easy for Ned to make this kind of error, and consequently decide not to reuse, which may make the replacement decision wrong. It can be summed up in this way.

1. The *cost* of continuing something in-service, or reusing it, is the salvage foregone by not disposing of it, plus any cost of preparing it for a possibly new function.

2. The *value* of something which may be replaced is the greater of the salvage which can be obtained or the present worth of the costs which can be avoided if it is reused.

Finally, it should be obvious, that if reuse is assumed in a replacement study, the next time similar equipment is considered for replacement, the alternative of actually reusing available equipment must also be considered. Otherwise, the subsequent reuse will never come to pass, and the removed equipment actually would be junked, because reuse was not considered. As a result, it is usually best to justify a single replacement based upon salvage value alone, unless a specific reuse is known. Then the reuse, if it actually occurs, is a bonus. The consistent application of a reuse value, when reuse does not actually take place, may even appear to justify economically a new technology which is not economic. Even if the new technology is economic in an overall sense, the reuse value may encourage its introduction at a more rapid pace than would be supported by a study of the overall effect. While a reuse value is intuitively appealing, it should be applied with caution, and the validity of the assumed reuse should be tracked as further studies are done.

NON-COTERMINATED ALTERNATIVES

In many studies, and often in replacement studies, the various items of plant may have different economic lives. Because a valid comparison requires that every alternative provide for the same time period, one or more alternatives may need to be supplemented. This is normally done by adding assets to the plans with shorter-lived plant. However, the supplemental assets may not have economic lives that will cause the plans to be coterminated either. This dilemma can be handled in one of three ways, depending upon the circumstances. *First,* the repeated plant assumption can be made if the costs are expected to remain constant over time. *Second,* if constant costs are not a reasonable expectation (and typically they are not), a study life may be selected which coincides with the retirement of plant in one alternative. A value analogous to salvage may then be found for each other alternative, to reflect the value of the continuation in service of the longer-lived plant. *Third,* if the difference in lives is small, the estimated lives can be adjusted so that the alternatives are coterminated at some point in time and the difference between them neglected. Of course, these approaches are not without problems, since estimated costs become less and less reliable as they are projected farther and farther into the future. However, these problems can be reduced if study results are treated for their sensitivity to the estimates.

All the examples in this chapter have been developed with the *PWE* technique. Of course, either the *ROR* or *PWAC* technique could have been substituted without any significant effect on the result.

The Repeated Plant Assumption

Example 13-2 illustrates the application of the repeated plant assumption to the replacement study.

Example 13-2. Suppose equipment A is now in service and could be kept for 5 more years. Equipment A could be sold now for $1,200, but in 5 years would net only $150. Operations costs are expected to be $500, $520, $540, $565, and $590 during these 5 years. The challenger, equipment B, has an economic life expectancy of 7 years. It will cost $2,500 now and have a gross salvage value of $1,150, with no cost of removal at the end of 7 years. It would qualify for ADR hybrid tax depreciation, but assume no investment tax credit. Because of trends in the manufacturing costs of equipment B, its purchase price 5 years from now is expected to be the same. Also, because technological improvements for this equipment have been solely in the manufacturing process, salvage values are not expected to change over time. Operations costs for equipment B are expected to remain constant at about $250/year. The company's cost of money is approximately 12%; income tax rate, 48%; debt ratio, 45%; debt interest rate, 9%; and $\phi = 0.61$.

The present worth of the expenditures for these two alternatives over their respective lives would be found as follows.

PWE for A:

$$PW(C_{rec}) = S_0 - S_n(P/F, 12\%, 5) \tag{13.1}$$
$$= 1{,}200 - 150(0.5674) = \$1{,}115$$

$$PW(T) = \phi\left[S_0 - S_n(P/F, 12\%, 5) - \frac{S_0 - S_n}{5}(P/A, 12\%, 5)\right]$$
$$- t(1 + \phi)\left[S_0 - S_n(P/F, 12\%, 5) - \frac{S_0 - S_n}{5}(P/A, 12\%, 5)\right] \tag{13.2}$$

$$\frac{S_0 - S_n}{5}(P/A, 12\%, 5) = \frac{1{,}200 - 150}{5}(3.605) = \$757$$

$$PW(T) = 0.61(1{,}115 - 757) - 0.48(1.61)(1{,}115 - 757)$$
$$= 218 - 277 = -\$59$$

$$PWE(\text{capital cost}) = 1{,}115 - 59 = \$1{,}056$$

$$PW(\text{operating costs}) = 500(P/F, 12\%, 1) + 520(P/F, 12\%, 2)$$
$$+ 540(P/F, 12\%, 3) + 565(P/F, 12\%, 4) + 590(P/F, 12\%, 5)$$
$$= 500(0.8929) + 520(0.7972) + 540(0.7118)$$
$$+ 565(0.6355) + 590(0.5674)$$
$$= \$1{,}939$$

$$PWE(A) = 1{,}056 + 1{,}939 = \$2{,}995$$

PWE for B:

$$PW(C_{rec}) = 2{,}500(P/F, 12\%, 0) - 1{,}150(P/F, 12\%, 7)$$
$$= 2{,}500(1.0) - 1{,}150(0.4523) = \$1{,}980$$
$$PW(T) = \phi[PW(C_{rec}) - PW(D_b)]$$
$$- t(1 + \phi)[PW(D_t) - PW(D_b)] \tag{9.1}$$

$$PW(D_b) = \frac{2{,}500 - 1{,}150}{7}(P/A, 12\%, 7)(P/F, 12\%, 0)$$
$$= 193(4.564)(1.0) = \$881$$

$$\% \text{ gross salvage} = \frac{1{,}150}{2{,}500} = 0.46 = 46\%$$
$$PW(D_t) = (\text{first cost})(P/D_t)(P/F, 12\%, 0)^*$$
$$(P/D_t) = \text{gross salvage factor}$$
$$= 0.53729 - (0.53729 - 0.46611)0.6 = 0.49458$$
$$PW(D_t) = 2{,}500(0.49458)(1.0) = \$1{,}236$$
$$PW(T) = 0.61(1{,}980 - 881) - 0.48(1.61)(1{,}236 - 881)$$
$$= 670 - 274 = \$396$$
$$PW(\text{capital cost}) = 1{,}980 + 396 = \$2{,}376$$
$$PW(\text{operating costs}) = 250(P/A, 12\%, 7)$$
$$= 250(4.564) = \$1{,}141$$
$$PWE(B) = 2{,}376 + 1{,}141 = \$3{,}517$$

*See Figure 9-5, page 200.

Providing service for 5 years with the defender (A) has a present worth of $2,995, and for 7 years with the challenger (B), $3,517. Since the defender is likely to be replaced with challenger-type equipment in 5 years, the same equipment will be in service in either alternative after that time. Also, since the assumption was made that the price of equipment A will not change, the annualized costs of the replacing plant will be the same as that of the challenger.

In a 7-year study, the 2-year difference in lives can be recognized by adding an allocation of the costs of a replacement to the defender. The levelized equivalent annual cost (AC) of the challenger for 2 years is then added to the costs of the defender to represent the cost over 7 years under the repeated plant assumption.†

$$AC = 3,517(A/P, 12\%, 7)$$
$$= 3,517(0.2191) = \$771$$
$$PWE(A) = 2,995 + 771(P/A, 12\%, 2)(P/F, 12\%, 5)$$
$$= 2,995 + 771(1.6901)(0.5674) = \$3,734$$

Therefore, a 7-year study under the repeated plant assumption would show that the costs of keeping the defender are $3,734, versus the challenger costs of $3,517. The result is an advantage for the challenger.

Another, often useful, way to look at this situation is that keeping the defender (A) allows the purchase of the challenger to be deferred for 5 years. On this basis, the cost of the defender for 5 years, $2,995, is compared with the present worth of 5 years of the AC of the challenger.†

$$PWE(B) = 771(P/A, 12\%, 5)$$
$$= 771(3.6048) = \$2,779$$

This calculation is consistent with the result of the 7-year study; it also indicates that keeping the defender is not justified because it would cost more than could be saved by deferring the cost of the challenger.

Continuation Value

One frequently used solution to unequal lives is to determine a salvage value based either on net investment or on the estimated junking price and to assume that the longer-lived asset is junked. That is an adequate approach for long study periods, but just how long the study period must be is questionable. Another solution would be to make the study over an infinite study period. However, in many studies an infinite study period is not practical, because the problem may only be one element of a larger economy study with a finite period. However, to make a finite study of a problem with unequal lives, a value analogous to salvage can be found which represents the value to the business of having plant in service for a longer period in one alternative than another. This value is called the *continuation value* and is the difference between the present worth of all future costs for the two alternatives for the infinite period beyond the end of the finite study period. Using the continuation value as the effective salvage value will permit any finite study period, because the results are insensitive to the period. In fact, the *PWE* values found are a proportional allocation of the infinite *PWE* values.

Example 13-3 illustrates the application of a continuation value to the problem of unequal lives. Even though the illustration is for a replacement study, this concept could be used in any type of study with non-coterminated plant.

---

*Because operations costs are constant, they can be combined with capital costs in this allocation. If operations costs are changing, they must be treated separately.

†See footnote p 323.

Example 13-3. Suppose equipment D is now in service and could be kept for 7 years more. It could be sold for $20,000 now, but in 7 years it would net only $5,000. Operations costs for the past year were $7,500, and are expected to grow at 5% a year. The challenger, equipment C, has an economic life expectancy of 25 years. It will cost $40,000 now, and will have a gross salvage value of $2,000 and a negligible cost of removal in 25 years. The purchase price, as well as the salvage value, are expected to increase at 3% per year. The operations costs for the challenger at present price levels are $5,000 per year; like those of the defender, they are expected to increase at 5% per year. The company's cost of money is 12%, and the other factors are the same as in Example 13-2.

The repeated plant assumption does not hold in this problem, because costs are expected to be changing throughout. As a result, a continuation value is needed for one of the alternatives at some point in time in order to do a proper economy study (i.e., where the alternatives will provide service over the same period of time). The values in the time-cost diagrams which follow reflect the costs described in the preceding paragraph, plus those reasonably expected in the future. Four different finite study periods have been used to demonstrate the effect of the life assumption.

The defender must be replaced in 7 years—presumably, with challenger-type equipment. If the challenger is installed now, it would still have 18 (that is, 25 − 7) years of service remaining at the time coincident with the eventual replacement of the defender. If it is reasonable to assume that subsequent replacements would also be made with challenger-type equipment, the capital expenditures for the alternatives are as shown in Figure 13-3.

The operations costs for each alternative are geometric progressions increasing at 5% per year. Since they were stated in year 0 dollars, the convenience-rate concept can be used to find their present worths. This convenience rate is then approximately 12% minus 5%, or 7%. However, after year 7, the same amount and type of plant exist. Therefore, the operations costs beyond year 7 can be considered common and are excluded from the study.

The study life can be selected as any time coincident with a retirement in either alternative. In this case, the study could be terminated at years 7, 25, 32, 50, and so on. If a study period of 7 years is selected, the actual retirement and junking of the challenger in 7 years is unrealistic; and any possible net salvage at that point is of no relevance. The situation is much like the repeated plant assumption, except that costs are increasing.

The choice of the challenger now means that the purchase of challenger-type equipment in 7 years is avoided. The present value of the difference between the two expenditure series, beyond the study period, is the incremental cost avoided by having the challenger in place 7 years from now. Therefore, it is the value of having the challenger in service rather than buying one in the defender plan. In an incremental analysis, this amount could either be added to the cost of the plan which actually retires

FIGURE 13-3
Capital Expenditures

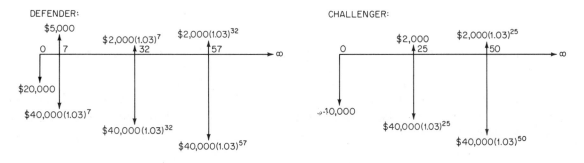

plant at the end of the study period, or it could serve a purpose similar to salvage value in the plan that retains plant. When the present value of this difference is used like a salvage value, the present worth of capital costs represent those incurred during the study period by the plan which actually retires plant at the end and an allocation of costs for the plan which does not.

To determine the continuation value, it is necessary to use the effective interest-rate concept from Chapter 5, as well as the convenience-rate concept (also from Chapter 5). In this example the capital expenditures occur every 25 years, and the annual cost of money is 12%. Compounding annually at 12% for 25 years is equivalent to compounding every 25 years at a rate found in the same way as a monthly-to-yearly conversion.

$$(1.12)^{25} = (1 + i_{25})^1$$
$$i_{25} = 17.00 - 1 = 16.00 = 1600\%$$

Likewise, the cost-trend rate for 25 years that is equivalent to 3% per year can be found as follows:

$$(1.03)^{25} = (1 + I_{25})^1$$
$$I_{25} = 2.09 - 1 = 1.09 = 109\%$$

The convenience rate for a 25-year period is found from the relationship demonstrated in Chapter 5:

$$i_g = \frac{i - I}{1 + I}$$
$$\text{25-year } i_g = \frac{16.00 - 1.09}{1 + 1.09} = 7.13 = 713\%$$

It will be necessary to use the (P/A) factor at 713%, which is not tabulated. However, the factor needed is for an infinite flow and is readily calculated.

Recall from Chapter 5 that $(P/A, i\%, \infty) = 1/i$. Then

$$(P/A, 713\%, \infty) = \frac{1}{7.13}$$

Therefore, the present worth of an infinite series of costs occurring every 25 years, growing at 3%/year, and discounted at 12%/year, is the base year dollar value of one payment divided by 7.13.

SEVEN-YEAR STUDY PERIOD. For a 7-year study, the capital expenditures assumed are as shown here.

Since the continuation value is not an actual receipt of cash, it has no effect on income taxes. Taxes during this 7-year period for the challenger will be computed as a portion of the total taxes over the life of the challenger. The amount of income taxes not allocated to the 7-year study period will influence the amount of the continuation value. After the initial 7 years, the future costs consist of those costs resulting from repeated installations of challenger-type equipment. The present worth of the expenditures for the first installation in the challenger plan is found first, as follows.

$$PW(C_{rec}) = PW(\text{first cost}) - PW(\text{salvage})$$
$$= 40,000(P/F, 12\%, 0) - 2,000(P/F, 12\%, 25)$$
$$= 40,000(1.0) - 2,000(0.0588) = \$39,882$$
$$PW(T) = \phi[PW(C_{rec}) - PW(D_b)] - t(1 + \phi)[PW(D_t) - PW(D_b)]$$
(neglecting investment tax credit)
$$PW(D_b) = \frac{40,000 - 2,000}{25}(P/A, 12\%, 25)$$
$$= 1520(7.843) = \$11,921$$
$$\% \text{ gross salvage} = \frac{2,000}{40,000} = 0.05 = 5\%$$
$$PW(D_t) = (\text{first cost})(P/D_t)(P/F, 12\%, 0)^*$$
$$(P/D_t) = \frac{0.49933 + 0.49344}{2} = 0.49638$$
$$PW(D_t) = 40,000(0.49638)(1.0) = \$19,855$$
$$PW(T) = 0.61(39,882 - 11,921) - 0.48(1.61)(19,855 - 11,921)$$
$$= 17,055 - 6,131 = \$10,924$$
$$PW(\text{capital cost}) = 39,882 + 10,924 = \$50,806$$

Because both salvage and first-cost estimates are increasing at the same rate (3%), each subsequent installation in either plan will have a PW(capital cost) of $50,806(1.03)^n$, where n is the year of installation. If the salvage and first cost were changing at different rates, the PW(capital cost) would have to be separated into a component for first cost and a component for salvage, because the combined PW(capital cost) would not change at a constant rate.

For the defender, the present worth at year 7 of all future costs, beyond year 7, is the present worth of an infinite geometric progression of PW(capital cost) amounts occurring every 25 years beginning with $50,806(1.03)^7$.

$$PWE(FD) = PWE(\text{future costs of defender plan})$$
$$PWE(FD) = 50,806(1.03)^7 + 50,806(1.03)^7 (P/A, 713\%, \infty)$$
$$= 50,806(1.23) + \frac{50,806}{7.13}(1.23) = \$71,256$$

NOTE: With the convenience-rate method, the present worth of the geometric series is found from the base-year amount, but the series does not include a base-year amount. In this case an amount does occur in the base year and it must be added in.

For the challenger, the present worth of all future costs beyond year 7 must include an allocation of the income tax paid during the entire life of first installation. Since 18 years of life are remaining, for simplicity, this allocation will be 18 years of the levelized equivalent annual cost of the total taxes. The present worth of the income tax for the first installation was previously found to be $10,924. The AC of the taxes is then

$$AC(T) = PW(T)(A/P, 12\%, 25)$$
$$= 10,924(0.12750) = \$1,393$$

For the challenger, the present worth, at year 7, of the expenditures, beyond year 7, is then the present worth of a geometric progression of PWE amounts beginning in year 25, with a base of $50,806(1.03)^0$, translated to year 7, minus the present worth of the actual net salvage in 18 years, plus the present worth of 18 years of AC of income tax for the first installation.

$$PWE(FC) = PWE(\text{future costs of challenger plan})$$
$$PWE(FC) = 50,806(P/A, 7\tfrac{1}{3}\%, \infty)(F/P, 12\%, 7)$$
$$\qquad -2,000(P/F, 12\%, 18) + 1,393(P/A, 12\%, 18)$$
$$\qquad = \frac{50,806}{7.13}(2.2107) - 2,000(0.1300) + 1,393(7.250)$$
$$\qquad = \$25,592$$

At this point, the capital costs of the infinite study period for each alternative are as shown in the following diagrams.

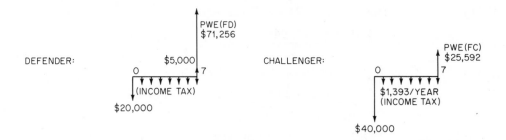

At 7 years then, the value of having a challenger with 18 more years of life is that future expenditures have a present worth of $25,592 rather than $71,256. This is the continuation value (CV), which is

$$CV = PWE(FD) - PWE(FC)$$
$$\qquad = 71,256 - 25,592$$
$$\qquad = \$45,664$$

Therefore, the 7-year capital costs can be considered as in the following diagrams.

The PWE values for the two plans can now be found as follows.
$\underline{PWE\ \text{for the defender:}}$

$$PW(C_{rec}) = S_0 - S_n(P/F, 12\%, 7) \tag{13.1}$$
$$\qquad = 20,000 - 5,000(P/F, 12\%, 7)$$
$$\qquad = 20,000 - 5,000(0.4523)$$
$$\qquad = \$17,738$$

$$PW(T) = \phi[S_0 - S_n(P/F, 12\%, 7) - \frac{S_0 - S_n}{7}(P/A, 12\%, 7)] \tag{13.2}$$

$$\qquad -t(1 + \phi)[S_0 - S_n(P/F, 12\%, 7) - \frac{S_0 - S_n}{7}(P/A, 12\%, 7)]$$

$$\frac{S_0 - S_n}{7}(P/A, 12\%, 7) = \frac{20,000 - 5,000}{7}(4.564)$$

$$\qquad = \$9,780$$

$$PW(T) = 0.61(17,738 - 9,780) - 0.48(1.61)(17,738 - 9,780)$$
$$= -\$1,296$$

PW(capital cost) $= 17,738 - 1,296$ $\qquad\qquad = \$16,442$

PW(operating cost) $= 7,500(P/A, 7\%, 7)$

$$= 7,500(5.389) \qquad\qquad\qquad = \$40,418$$

$$\text{Total } PWE(D) \qquad\qquad = \$56,860$$

PWE for the challenger:

$$PW(C_{rec}) = PW(\text{first cost}) - PW(CV)$$
$$= 40,000(P/F, 12\%, 0) - 45,665(P/F, 12\%, 7)$$
$$= 40,000(1.0) - 45,664(0.4523) = \$19,346$$
$$PW(T) = AC(T)(P/A, 12\%, 7)$$
$$= 1,393(4.564) = \$6,358$$

PW(capital cost) $\quad= 19,346 + 6,358 \qquad\qquad = \$25,704$

PW(operating cost) $= 5,000(P/A, 7\%, 7)$

$$= 5,000(5.389) \qquad\qquad\qquad = \$26,945$$

$$\text{Total } PWE(C) \qquad = \$52,649$$
$$PWE(D) \qquad = \underline{\;\;56,860}$$
$$\text{Difference in favor of challenger} \qquad \$\;\,4,211$$

The cost of keeping the defender in service for seven more years therefore exceeds the allocated costs of the challenger, when the value of the remaining life of the challenger after seven years is determined.

TWENTY-FIVE-YEAR STUDY PERIOD.   For a 25-year study, the defender is the alternative which has plant with remaining life and a continuation value. The following sketch shows the effective capital expenditures.

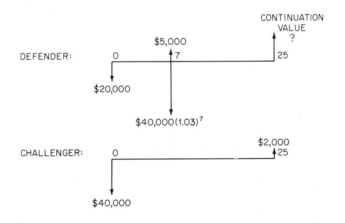

This time, the present worth of the future costs of the defender must include an allocation of income tax; but this tax is based on the challenger-type equipment placed in year 7.

$$AC(T) = PW(T)(1.03)^7 (A/P, 12\%, 25)$$
$$= 10,924(1.23)(0.12750)$$
$$= \$1,713$$

In this study, the present worth of the future defender costs beyond year 25 includes the present worth of the geometric progression, at year 25, minus the salvage, plus the tax allocation.

$$PWE(FD) = 50{,}806(1.03)^7(P/A,\ 713\%,\ \infty)(F/P,\ 12\%,\ 18)$$
$$-2{,}000(1.03)^7(P/F,\ 12\%,\ 7) + 1{,}713(P/A,\ 12\%,\ 7)$$

$$= \frac{50{,}806}{7.13}(1.23)(7.690) - 2{,}000(1.23)(0.4523) + 1{,}713(4.564)$$

$$= \$74{,}105$$

The present worth of the future challenger costs is just the present worth of the geometric progression, beginning in year 25.

$$PWE(FC) = 50{,}806(1.03)^{25} + 50{,}806(1.03)^{25}\ (P/A,\ 713\%,\ \infty)$$

$$= 50{,}806(2.09) + \frac{50{,}806(2.09)}{7.13}$$

$$= \$121{,}077$$

Now the continuation value is the value of having challenger-type equipment, in the defender plan, with life remaining. This continuation value is then the $PWE(FC)$ less the $PWE(FD)$.

$$CV = PWE(FC) - PWE(FD)$$
$$= 121{,}077 - 74{,}105 = \$46{,}972$$

PWE for the defender:

PW(capital cost) =	7-year PW(capital)	= \$16,442
$PW(C_{\text{rec}})$	= PW(first cost) − $PW(CV)$	
	= $40{,}000(1.03)^7\ (P/F,\ 12\%,\ 7)$	
	− $46{,}972(P/F,\ 12\%,\ 25)$	
	= $40{,}000(1.23)(0.4523)$	
	− $46{,}972(0.0588)$	= \$19,491
$PW(T)$	= $AC(T)(P/A,\ 12\%,\ 18)(P/F,\ 12\%,\ 7)$	
	= $1{,}713(7.250)(0.4523)$	= \$ 5,617
	Total PW(capital cost)	= \$41,550
PW(operating cost) =	7-year PW(operating cost)	= \$40,418
	Total $PWE(D)$	= \$81,968

TABLE 13-1
COSTS OVER DIFFERENT STUDY PERIODS

	Study period, years				
	7	25	32	50	∞
Defender					
PW(capital cost)	\$16,400	\$41,500	\$44,700	\$47,800	\$48,700
PW(operating cost)	40,400	40,400	40,400	40,400	40,400
$PWE(D)$	\$56,800	\$81,900	\$85,100	\$88,200	\$89,100
Challenger					
PW(capital cost)	\$25,700	\$50,800	\$54,000	\$57,100	\$58,000
PW(operating cost)	26,900	26,900	26,900	26,900	26,900
$PWE(C)$	\$52,600	\$77,700	\$80,900	\$84,000	\$84,900
Difference $[PWE(D) - PWE(C)]$	\$ 4,200	\$ 4,200	\$ 4,200	\$ 4,200	\$ 4,200

FIGURE 13-4
A Plot of the Data in Table 13-1

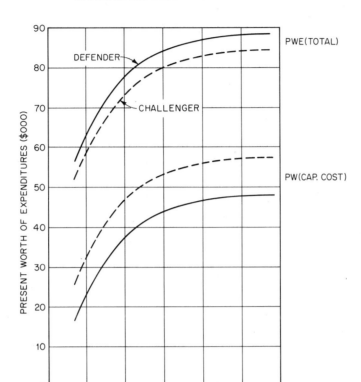

PWE for the challenger:

PW(capital cost) = initial *PWE* (time 0)	=	$50,806
PW(operating cost) = 7-year *PW* (operating cost) =		26,945
Total *PWE(C)* =		$77,751
PWE(D) =		81,968
Difference in favor of the challenger		$ 4,217

Even though the total costs of each alternative are higher for the 25-year study, the difference in costs still shows that the challenger is favored—and by approximately the same amount. This result is also borne out by further analysis, using study periods of 32 and 50 years and an infinite study. Calculations similar to those illustrated result in the costs for each study period shown in Table 13-1. Since the results can be accurate only to three significant digits, these results have been rounded accordingly.

These amounts are also plotted on Figure 13-4, illustrating the results of the use of a continuation value. With a continuation value, the study can be of any length considered appropriate without the distortion which can result from an estimated premature salvage value.

Future Challengers

In both Example 13-2 and Example 13-3, a presently available challenger was pitted against the defender; and in both cases, the challenger was more desirable, given the results of the *PWE* analysis. However, an economical present challenger could be rejected in favor of keeping the defender in service and waiting for a future technologically improved challenger which would be even more economical than the present challenger.

Suppose a defender could last for 10 more years, or be replaced now with a present challenger. Further suppose that an improved challenger will be available in 5 years. Now there may be several alternatives. At some point in all alternatives it should be assumed that the then current technology is common to all. The continuation value can be used to decide which of several alternatives is the most economic.

For example, three alternatives which could result from the above situation are the following, where

$$C_1 = \text{cost of first challenger}$$
$$N_1 = \text{life of first challenger}$$
$$C_2 = \text{cost of second challenger (present dollars)}$$
$$N_2 = \text{life of second challenger}$$
$$S_n = \text{salvage in year } n$$

1. Keep defender for 10 years, replacing with the improved challenger.

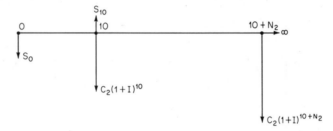

2. Replace with present challenger now, future replacements with improved challenger.

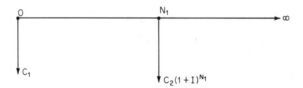

3. Replace with improved challenger in 5 years.

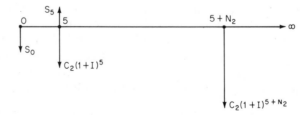

A study period can be selected coincident with the retirement of plant in any one of the three alternatives. For instance, a 10-year study period is coincident with retirement of the defender in plan 1. The continuation values for the other two plans are then the present worth of future costs for plan 1 (the plan with a coincident retirement) less the present worth of future costs for each of the other plans. The *PWE* of each of the three plans can then be found as in Example 13-3, and the results compared. In this case the operations costs beyond the study period may not be common and any difference should also be determined and included.

REPLACEMENT OF JOINTLY OWNED ASSETS

Many times in the Bell System, assets are jointly owned by two (or more) Bell System companies. The most common cases involve interstate plant. Typical examples are the joint ownership of cables and toll switching offices. This section of the chapter treats the economics of retiring such assets.

The consideration of a possible retirement of jointly owned assets usually arises when some new technology is available to one of the owners. If this company makes an independent economic analysis which demonstrates cost savings with the new technology, it may be desirable to retire the existing asset (the defender). In an independent study of this type, the existing asset has often been treated as if the studying company had complete control of the asset's fate. This approach overstates the advantages of the new technology. Usually, most of the ad valorem tax savings, salvage, and often some of the operations cost savings cannot be realized unless both parties agree to the retirement. Similarly, an independent study by the other owner will understate the costs for continued use of the defender.

A simple way out of the dilemma would be for one of the companies to sell its portion to the other; but even if some fair transfer price is found and included in both analyses, the unilateral studies may not indicate the most economic course of action. The only consistently correct approach is to take the Bell System viewpoint. This does not mean that Bell System average costs should be substituted for the actual costs of the companies involved, but rather that the study should be made to see whether the System as a whole (and not just an individual company) will benefit or lose by the retirement.

When the Bell System viewpoint is adopted, several facts become apparent.

1. The total number of alternatives available to the Bell System is equal to the product of the number of alternatives available to each joint owner. In the simplest case with two owners, each has two alternatives (i.e., continue the present operation or retire the existing asset and replace it with something else). From the Bell System viewpoint, there would be four (2 × 2) alternatives:

   a. Both companies could continue the present operation.
   b. Company A could continue the present operation while Company B acquires a new asset.
   c. Company A could acquire a new asset while Company B continues the present operation.
   d. Both companies could discontinue the present operation in favor of new assets.

   Usually, one or more of the alternatives can be eliminated by examination. For example, if it were clear that Company A would benefit from a new asset while Company B would not, alternative "b" would be an unlikely prospect.

2. The Bell System decision among its alternatives is completely independent of any transfer of funds between the two companies resulting from a sale of assets from one joint owner to the other. In alternative "b" or "c," one company might assume 100% ownership of the existing asset, and there would be an appropriate transfer of funds. Regardless of the amount

of this transfer, or how it was arranged, it represents an expenditure by one company that is exactly offset through a receipt by the other company. The Bell System net cash flow resulting from the transaction is 0; therefore, the transaction itself would not be a factor in deciding between the alternatives.

3. To evaluate costs, various alternatives available to the Bell System can be grouped into three categories:

   Category 1: All companies continue the present operation to a common retirement date.
   Category 2: One or more companies continue the present operations, while the others acquire new assets.
   Category 3: All companies discontinue the present operation, replacing it with something else.

The appropriate costs for comparing these categories should be assembled as follows:

Category 1: Sum of the *PWE* for

   a. Total operations cost savings foregone by the continuation of the present operation.

   b. Total ad valorem tax savings foregone by the continuation of the present operation.

   c. Total capital costs attributable to loss in salvage or reuse value by the continuation of the present operation.

   d. Total costs related to replacing assets after the common retirement date (if required).

Category 2: Sum of the *PWE* for:

   a. Total operating costs savings foregone by the continuation of the present operation.

   • Less the operations cost savings of the companies which are discontinuing the present operation

   • Plus the operating cost increases experienced by the companies which are continuing the present operation.

   b. Total ad valorem tax savings foregone by the continuation of the present operation.

   • Less the ad valorem tax savings of the companies which are discontinuing the present operation.

   • Plus the ad valorem tax increases experienced by the companies which are continuing the present operation.

   c. Total capital costs attributable to loss in salvage or reuse value by the continuation of the present operation.

   d. Total capital costs attributable to new assets acquired by the companies which are discontinuing the present operation.

   e. Total costs related to replacing defender assets after the retirement date (if this is required).

Category 3: Sum of the *PWE* for total costs related to new assets assuming zero current salvage for the defender.

The most economic alternative for the Bell System will be the one with the smallest combined *PWE*. Very often, this alternative will not be the most economic alternative for all the companies involved. However, this situation is not really any different from one in which an individual company adopts a plan requiring one of its departments to take on extra costs in order to save a larger amount in other departments. To do anything else would result in a situation that would be economically detrimental to the company as a whole.

SUMMARY

The continual evolution of technology means that many economy studies consider the replacement of existing plant with newly developed plant. Just as in all economy studies, the critical economic factor is the incremental cost differences among alternatives. However, when replacement is considered, the remaining life, value, and costs associated with existing plant must be estimated. An existing asset is frequently called the defender, whereas the potential replacing asset is the challenger.

The most significant aspect of defender-associated costs is that the capital cost of keeping a defender in service is dependent on the current salvage foregone or the cost which can be avoided if it is kept and reused. This capital cost may be as low as the net salvage value foregone by keeping the defender, and can be as high as the cost of a new defender-type asset if reuse of the asset is planned. The original cost of the defender, and any past costs associated with its past operation, are common to all alternatives. These are sunk costs and should not be a factor in the economy study. This is so even though the accounting allocations of some of the costs may or may not have been completed. When an asset is retired, the only costs which cease are those which are associated with its operations, such as ad valorem taxes and maintenance.

In many replacement studies, as well as in other types of studies, the estimates of the economic lives of plant may not permit the various alternative plans to have a coincident termination time in the future. When plans are not coterminated, there are three basic ways to make an economy study over a comparable period. The first is through the use of the repeated plant assumption; but this simple expedient is inadequate when future costs are expected to change. Second, a continuation, or reuse in place, value can be determined for the asset with remaining life; and the study can be terminated by assuming that this value is analogous to salvage. Finally, if the life differences are small, the estimated lives can be revised and the difference neglected.

The retirement of jointly owned assets should be investigated from a Bell System viewpoint. Often, a replacement will be economically desirable for one of the owners, but not for the others. When the combined effect is considered from a system viewpoint, it may be beneficial for one of the companies to make an economic sacrifice for the benefit of the whole.

14

DISPERSED RETIREMENTS

INTRODUCTION

A single asset will have some finite service life. Yet this life must be estimated for an economy study because it will not be known with certainty until the asset is retired. When the economy study is considering large groups of many assets, the problem of service life becomes more complex. With large numbers of similar but independent items, it would be impractical to estimate a specific life for each item. It would also be unrealistic to assume that all the items will live exactly the same life. Random events will cause variations in the lives of similar units, like telephone sets, just as they do in the life expectancies of people.

This chapter describes the *dispersion of retirements*, which represents the realistic variations in the lives of items in a group. The chapter also presents guidelines for reflecting these variations in an economy study.

EFFECT OF A GROUP AVERAGE

Because it is impractical to account individually for each of the millions of units of plant necessary to provide telephone service, the Bell System uses group depreciation techniques. As Chapter 7 explained, depreciation accruals are based on an average service life and average salvage for a group of telephone plant items. This approach recognizes that in any large group of property items certain things will happen which will force some items to retire before the average life, and some items after it. Nevertheless, if a depreciation rate based on the average life of all the units in the group is applied to each unit, the original cost of that group will be completely recovered—provided all the units live according to the estimate.

When large numbers of items are being considered in an economy study, the question might be asked whether a capital-recovery factor based on the group average life would not also result in complete capital recovery, including return. In general, the answer is "no," because an economy study is affected by the time value of capital recovery. When time-value weighting is introduced, the effect of dispersed retirements on capital recovery does not "average out." Example 14-1 will illustrate this time-value impact of dispersed retirements.

Example 14-1. Suppose a company spends $100 to buy an asset group of many units, with an average life of 4 years. For simplicity, assume further that this company has a cost of money of 12%, that it exists in a tax-free environment, and that it incurs no non-capital-related expenses because of these assets. If all the items in the group retire at the same time, each having lived the "average" life, the capital recovery process could be illustrated as in Table 14-1A:

TABLE 14-1A

Year (N) (A)	Plant in service (B)	Revenue requirement B × (A/P, 12%, 4) (C)	Capital repayment (B × 0.25) (D)	Available for return (C − D) (E)	(P/F, 12%, N) (F)	PW(C_{rec}) (C × F) (G)
1	$100	$32.92	$ 25.00	$7.92	0.8929	$29.39
2	100	32.92	25.00	7.92	0.7972	26.24
3	100	32.92	25.00	7.92	0.7118	23.43
4	100	32.92	25.00	7.92	0.6355	20.92
Totals			$100.00			$99.98

Except for a rounding difference, the present worth of the capital recovery is equal to the original investment, thus proving that the revenues were sufficient.

Now suppose the average life of the asset group is the same 4 years, but retirements are dispersed in such a way that 50% of the assets are retired after 2 years and 50% at 6 years. In that case, the revenue for capital recovery found in the same way as in Table 14-1A will not be sufficient, as Table 14-1B illustrates.

The difference between the capital repayment and the present worth of capital recovery is not large in Table 14-1B, because small numbers were used in this example for convenience. In practice, though, this difference can be substantial. Table 14-1A illustrates that this revenue is too low to allow complete capital recovery for the dispersed assets, even though it is high enough to provide for the capital repayment. As Table 14-1C shows, the proper revenue level is actually $34.47 for the first 2 years and $17.24 for the remaining 4. Just how these amounts are found will be demonstrated in the next section.

Example 14-1 demonstrated (in Table 14-1A) that when the (A/P) factor is applied to a constant amount of surviving plant, the result is the correct amount of capital recovery. But the original capital is not completely recovered when the (A/P) factor is applied to diminishing plant survivors (as in Table 14-1B). However, there is some total revenue amount (given in Table 14-1C) which will provide complete capital recovery. In fact, the example proves that these total annual revenue amounts are adequate

TABLE 14-1B

Year (N) (A)	Plant in service (B)	Revenue requirement B × (A/P, 12%, 4) (C)	Capital repayment (B × 0.25) (D)	Available for return (C − D) (E)	(P/F, 12%, N) (F)	PW(C_{rec}) (C × F) (G)
1	$100	$32.92	$ 25.00	$7.92	0.8929	$29.39
2	100	32.92	25.00	7.92	0.7972	26.24
3	50	16.46	12.50	3.96	0.7118	11.72
4	50	16.46	12.50	3.96	0.6355	10.46
5	50	16.46	12.50	3.96	0.5674	9.34
6	50	16.46	12.50	3.96	0.5066	8.34
Totals			$100.00			$95.49

TABLE 14-1C

Year (N) (A)	Plant in service (B)	Revenue requirement (C)	Capital repayment (B × 0.25) (D)	Available for return (C − D) (E)	(P/F, 12%, N) (F)	PW(C_rec) (C × F) (G)
1	$100	$34.47	$ 25.00	$9.47	0.8929	$ 30.78
2	100	34.47	25.00	9.47	0.7972	27.48
3	50	17.24	12.50	4.74	0.7118	12.27
4	50	17.24	12.50	4.74	0.6355	10.96
5	50	17.24	12.50	4.74	0.5674	9.78
6	50	17.24	12.50	4.74	0.5066	8.73
Totals			$100.00			$100.00

because the sum of their present worths equals the initial investment. This relationship must hold if the revenue is to result in time-value equivalence; and it then is the basis for determining suitable annuity factors which will produce the proper revenue amounts when retirements are dispersed.

CALCULATION OF DISPERSED ANNUITY FACTORS

The first step in deriving suitable annuity factors for dispersed retirements is to define a constant (α/P), which is the *dispersed annuity from a present amount factor*. This factor is like the (A/P) factor discussed in Chapter 5. The revenue required in each year is therefore (α/P) times the investment remaining in the survivors of that year. The sum of the present worths of those annual revenue requirements must then be equal to the original investment. The equation is

$$(\alpha/P)S_1(P/F)_1 + (\alpha/P)S_2(P/F)_2 + \cdots + (\alpha/P)S_N(P/F)_N = S_0 \qquad (14.1)$$

Here, S_0 is the initial investment; and $S_1, S_2, \ldots, S_N$ are the average investment surviving during years 1, 2, and so on, through the last year N that any plant from that group is still not retired.

Since (α/P) is a constant, equation (14.1) may be simplified as

$$(\alpha/P)[S_1(P/F)_1 + S_2(P/F)_2 + \cdots + S_N(P/F)_N] = S_0$$

To find the (α/P) factor, the initial investment is divided by the sum of the present worths of each year's survivors.

$$(\alpha/P) = \frac{S_0}{S_1(P/F)_1 + S_2(P/F)_2 + \cdots + S_N(P/F)_N} \qquad (14.2)$$

Without Dispersed Retirements

The situation in which retirements are not dispersed may now be considered as a special case of equation (14.2), in which $S_0 = S_1 = S_2 = \cdots = S_N$. Without dispersion, then

$$(\alpha/P)_N = \frac{S_0}{S_0 \sum_{n=1}^{N} (P/F)_n} = \frac{1}{\sum_{n=1}^{N} (P/F)_n} = \frac{1}{(P/A)_N} = (A/P)_N$$

Chapter 5 showed that the (P/A) factor is just the summation of successive (P/F) factors.

TABLE 14-1D

Year	Plant surviving	Present worth Factor	Present worth Amount
1	$100	0.8928	$ 89.29
2	100	0.7972	79.72
3	50	0.7118	35.59
4	50	0.6355	31.78
5	50	0.5674	28.37
6	50	0.5066	25.33
Total			$290.08

With Dispersed Retirements

The data of Table 14-1B may also be used to show the development of an (α/P) annuity factor with dispersed retirements, as illustrated in Table 14-1D.

$$(\alpha/P) = \frac{\text{initial investment}}{\text{sum of } PW(\text{survivors})} = \frac{\$100.00}{\$290.08} = 0.3447$$

Table 14-1C demonstrated that this factor would produce the appropriate revenue requirement. With the amounts used in that table, the calculation is as follows:

$$\$100 \times (\alpha/P, 12\%, 4) = 100(0.3447) = \$34.47 \qquad \text{for the first 2 years}$$

and

$$\$50 \times (\alpha/P, 12\%, 4) = 50(0.3447) = \$17.24 \qquad \text{for the next 4 years}$$

Relation to Certain Costs for Pricing Studies

There is a seemingly different way to account for dispersion when costs are computed for a pricing study. The procedure is to calculate a uniform annual cost factor per unit which can be applied when the demand for units changes with time. In fact, the purpose of a pricing study often is to find this uniform annual cost factor. Because the number of survivors of all vintages must equal the demand for units at any time, the factor is

$$\text{Annual cost factor} = \frac{PW(\text{all costs})}{\text{sum of } PW(\text{survivors or demand})}$$

This ratio is what might be considered a time-value-weighted average cost factor. For example, consider just that part of this expression which is the capital-recovery cost.

$$\text{Annual capital-recovery cost factor} = \frac{PW(\text{capital recovery})}{\text{sum of } PW(\text{survivors})}$$

If the period under study is the total time that there are any survivors, then

$$PW(\text{capital recovery}) = \text{initial investment}$$

and $$\text{annual capital-recovery cost factor} = \frac{\text{initial investment}}{\text{sum of } PW(\text{survivors})} = (\alpha/P)$$

This costs-for-pricing technique is therefore consistent with the dispersed annuity approach. The difference is that in costs-for-pricing studies, the unit annual cost factor is usually found over a planning period different from the total time that there are any survivors.

SURVIVOR CURVES

Example 14-1 illustrated the concept of dispersed retirements by assuming that half the plant retired at 50% of the average life; the remaining plant, at 150% of the average life. Of course this example was simplified for the illustration. In reality, when a large group of items of telephone plant is put into service, retirement activity will start very soon after the installation and will continue as long as that group exists.

The Bell System has gathered vast amounts of data based on years of actuarial analyses of the characteristics of telephone plant life. Using these data, it is possible to estimate with accuracy the rate of retirement of a particular class of property at each age of life. These data are usually represented in Bell System depreciation studies as *survivor curves* (also called "life tables"), which show graphically what proportion of the initial investment is expected to survive at various ages. (In some other industries and in the life insurance field, comparable data are exhibited as "mortality distributions"—either the rate of retirement of property, or the human death rate by ages. These are simply different ways to express the same information.)

Figure 14-1 illustrates seven survivor curves. Curves numbered 1 to 5 plus the negative exponential are typical of those used with telephone plant. These curves show varying degrees of dispersion. Since the vertical change in a survivor curve during a year represents the retirements in that year, the steeper the initial slope of the survivor curve, the greater the rate of retirements at an early age. So, the negative exponential has the greatest dispersion, and curve No. 5 has the least except for the rectangular

FIGURE 14-1
Seven Survivor Curves

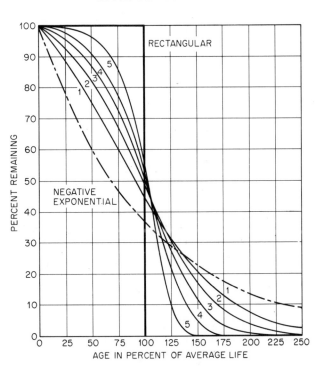

curve shown in Figure 14-1. This rectangular curve would not normally be representative of the life characteristics of a large mass of plant, because it assumes that all retirements occur at one point in time. The rectangular survivor curve is usually assumed in an economy study when a single unit of plant is involved, or whenever dispersion is not significant.

Survivor curve No. 5 shows that most of the vintage is retired in a relatively short period of time, which is centered near the average life. Such experience would be typical of motor vehicles, for example. A few might be retired early because of accidents or other reasons, but most would be retired within a short time of each other as

TABLE 14-2

PERCENTAGE SURVIVING AT VARIOUS PERCENTAGES OF AVERAGE LIFE FOR THE SEVEN SURVIVOR CURVES

Age as % of average life	Curve						
	Negative exponential	No. 1	No. 2	No. 3	No. 4	No. 5	Rectangular
0	100.0	100.0	100.0	100.0	100.0	100.0	100.0
10	90.5	96.1	97.7	99.2	99.6	99.9	100.0
20	81.9	91.6	94.6	97.8	98.7	99.7	100.0
30	74.1	86.5	90.9	95.3	97.4	99.3	100.0
40	67.0	81.1	86.4	92.2	95.3	98.5	100.0
50	60.7	75.3	81.3	87.8	92.2	96.9	100.0
60	54.9	69.3	75.5	82.4	88.0	94.3	100.0
70	49.7	63.2	69.3	75.7	82.3	89.8	100.0
80	44.9	57.0	62.6	68.2	74.8	82.6	100.0
90	40.7	51.0	55.7	59.7	65.4	71.6	100.0
100	36.8	45.2	48.7	50.7	54.3	56.1	100.0
110	33.3	39.6	41.8	41.5	41.8	37.0	0
120	30.1	34.4	35.1	32.5	29.2	18.2	
130	27.3	29.6	28.7	24.5	17.7	5.4	
140	24.7	25.1	22.9	17.2	8.9	0.7	
150	22.3	21.1	17.8	11.3	3.4	0	
160	20.2	17.6	13.4	6.8	0.9		
170	18.3	14.5	9.7	3.7	0.1		
180	16.5	11.8	6.8	1.8	0		
190	15.0	9.4	4.5	0.8			
200	13.5	7.5	2.9	0.3			
210	12.3	5.9	1.8	0.1			
220	11.1	4.6	1.0	0			
230	10.0	3.5	0.5				
240	9.1	2.6	0.3				
250	8.2	2.0	0.1				
260	7.4	1.4	0.1				
270	6.7	1.0	0				
280	6.1	0.7					
290	5.5	0.5					
300	5.0	0.4					

they reached an age when replacement would be appropriate. In such a pattern, age appears to be the predominant reason for retirement. By contrast, the negative exponential curve is assumed typical of the life of the station connection account. In that account, age plays little part in the retirement, because a customer may decide to move at any time after taking service, thus retiring a telephone connection.

Table 14-2 supports Figure 14-1 by showing the numerical values for the curves. Dispersed annuity factors for these curves are included among the tables (in Appendixes D and E at the end of the book) accompanying this text. However, the analyst who might need to use these curves in an economy study should consult the company depreciation engineer for guidance about the appropriate curve shape to represent a particular class of plant.

Impact of Dispersion

The factors below illustrate the effect of the retirement pattern, or dispersion, on annuity factors for plant having an average life of 20 years and assuming a 12% cost of money.

Survivor curve	Dispersed annuity from a present amount (α/P)
Rectangular	0.1339
No. 5	0.1364
No. 4	0.1391
No. 3	0.1422
No. 2	0.1476
No. 1	0.1535
Negative exponential	0.1729

As these factors show, the impact of dispersion may be quite significant, and probably should be considered—unless just a few units of plant are being studied, or unless the effects of dispersion are believed to influence the competing alternatives in similar ways.

Dispersion Factors in Computer Programs

The survivor curves typical of telephone plant are based on a mathematical formula which is easily represented in computer languages. If an analyst is using the computer as an aid in an economy study, it might be more efficient to represent the survivor curve mathematically than to refer to tabulated factors.

Gompertz-Makeham Formula for the Numbered Survivor Curves

Equation (14.3) is the underlying relationship expressing the five numbered survivor curves in Figure 14-1. Called the Gompertz-Makeham formula, it was originally developed for studies of human mortality. Thus,

$$l_x = l_0\, s^x\, g^{c^x-1} \tag{14.3}$$

Here, l represents survivors, with x denoting age (l_0 is the initial amount); s, g, and c are constants which define a particular curve. For convenience, equation (14.3) is usually represented in logarithmic form, with capital letters denoting logarithms:

$$L_x = L_0 + xS + G(c^x - 1) \qquad (14.4)$$

Equation (14.4) may also be stated as:

$$L_{x-1/2} = L_0 + (x - \tfrac{1}{2})S + G(c^{x-1/2} - 1)$$

This latter form is appropriate if data are to be represented in phase with calendar years rather than years of age. The assumption is that plant is installed at a uniform rate throughout a calendar year. Therefore, the survivors are on the average ½-year-old at the end of the first calendar year of service; 1½ years old at the end of the second year; and so on.

However, it is important to keep two points in mind when representing these expressions in computer programs.

- Since L is expressed in common logarithms, it will be necessary to convert to the corresponding antilog by using a built-in program, or by expressing the result as a power of 10.

- Some arbitrary cutoff value should be established to avoid unnecessary calculations. This might be some dollar value of surviving plant beyond which data would be trivial; or it might be a number of years which is equal to some multiple of average life.

Values of S, G, and c for specific curves are available from the company depreciation engineers.

Expression for the Negative Exponential Curve

Equation (14.3) and its logarithmic forms are related to the numbered curves in Figure 14-1, as the previous section pointed out. The negative exponential curve may be represented by equation (14.5).*

$$l_x = l_0 e^{-x/N} \qquad (14.5)$$

where l_x = survivors at age x
l_0 = initial amount
N = average life
e = base of natural system of logarithms, or 2.71828 . . .

Equation (14.5) is handled conveniently with integral calculus if continuous compounding is assumed. (See Appendix 14A at the end of the chapter for the derivation of the present worth of the survivors and Appendix 14B for additional observations.) The present worth of survivors for an initial placement of 1.0 will then be

$$PW(\text{survivors}) = \frac{N}{1 + rN}$$

where N = average life
r = force of interest

---

*The negative exponential curve may also be represented by equation (14.3), where $c = 0$, $g = 1$, and $s = e^{-1/N}$.

Negative Exponential Factor with Continuous Compounding

This relationship can now be used to express the dispersed annuity factor for a negative exponential curve when continuous compounding is assumed.

$$(\bar{\alpha}/P) = \frac{\text{initial investment}}{\text{sum of } PW(\text{survivors})}$$

where $\bar{\alpha}$ denotes a continuous dispersed annuity.

$$(\bar{\alpha}/P) = \frac{1.0}{\dfrac{N}{1+rN}}$$

$$= \frac{1+rN}{N}$$

$$= \frac{1}{N} + r \tag{14.6}$$

Negative Exponential Factor with Annual Compounding

Since annual compounding is used more often than continuous compounding in economy studies, it would be convenient to restate equation (14.6) in terms of annual compounding.

The relationship between a periodic annuity and a continuous annuity is expressed by the ratio of the effective annual rate of interest (i) to the nominal rate of interest (r),* or

$$\frac{(A/P)}{(\bar{A}/P)} = \frac{i}{r}$$

This relationship is also true for dispersed annuity factors.

$$\frac{(\alpha/P)}{(\bar{\alpha}/P)} = \frac{i}{r} \qquad \text{or} \qquad (\alpha/P) = (\bar{\alpha}/P)\left(\frac{i}{r}\right)$$

For the negative exponential,

$$(\alpha/P) = \left(\frac{1}{N} + r\right)\left(\frac{i}{r}\right)$$

$$= \frac{i}{Nr} + i$$

---

*Referring to Chapter 5,

$$\frac{(A/P)}{(\bar{A}/P)} = \frac{\dfrac{i(1+i)^N}{(1+i)^N - 1}}{\dfrac{re^{rN}}{e^{rN} - 1}}$$

But $\qquad e^r = 1 + i$

Therefore,

$$\frac{(A/P)}{(\bar{A}/P)} = \frac{\dfrac{i(1+i)^N}{(1+i)^N - 1}}{\dfrac{r(1+i)^N}{(1+i)^N - 1}} = \frac{i}{r}$$

But Chapter 5 pointed out that $r = \ln (1 + i)$. Therefore, for annual compounding,

$$(\alpha/P) = \frac{i}{N \ln (1 + i)} + i$$

If this expression is evaluated at $i = 12\%$ and $N = 20$ years, the result is a factor of 0.1729. This is the value shown in the data presented on page 351 to illustrate the relative impact of retirement patterns.

CONTINUING PLANT

So far, the discussion has focused on the application of the dispersed annuity factor to a single placement of many units of plant which are expected to be retired bit by bit until none is left. This might be called the *decaying plant concept*. In practice, however, many studies will require that plant which is retired be replaced in order to keep the plant in service at a certain capacity. This may be called the *continuing plant concept*.

The continuing plant concept assumes that plant and capacity remain constant. It also assumes that as individual units are retired, they are replaced by units having the same first cost and life and survivor pattern as those of the initial placement of plant. The dispersed annuity factor may then be used very effectively. Under these assumptions, the dispersed annuity factor is applied each year to the surviving plant balance of an initial placement; it is also applied to the surviving plant balance of each replacement. But if plant remains constant, the initial placement equals the total of its surviving plant balances plus the surviving plant balances of replacements. Therefore, the total amount of the annuity is the same as it would be if the dispersed annuity factor were applied to the initial investment. The data of Example 14-1 can be used to illustrate this fact.

Consider how the pattern of retirements for Example 14-1 might be portrayed graphically as a survivor curve.

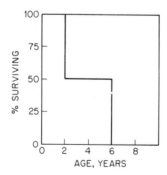

The age has been represented here in years, rather than as a percentage of average life; 50% of the investment retires after 2 years, and it must be replaced in order to maintain the total plant at 100%. Accordingly, 50% of this replacement retires in another 2 years, and it too must be replaced to keep the total plant capacity at a constant level. This process continues as long as the plant is required to provide service.

Therefore, sufficient new plant would be added to maintain the total plant at $100, and the total revenue requirements could be derived by applying the (α/P) factor to survivors from the initial placement and from all subsequent placements, as shown in Table 14-3.

TABLE 14-3
REVENUE REQUIREMENT BY YEARS FOR THE SURVIVING INVESTMENT IN CONTINUING PLANT

Placement	1	2	3	4	5	6	7	8	9
Initial	$100\alpha/P$	$100\alpha/P$	$50\alpha/P$	$50\alpha/P$	$50\alpha/P$	$50\alpha/P$			
1st replacement			$50\alpha/P$	$50\alpha/P$	$25\alpha/P$	$25\alpha/P$	$25\alpha/P$	$25\alpha/P$	
2d replacement					$25\alpha/P$	$25\alpha/P$	$12.50\alpha/P$	$12.50\alpha/P$	$12.50\alpha/P$
3d replacement							$62.50\alpha/P$	$62.50\alpha/P$	$31.25\alpha/P$
4th replacement									$56.25\alpha/P$
Total	$100\alpha/P$	$100\alpha/P$	$100\alpha/P$	$100\alpha/P$	$100\alpha/P$	$100\alpha/P$	$100\alpha/P$	$100\alpha/P$	$100\alpha/P$

Relationship to Continuing Plant with Rectangular Survivor Curve

The annual revenue requirement for capital recovery can be found for plant with a rectangular survivor curve (e.g., a single unit) by multiplying the first cost of that plant by the (A/P) factor. If the continuing plant concept is assumed, this (A/P) factor would also apply to each subsequent addition. Underlying this concept of continuing plant, of course, is the assumption that there were no cost changes, and that there is a constant level of total plant investment. Therefore, the product of the first cost of the initial placement and the (A/P) factor would represent the annual cost for capital recovery at any time during the life span of the succession of plant units. With this point of view, it is not necessary to consider explicitly the expenditures for future plant additions. Furthermore, when the continuing plant concept is applied to the rectangular survivor curve, it is in fact the repeated plant assumption.

When these assumptions are changed to recognize the dispersion of retirements, however, the annual cost figure found by multiplying the initial amount by the (α/p) factor plays the same role. In other words, it provides the levelized equivalent annual cost of capital recovery; and it becomes unnecessary to consider the expenditures for plant replacements—even though those expenditures may be occurring in varying amounts during each year.

TABULATED FACTORS

The (α/P) factors for the seven typical survivor curves are tabulated in Appendix D at the end of this book for various rates of interest. Although the (α/P) factor is used most often, the (α/F) factor is sometimes useful too. It is not tabulated, though, because it can easily be derived from the (α/P) factor.

Recall from Chapter 5,

$$(A/F) + i = (A/P)$$

Similarly,

$$(\alpha/F) + i = (\alpha/P)$$

Therefore, the (α/F) factor is found by simply subtracting the interest rate (i) from the tabulated (α/P) factor. Both factors may be used in a similar way to find the annuity equivalent to the purchase of a group of units and the continuing replacement of those units as they retire according to some survivor curve.

For a rectangular survivor curve, with no replacements assumed during the life of the units,

$$AC(C_{\text{rec}}) = \text{first cost}(A/P) - \text{net salvage}(A/F)$$

For a dispersion of retirements,

$$AC(C_{\text{rec}}) = (\text{first cost})(\alpha/P) - (\text{net salvage})(\alpha/F)$$

This equation now assumes continuous plant and the replacements required by the survivor curve.

APPLICATION OF DISPERSION TO ECONOMY STUDIES

When the concept of dispersion of retirements is combined with the income tax effects of accelerated tax depreciation, the problem of calculating the annual costs becomes enormously complex. The additional considerations in a study when the effects of price trends must also be included make a simple technique a practical impossibility. If such a study is required, the only practical solution is a rather sophisticated computer program which can find the year-by-year costs of vintage after vintage of plant with changing costs.

The most important point for the analyst to remember about the dispersion of retirements is that it will increase capital costs. Many economy studies can be done without recognizing dispersion, because the study results would not be sensitive to the effects of dispersion. An example is an economic selection study with a common survivor curve for each plan and with approximately equal capital costs. The reason is that if the amount of dispersion, as reflected by the assumed survivor curves, is approximately the same for each alternative plan in a set, the capital cost for each plan is greater than it would be without dispersion by a constant percentage. Also, if the most economic plan also has the lowest proportion of capital costs, it is not necessary to recognize dispersion because dispersion will only emphasize the economic superiority of that plan.

However, if dispersion influences the alternatives differently, and especially if the most economic plan is most affected by dispersion, the analyst may have to recognize its effect. Example 14-2 illustrates how the effects of dispersion can be approximated in a simple study.

Example 14-2. Suppose there is a proposal that the company equip its installation and repair people with portable paging devices, which would allow rapid communication between the dispatcher and the field forces. These devices clip to a belt or pocket, and would be carried at all times. The use of the pagers saves an estimated $54,000 per year in clerical expense in trying to locate the field people. However, 1,000 paging devices, costing $200 each, are needed, with annual maintenance and ad valorem taxes of $10 per set. If the pagers are expected to have an average life of 10 years, no salvage, and no cost of removal, the study might be done as follows, assuming a rectangular life table.

Cost of money	12%
Composite tax rate	50%
Debt ratio	45%
Cost of debt	9%

$$\phi = \frac{0.5}{1 - 0.5}\left(1 - \frac{0.45 \times 0.09}{0.12}\right) = 0.6625$$

ADR hybrid tax depreciation applies over a period of 8 years (10×0.8). And other tax effects may be neglected.

Therefore,

$$AC(C_{rec}) = AC(\text{first cost}) - AC(\text{net salvage})$$
$$AC(C_{rec}) = \$200(A/P, 12\%, 10) - 0(A/F, 12\%, 10)$$
$$= \$200(0.17698) = \$35.40$$
$$AC(T) = \phi\,[AC(C_{rec}) - AC(D_b)] - t(1 + \phi)[AC(D_t) - AC(D_b)]$$
$$AC(D_b) = \frac{200 - 0}{10} = \$20.00$$
$$AC(D_t) = PW(D_t)(A/P, 12\%, 10)$$
$$PW(D_t) = 200(P/D_t)^*$$
$$= 200(0.70875) = \$141.75$$
$$AC(D_t) = 141.75(0.17698) = \$25.09$$
$$AC(T) = 0.6625(35.40 - 20.00) - 0.5(1.6625)(25.09 - 20.00)$$
$$= 10.20 - 4.23 = \$5.97$$

$AC(\text{capital cost}) = 35.40 + 5.97 =$	\$41.37
$AC(\text{operating cost}) =$	10.00
Unit $AC =$	\$51.37
Total $AC = \$51.37 \times 1,000 =$	\$51,370

Because the levelized equivalent annual costs for the purchase of the pagers is apparently less than the \$54,000 savings they allow, the use of the pagers appears to be economically justified.

However, these pagers will live a hazardous life indeed. They will be lost, run over, stepped on, and, in general, be subjected to many of the causes of dispersion in their lives. If it is recognized that some will not last 10 years and must be replaced, whereas others will live more than the average, the costs will change. Suppose it is estimated that these pagers will live according to a No. 2 survivor curve. An approximation of the annual cost of maintaining the 1,000-unit level of continuing plant could be found as follows:

$$\text{Initial investment} = \$200 \times 1,000 = \$200,000$$
$$AC(C_{rec}) = AC(\text{first cost}) - AC(\text{net salvage})$$
$$= 200,000(\alpha/P, 12\%, 10)_{No.2} - 0(\alpha/F, 12\%, 10)_{No.2}$$
$$= 200,000(0.19284) = \$38,568$$
$$AC(T) = \phi[AC(C_{rec}) - AC(D_b)] - t(1 + \phi)[AC(D_t) - AC(D_b)]$$
$$AC(D_b) = 200,000 \times \frac{1}{10} = \$20,000\dagger$$
$$AC(D_t) = PW(D_t)(\text{initial investment})(\alpha/P, 12\%, 10)_{No.2}$$
$$= \$141.75 \times 1,000(0.19284) = \$27,335$$
$$AC(T) = 0.6625(38,568 - 20,000) - 0.5(1.6625)(27,335 - 20,000)$$
$$= 12,301 - 6,097 = \$6,204$$

---

*See Chapter 9, page 200.

†This assumes SLVG book depreciation because, with continuing plant, the $PW(D_b)$ can be shown to be depreciation rate $\times PW(\text{survivors})$. The $AC(D_b)$ is then $PW(D_b)(\alpha/P, 12\%, 10)_{No.2}$. Since

$$(\alpha/P) = \text{initial investment}/PW(\text{survivors})$$
$$AC(D_b) = \text{depreciation rate} \times PW(\text{survivors}) \times \frac{\text{initial investment}}{PW(\text{survivors})}$$
$$AC(D_b) = \text{depreciation rate} \times \text{initial investment}$$

With SLELG the $PW(D_b)$ is a complex series of terms which defy simplification. It can be approximated as $PW(D_b)_{ELG} = \text{depreciation rate} \times \text{initial investment} \times (P/A, 12\%, 10)$, which is the $PW(D_b)$ for the initial investment as if all units lived the average life (rectangular). The $AC(D_b)$ is then $PW(D_b)_{rectangular} \times (\alpha/P, 12\%, 10)_{No.2}$.

$$AC(\text{capital cost}) = 38,568 + 6,204 = \$44,772$$
$$AC(\text{operating cost}) = 10 \times 1,000 = \$10,000$$
$$\text{Total } AC = \$44,722 + \$10,000 = \$54,772$$

Because the costs of maintaining a continuous level of 1,000 paging devices are somewhat greater than the annual savings of $54,000, the proposal for buying pagers may not be justified.

To summarize, even in this simple example, the consideration of dispersion effects can influence the decision. If dispersion is not considered, the company appears to save money by using the pagers. However, when dispersion is considered, the cost of the pagers is increased and pagers are not the economic choice. Now the final decision rests on whether the pagers have other intangible benefits.

CONTINUING PLANT WITH GROWTH

Up to now, the discussion of the continuing plant concept has been limited to the assumption of a constant level of plant investment. This limitation may be restrictive if service needs require changing plant capacity, or if the price levels for plant are changing. Under these conditions, the plant or unit cost may be growing larger with each addition. Nevertheless, the (α/P) factor may still be used provided it is applied appropriately.

To review, the logic which was examined earlier held that:

1. The (α/P) factor may be applied to the initial plant placement.

2. The (α/P) factor may also be applied to subsequent plant placements provided they have the same average life and their survivor curve has the same shape.

3. According to the distributive law of mathematics, the result will be the same if the factor is applied to the sum of the survivors as if it is applied separately to survivors from each placement and their products are summed. In other words, the (α/P) factor may be applied to total survivors.

4. If the objective of the continuing plant concept is to maintain the total plant at the level of the initial placement, then the sum of the survivors will always equal the initial placement, and the appropriate annual cost figure can be found by multiplying the (α/P) factor by the initial cost.

The only statement that is not valid with growing plant, or a growing cost of plant, is number 4. Statements 1 to 3 still hold, and the (α/P) factor may be applied to the sum of the investment in survivors at each point.

With this approximation for SLELG effects, then, all quantities in the capital costs expression are proportionally greater due to dispersion.

The annual capital costs for SLELG are, therefore,

$$AC(\text{capital cost})_{\text{dispersion}} = AC(\text{capital cost})_{\text{rectangular}} \frac{(\alpha/P, 12\%, 10)}{(A/P, 12\%, 10)}$$

And in this example,

$$AC(\text{capital cost}) = 41.37 \frac{0.19284}{0.17898} = \$45.08$$
$$AC(\text{operating cost}) = \$10.00$$
$$\text{Unit } AC = \$45.08 + \$10.00 = \$55.08$$
$$\text{Total } AC = \$55.08 \times 1,000 = \$55,080$$

SUMMARY

When the lives of units within a group of assets vary, or are dispersed, the annual capital costs of the group are greater than if all the units lived the average life. Sometimes, dispersion can be ignored because it has the same effect on all competing alternatives, or because only a single unit of plant is being used. However, the analyst may need to recognize the effects of dispersion when competing alternative plans in an economy study include many units of plant which are expected to be influenced differently by dispersion.

The way to recognize the effects of dispersion is to use mathematically developed factors similar to the time-value factors. These factors are tabulated for many different survivor curves, representing different amounts of dispersion. The more dispersion anticipated, the greater the capital costs.

When dispersion, the various tax effects, and the changing price levels are all considered, economy studies can become exceedingly difficult. Such complex studies almost always require the use of sophisticated computer programs.

Present Worth of Survivors with Negative Exponential Survivor Curve

The survivors at a particular age can be found from the expression for the negative exponential survivor curve:

$$\text{Survivors at age } x = l_x = l_0\, e^{-x/N}$$

For convenience, the proportion of the initial investment surviving at age x can be expressed as follows:

$$\frac{l_x}{l_0} = e^{-x/N}$$

With continuous compounding, the present worth of the proportion surviving at age x would be

$$PW\left(\frac{l_x}{l_0}\right) = (e^{-rx})(e^{-x/n})$$

The total present worth of survivors along the entire curve would then be

$$\text{Total } PW(\text{survivors}) = \int_0^\infty (e^{-rx})(e^{-x/N})\, dx$$

$$= \int_0^\infty e^{-(1/N + r)x}\, dx$$

$$= \frac{1}{\dfrac{1}{N} + r}$$

$$= \frac{N}{1 + rN}$$

Some Observations About the Negative Exponential Curve

The negative exponential survivor curve is represented by the following expression:

$$l_x = l_0\, e^{-x/N}$$

where l_x = survivors at time x
l_0 = initial placement (survivors at time 0)
e = base of natural logarithms
N = average life

This expression can be used to find the retirement rate, the depreciation rate, the depreciation reserve, and the relationship between the negative exponential (α/P) factor and other (α/P) factors. These relationships are explained in this appendix.

RETIREMENT RATE

The first derivative of the negative exponential curve is

$$\frac{d(l_x)}{dx} = \frac{-l_0}{N}\, e^{-x/N}$$

This derivative represents the instantaneous rate of change of the survivor curve. It is negative because the curve is decreasing as time, x, increases. This is expected because a survivor curve cannot increase. It can only decrease, and this decrease, of course, is due to retirements. The derivative has therefore provided a measure of the rate at which the original placement is decreasing—in other words, the rate at which retirements of l_0, are occurring.

It would also be of interest to consider the rate at which the survivors, l_x, are retiring. To do this, the retirements will be related to survivors:

$$\text{Rate of retirement relative to survivors} = \frac{\dfrac{l_0}{N}\, e^{-x/N}}{l_0\, e^{-x/N}} = \frac{1}{N}$$

This expression reveals one unusual feature of the negative exponential curve: The rate of retirement for survivors is a constant, and it is independent of age.

DEPRECIATION RATE

Under straight-line vintage group depreciation, the depreciation rate (as a ratio) is equal to 1.0/average life, ignoring salvage. The average life for a group of assets may be found by dividing the area under the survivor curve by the initial placement. The area under the negative exponential curve may be found by integration, as follows:

$$\text{Area} = \int_0^\infty l_0 e^{-x/N}\, dx$$
$$= -N l_0 e^{-x/N}\Big|_0^\infty$$
$$= N l_0$$

The average life, then, would be

$$\text{Average life} = \frac{\text{area}}{\text{initial placement}}$$
$$= \frac{N l_0}{l_0}$$
$$= N$$

The depreciation rate therefore equals $1/N$. If this rate is applied to plant surviving throughout the existence of the group, all of the cost will be recovered. In other words,

$$\text{Depreciation accrual at age } x = \frac{1}{N} l_0 e^{-x/N}$$
$$\text{Total depreciation} = \int_0^\infty \frac{1}{N} l_0 e^{-x/N}\, dx$$
$$= l_0$$

DEPRECIATION RESERVE

The depreciation rate equals the retirement rate for the negative exponential survivor curve. This means that credits to the depreciation reserve (accruals) equal the debits (retirements, since salvage is being ignored here), and no depreciation reserve is accumulated for the group.

RELATION TO (α/P) FACTOR

In the text of Chapter 14, the (α/P) factor for the negative exponential curve was found to be $(1/N) + r$. However, this expression is simply the sum of the straight-line depreciation rate and the force of interest.

Readers who have examined the (A/P) factor may have observed that for normal values of life and the interest rate, the factor itself is always less than the sum of straight-line depreciation and the interest rate. This is because a depreciation reserve accumulates over the life of the plant, and the levelized return requirement is lowered to reflect it. (This was apparent in the procedure recommended in the 1952 edition of *Engineering Economy*, in what was called the "Equated Cost of Money.")

This chapter has pointed out that dispersion requires the use of an annuity factor which is greater than that associated with a rectangular survivor curve. This difference could be explained in terms of the lower depreciation reserve level which is inherent in the increased ratio of reserve debits to reserve credits when retirements occur at early ages.

The negative exponential represents a limiting case among the curves explored in this text. It results in reserve debits equal to reserve credits, and consequently no accumulated depreciation reserve. The (α/P) factor for the negative exponential curve is therefore the maximum among those curves considered in this text. It is equal to depreciation plus interest on the full cost.

15

OWNING VERSUS LEASING

INTRODUCTION

Since World War II, the growth in leasing has been dramatic. This growth is partly due to the continually increasing demand for capital; and since this increasing demand must be expected to continue, the appeal of leasing is likely to become even greater.

The appeal of leasing, of course, is that a business can get the use of some asset without actually having to get the capital necessary to buy it outright. This point is emphasized often by the various companies and individuals who specialize in making leases (lessors). However, the fact is that someone must provide the capital—whether it is an investor in a leasing company or an investor who buys the asset and leases it as an individual. For investors, an investment in assets which will then be leased is an alternative investment opportunity to buying stocks or bonds in the companies which will use the assets. They may find that leasing is a more attractive investment because they actually own a tangible asset, which has some value; and they are not relying entirely on the management of a business to protect their investment. On the other hand, the managers of businesses which lease assets (lessees) do not have the same control over these leased assets that ownership would give.

Just as investors consider leasing an alternative investment opportunity, a firm which leases must consider leasing an alternate method of financing assets. Leased assets are essentially like assets financed with mortgage bonds except that they are not merely pledged to investors—they are actually owned by the lessors. As Chapter 4 indicated, leases are debt-type financing, because they impose essentially the same obligation on a firm as mortgage bonds or debt.

Leasing can be an extremely complex topic because it is viewed from so many different vantage points. This chapter is limited to the point of view of the management of a business, which may consider obtaining the use of assets by leasing. It is therefore primarily concerned with the viewpoint of *lessees;* it is not concerned with the viewpoint of *lessors* and that of others associated with them.

FINANCIAL ASPECTS OF LEASING

For many years, accountants treated lease payments as operating expenses, and they did not recognize the financial aspects of leasing. However, as Chapter 4 pointed out, financial people have always considered leasing as an alternative source of external debt-type financing. This apparent difference in point of view between the financial and accounting communities has experienced continual change. Before 1974, the Accounting Principles Board (APB) of the American Institute of Certified Public Accountants established generally accepted accounting principles. Then in 1974, the APB was replaced by the Financial Accounting Standards Board (FASB). Over the years, these two groups have issued a series of opinions on leasing which have gradually changed the accountants' treatment of leases so that they now give greater recognition to the financial aspects of leasing. The most recent FASB opinion, in fact, would require the disclosure of future lease commitments on financial statements.

Today, engineering economists are in general agreement that leasing is a form of debt financing, whether or not the value of the assets leased appears on the balance sheet as debt. This is because a firm which takes a lease has incurred an obligation to make lease payments, and this obligation is just like the obligation to make loan payments. The financial impact is therefore the same as if the firm had actually taken a loan. This means that leasing must be appraised as a form of debt financing; and it must stand or fall, compared with the traditional forms of financing with debt.

TYPES OF LEASES

However, not all rental agreements are appropriately considered alternative forms of debt financing. In recognition of this fact, engineering economists often divide rental agreements into two broad categories: *operational* rentals (often called operational leases) and *financial* leases.

Operational Leases

Usually, leases are considered operational if they are arranged to meet temporary peak requirements, to defer major expenditures, or to fill a temporary need while something more permanent is being developed or built. They are a different way to provide a service or a function; they are not alternative financing.

An example of an operational lease might be the occasional rental of automobiles to meet a peak demand for cars in a motor pool. Another example could be the temporary rental of building space while a new building is under construction.

Financial Leases

Financial leases usually bind a lessee to fixed lease payments over a considerable period of time—often equal to the total life of the plant. Examples of financial leases are the leasing of a headquarters building, a motor vehicle for continuous service, or computers or reproduction equipment over their technologically satisfactory life.

Because every new financial lease increases long-term fixed charges, the cumulative effect of all leasing decisions could cause a major impact on a firm's financial position, just as if the firm had a larger amount of debt. As a result, other debt obligations must be curtailed, to offset the financial leases and to stay within the desired limit on long-term fixed charges. Otherwise, the risks of the business would be unintentionally increased beyond what is considered prudent for financial safety. In other words, the apparent ratio of debt capital to equity capital should be the same regardless of whether the asset is owned directly or leased from someone else. This is because the ratio of debt obligations to the total capital influences a firm's ability to raise both debt and equity. It is important to recognize the fixed obligation which a financial lease represents in order to keep fixed charges within the limit required if the capital structure is to be managed successfully.

Treatment of Operational and Financial Leases

Operational and financial leases are treated differently in economy studies because operational leases are not considered to have any impact on the capital structure of the business. The lease payments are simply treated as operations costs, like the costs of maintenance or electric power. On the other hand, financial leases are considered as having a financial impact, which means that lease payments for financial leases cannot be treated as operations costs. They must be considered as the repayment of capital with return to the lessor; and for this reason it is necessary to recognize their effect on the financial structure of the business.

From their definitions, the distinction between the two basic types of leases would appear to be clear; and in most cases it is clear. However, in some situations, whether to consider a specific lease as operational or financial is not obvious; and the decision must be left to the analyst's judgment.

The question of whether a lease is financial or operational often arises because the lessors would usually prefer that their leases be considered operational. However, if a lease in an economy study is treated as operational when it is actually financial, the costs of that lease will be significantly understated because they do not include the financial effects of the lease. If the purpose of a study is to decide between owning or leasing some asset, the lease must be financial because it is an alternative way of financing. If the purpose of the study is to determine the costs of an undertaking which includes a lease, the lease might be considered operational. However, if that lease has some reasonable buy alternative, it too should be considered a financial lease, even if the buy alternative is not being studied. The decision of whether a lease is financial or operational can be avoided by treating all leases as financial for study purposes. With this generalization, the results of the studies would not be wrong because leases which are truly operational would not be significantly penalized.

THREE APPROACHES TO FINANCIAL LEASES

Because this chapter is primarily concerned with owning versus leasing studies, it is concerned with the analysis of financial leases only. Operational leases are treated as operations costs, which have been covered in the previous chapters.

This chapter explains three increasingly sophisticated methods for analyzing financial leases:

- The implicit financing rate
- The apparent first cost
- The effective capital cost

The *implicit-financing-rate* method measures the approximate cost of capital secured through lease financing. Its chief value is that it is a convenient way to eliminate highly uneconomical leases which do not warrant further analysis. The second technique is the *apparent-first-cost* method. This method compares the first cost of the purchased asset, adjusted for the major income tax advantages of ownership, with the apparent first cost that the lease payments are repaying. This is a traditional approach, and is more precise than the simple implicit financing rate. The third technique is the *effective-capital-cost* method. It compares the complete capital costs for owning, as developed in the previous chapters, with the effective capital costs of the lease, including the financial impact leasing causes. This last method is the most complex of the three, but it is also the most accurate and versatile. All three of these methods are developed consistent with a basic principle, however; and to understand the methods, it is necessary to understand this principle.

Basic Principle Common to All Methods

Because taking a lease is equivalent to increasing debt-type financing, a firm will move towards a different financial structure when it leases an asset rather than buying it. If this firm buys an asset with composite capital, the debt ratio and the earnings percentage will be the same as they were before ownership. However, if the firm leases the asset, it is incurring a fixed obligation which is equivalent to an increase in debt financing. Since this increase is not accompanied by a concurrent increase in equity financing, the apparent debt ratio will be greater than it was before the leasing arrangement was made. From the investors' point of view, this would be an increase in risk, which would require an increase in their return and an increase in the cost to the firm. Because of constantly changing market conditions, it is not practical to try to predict these increases, and the own versus lease analyst is in no position to evaluate the various financial structures that a firm may wish to have at any particular time. Therefore, the basis of any own versus lease analysis is that the overall financial characteristics must be comparable regardless of how the asset is financed.

Implicit Financing Rate

Overall financial characteristics comparable to those resulting from a lease can be achieved by borrowing the funds. As a result, the major economic difference between owning and leasing is the relative financing cost which the firm would pay for either debt or lease financing. As Chapter 5 pointed out, the series of payments which repays a loan with interest is equivalent to the original loan at the interest rate charged. With an outright purchase, the price of an asset, or the loan required, is a lump sum at the time of the purchase. For a lease of the same asset, the price amounts to payments spread over the term of the lease. The interest rate at which these two payment schedules are equivalent is the *implicit financing rate* of the lease.

If some future salvage is estimated for ownership, the implicit financing rate is the interest rate at which the lease payments are equivalent to the capital expenditure for the first cost and the receipt for future salvage. The implicit financing rate, then, is the incremental internal rate of return (*IROR*), as covered in Chapter 12, on the difference between the "buy" cash flow and the "lease" cash flow. There is an important difference, however.

The lease cash flow usually looks like this.

A typical buy cash flow is

Then, the lease cash flow minus the buy cash flow is

Because an up arrow represents a receipt, the lease minus buy cash flow represents a source of funds and its repayment. The *IROR* is then the cost of these funds, rather than the rate of return on an additional investment, and it is the cost of the capital that leasing makes available.

In the absence of salvage and with equal lease payments, the implicit financing rate is found directly because the equation has only one unknown. If there is some salvage or the lease payments are unequal, however, the rate must be found by trial and error. Example 15-1 illustrates the calculation for equal lease payments but with future salvage.

Example 15-1.
Own: Purchase for $6,000; 10-year life; $1,000 gross salvage, $100 cost of removal.
Lease: Lease for 10 years at $1,000 per year.

The lease minus own cash flow is then

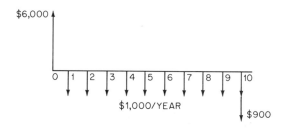

The implicit financing rate (k) is the interest rate which will cause the present worth of the cash flow to be zero. [See equation (12.1), page 271.]

$$\$6,000 = \$1,000(P/A, k\%, 10) + (\$1,000 - \$100)(P/F, k\%, 10)$$

Try 10%: $1,000(6.145) + $900(0.3855) = $6,492
Try 12%: $1,000(5.650) + $900(0.3220) = $5,940
Try 11%: $1,000(5.889) + $900(0.3522) = $6,206

$$k = 11\% + \left[\left(\frac{\$6,206 - \$6,000}{\$6,206 - \$5,940} \right) \times 1\% \right] = 11.8\%$$

The implicit financing rate is then 11.8%. This means that if debt-type capital usually costs less than 11.8%, the lease should not be taken; but if debt capital usually costs more, the lease *may* be economically justified.

Keep in mind that the implicit financing rate may be used as a quick way to identify very poor leases. However, the calculations do not include any income tax advantages which normally apply to a purchase plan. Therefore, if the implicit financing rate approaches the firm's cost of debt—or if it is less than the firm's cost of debt—it is important to analyze the situation further before making a final decision.

A rough approximation of the maximum acceptable lease payment can be found by determining the series of payments which is equivalent to the first cost (offset by any salvage) at the firm's cost of debt. Example 15-2 illustrates this approximation for a break-even lease payment.

Example 15-2. The data from Example 15-1 are used to determine the break-even lease payment if the cost of debt is 9%.

$$PW(\text{first cost}) - PW(\text{net salvage}) = PW(\text{lease payments})$$
$$\$6,000 = \$900(P/F, 9\%, 10) = (\text{lease payment})(P/A, 9\%, 10)$$
$$\$6,000 - \$900(0.4224) = (\text{lease payment})(6.418)$$
$$\text{Break-even lease payment} = \$876$$

This means that if the lease payments stipulated in the lease are greater than $876, it is cheaper to buy the asset than to lease it. Since this lease proposal stipulates payments of $1,000, it should be rejected. A lease payment of less than $876 might be acceptable, but further analysis of the lease proposal would be required to reach an economic decision.

Apparent First Cost

The implicit-financing-rate method is similar to the *IROR* technique used for other economy studies. However, leases can also be evaluated using present-worth techniques. For an own versus lease study, the first cost of the own alternative is usually known, or it can be estimated. For the lease alternative, the lease payments are the cost, since there is no actual first cost. However, since the lease payments are the repayment of a debt obligation, they are equivalent to the annual capital costs which would have resulted from the expenditure for some first cost. This apparent first cost of the lease is then the amount of capital that the lease payments would repay. Since the lease is equivalent to debt, the apparent first cost of the lease is the present worth of the lease payments at the cost of debt.

A simple comparison of the first cost to buy with the apparent first cost of the lease is another way to make an approximate evaluation. However, it is not any more accurate than the implicit-financing-rate method, because it does not recognize income tax effects. A more meaningful comparison can be made if the first cost of buying is adjusted for the major income tax effects. To make the comparison, a common financial structure must be maintained between owning and leasing.

A traditional way to maintain this common financial structure, and to determine the income tax effects, is to assume that the owned asset is purchased with 100% debt capital. This does not imply that the asset will *actually* be purchased with 100% debt; it implies only that if the asset is purchased, the financial structure of the firm could be modified so it would be common with the financial structure under a lease. The assumption of the 100% debt purchase approximates the cost the firm would incur if it bought the asset with composite capital and then adjusted its financial structure so it would be common with the financial structure under the lease. The important point is that the 100% debt purchase assumption makes the lease and purchase comparable. It also allows income tax effects to be included that are due to accelerated depreciation and investment tax credit, which usually apply to owned assets but not to leased assets.

The present worth of income tax, with normalization of the accelerated tax benefit and with service-life flow through of the investment tax credit, was shown in Chapter 9 to be

$$PW(T) = \phi[PW(C_{\text{rec}}) - PW(D_b)] - t(1 + \phi)[PW(D_t) - PW(D_b)]$$
$$- \frac{1}{1 - t} PW(ATC) - (1 + \phi)[PW(TC) - PW(ATC)]$$

by combining equations (9.1) and (9.3).

If 100% incremental debt financing is assumed, the debt ratio (r) is 100% and the cost of capital (i) equals the cost of debt (i_d). The tax factor (ϕ) is therefore zero, which means that the basic tax liability is zero. The adjustments for accelerated depreciation and investment tax credit, however, do not disappear with 100% debt financing. Even though the basic tax liability is zero, these other tax effects must be computed for the owning option. These adjustment terms are negative, indicating the relative economic benefits of acquiring an asset which would be eligible for them. They usually do not need to be computed for the leasing option, because assets which are obtained through leasing are not usually eligible for accelerated tax depreciation or investment tax credit. Cases in which investment tax credit may be passed through from the lessor to the lessee will be covered later in this chapter.

The apparent first cost of owning is the present worth of the capital costs of owning, assuming 100% debt financing. The apparent first cost of the lease is the present worth of the lease payments at the cost of debt. These apparent-first-cost values are similar to PWE values used in other economy studies; but they cannot be compared directly to other costs because these amounts are basically adjusted first-cost values. They have been found using an assumption of incremental debt financing which would cause a different financial structure from that assumed for other studies. Example 15-3 illustrates the apparent-first-cost method of an own versus lease analysis.

Example 15-3. Compare the alternatives described in Example 15-1 on the basis of the apparent first cost.

The purchased equipment is eligible for accelerated depreciation using the hybrid method (i.e., DDB with a switch to SYD after the first 2 years). The purchased equipment is also eligible for an investment tax credit of 10% of the first cost.

$$\text{ADR tax life} = 80\% \text{ of service life} = 8 \text{ years}$$

$$\text{Cost of debt} = 9\%$$

$$\text{Composite tax rate} = 48\%$$

$$\phi = 0 \text{ for a } 100\% \text{ debt purchase}$$

Buy Alternative:

$$\text{Apparent first cost} = PW(\text{capital costs}) = PW(C_{\text{rec}}) + PW(T)$$
$$PW(C_{\text{rec}}) = (\text{first cost})(P/F, 9\%, 0) - (\text{net salvage})(P/F, 9\%, 10)$$
$$= \$6,000(1) - (\$1,000 - \$100)(0.4224)$$
$$= \$5,620$$

$$PW(T) = -t(1 + \phi)[PW(D_t) - PW(D_b)] - \frac{1}{1 - t} PW(ATC)$$
$$- (1 + \phi)[PW(TC) - PW(ATC)]$$

$$D_b = \frac{\text{first cost} - \text{net salvage}}{\text{life}}$$

$$= \frac{\$6,000 - (\$1,000 - \$100)}{10}$$

$$= \$510$$

$$PW(D_b) = D_b(P/A, 9\%, 10)$$
$$= \$510(6.418)$$
$$= \$3,273$$

$$\% \text{ gross salvage} = \frac{\$1,000}{\$6,000} = 16.7\%$$

$$\% \text{ cost of removal} = \frac{\$100}{\$6,000} = 1.67\%$$

$$PW(D_t) = \text{ first cost } (P/D_t)$$

The P/D_t factor can be determined from the present worth of tax depreciation tables.

$$(P/D_t) = [0.72421 - \frac{0.167 - 0.100}{0.200 - 0.100}(0.72421 - 0.66975) + 0.0167(0.42241)] = 0.69478$$

$$PW(D_t) = \$6,000(0.69478)$$
$$= \$4,169$$

$$TC = \$6,000 \times 10\% = \$600$$
$$PW(TC) = TC(P/F, 9\%, 1) = \$600(0.9174)$$
$$= \$550$$

$$ATC = \frac{TC}{\text{life}} = \frac{\$600}{10}$$
$$= \$60$$

$$PW(ATC) = ATC(P/A, 9\%, 10) = \$60(6.418)$$
$$= \$385$$

$$PW(T) = -0.48(1 + 0)(\$4,169 - \$3,273) - \frac{1}{1 - 0.48}(\$385)$$
$$- (1 + 0)(\$550 - \$385)$$
$$= -\$430 - \$740 - \$165$$
$$= -\$1,335$$

$$\text{Apparent first cost} = PW(\text{capital costs}) = PW(C_{\text{rec}}) + PW(T)$$
$$= \$5,620 - \$1,335$$
$$= \$4,285$$

Lease Alternative:

$$\text{Apparent first cost} = PW(\text{lease payments})$$
$$PW(\text{lease payments}) = \$1,000(P/A, 9\%, 10)$$
$$= \$1,000(6.418)$$
$$= \$6,418$$
$$\text{Apparent-first-cost difference (lease} - \text{buy)} = \$6,418 - \$4,285$$
$$= \$2,133$$

The difference in the apparent first costs indicates that leasing is about 50% more expensive than purchasing the asset directly.

Interpretation of the Apparent First Cost

The apparent first cost of the lease is the present worth of the lease payments at the debt interest rate, because that is the amount which could be borrowed if the repayment were the lease payments. The actual first cost of buying is reduced, first, to reflect the fact that owning will result in some future salvage. The revenue required for owning is then the same as if a lower price were paid and there were no salvage. Because income tax effects are different if the asset is owned rather than leased, and because these effects encourage owning, this difference is reflected in a further reduction in the apparent first cost of owning. The apparent first cost of owning is the first cost which would have caused the revenue requirements if there were no salvage and no tax effects. This apparent first cost is then compared with the apparent first cost of the lease, for which the salvage and tax effects actually do not occur. Therefore, the results of Example 15-3 can be interpreted as follows:

- The revenue requirements for leasing are the same as if the firm spends $6,418 of debt capital and has no salvage or tax effects.

- The revenue requirements for owning are the same as if the firm spends $4,285 of debt capital and has no salvage or tax effects.

The apparent-first-cost method also permits a refinement in the calculation of a maximum acceptable, or break-even, lease payment. The break-even lease payment is the annual amount which would have a present worth equivalent to the apparent first cost of owning at the cost of debt.

$$\text{Apparent first cost of owning} = \text{lease payment}(P/A, 9\%, 10)$$
$$\$4,285 = \text{lease payment}(6.418)$$
$$\text{Break-even lease payment} = \$668$$

Here, the break-even lease payment is $208 less than was estimated by the implicit-financing-rate method. The reason for this difference is that the apparent-first-cost method recognizes most of the income tax differences between owning and leasing that the implicit-financing-rate approach does not reflect.

Effective Capital Cost

The implicit-financing-rate and the apparent-first-cost methods of owning versus leasing comparisons are usually adequate for most leasing decisions. However, it is important to recognize what they really are: They are not engineering economy studies; they are financial studies. They answer the question of whether to buy or to lease; an economy study is needed to answer the question of whether to acquire an asset or not. If an economy study finds that the acquisition is justified, the question of how to finance it can then be answered with an owning versus leasing study.

Because these two methods answer financial questions alone, they have some drawbacks for many engineering economy applications. These drawbacks are as follows:

1. The apparent-first-cost method presumes specific (debt) financing for specific assets. This appears in conflict with other economy studies, in which a pool of composite capital is assumed. As a result, the apparent-first-cost method is easily misunderstood.

2. In the apparent-first-cost method, only the major income tax effects are taken into account. (Because 100% debt is assumed, ϕ is zero; therefore, the component of income tax which is proportional to return is zero.) The lease capital is repaid differently from actual debt capital, and there are further tax effects as a result.

3. The results do not reflect the full revenue requirements of either owning or leasing under the firm's desired capital structure. Therefore, these owning versus leasing studies are not compatible with other revenue-requirement studies.

4. Because separate acquisition studies are needed, the question of leasing versus the status quo is difficult to answer. For example, assume a new

purchase is not economically justified, but that leasing is less expensive than the new purchase. Is leasing justified? The results of the own versus lease study and the status quo versus new purchase study cannot be compared directly because they have been made assuming different financing.

5. When no reasonable buy alternative exists for a lease which is embedded in an otherwise composite capital economy study, it is difficult to account for the leasing costs. For example, suppose an analyst is performing a composite capital economy study between buying two different trouble locating and analysis systems. If a minicomputer must be leased for one alternative, the apparent-first-cost concept does not allow the costs of the lease to be added to the other costs for that alternative.

The goal of the 100% debt purchase assumption is to maintain a common financial structure whether the firm buys or leases. The apparent-first-cost approach is actually an adjustment to the own alternative so that its incremental financing will be comparable to the lease financing; but it is just as reasonable to adjust the lease alternative so that its adjusted financing is common with that of the own alternative.

Previous chapters in this text have shown how to find the present worth of the capital costs of making a composite capital investment. To be compatible with other types of economy studies, this is the amount which must represent the own alternative. As the apparent-first-cost approach has illustrated, the lease is equivalent to an investment of an amount equal to the present worth of the lease payments at the cost of debt. However, it is an investment of debt-type capital only, and the terms of the lease specify the tax deductions and book expense.

If the same financial structure that exists under owning is maintained during a lease, other debt and equity financing must be balanced with the lease financing. When something is leased, the total amount of debt financing is increased; therefore, other debt financing must be adjusted down, and other equity adjusted up, to compensate for the lease.

Appendix 15A demonstrates that the net result of the lease plus these financial adjustments is equivalent to the firm investing an amount of composite capital equal to the present worth of the lease payments. In other words, the effective capital costs of the lease are equivalent to the capital costs of a composite investment equal to the apparent first cost of the lease.

Effective Depreciation

Each year of the lease, the present worth of the remaining lease payments represents the amount of capital which has yet to be recovered. The difference between the capital at the beginning and end of a year is the amount which was recovered during that year. Therefore, the sum total of those amounts is the capital repayment, or effective depreciation, during the lease. Since very few leases are capitalized for tax purposes, this effective depreciation can be considered the same for both book and tax depreciation. This means that the simple capital-cost equations from Chapter 8 can be used to find the effective capital costs for the lease.

$$PW(C_{rec}) = PW(\text{first cost}) - PW(\text{net salvage})$$

For leasing:　　net salvage $= 0$

$$PW(C_{rec}) = PW(\text{apparent first cost})$$
$$PW(T) = \phi[PW(C_{rec}) - PW(D_b)]$$
$$PW(D_b) = PW(\text{effective depreciation})$$
$$PW(T) = \phi[PW(\text{apparent first cost}) - PW(\text{effective depreciation})]$$

$$PW(\text{capital cost}) = PW(C_{\text{rec}}) + PW(T)$$
$$= PW(\text{apparent first cost}) + \phi[\,PW(\text{apparent first cost})$$
$$- PW(\text{effective depreciation})]$$

Remember that each year throughout the life of the lease, the present worth (at the cost of debt) of the remaining lease payments is the amount of outstanding capital. This amount is analogous to the accountants' net plant for owned assets. However, net plant reduces equally each year because of straight-line depreciation. As can be seen in Figure 15-1, the present value of future lease payments (the outstanding capital) decreases in a geometric progression, depending on the amount of the lease payments and the interest rate.

Each lease payment consists of an effective depreciation component and a return component. The return component is equal to the debt interest rate times the capital outstanding *before the lease payment is made.* For now we will assume that the lease payment is made at year end.

$$\text{Interest portion}_n = \text{outstanding capital}_n \times i_d$$

The capital repayment portion (effective depreciation) is equal to the lease payment minus this interest portion.

FIGURE 15-1
(Effective) Net Plant

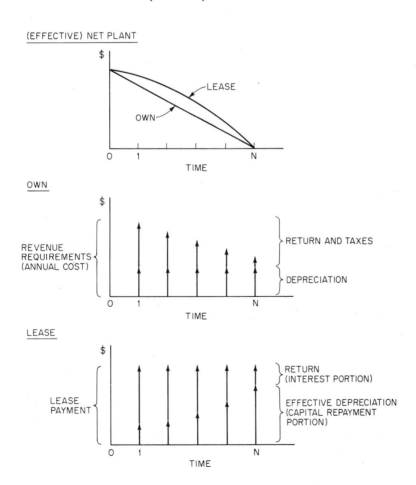

$$\text{Effective depreciation}_n = \text{lease payment}_n - \text{interest portion}_n$$
$$= \text{lease payment}_n - (\text{outstanding capital}_n \times i_d)$$
$$= \text{lease payment}_n - [PW(\text{remaining lease payments})] \times i_d$$

Example 15-4 illustrates the effective depreciation calculation for a simple case with only three lease payments.

Example 15-4. Find the year-by-year effective depreciation for the following lease agreement and compare the total accumulated effective depreciation to the apparent first cost.

Year	Contracted lease payment
1	$1,200
2	$1,100
3	$1,000

Debt interest = 9%

Composite cost of capital = 12%

Year	Depreciation	
1	$1,200 − [$1,200(P/F, 9%, 1) + $1,100($P/F$, 9%, 2) + $1,000($P/F$, 9%, 3)] 0.09	= $ 948
2	$1,100 − [$1,100(P/F, 9%, 1) + $1,000($P/F$, 9%, 2)] 0.09	= $ 933
3	$1,000 − [$1,000(P/F, 9%, 1)] 0.09	= $ 917
	Total accumulated effective depreciation	= $2,798

$$PW(\text{lease payments}) = \text{apparent first cost}$$
$$= \$1,200(P/F, 9\%, 1)$$
$$+ \$1,100(P/F, 9\%, 2)$$
$$+ \$1,000(P/F, 9\%, 3) = \$2,798$$

Since the sum of yearly effective depreciation, $2,798, is also the present value of all future lease payments, all capital (effective investment) has been recovered.

The effective depreciation for a lease is analogous to the repayment of principal on a home mortgage loan. Suppose a person takes out a $40,000 mortgage at an interest rate of 7% annually, or $7/12$% monthly, for 40 years. The monthly mortgage payment will be $40,000($A/P$, $7/12$%, 480) = $248.57. Of the first $248.57 payment, $40,000 × $7/12$% = $233.33 will be interest and $248.57 − $233.33 = $15.24 will be the repayment of principal. This repayment reduces the outstanding capital balance to $40,000 − $15.24 = $39,984.76. In the second month, the person will pay $39,984.76 × $7/12$% = $233.24 in interest and $248.57 − $233.24 = $15.33 in principal. The relationship between principal and interest payments over the loan period is illustrated by Figure 15-1.

For a lease, the effective first cost is the present worth of lease payments at the cost of debt. Suppose a firm having a 9% cost of debt leases an asset for 5 years at $257.07 per year. The equivalent first cost is $257.07($P/A$, 9%, 5) = $1,000. In the first year, the effective interest payment is $1,000 × 9% = $90. The effective depreciation is $257.07 − $90 = $167.07. In the second year, the effective outstanding balance is $1,000 −

$167.07 = \$832.93$. The effective interest payment in year 2 is $\$832.93 \times 9\% = \$74.96$. The effective capital recovery is $\$257.07 - \$74.96 = \$182.11$.

The difference between the mortgage and the lease is that the monthly payment for the mortgage is determined from the first cost and the interest rate; for the lease, from the viewpoint of the lessee, the lease payments determine the effective first cost at the lessee's debt interest rate.

Equal Periodic Payments

In the familiar case of *equal* periodic lease payments over the life of the lease, the periodic effective depreciation is a geometric progression increasing at a constant rate equal to the debt interest rate. The base of this geometric progression is the constant lease payment times the (P/F) factor at the debt interest rate for one period more than the lease. This relationship is derived in Appendix 15B at the end of the chapter, but it may be easier to visualize from the following explanation. The last payment includes an amount of interest. The capital repayment portion of the payment is the total payment less the interest portion. If the number of terms in the series of capital repayment amounts were increased to $N + 1$, the amount of that next term would be the lease payment. (See Figure 15-1.) The amount of the effective depreciation at time N is then [payment/$(1 + i_d)$]. However, the last amount in a geometric progression is also [base $(1 + i_d)^N$], if i_d is the rate of change. Therefore,

$$\text{Base}(1 + i_d)^N = \frac{\text{payment}}{1 + i_d}$$

$$\text{Base} = \frac{\text{payment}}{(1 + i_d)^{N+1}}$$

$$\text{Base} = \text{payment}(P/F, i_d\%, N + 1) = A_0$$

As shown in Chapter 5, the present worth of a geometric series having a base A_0 and a constant rate of change i_d is

$$PW(\text{geometric progression}) = A_0(P/A, i_g\%, N)$$

where $i_g = \dfrac{i - i_d}{1 + i_d}$

The present worth of the effective depreciation for a lease with equal payments is then

$$PW(\text{effective depreciation}) = \text{lease payment}(P/F, i_d\%, N + 1)(P/A, i_g\%, N)$$

Example 15-5 illustrates the use of this effective depreciation formula in the effective-capital-cost method of own versus lease analysis. The own versus lease comparison is the same as in Examples 15-1, 15-2, and 15-3.

Example 15-5. Solve Example 15-3 by comparing the effective capital costs.

Composite cost of capital = 12%

Debt interest rate = 9%

Debt ratio = 45%

Tax rate = 48%

$$\phi = \frac{0.48}{1 - 0.48}\left(1 - \frac{0.45 \times 0.09}{0.12}\right) = 0.6115$$

Own Alternative:

This computation for present worth of expenditures is done exactly the same as other present worth of expenditures computations in this text.

$$PW(\text{Capital Costs}) = PW(C_{\text{rec}}) + PW(T)$$

$$PW(C_{\text{rec}}) = FC(P/F, 12\%, 0) - (\text{net salvage})(P/F, 12\%, 10)$$

$$= \$6{,}000(1) - (\$1{,}000 - \$100)(0.3220)$$

$$= \$5{,}710$$

$$PW(T) = \phi[PW(C_{\text{rec}}) - PW(D_b)] - t(1 + \phi)[PW(D_t) - PW(D_b)]$$

$$- \frac{1}{1 - t} PW(ATC) - (1 + \phi)[PW(TC) - PW(ATC)]$$

$$D_b = \frac{\text{first cost} - \text{net salvage}}{\text{life}}$$

$$= \frac{\$6{,}000 - (\$1{,}000 - \$100)}{10}$$

$$= \$510$$

$$PW(D_b) = D_b(P/A, 12\%, 10)$$

$$= \$510(5.650)$$

$$= \$2{,}882$$

$$\% \text{ gross salvage} = \frac{\$1{,}000}{\$6{,}000}$$

$$= 16.67\%$$

$$\% \text{ cost of removal} = \frac{\$100}{\$6{,}000}$$

$$= 1.67\%$$

$$PW(D_t) = (\text{first cost}) (P/D_t)$$

$$= \$6{,}000\left[0.67656 - \frac{0.167 - 0.100}{0.200 - 0.100}(0.67656 - 0.63155) \right.$$

$$\left. + 0.0167(0.32197) \right]$$

$$= \$3{,}911$$

$$TC = \$6{,}000 \times 10\%$$

$$= \$600$$

$$PW(TC) = TC(P/F, 12\%, 1)$$

$$= \$600(0.8929)$$

$$= \$536$$

$$ATC = \frac{TC}{\text{life}} = \frac{\$600}{10}$$

$$= \$60$$

$$PW(ATC) = ATC(P/A, 12\%, 10)$$

$$= \$60(5.650)$$

$$= \$339$$

$$PW(T) = 0.6115(5{,}710 - 2{,}882)$$

$$- 0.48 (1 + 0.6115)(\$3{,}911 - \$2{,}882)$$

$$- \frac{1}{1 - 0.48} (\$339) - (1 + 0.6115)(\$536 - \$339)$$

$$= \$1{,}729 - \$796 - \$652 - \$317$$

$$= -\$36$$

$$PW(\text{capital costs}) = \$5{,}710 - \$36 = \$5{,}674$$

<u>Lease Alternative:</u>

$$PW(\text{effective capital cost}) = PW(C_{rec}) + PW(T)$$
$$PW(C_{rec}) = \text{apparent first cost} = (\text{lease payment})(P/A, 9\%, 10)$$
$$= \$1,000(6.418)$$
$$= \$6,418$$
$$PW(T) = \phi[PW(C_{rec}) - PW(\text{effective } D_b)]$$
$$PW(\text{effective } D_b) = (\text{lease payment})(P/F, i_d\%, N+1)(P/A, i_g\%, N)$$
$$i_g = \frac{12 - 9}{1.09} = 2.75\%$$
$$PW(\text{effective } D_b) = \$1,000(P/F, 9\%, 11)(P/A, 2.75\%, 10)$$
$$= \$1,000(0.3875)(8.640)$$
$$= \$3,348$$
$$PW(T) = 0.6115(\$6,418 - \$3,348)$$
$$= \$1,877$$
$$PW(\text{effective capital cost}) = \$6,418 + \$1,877 = \$8,295$$
$$\text{Difference in capital cost (lease} - \text{own)} = \$8,295 - \$5,674 = \$2,621$$

This difference in the present worth of effective capital costs indicates that the asset should be purchased rather than leased. The PW(capital cost) for owning is \$5,674, whereas the comparable value for leasing is \$8,295.

This comparison is based on the effective capital cost only. However, any operations costs which are different between the two plans must also be considered in a complete study. Therefore, the amount of the lease payment used to determine the effective first cost should be reduced by the amount of maintenance, executory (expenses of executing the lease), and other operations costs to be paid by the lessee. For study purposes, these operations costs should then be treated separately, just as any other expense items, and in the absence of more specific information, they may be considered to have the same cost whether they are paid by the lessor or the lessee.

Interpretation of the Effective Capital Cost

The present worth of the capital cost for owning is simply the present worth of capital costs discussed throughout this book. This means that it would be the same quantity which would be used in an acquisition study of this investment. The present worth of the effective capital cost for leasing is the cost of leasing plus the financial adjustments necessary to restore the firm to the pre-lease financial conditions. Unlike the apparent-first-cost quantity, it can be added to or compared with other present worth of capital costs amounts.

This means that although the effective-capital-cost method can be used to make the financial decision between owning and leasing, it is not restricted to that: It can be interpreted and used in an economy study as well.

Rate-of-Return Technique

The effective-capital-cost method is basically the PWE technique. As Chapter 12 pointed out, the rate of return (ROR) technique is equivalent to the PWE technique. It therefore follows that the ROR technique should also be valid for own versus lease studies. Since Chapter 12 covers the rate-of-return analysis in detail, a full explanation will not be presented here. However, Example 15-6 illustrates the application of the

ROR technique to the owning versus leasing case illustrated in Example 15-5 and others.

Example 15-6. Table 15-1 is the cash-flow table for the own alternative assuming $1,200 per year in revenue and $200 per year in operations costs. Table 15-2 is the effective cash-flow table for the lease alternative as if it were an equivalent purchase, assuming the same revenue and operations costs as in Table 15-1. Table 15-3 is the incremental cash flow for the difference between leasing and owning.

The tax-depreciation amounts were found as follows:

$$\text{ADR tax life} = 80\% \text{ of 10-year service life} = 8 \text{ years}$$

$$
\begin{aligned}
\text{Maximum allowable} \\
\text{accumulated tax depreciation} &= \text{first cost} - (\text{gross salvage} - 10\% \text{ of first cost})^* \\
&= \$6,000 - (\$1,000 - \$600) \\
&= \$6,000 - \$400 \\
&= \$5,600
\end{aligned}
$$

$$
\begin{aligned}
\text{Taxable gain} &= \text{gross salvage} - (\text{gross salvage} - 10\% \text{ of first cost}) \\
&= \$1,000 - (\$1,000 - \$600) \\
&= \$1,000 - \$400 \\
&= \$600
\end{aligned}
$$

$$
\begin{aligned}
\text{Book depreciation} &= \frac{\text{first cost} - \text{net salvage}}{\text{life}} \\
&= \frac{\$6,000 - \$900}{10} \\
&= \frac{\$5,100}{10} \\
&= \$510
\end{aligned}
$$

$$
\begin{aligned}
\text{Effective retirement} &= \text{cost of removal} - \text{taxable gain} \\
&= \$100 - \$600 \\
&= -\$500
\end{aligned}
$$

Tax depreciation in each year is shown in the following table.

Year	ADR balance	DDB depreciation	SYD rate	SYD depreciation	Tax depreciation	Accumulated tax depreciation
1	$6,000	$1,500	8/36	$1,333	$1,500	$1,500
2	4,500	1,125	7/28	1,125	1,125	2,625
3	3,375	844	6/21	964	964	3,589
4	2,411		5/21	804	804	4,393
5	1,608		4/21	643	643	5,036
6	966		3/21	482	483	5,518
7	484		2/21	321	82	5,600
8						5,600
9						5,600
10					(500)	5,100

*Gross salvage is reduced by 10% of the tax basis, which is the first cost in this example.

TABLE 15-1
CASH FLOW FOR OWN

	0	1	2	3	4	5	6	7	8	9	10
1. Revenue		1,200	1,200	1,200	1,200	1,200	1,200	1,200	1,200	1,200	1,200
2. Operations expenses (excluding depreciation)		200	200	200	200	200	200	200	200	200	200
3. Tax depreciation		1,500	1,125	964	804	643	482	82	0	0	(500)
4. Book depreciation		510	510	510	510	510	510	510	510	510	510
5. Net operating revenue $(1 - 2 - 4)$		490	490	490	490	490	490	490	490	490	490
6. F.I.T. without TC $\{(1 - 2 - 3 - 11)t\}$		(357)	(147)	(55)	35	123	211	411	455	459	703
7. Investment tax credit		600									
8. Amortized TC $(7 \div \text{life})$		60	60	60	60	60	60	60	60	60	60
9. F.I.T. deferred $[(3 - 4)t]$		475	295	218	141	64	(13)	(205)	(245)	(245)	(485)
10. Operating income $(5 - 6 + 8 - 9)$		432	402	387	374	363	352	344	340	336	332
11. Interest on debt $(17_{n-1} \times r \times i_d)$		243	181	151	124	100	79	62	52	43	35
12. Net income $(10 - 11)$		189	221	236	250	263	273	282	288	293	297
13. Required return $(17_{n-1} \times i_c)$		720	537	448	367	297	235	186	153	128	104
14. ATCF $(1 - 2 - 6 + 7)$		1,957	1,147	1,055	965	877	789	589	545	541	297
15. New capital/salvage	6,000	0	0	0	0	0	0	0	0	0	(900)
16. Net cash flow $(14 - 15)$	(6,000)	1,957	1,147	1,055	965	877	789	589	545	541	1,197
17. Remaining capital balance $(17_{n-1} + 10 - 16)$	6,000	4,475	3,730	3,062	2,471	1,957	1,520	1,275	1,070	865	0

TABLE 15-2
EFFECTIVE CASH FLOW FOR LEASE

	0	1	2	3	4	5	6	7	8	9	10
1. Revenue		1,200	1,200	1,200	1,200	1,200	1,200	1,200	1,200	1,200	1,200
2. Operations expenses (excluding depreciation)		200	200	200	200	200	200	200	200	200	200
3. Tax depreciation = book depreciation		422	460	502	547	596	650	708	772	842	917
4. Book depreciation = lease payment − i_d [PW(future lease payments)]		422	460	502	547	596	650	708	772	842	917
5. Net operating revenue (1 − 2 − 4)		578	540	498	453	404	350	292	228	158	83
6. F.I.T. without TC [(1 − 2 − 3 − 11)t]		153	143	132	120	107	92	77	60	42	22
7. Investment tax credit		0									
8. Amortized TC (7 ÷ life)		0	0	0	0	0	0	0	0	0	0
9. F.I.T. def. [(3 − 4)t]		0	0	0	0	0	0	0	0	0	0
10. Operating income (5 − 6 + 8 − 9)		425	397	366	333	297	258	215	168	116	61
11. Interest on debt ($17_{n-1} \times r \times i_d$)		260	243	224	204	182	158	131	103	71	37
12. Net income (10 − 11)		165	154	142	129	115	100	84	65	45	24
13. Required return ($17_{n-1} \times i_c$)		770	720	664	604	538	467	389	304	211	110
14. ATCF (1 − 2 − 6 + 7)		847	857	868	880	893	908	923	940	958	978
15. New capital/salvage = PW(lease payments)	6,418	0	0	0	0	0	0	0	0	0	0
16. Net cash flow (14 − 15)	(6,418)	847	857	868	880	893	908	923	940	958	978
17. Remaining capital balance ($17_{n-1} + 10 − 16$)	6,418	5,996	5,536	5,034	4,487	3,891	3,241	2,533	1,761	919	0

TABLE 15-3
INCREMENTAL CASH FLOW FOR LEASE MINUS OWN

	0	1	2	3	4	5	6	7	8	9	10
1. Revenue		0	0	0	0	0	0	0	0	0	0
2. Operating expenses (excluding depreciation)		0	0	0	0	0	0	0	0	0	0
3. Tax depreciation		(1,078)	(665)	(462)	(257)	(47)	168	626	772	841	1,417
4. Book depreciation		(88)	(50)	(8)	37	86	140	198	262	332	407
5. Net operating revenue (1 − 2 − 4)		88	50	8	(37)	(86)	(140)	(198)	(262)	(332)	(407)
6. F.I.T. without TC [(1 − 2 − 3 − 11)t		510	290	187	85	(16)	(119)	(334)	(395)	(417)	(681)
7. Investment tax credit		(600)									
8. Amortized ITC (7 ÷ life)		(60)	(60)	(60)	(60)	(60)	(60)	(60)	(60)	(60)	(60)
9. F.I.T. deferred [(3 − 4)t]		(475)	(295)	(218)	(141)	(64)	13	205	245	245	485
10. Operating income (5 − 6 + 8 − 9)		(7)	(5)	(21)	(41)	(66)	(94)	(129)	(172)	(220)	(271)
11. Interest on debt ($17_{n-1} \times r \times i_d$)		17	62	73	80	82	79	69	51	28	2
12. Net income (10 − 11)		(24)	(67)	(94)	(121)	(148)	(173)	(198)	(223)	(248)	(273)
13. Required return ($17_{n-1} \times i_c$)		50	183	216	237	241	232	207	151	83	6
14. ATCF (1 − 2 − 6 + 7)		(1,110)	(290)	(187)	(85)	16	119	334	395	417	681
15. New capital/salvage	418	0	0	0	0	0	0	0	0	0	900
16. Net cash flow (14 − 15)	(418)	(1,110)	(290)	(187)	(85)	16	119	334	395	417	(219)
17. Remaining capital balance (17_{n-1} + 10 − 16)	418	1,521	1,806	1,972	2,016	1,934	1,721	1,258	691	54	0

385

Net present value:

	Factor	PW
−418(*P/F*, 12%, 0)	1.0000	−418
−1,110(*P/F*, 12%, 1)	0.8929	−991
−290(*P/F*, 12%, 2)	0.7972	−231
−187(*P/F*, 12%, 3)	0.7112	−133
−85(*P/F*, 12%, 4)	0.6355	−54
16(*P/F*, 12%, 5)	0.5674	9
119(*P/F*, 12%, 6)	0.5066	60
334(*P/F*, 12%, 7)	0.4523	151
395(*P/F*, 12%, 8)	0.4039	160
417(*P/F*, 12%, 9)	0.3606	150
−219(*P/F*, 12%, 10)	0.3220	−71

$$NPV = -\$1,336$$

The negative net present value for the cash flow of leasing minus owning indicates that owning is the better economic choice.

The *IROR* for this cash flow is found by trial and error, as Chapter 12 demonstrated. For this cash flow of lease minus own, it is approximately −10.5%, indicating that leasing is a very poor investment compared with owning.

The *IROR* of the reversed cash flow (that is, with all signs reversed or for own minus lease) would be this same, 10.5%. This means that owning is a source of funds with a low cost compared with that of the lease.

LEASE OPTIONS

A type of lease which may be difficult to analyze is one in which the lessee is given the opportunity to buy the leased asset for a fixed price at some future time. Often, the lease will continue even if the lessee elects not to buy, but there will be some change in the lease payment. Sometimes, the lease can be terminated if the buy option is not exercised.

In analyzing own and lease alternatives, it is important to make certain that conditions are comparable. Both the own and the lease alternatives must provide essentially the same service function over the same period of time. If a lease with an option to buy clause is expected to be terminated with the buy option *not* exercised, the own alternative must also be terminated, for study purposes, at the same point in time; and the owned property must then be assumed to be sold. If the plant is required for a period longer than the lease, the cost of some substitute for the lease must be included after the lease ends.

Each leasing alternative should be considered as a separate economic plan and analyzed accordingly. Barring exceptional conditions, outright ownership should be cheaper than leasing, and exercising a lease-purchase option should be cheaper than continuing to lease.

Income tax calculations for a lease with an option to buy may be particularly difficult to quantify. The income tax component, which is a function of return, is proportional to the amount of invested capital and its schedule of repayment. With owned property, the investment is considered to be made at the time of the purchase, and capital repayment (and associated return and taxes) occurs throughout its service life. If a lease has an option to buy, it is just as reasonable to consider that the composite capital required for the purchase is committed when the option is exercised; and repayment occurs over the additional service period of the then owned property.

A lease is viewed as a capital commitment equivalent to 100% debt-type financing. For a lease without renewal or purchase options, the capital commitment starts at the beginning of the lease; throughout its term, then, the firm must raise additional equity capital for other projects in order to maintain its desired financial ratios. With a renewal option, the firm does not have a real financial commitment for the additional capital required until the option is actually exercised. Of course, the firm does have a commitment to supply the service that the leased property may provide. The firm need not recognize the need for other equity financing in order to keep its desired financial ratios until it has a legal commitment to continue the lease. It is therefore reasonable to reflect capital repayment, return, and taxes associated with a lease renewal over the term of the renewed lease, and not throughout the period of the original lease.

Leases with a renewal or purchase option allow partial or incremental financing. This is equivalent to raising capital at some point in time in order to satisfy an initial need, and raising supplemental capital at a later time to satisfy the continuing need. Therefore, leases with renewal or purchase options may have an advantage over an outright purchase in that they delay capital requirements. Example 15-7 illustrates the analysis of leases with options.

Example 15-7. Compare the following alternatives:
  Plan A—Own: Same as Example 15-5.
  Plan B—Financial Lease: Same as Example 15-5.
  Plan C—Financial Lease with Purchase Option:

  Lease for 5 years at $1,200 per year

  Option to purchase at end of 5th year for $3,000

  Gross salvage at end of 10th year is $1,000

  Cost of removal at end of 10th year is $100

  Plan D—Financial Lease with Renewal Option:

  Lease for 5 years at $1,200 per year

  Option to renew for 5 additional years at $800 per year.

The capital cost of each plan is calculated as follows.
  Plan A—Own: PWE(capital costs) = $5,674
  Plan B—Financial Lease: PWE(effective capital cost) = $8,295
  Plan C—Financial Lease with Purchase Option:
    Lease portion:

$$PW(C_{rec}) = \$1,200(P/A, 9\%, 5)$$
$$= \$1,200(3.890)$$
$$= \$4,668$$

$$i_g = \frac{0.12 - 0.09}{1.09} = 2.75\%$$

$$PW(\text{effective } D_b) = \$1,200(P/F, 9\%, 6)(P/A, 2.75\%, 5)$$
$$= \$1,200(0.5963)(4.613)$$
$$= \$3,301$$

$$PW(T) = 0.6115(\$4,668 - \$3,301)$$
$$= \$836$$

$$PW(\text{effective capital cost}) = \$4,668 + \$836$$
$$= \$5,504$$

Purchase portion:

$$PW(C_{rec}) = \$3,000(P/F, 12\%, 5) - (\$1,000 - \$100)(P/F, 12\%, 10)$$
$$= \$3,300(0.5674) - \$900(0.3220)$$
$$= \$1,412$$

$$(D_b) = \frac{\text{first cost} - \text{net salvage}}{\text{life}}$$

$$= \frac{\$3,000 - (\$1,000 - \$100)}{5}$$

$$= \$420$$
$$PW(D_b) = \$420(P/A, 12\%, 5)(P/F, 12\%, 5)$$
$$= \$420(3.605)(0.5674)$$
$$= \$859$$

$$\% \text{ gross salvage} = \frac{\$1,000}{\$3,000} = 33.3\%$$

$$\% \text{ cost of removal} = \frac{\$100}{\$3,000} = 3.3\%$$

This is a purchase of used equipment eligible for accelerated depreciation under the 1.5 DB with a switch to the straight-line method. The ADR life is 80% of the service life.

$$PW(D_t) = \$3,000(P/D_t)(P/F, 12\%, 5)$$
$$= \$3,000\left[0.59724 - \frac{0.333 - 0.30}{0.40 - 0.30}(0.59724 - 0.52943)\right.$$
$$\left. + 0.033\,(0.56743)\right]0.5674$$
$$= \$1,010$$

$$PW(T) = 0.6115\,(\$1,412 - \$859) - 0.48(1 + 0.6115)(\$1,010 - \$859)$$

$$= \$221$$
$$PW(\text{capital cost}) = \$1,412 + \$221 = \$1,633$$
$$PWE(\text{financial lease with purchase option}) = \$5,504 + \$1,633 = \$7,137$$

Plan D—Financial Lease with Renewal Option:

Original Lease:

$$PW(\text{effective capital cost}) = \$5,504$$

(Note that results are the same as the lease with purchase option.)

Renewal:

$$PW(C_{rec}) = \$800(P/A, 9\%, 5)(P/F, 12\%, 5)$$
$$= \$800(3.890)(0.5674)$$
$$= \$1,766$$
$$PW(\text{effective } D_b) = [\$800(P/F, 9\%, 6)(P/A, 2.75\%, 5)]\,(P/F, 12\%, 5)$$
$$= \$800(0.5963)(4.613)(0.5674)$$
$$= \$1,249$$
$$PW(T) = 0.6115(\$1,766 - \$1,249) = \$316$$
$$PW(\text{effective capital cost}) = \$1,766 + \$316 = \$2,082$$
$$PWE(\text{lease with renewal option}) = \$5,504 + \$2,082 = \$7,586$$

Alternative	*PWE*
Plan A—Own	$5,674
Plan B—Financial lease	$8,295
Plan C—Financial lease with purchase option	$7,137
Plan D—Financial lease with renewal option	$7,586

The straight-purchase alternative (Plan A) logically should be more economical than the straight-lease alternative (Plan B), since the lessor can probably get the plant at the same cost available to the lessee, but must also realize a profit. The cost of a lease should therefore be higher than the cost of a direct purchase. The lease with a purchase option (Plan C) has an economic advantage over the lease with a renewal option (Plan D) for the same reasons that a straight purchase is usually more attractive than renting. The lease with a renewal option (Plan D) is more economical than the straight lease (Plan B), because the option to renew has deferred a portion of the firm's financial commitment until the end of year 5. However, this economic ordering of the alternatives could be reversed in some cases, depending upon the terms of the lease and the cost of money. As a result, it is important not to prejudge any of the alternatives without a careful economic analysis.

The decision about which lease option to exercise does not have to be made when the lease is taken. The current analysis favors owning (Plan A); but if owning is not practical for some other reason, the most attractive alternative is the lease with the option to buy. However, if this alternative is chosen, it does not mean that the firm should buy in 5 years. A change in the cost of money during the lease period could alter this decision. For this reason, the lease options should be evaluated separately when a decision needs to be made.

SPECIAL INCOME TAX CONSIDERATIONS FOR LEASES

The previous examples have treated the income tax effects of leasing in a very simple way. For most leases, this is usually adequate; but there are circumstances where some additional effects should be considered. These effects are the result of a lessor's option about investment tax credit and the possible capitalization of certain leases for tax purposes.

Investment Tax Credit for Leased Property

The federal income tax regulations permit the original buyer of property eligible for investment tax credit to transfer this credit to a lessee of that property, subject to certain restrictions. In effect, the lessor can elect to treat the lessee as the one who has made the investment in the property. If this option is elected, the lessor must provide the lessee with an election statement. This election statement may take the form of a blanket statement covering all property which the lessor has leased to the lessee during the taxable year, or it may be a property-by-property election for the individual units under the lease. This election is called *pass through* of investment tax credit.

Before a property can qualify for this pass through, it must be new eligible property and the lessee must be the original user of that property. The amount of credit is based on the fair market value of that property which the lessor has provided and its estimated useful life to the *lessor*, regardless of the terms of the lease or the ADR guideline life if the property were owned by the lessee.

In most cases, the amount of investment tax credit available under a lease should be the same as would be available if the asset were purchased. Therefore, in lease versus own comparisons of the same property, credit which is passed through can usually be excluded from both the owning and leasing alternatives. However, if the lessor does not elect to pass the credit through, the effect of the credit should be computed for the own option. Presumably, the lessor has included any credit effect in the computation of lease payments. In economic selection studies having an embedded lease that is not common to other considered alternatives, the investment credit passed on to the lessee by the lessor must be included. In the absence of additional information, the present worth of all lease payments may be used as a fair market value of the leased property.

Investment credit is realized in the year that the lessee takes physical possession, or control, of the property.

Because the internal revenue code may change from time to time, the analyst should check with the locally responsible organization to determine whether the leased property is eligible for investment tax credit—and if so, what investment-tax-credit rate applies. Example 15-8 illustrates an own versus lease study if the lessor elects to pass through investment tax credit.

Example 15-8. Solve Example 15-5 if the lessor passes investment-tax-credit benefits through to the lessee and raises the lease payment to $1,100 per month. The TC rate is 10%, based on the full $6,000 fair market value.

In Example 15-5 the lessor has already considered the effect of the investment tax credit in setting the $1,000 lease payment. Because of the election to pass the investment tax credit through to the lessee, the annual lease payments must be increased to reflect his loss of the tax credit.

This study could have been done by ignoring the common credit in the own and lease capital costs. However, if the full capital costs must be found, the credit effect can be included as below.

Own:

$$PWE = \$5,674$$

Lease:

$$PW(\text{effective capital cost}) = PW(C_{\text{rec}}) + PW(T)$$
$$PW(C_{\text{rec}}) = (\text{lease payment})(P/A, 9\%, 10)$$
$$= \$1,100(6.418)$$
$$= \$7,060$$

$$PW(T) = \phi[PW(C_{\text{rec}}) - PW(D_b)] - \frac{1}{1-t} PW(ATC) - (1 + \phi)[PW(TC) - PW(ATC)]$$

$$i_g = \frac{0.12 - 0.09}{1.09} = 2.75\%$$

$$PW(\text{effective } D_b) = (\text{lease payment})(P/F, i_d\%, N + 1)(P/A, i_g\%, N)$$
$$= \$1,100(0.3875)(8.640)$$
$$= \$3,683$$
$$TC = \$6,000 \times 10\%$$
$$= \$600$$
$$PW(TC) = TC(P/F, 12\%, 1)$$
$$= \$600(0.8929)$$
$$= \$536$$
$$ATC = \frac{TC}{\text{life}}$$
$$= \frac{\$600}{10}$$
$$= \$60$$
$$PW(ATC) = ATC(P/A, 12\%, 10)$$
$$= \$60(5.650)$$
$$= \$339$$

$$PW(T) = 0.6115(\$7,060 - \$3,683)$$
$$- \frac{1}{1 - 0.48}(\$339) - (1 + 0.6115)(\$536 - \$339)$$
$$= \$2,065 - \$652 - \$317$$
$$= \$1,096$$

$$PW(\text{effective capital cost}) = \$7,060 + \$1,096 = \$8,156$$
$$PW(\text{lease}) - PW(\text{own}) = \$8,156 - \$5,674$$
$$= \$2,482$$

As can be seen from the results of this example and Example 15-5, the difference in capital cost between owning and leasing is approximately the same whether the ITC is passed through or embedded in the lease payments.

Capitalized Leases

Accounting and income tax regulations require that certain kinds of financial leases be capitalized. Generally, this requirement applies to a lease which is considered to be an installment purchase or a conditional sales contract, rather than a normal lease. Very few leases are actually capitalized, and just because a lease has an option to purchase does not necessarily subject it to capitalization. The usual reason for doing so is that the lease payments are considered to be too rapid an amortization of the cost of an asset. This situation occurs if most of the lease payments are made over a period considerably shorter than the useful life, yet the lessee has the use of the property during most or all of the life. The tax deductions may then be restricted to the depreciation allowed for an owned asset. Analysts should get the advice of the company tax attorneys and accounting experts whenever there is any doubt in these matters, because accounting and tax treatments are not engineering decisions.

If a lease is capitalized, it is eligible for accelerated tax depreciation. These tax-depreciation calculations are the same as for owned assets, and are computed as described in Chapter 9; but book depreciation should be computed as indicated in this chapter—with a portion of the lease payment actually treated like book depreciation, and the remainder treated as if it were interest on debt. The equation is

$$PW(T) = \phi[PW(\text{lease payments}) - PW(\text{effective } D_b)]$$
$$- t(1 + \phi)[PW(D_t) - PW(\text{effective } D_b)]$$

Example 15-9 illustrates how to analyze a capitalized lease.

Example 15-9
    Own:

Same as in Example 15-5.

$$PW(\text{capital costs}) = \$5,674$$

Lease:

  Lease for 3 years at $2,450 per year

  Ownership transferred to lessee at end of 3d year

  $1,000 gross salvage at end of 10th year

  $100 cost of removal at end of 10th year

$$PW(\text{effective capital cost}) = PW(C_{\text{rec}}) + PW(T)$$
$$PW(C_{\text{rec}}) = PW(\text{effective first cost}) - PW(\text{net salvage})$$
$$= \$2,450(P/A, 9\%, 3) - (\$1,000 - \$100)(P/F, 12\%, 10)$$
$$= \$2,450(2.531) - \$900(0.3220)$$
$$= \$5,911$$
$$PW(T) = \phi[PW(C_{\text{rec}}) - PW(\text{effective } D_b)] - t(1 + \phi)[PW(D_t) - PW(\text{effective } D_b)]$$

$$i_g = \frac{0.12 - 0.09}{1.09} = 2.75\%$$

$$PW(\text{effective } D_b) = PW(\text{effective lease depreciation}) - PW(\text{effect of salvage})$$

$$= (\text{lease payment})(P/F, i_d\%, N + 1)(P/A, i_g\%, N)$$

$$-\frac{\text{salvage}}{\text{life}} (P/A, i\%, 10)$$

$$= \$2,450(P/F, 9\%, 4)(P/A, 2.75\%, 3)$$

$$-\frac{\$900}{10} (P/A, 12\%, 10)$$

$$= \$2,450(0.7084)(2.842) - \frac{\$900}{10} (5.650)$$

$$= \$4,424$$

$$\% \text{ gross salvage} = \frac{\text{gross salvage}}{\text{effective first cost}}$$

$$= \frac{\$1,000}{\$2,450(P/A, 9\%, 3)}$$

$$= \frac{\$1,000}{\$6,201}$$

$$= 16.1\%$$

$$\% \text{ cost of removal} = \frac{\text{cost of removal}}{\text{effective first cost}}$$

$$= \frac{10}{\$2,450(P/A, 9\%, 3)}$$

$$= \frac{\$100}{\$6,201}$$

$$= 1.61\%$$

$$PW(D_t) = \text{effective first cost } (P/D_t)$$

$$= \$2,450(P/A, 9\%, 3)\left[0.67656 - \frac{0.161 - 0.100}{0.200 - 0.100} (0.67656 - 0.63155) \right.$$

$$\left. + 0.0161(0.32197) \right]$$

$$= \$4,057$$

$$PW(T) = 0.6115(\$5,911 - \$4,424) - 0.48(1 + 0.6115)(\$4,057 - \$4,424)$$

$$= \$1,193$$

$$PW(\text{effective capital cost}) = \$5,911 + \$1,193 = \$7,104$$

If this lease were not capitalized, the tax depreciation would be the same as the book depreciation and the effective capital costs would be

$$PW(\text{effective capital cost}) = PW(C_{\text{rec}}) + PW(T)$$

$$= \$5,911 + 0.6115(\$5,911 - \$4,424)$$

$$= \$6,820$$

FLOW OF CASH AND ACCOUNTING PROCEDURES

Beginning of Period Payments

The formulas, financial tables, and all the calculations used in this chapter are based on the assumption that all transactions occur at the end of the year. However, lease terms usually call for each lease payment to be made in advance at the beginning of each period. Payments at the beginning of each year may be reflected by decreasing by one the number of lease payments applied in the formulas from N to $N - 1$, and then adding on the first lease payment.

This treatment can be applied to Example 15-5. For a lease with beginning of period payments of $1,000 per year for 10 years starting now,

$$PW(\text{apparent first cost}) = \$1,000(P/A, 9\%, 9) + \$1,000$$
$$= \$1,000(5.995) + \$1,000$$
$$= \$6,995$$
$$PW(\text{effective } D_b) = \$1,000(P/F, 9\%, 10)(P/A, 2.75\%, 9) + \$1,000$$
$$= \$1,000(0.4224)(7.878) + \$1,000$$
$$= \$4,328$$
$$PW(T) = 0.6115(\$6,995 - \$4,328)$$
$$= \$1,631$$
$$PW(\text{effective capital cost}) = \$6,995 + \$1,631$$
$$= \$8,626$$

The PW(effective capital cost) determined by using the end-of-year convention was $8,295. For this lease the difference in answers is only about 4%, which demonstrates that the assumption of end-of-year payments is sufficiently accurate for most study purposes.

Monthly Versus Yearly Compounding

Lease terms often require equal monthly payments. This would suggest that a monthly compounding period and a monthly rate of interest should be used in the own versus lease analysis. In fact, a firm accounts for most cash flows monthly—both inward flows from revenues, and outward flows for expense. Although cost studies based on these actual monthly flows might be more accurate, calculations based on monthly compounding are cumbersome. Annual rates will usually be accurate enough, and are therefore recommended. When annual compounding is used on a lease with monthly lease payments, the yearly rental is considered to be the sum of the actual monthly payments during the year. In any case, the same compounding period should be used for both the own and lease alternatives.

ADVANTAGES OF OWNERSHIP

Common sense dictates that leasing an asset should cost more than owning it, unless

- The lessor can buy or build at less cost than the owner.
- The lessor can borrow money at a lower rate, or is willing to finance at a lower rate, than the lessee must pay for debt.
- The lessor can take advantage of income tax "deductions" not usually available to the lessee.

Even when the costs of leasing compare favorably with ownership, owning has advantages which should be considered before a decision is made. Some of these are:

- The owner has full control of the property concerning such matters as rearrangements, additions, and retirements.

- The owner is free from uncertainty about future rental rates after the original agreement expires, and might avoid a costly relocation or other rearrangement.

- The owner is a part of the community. This often has advantages such as a voice in zoning or improvement proposals which cannot be reduced simply to dollars, but are nevertheless very important.

Some Exceptions

There are occasions when leasing is practical and economically sound, such as the leasing of temporary quarters or the leasing of certain types of equipment which may be subject to rapid obsolescence because of technological improvements. In these cases, leasing might be an economical way of avoiding unusual risk. But presumably a lessor will charge enough to provide protection from the risk of obsolescence, unless the lessor has access to secondary markets and in effect anticipates a higher salvage.

Another case where a lease might be justified would be when the usual sources of capital cannot, or will not, provide sufficient funds. Leasing can then be used as a supplementary source of capital. However, if leases are available, there probably is not a shortage of capital. The firm is just unwilling to pay the price asked by its usual sources; yet lessors often use the credit standing and financial reputation of the lessee to get the capital they need.

This situation is very difficult to quantify because lessors may be obtaining capital from sources that the firm would not normally use. The firm may wish to lease to increase its sources of capital, but this decision should be made by the financial experts. It is not a decision to be made by the analyst doing an engineering economy study.

SUMMARY

Leases fall into two basic categories for the purpose of making economic comparisons—operational and financial leases. Operational leases are a convenient way to meet temporary requirements, or to delay major commitments until it is possible to determine which mode of operation is actually appropriate. Operational lease payments are treated as operations costs. On the other hand, financial leases bind lessees to fixed payments over a considerable period of time, and are treated as an optional financing method. The fact that an own alternative exists for some lease defines the lease as financial.

This chapter presented three methods for analyzing a financial lease: by finding the implicit financing rate; by finding the apparent first cost through adjustments to the own option; and by finding the effective capital cost, with financial adjustments to the lease option.

In comparing financing rates, the measurement criterion is the interest rate which equates the present worth of the lease payments with the current cost of the purchase. Although the implicit financing rate is the simplest of the three methods to apply, it actually gives the analyst the least amount of economic information. Nevertheless, it is useful for eliminating leases which obviously cannot compete economically.

If the implicit cost of financing approach does not provide a clear economic choice, the apparent-first-cost approach can be used for further analysis. In this second approach, a purchase with 100% debt capital is assumed in order to keep a uniform financial structure between the own and lease alternatives. This approach allows the inclusion of the major income tax effects applicable to the own alternative.

The third and most versatile method is to compute the cost of ownership as in other studies, but to compute the lease costs as if the lease were a composite capital investment equal to the apparent first cost of the lease. Although this method is the most complex of the three approaches, it is also the most accurate because it takes into account income tax effects which the other two methods ignore.

THE FINANCIAL IMPACT OF INCURRING A LEASE

This appendix illustrates that the effective capital cost of a lease is equal to the present worth of the capital costs for a composite capital purchase equal to the apparent first cost of the lease (i.e., the present worth of lease payments at the cost of debt). The reasoning that leads to this conclusion is as follows:

- A lease commitment has the same priority for repayment (risk) and impact on a firm's financial structure as debt-type financing. (See Chapter 4.)

- Undertaking a lease is therefore equivalent to making a 100% debt capital purchase (P) equivalent to the present worth (at the cost of debt) of the lease payments.

- Consistent comparisons between alternatives require the assumption that a firm will maintain a constant debt ratio (r) regardless of its investment decisions.

- In order to maintain this debt ratio after committing to a lease, the firm will increase its level of equity financing (and thus decrease its level of debt financing) in other projects. The lease plus the financial adjustment is equivalent to a composite financing of an amount P. This conclusion is illustrated in the following development:

Present worth of lease payments at debt interest rate = effective capital investment
$$= P$$
Total capital invested in firm before undertaking lease
$$= \text{(debt capital before lease)} + \text{(equity capital before lease)}$$
$$= C$$
Debt capital before lease = (total capital before lease) × (debt ratio)
$$= D = C \times r$$
Equity capital before lease = (total capital before lease) × (1 − debt ratio)
$$= E = C \times (1 - r)$$
Total capital invested in firm after lease
$$= \text{(present worth of lease payments)} + \text{(capital invested in firm before lease)}$$
$$= C + P$$
Total allowed debt after lease to maintain desired debt ratio
$$= \text{(total capital invested after lease)} \times \text{(debt ratio)}$$
$$= (C + P) \times r$$
Total allowed equity after lease to maintain desired debt ratio
$$= \text{(total capital invested after lease)} \times \text{(1 − debt ratio)}$$
$$= (C + P) \times (1 - r)$$

Change in debt capital allowed = (allowed debt) − (debt before lease)
$$= (C + P) \times r - C \times r = P \times r$$
Change in equity capital allowed = (allowed equity) − (equity before lease)
$$= (C + P) \times (1 - r) - C \times (1 - r) = P \times (1 - r)$$
Change in total capital after lease = effective capital investment
$$= \text{(change in equity capital)} + \text{(change in debt capital)}$$
$$= [P \times (1 - r)] + (P \times r) = P$$
Debt ratio of effective capital investment
$$= \text{(total change in debt capital)} \div \text{(total change in capital)}$$
$$= (P \times r) \div P = r$$

- The total revenue requirements for a lease are equal to the revenue requirements for a composite capital purchase, which is equal in size to the present worth of lease payments at the cost of debt.

- With a lease, the capital repayment schedule is specified by the lease payments; and it must be used instead of the straight-line depreciation normally assumed for composite capital purchases.

The following numerical example illustrates the above points.

Example 15A-1. Assume that a firm with the following characteristics undertakes a lease of $1,164.14 per year for 3 years,

Equity required rate of return	16%
Debt interest rate	8%
Desired debt ratio	50%
Tax rate	50%

If the firm leases the asset, the revenue required for the lease is $1,164.14 in each of the 3 years. However, the firm has committed itself to a debt-type obligation equivalent to borrowing. At its present debt interest rate, this obligation is

$$\$1,164.14(P/A, 8\%, 3) = \$3,000$$

To maintain its debt ratio, the firm must revise its financing elsewhere. For this illustration, "elsewhere" can be assumed to be a single project with a life equal to the lease. Actually, it would be the combined operation of the total firm over time.

Suppose the firm has to make a separate $11,000 capital expenditure. Normally, it would finance 50% debt and 50% equity. Because of the lease, however, it must do otherwise in order to restore the 50% *overall* debt ratio, including the lease:

Total outstanding capital = $11,000 + $3,000 =	$14,000
Debt ratio	× 0.50
Total allowed debt	$ 7,000
Lease incurred debt	−3,000
New debt permitted in $11,000 purchase	$ 4,000

Therefore, this investment must be financed with $4,000 of debt capital and $11,000 −
$4,000, or $7,000, of equity capital. In the first year, the revenue required for the
$11,000 project is

16% of $7,000	$1,120.00 for equity
8% of $4,000	320.00 for debt
0.5/(1 − 0.5) × 1,120.00	1,120.00 for taxes
11,000/3	3,666.67 for depreciation
Total	$6,226.67

(See Table 15A-1 on the next page.)

At the beginning of the second year, two more lease payments must still be made.
They now represent $1,164.14(P/A, 8%, 2) = $2,075.66 worth of debt commitment. The
other project has repaid $3,666.67 of its capital, so that $11,000 − $3,666.67 =
$7,333.33 remains unrecovered. To maintain the desired debt ratio, the total of
$7,333.33 + $2,075.66 = $9,408.99 must contain 50% debt; that is, $9,408.99 × 0.50 =
$4,704.50 of total debt. Therefore, the amount of debt left unrecovered in the second
project is ($4,704.50 − $2,075.66) = $2,628.84. The outstanding equity capital is
($7,333.33 − $2,628.84) = $4,704.49. Now the revenue required is

16% of $4,704.49	$ 752.72 for equity
8% of $2,628.84	210.31 for debt
0.5/(1 − 0.5) × $752.74	752.72 for taxes
11,000/3	3,666.67 for depreciation
Total	$5,382.42

By applying the same reasoning to the third year, the financial situation is:

Lease commitment = $1,164.14(P/A, 8%, 1)	= $1,077.99
Unrecovered capital in $11,000 purchase = $7,333.33 − $3,666.67 =	3,666.66
Total	$4,744.65
Debt ratio	× 0.50
Debt portion	$2,372.33
Equity portion = $4,744.65 − $2,372.33	$2,372.32

This leaves $2,372.33 − $1,077.99 = $1,294.34 unrecovered debt in the second project.
Revenue required in the final year for the $11,000 project is

16% of $2,372.32	$ 379.57 for equity
8% of 1,294.34	103.55 for debt
0.5/(1 − 0.5) × 379.57	379.57 for taxes
11,000/3	3,666.67 for depreciation
Total	$4,529.36

Total revenues required are found by adding the revenue required for lease payments to
the revenue required for the adjusted second project, as calculated above. This is
shown in Table 15A-1 (columns A to L). If the second project had been financed

TABLE 15A-1

REVENUE REQUIRED FOR $11,000 PURCHASE + LEASE + ADJUSTMENT TO MAINTAIN A 50% DEBT RATIO

Year	PW of outstanding lease commitments[a] A	Unrecovered capital from $11,000 purchase[b] B	Total outstanding capital $C = A + B$	Total allowed debt[c] $D = 50\%$ of C	Allowed debt in $11,000 purchase[a,c] $E = D - A$	Required equity capital in $11,000 purchase[a] $F = B - E$	Return on equity capital[d] $G = 16\%$ of F	Return on debt capital[e] $H = 8\%$ of E	Taxes[f] $I = \dfrac{0.5}{1 - 0.5}\,G$	Depreciation of $11,000 purchase[b] $J = \$11,000/3$	Lease payment K	Total revenue required $L = K + G + H + I + J$
1	3,000.00	11,000.00	14,000.00	7,000.00	4,000.00	7,000.00	1,120.00	320.00	1,120.00	3,666.67	1,164.14	7,390.81
2	2,075.66	7,333.33	9,408.99	4,704.50	2,628.84	4,704.49	752.72	210.31	752.72	3,666.67	1,164.14	6,546.56
3	1,077.99	3,666.66	4,744.65	2,372.33	1,294.34	2,372.32	379.57	103.55	379.57	3,666.67	1,164.14	5,693.50

REVENUE REQUIRED FOR $11,000 PURCHASE + LEASE WITH NO ADJUSTMENT TO REFLECT FINANCIAL LEASE

Year	Outstanding capital[b] M	Total return[g] $N = 12\%$ of M	Taxes[h] $O = \phi \times N$	Depreciation[b] $P = \$11,000/3$	Lease payment Q	Total revenue required $R = N + O + P + Q$
1	11,000.00	1,320.00	880.04	3,666.67	1,164.14	7,030.85
2	7,333.33	880.00	586.70	3,666.67	1,164.14	6,297.51
3	3,666.66	440.00	293.35	3,666.67	1,164.14	5,564.16

FINANCIAL IMPACT OF LEASE

Year	Adjustment to maintain 50% debt ratio $S = L - R$	Total annual cost attributed to lease $T = S + Q$
1	359.96	1,524.10
2	249.05	1,413.19
3	129.34	1,293.48

[a] Lease considered equivalent to 100% debt obligation

[b] Straight-line depreciation

[c] 50% debt ratio

[d] Required equity return = 16%

[e] Debt interest rate = 8%

[f] $\text{Tax} = \dfrac{t}{1 - t} \times$ equity return

[g] Composite cost of capital = 12%

[h] $\phi = \dfrac{(i - i_d r)t}{(1 - t)i} = \dfrac{[0.12 - (0.08 \times 0.5)]0.5}{(1 - 0.5)\,0.12} = 0.6667$

TABLE 15A-2
LEASE CALCULATIONS USING THE EFFECTIVE-CAPITAL-COST METHOD

Year	Outstanding capital[a] A = PW (future lease payments)	Return[b] B = 12% of A	Depreciation[c] C = (1,164.10 − 0.08A)	Taxes[d–g] $D = \phi B$	Total annual cost attributed to lease $E = B + C + D$
1	3,000.00	360.00	924.14	240.01	1,524.15
2	2,075.66	249.08	998.09	166.06	1,413.23
3	1,077.99	129.36	1,077.90	86.24	1,293.50

[a]Cost of debt = 8%

[b]Composite cost of capital = 12%

[c]Lease payment − (debt interest rate × outstanding capital)

[d]$\phi = \dfrac{(i - i_{dr})t}{(1 - t)i} = \dfrac{(0.12 - 0.08 \times 0.5)0.5}{(1 - 0.5)0.12} = 0.6667$

[e]Debt ratio = 50%

[f]Income tax rate = 50%

[g]Tax = total return × ϕ

normally, the revenue required would have been as calculated in columns M to R of the table. Subtracting the revenue requirement for a normally financed second project and the lease payments (column R) from the total derived in columns A to L leaves the revenue requirements representing the financial adjustments necessary to maintain the desired debt ratio (column S). The total of the annual revenue requirements attributed to the lease is the lease payment plus this cost of financial readjustment (column T). These same revenue requirements are computed using the effective-capital-cost method in Table 15A-2.

EFFECTIVE DEPRECIATION UNDER EQUAL PERIODIC LEASE PAYMENTS

In the familiar case of *equal* periodic lease payments over the life of the lease, the periodic effective depreciation (the capital portion of each lease payment) takes on a convenient form for calculation.

Each lease payment consists of two components: a capital repayment portion (repayment of the lessor's investment) and a return component (the lessor's return on the investment). The interest portion *decreases* each period in proportion to the capital repaid in the previous period. With equal periodic lease payments, the capital recovery portion must *increase* each period in the following way:

$$\text{Capital portion in Period 1} = C_1$$
$$\text{Capital portion in Period 2} = C_2 = C_1 + C_1 i_d = C_1(1 + i_d)$$
$$\text{Capital portion in Period 3} = C_3 = C_2 + C_2 i_d = C_1(1 + i_d)^2$$
$$\text{Capital portion in Period } n = C_n = C_1(1 + i_d)^{n-1}$$
$$\text{or: } C_n = C_0(1 + i_d)^n$$

where C_0 is the value at time base zero of a geometric progression increasing geometrically at the rate i_d. (Chapter 5 discussed geometric progressions.) This is also the identical form of the annuity or sinking-fund depreciation.

The value of C_0 may be determined from what is known about the last lease payment; and, knowing this value at time base zero, the present worth of the geometric progression (the present worth of the total effective depreciation) can be determined.

The *last* lease payment, occurring in period N, will be just enough to provide for the remaining unrecovered capital and the interest on that capital.

For the *last* payment in period N,

$$\text{Interest portion}_N = \text{outstanding capital}_N \times i_d$$

The capital repayment (effective depreciation) is equal to the lease payment minus the interest portion.

$$\text{Effective depreciation}_N = \text{lease payment}_N - \text{interest portion}_N$$
$$= \text{lease payment}_N - (\text{outstanding capital}_N \times i_d)$$

But since the effective depreciation in period N equals the capital outstanding in period N:

$$(1 + i_d) \times \text{effective depreciation}_N = \text{lease payment}_N$$
$$\text{Effective depreciation}_N = \frac{\text{lease payment}_N}{(1 + i_d)}$$

The final term of the geometric progression, then, has a value of

$$C_N = \frac{\text{periodic lease payment}}{(1 + i_d)}$$

By using the expression just developed for C_N, this term may also be expressed as

$$C_0(1 + i_d)^N = \frac{\text{periodic lease payment}}{(1 + i_d)}$$

$$C_0 = \frac{\text{periodic lease payment}}{(1 + i_d)(1 + i_d)^N}$$

Recall that

$$\frac{1}{(1 + i_d)^{N+1}} = (P/F, i_d\%, N + 1)$$

The base of the geometric progression is therefore

$$C_0 = (\text{lease payment})(P/F, i_d\%, N + 1)$$

The effective depreciation over the term of the lease is a geometric progression having a base of C_0. According to Chapter 5, the present value at interest rate i of a geometric progression increasing at the periodic rate i_d for N periods is the base value times $(P/A, i_g\%, N)$, where i_g is defined as

$$i_g = (i - i_d)/(1 + i_d).$$

Therefore, the present worth (at the composite cost of capital, i) of the total effective depreciation for a lease having equal periodic payments for N periods, is

$$PW(\text{effective depreciation}) = (\text{lease payment})(P/F, i_d\%, N + 1) \times (P/A, i_g\%, N)$$

16

PRESENTING AND EVALUATING
AN ECONOMY STUDY

INTRODUCTION

Managers are often asked to make recommendations or investment decisions based upon economy studies which they themselves have not made, but which their subordinates or other analysts have performed. These studies should give the manager whatever information is required for making a valid choice among alternatives involving different amounts of capital expenditures, expenses, and revenues, and the presentation of these results should be in a format that is as helpful as possible. However, in order to have confidence in the recommendations and to reach a proper decision, the manager must then objectively evaluate the study according to consistent criteria.

An economy study consists of six basic elements:

1. A definition of the objective of the study

2. A description of the alternatives considered

3. A description of the data used

4. The study techniques employed

5. The study results

6. The plan recommended

The integrity of the results of any study therefore depends on how thorough a job the analyst has done in considering all viable alternatives, on the validity of the data used, and on whether the study technique was correctly applied. In addition, the plan recommended must be compatible with corporate goals and the prevailing economic, regulatory, social, and technological environment.

This chapter explains how the manager evaluates all six elements of an economy study. It also gives guidelines for presenting the results. In the Summary, there is a set of questions which a manager will find helpful to use as a checklist when reviewing a study, to make sure that all the relevant factors were adequately considered.

THE OBJECTIVE

The first step in evaluating an economy study is to identify the reason for the study. What is the problem? Where is the problem? Why is there a problem? What is the extent of the problem?

The analyst should have answered all of these questions concisely in the preface of the study. Once the purpose of the study has been established, subsequent analysis will determine if the problem can be solved, and, if so, at what cost. The objective should also establish the limits on the scope of the study.

ALTERNATIVES CONSIDERED

Before any courses of action can be identified, it is necessary to define the total environment in which the alternatives will be evaluated. Then, those plans which are compatible with this environment, and which will also perform the same function, can be compared on the basis of their relative economics. Courses of action that are not compatible with the environment can be rejected outright.

Once the relevant study environment has been defined, it is important that the analyst consider all viable courses of action. Each reasonable plan which is discarded for any reason should be backed up with adequate documentation explaining why it was rejected. Such comprehensiveness adds credibility to the final recommendation. Furthermore, it often happens that an alternative which at the outset does not appear to be economically competitive will, after detailed analysis, prove to be very attractive indeed. In those cases where there are too many viable courses of action to evaluate separately, the rationale should be stated, explaining why certain of these alternatives were selected for detailed analysis. Nothing can discredit a study more than for the manager to suggest reasonable alternatives to the recommended plan which the analyst had never considered.

It is also very important that all the alternatives which are considered be truly comparable. Each alternative in an economy study should perform essentially the same service function over the same period of time. For example, it is not reasonable to compare the cost of buying a minicomputer that has a 5-year service life and that will perform a specific accounting function with the full costs of buying a large-scale computer that has a 7-year service life and that will serve a multiplicity of corporate computer needs.

To summarize, in evaluating an economy study, the manager must know that all viable alternatives have been considered, and that those courses of action which are presented are indeed comparable.

DATA

As this book has emphasized, the results of an economy study are only as good as the data used in the study. Even though the analyst has a large degree of control over the study technique and the alternatives, often the data to be included in the study will come from numerous sources, not all having the same degree of reliability. In reviewing a study, the manager must be confident that the data used are reasonable, reliable, and consistent with other known facts. To help assure this confidence, sources of data should be documented and all assumptions stated and justified.

Specific data about future events are usually not available, and an estimate or projection is needed. However, reliable forecasting is a sophisticated technique which may require the help of qualified specialists.

In short, the integrity of the entire study depends on using the most appropriate and accurate data for the specific circumstances and expenditures being considered. Of course, the collection and verification of the data can be time-consuming, and it may be very attractive to shortcut the study process by using data from the plant account average. However, this cannot be done if the study is to have meaningful results. As its name implies, the *account average* is the composite experience of an account. It would be only a coincidence if the plant being considered required average maintenance, lived an average life, and was average in every other way. Studies are usually made of specific plant, with specific plans for service. Specific placing and replacing are therefore expected. Many accounts include quite different kinds of plant; as an example, the only thing that a repeater hut and a large toll building have in common is the plant account they share. Similarly, although a small car and a large truck are very different, both are included in the same account, the motor vehicle account. Even in an account with a more homogeneous composition, lives may vary widely. The account average unit cost may not be useful because in the study the placement of a single unit may require a higher cost per unit. Or, if a very large addition is being planned, a single unit may well have a lower-per-unit cost than the average. The analyst should investigate details about the life, salvage, cost, and maintenance of the specific plant being studied until these parameters are reasonably well established. A thorough familiarity with the equipment and the plans for the future will help establish these parameters.

The study data may be influenced by concurrent projects or events which are not included in the study. Even though projects are usually studied in isolation, the analyst must consider whether concurrent projects are actually independent—and, if they are not independent, how they interact. For example, there may be conflicts in the availability of space, facilities, capital, labor, and other resources.

Besides being confident that the data are reasonable, reliable, and consistent, the manager must determine that *all* relevant economic factors have been considered. Nothing should be disregarded without consideration. Factors which on the surface appear insignificant may, upon detailed examination, have a substantial economic impact. The analyst should specifically note any economic factors which are eliminated from the study because they have a common effect on the alternatives being considered.

Some of the major economic factors to be considered in a study are:

- Capital expenditures
- Revenues
- Income taxes
- Salvage
- Inflation
- Tax depreciation
- Book depreciation
- Rent
- One-time expenses
- Recurring expenses
- Maintenance expense
- Cost of removal
- Investment tax credit

- Debt interest
- Equity return
- State, local, and miscellaneous taxes
- Ad valorem taxes
- Interest during construction
- Deferred tax normalization and/or flow through
- Capital gains taxes
- Gross receipts taxes

Some of the many sources of data about these factors are

- General letters
- Continuing property records
- General traffic engineering forms
- Job record sheets
- General planning forecasts
- Broad-gauge manuals
- Supply catalogs
- AT&T staff organizations
- Engineering letters
- Other engineering economy texts
- Bell Laboratories letters
- Bell System practices
- Traffic facility charts
- General trunk forecasts
- Western Electric brochures and planning aids
- Other operating telephone company organizations
- Advertisements
- Manufacturer's representatives
- Company reports
- Trade journals

Organizational responsibilities will differ from company to company, and the availability of locally generated data will vary too. Where to go for what information is usually a matter of informed judgment, which can be developed only through experience. However, the study should make clear where and/or from whom the analyst has gathered the information for generating all the cost data described.

STUDY TECHNIQUES

The objective of an economy study is to describe the economic impact that alternatives will have on the firm. Alternatives can be analyzed and compared by the use of a mathematical model, because this is an inexpensive way to examine a large number of alternatives without actually making a commitment of resources. In fact, only a few algebraic expressions may be needed to represent the very complex economic structure of the firm. The choice of algebraic expressions depends on which objectives are to be evaluated. Several criteria are possible, depending upon the circumstances of a particular study, and they are probably not compatible. Generally, it is not possible to optimize two or more objectives simultaneously—for example, maximizing service quality versus minimizing cost.

Chapter 12 describes in detail three approaches for performing economic analysis in the Bell System:

- The internal rate of return

- The present worth of expenditures

- The present worth of annual costs

As Chapter 12 pointed out, for a given set of mutually exclusive alternatives, the same plan would be selected regardless of which approach is used. However, the numbers in the input and the output would normally appear different because each method represents a somewhat different viewpoint. When evaluating a study, then, the manager should first note which technique was used and then make sure that this technique was correctly applied. These approaches are outlined next, as a review.

The Internal Rate of Return

The *internal-rate-of-return* approach (*IROR*) views projects from the investor's point of view. Internal rate of return is the critical cost of money at which a project breaks even. However, an *incremental* internal rate of return is required before this method can be used to decide among alternatives.

In its simplest form, the incremental internal rate of return can be defined as the interest rate which causes the present worth of the incremental cash flow between two projects to be zero. If the incremental cash flow represents an investment followed by its recovery, the internal rate of return is then the break-even cost of money. If money costs less than the internal rate of return, the incremental investment is beneficial. If money costs more, the incremental investment is not beneficial. However, the internal rate of return should not be misinterpreted as the profitability of a project, or of an incremental investment, because it does not provide any information about the magnitude of the benefit at the firm's actual cost of money.

It is important to note that the *IROR* approach can be applied only in deciding between two mutually exclusive alternatives. When more than two alternatives must be evaluated, it is necessary to make sure that all the pairs were compared. Knowing the internal rate of return between Plan A and Plan B, for example, or the internal rate of return between Plan C and Plan B, tells little about the comparison between Plan A and Plan C. Plan A and Plan C must be compared directly. As the number of alternatives increases, the number of comparisons required increases geometrically.

When there is no differential revenue between two competing plans, the incremental internal rate of return can be used to compare the two investments on the basis of cost savings. However, when there is a differential revenue, this must be included in the analysis as well.

A critical problem which may arise from the use of the internal-rate-of-return technique is the possibility of multiple answers, or misleading ones. This is a consequence of the algebraic formulation arising when the annual cash flows occur in certain patterns, and the incremental cash flow does not meet the criteria necessary for a valid internal rate of return on an investment. Nevertheless, the problem still requires a solution, and the *modified-rate-of-return* (*MROR*) (described in Appendix B to Chapter 12) will usually provide a suitable answer. The modified-rate-of-return technique is a variation of the internal-rate-of-return (*IROR*) method, and the results found by using it may be viewed from the same perspective. If the problem does not require an internal rate of return type of measurement, the present-worth-of-expenditures or the present-worth-of-annual-costs techniques will also indicate the correct decision.

In summary, as Chapter 12 emphasized, managers must exercise extreme caution in using the *IROR* technique. The *IROR* is not a measure of profitability, two *IROR* numbers cannot be compared directly, and *IROR* is not a theoretically valid project-ranking tool. The method is complex to use, and the results may be more confusing than enlightening. In the hands of the uninformed or the misinformed, decisions based on *IROR* data can be dangerous. Consequently, the interpretation of *IROR* data should be left to experts.

The Present Worth of Expenditures

The *present-worth-of-expenditures* (*PWE*) technique approaches an economy study from the customer's viewpoint. This method will select the alternative with the lowest revenue requirements, because its objective is to find which alternative will meet the service objectives at the least cost. The decision is based on the lowest total present worth of expenditures over the life of the study.

The *PWE* technique is based on the requirement that the firm's long-run rate of return will be realized, and that all revenues required to pay capital and operating costs will be generated to support whichever alternative is selected. Of course, the results will be valid only for the cost of money used in the analysis.

The *PWE* may be calculated for incremental differences among alternatives when one is a null alternative, or when all have cash flows. It avoids the problems of multiple or misleading answers which can result from an *IROR* analysis. Furthermore, the absolute *PWE* values for each alternative can be compared directly, and incremental analysis is not required. However, absolute *PWE* values for various alternatives are influenced by the treatment of common costs, even though differences in the *PWE* values are not. As a result, the significance of *PWE* differences may be difficult to determine.

The Present Worth of Annual Costs

The *present-worth-of-annual-costs* (*PWAC*) method also approaches an economy study from the customer's viewpoint. However, with this technique, the actual expenditures are represented by a levelized equivalent annual amount which in effect represents the costs the customer sees when paying a constant monthly bill for service. The decision is based on the lowest present worth of equivalent annual costs over the life of the study. When the present worth of annual costs is calculated over the life of an investment, the *PWAC* approach is equivalent to the *PWE* approach and the same rationale and constraints apply. The difference is that if the study period and the various lives of investments are not coincident, the *PWAC* technique implies an allocation of costs over the study period, whereas the *PWE* technique requires explicit forecasts and cost allocations.

The *PWAC* technique has been very popular, and for two reasons. The first is that it encourages equipment to be grouped into specific categories; then average lives, salvage values, maintenance costs, operating costs, and ad valorem tax rates are determined for each category. Once these averages are determined, it is a relatively simple matter to calculate an annual cost rate as a percentage of the total installed cost for each different category. These all-inclusive annual cost rates can then be used to put together a *PWAC* study very quickly. However, as this text has emphasized, the use of data from plant account averages has certain pitfalls and should be exercised with caution.

The second reason for the popularity of the *PWAC* method is that working with annual costs often simplifies the treatment of noncoincident equipment placements and retirements. While this is often a convenient simplification, it may conflict with the necessity of including the effects of changing costs due to inflation.

Other Methods

The three approaches described in Chapter 12 have been recommended because they are consistent with Bell System objectives, and each other. However, several other methods can be used to analyze competing investments. Many of these are special application techniques, or variations of the methods discussed in Chapter 12. A complete discussion of all of these methods is beyond the scope of this text, and other publications should be consulted for detailed explanations.

Before accepting a recommendation based on any approach, however, the manager must understand the objectives of the approach and be able to interpret the results in terms of them. Some of the other popular approaches are described next.

Benefit-to-Cost Ratio

The ratio of the benefits of an alternative to its costs is a frequently used economic indicator. There are many forms of benefit-to-cost ratios based on various definitions of costs and benefits. In one form the benefit-to-cost ratio is consistent with Bell System objectives and it can be particularly useful for capital budgeting. This is because it permits the comparison of alternative solutions to different projects.

The benefit-to-cost ratio is a measure of overall efficiency. Efficiency is defined as

$$\text{Efficiency} = \frac{\text{output}}{\text{input}} = \frac{\text{benefit}}{\text{cost}}$$

For a given net cash flow, the output is the positive net flows and the input is the negative net flows. The benefit can be defined as the present worth of the positive net cash flows, and the cost as the present worth of the negative net cash flows. The benefit-to-cost ratio is then

$$\frac{\text{Benefit}}{\text{Cost}} = \frac{PW(\text{positive net cash flows})}{PW(\text{negative net cash flows})}$$

The Chapter 12 discussion of *IROR* pointed out that a net cash flow is only attractive if its *net present value* (*NPV*) is positive at the firm's cost of money. The *NPV* is the present worth of positive flows less the present worth of negative flows.

This relationship can be expressed mathematically:

$$NPV = PW \text{ (positive NCF)} - PW \text{ (negative NCF)}$$

or $$PW \text{ (positive } NCF) = NPV + PW \text{ (negative } NCF)$$

Therefore, $$\frac{\text{Benefit}}{\text{Cost}} = 1 + \frac{NPV}{PW \text{(negative NCF)}}$$

If this ratio is greater than 1.0, the alternative which results in the cash flow is generating enough cash to recover the invested funds with the required return, and to contribute additional funds to the business. The most efficient alternatives will have a ratio much greater than 1 and less efficient alternatives will have a lower ratio.

Similar to the *IROR*, this ratio can be used to indicate the most economic alternative of a set of mutually exclusive alternatives. However, the choice is not simply the alternative with the greatest benefit-to-cost ratio over some base alternative. Like *IROR*, the incremental benefit-to-cost ratios between pairs of alternatives must be used. As a result, the use of the benefit-to-cost ratio for alternative selection is almost as complex as the use of incremental *IROR*.

However, unlike *IROR*, benefit-to-cost ratios can be used to order the various alternative solutions of many different projects for capital budgeting. Through a rather complex procedure, those alternative solutions for each project can be selected which will maximize the overall efficiency of the use of the limited available capital. This ordering of project alternatives should not be considered a ranking list of the various alternatives. Depending on the amount of capital available, a high-ordered alternative may be superseded by a lower-ordered alternative for the same project. This occurs because the lower-ordered alternative is incrementally more efficient than the most efficient alternative for some other project further down the list. The result is a series of choices which may reject the most economic alternative of some projects because of capital limitations.

Payback (Payout) Period

The payback period is often used as a criterion for investment decisions, because it is relatively easy to calculate and because a manager instinctively would like to recover capital as quickly as possible. The payback (sometimes called the *payout*) period is the number of years required for the cost savings and receipts from an investment to equal the amount of the investment with no interest. It is a measure of the speed with which invested funds are returned to the firm.

When it is used as the sole criterion for evaluating a project, an uneconomical alternative may be chosen which has high cash flow in the early years. In calculating the payback period, the investments, savings, and earnings may be converted to present values. In that case, the payback period is called the *discounted payback period*. Payback, even discounted payback, ignores all events which occur after the payback itself.

Break-even Study

Often, one alternative will be more economical under one set of conditions; another alternative more economical under a different set of conditions. By altering one variable and holding all other factors constant, it is possible to find a value for the variable which will make the competing investments economically equivalent. This is called the *break-even point*. When it is known, the analyst can estimate on which side of this point the future conditions are *most likely* to occur, and can then choose the most attractive alternative. Break-even analysis can be helpful when the break-even variable is difficult to quantify. For example, it can be applied to such factors as cost savings, inflation, size of plant, revenues, and maintenance costs. However, evaluating alternatives by directly comparing them will be more illuminating than break-even analysis, because it will indicate the magnitude of the consequences of being on one side or the other of the break-even point. The break-even point may be helpful in a different way when the manager can in fact control some of the parameters of a study. For example, the break-even point for the time of placing, the size of plant, and similar factors may be helpful in formulating plans.

Probabilistic Approaches

Although some business decisions can be made under conditions which approach certainty, most decisions involve a certain amount of risk and uncertainty. Generally, probabilistic approaches require the manager to subjectively or objectively determine the probability of various events (input data); then, after considering the statistical relationships among these events, predict the relative likelihood of the possible outcomes (results). Some of the more common models used to reflect conditions of risk and uncertainty are

- Expected monetary value
- Conditional profit
- Loss-minimization
- Marginal analysis

The details of these approaches are beyond the scope of this text, and will be left to other publications. Furthermore, a comprehensive knowledge of probability and statistics is required to correctly apply and interpret the results. To the untrained, the results of probabilistic analysis can be misleading and confusing indeed.

Utility Functions and Values

In general, *utility* is defined as the benefit, or subjective value, to be derived from particular outcomes. In economics, utility functions are represented by mathematical expressions indicating the amount of risk an investor is willing to take in order to achieve a specific return. Of course, the utility value of an investment varies from investor to investor, and depends on many factors, including the environment, the alternatives considered, personal preferences, and the investor's economic position. It represents a *subjective* estimate of the investor's attitudes about risks and incremental costs and revenues. Sometimes, utility functions are used in an attempt to quantify noneconomic considerations, such as public opinion and environmental factors. However, generally they are applied using probabilistic analysis techniques, and involving complex statistical functions which are beyond the scope of this text.

STUDY RESULTS

The manager should be able to see that the study results are a logical conclusion based on the data. Since the results are transmitted by way of the presentation, the presentation may be the most important part of an economy study. It is described next.

Presentation

The purpose of the economy study is to help managers decide which plan is most economical, and the same care and planning should be given to the presentation as to the analysis. It is the presentation which gives the conclusions of the analysis in tabular, narrative, or graphic form, and with enough details to establish the integrity of the study.

Although many different formats can be used to present a study, there are a few basic principles to follow. The most important one is that a study should be presented in different levels of detail, depending on which managers will use the study. It is important to keep in mind that different managers have different goals, and therefore have a need for different degrees of detail.

The presentation should include a written narrative progressing from the most general to more and more specific. The most general level is the conclusion. It may be as simple as, "of all the plans considered, Plan A is the least costly to the firm over a 30-year period." Or perhaps the conclusion is to defer a decision.

Next, each conclusion needs to be qualified, but in just a simple statement. A short beginning paragraph can include the conclusions, any qualifications needed, as well as enough information to identify the project being considered. This one paragraph may be all that is required for some managers. Others will require greater and greater detail, possibly even including the working papers used in the study.

The narrative can be continued to explain the study conclusions in more detail for those managers who need it. Sufficient documentation should be included to support any assumptions and projections that were made in doing the study. This can include tables of data, as well as narrative discussion, but graphical illustrations are often a more effective way to present the conclusions. Some typical graphs that can be used are explained next.

Cash-Flow Diagram

The basic graph of all economy studies is the cash-flow diagram introduced in Chapter 5 and used throughout the text. This simple diagram gives the timing, quantity, and direction of the flow of funds in a project being analyzed. Often, it can be used to compare several projects, because it shows so clearly the magnitude and timing of the net cash outlays. Although these data can be presented in the narrative, they are certainly easier to grasp intuitively in cash-flow diagrams because the relationship between savings and expenses is available at a glance.

Sensitivity Graph

Analyzing various alternatives under future economic conditions will involve many variables having complex relationships, and it is often difficult to demonstrate the effect that each variable could have on the conclusions. However, a sensitivity graph allows readers to compare the effect of one variable on a number of plans at one time. For example, suppose A, B, and C are three switching technologies considered for an area with an expected growth rate of G_2. An analysis made at growth rate G_2 will require additions to plant over some period of time that will meet the expected demand for service. The conclusions of the *PWE* analysis are that Plan B has the lowest *PWE*. This is often the single conclusion of such a study. But if the growth rate is different the conclusion may change. This relationship can be described if the *PWE*s are found and plotted over a range of growth rates. Figure 16-1 is an example of a sensitivity graph of three plans, with the *PWE* compared with the growth rate.

FIGURE 16-1
Comparisons of PWE with Growth Rate

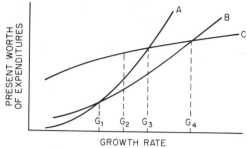

This type of graph shows plainly that the question of which technology is the most economic is sensitive to the growth rate to be experienced. Plan B is the economic choice only if growth lies between G_1 and G_4. For growth rates less than G_1, Plan A is the winner; if the growth is greater than G_4, Plan C is the best. Presented in this manner, the study does not have to substantiate G_2 as the growth rate, because it illustrates that the decision would be the same over a range of growth. Management can then better exercise its judgment as to which plan may be superior, based upon the most likely growth level.

This kind of analysis can be applied to all elements of a study to which decisions may be sensitive. It will also show when the decision is not sensitive to a variable, since the various plots would not then intersect. A sensitivity graph can be applied to inflation rates, discount rates, study lives, salvage values, maintenance levels, or anything else that the analyst may feel is significant.

Studies involving particular factors such as the date of placing or the size of job, can also benefit from a similar graph. The critical quantity (*PWE* in this case) can be plotted against the range of possible solutions; and the minimum point will be apparent, as in Figure 16-2.

Here, the sensitivity graph shows that the 2-year job has the lowest *PWE* and should be selected. The shape of the curve indicates how sensitive the *PWE* would be to a different-size job. A flat curve would indicate little sensitivity, while a sharply curving one indicates that the results are very sensitive to the variable.

Cumulative Present Worth of Expenditures

The present worth of expenditures (*PWE*) form of analysis was explained in detail in Chapter 12. One of the more descriptive interpretations of *PWE* data is a graphic

FIGURE 16-2
Minimum PWE

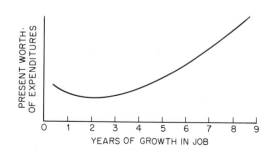

presentation of the cumulative PWE over the study period. Example 12-9 was a comparison of two plans over ten years which can be used to demonstrate the cumulative *PWE* graph. Table 16-1 has been developed here using details from that example.

The income tax portion of the capital costs actually occurs over the life of the investment. However, the actual tax payments are different for each year of the study. The equations for taxes in Example 12-9 give the present worth of the actual taxes over the life. In order to include income tax in the expenditures, as well as to simplify the graphing, a levelized annual equivalent of the present worth of taxes was assumed for each year.

In Alternative I, Unit A has a present worth of taxes of \$1,283 over the 10 years. An annuity from this tax liability at 12% is

$$\text{Levelized tax} = \$1,283(A/P, 12\%, 10) = \$227$$

For Unit B, the taxes are allocated over its 5-year life:

$$\text{Levelized tax} = \$182(F/P, 12\%, 5)(A/P, 12\%, 5) = \$89$$

The allocation of taxes for the first 5 years is \$227 per year, and for the second 5 years, it is \$316 (\$227 + \$89).

TABLE 16-1
COMPARATIVE DATA FROM EXAMPLE 12-9

Time point	Capital	Rearrangements	Operating expenses	Levelized income tax	Ad valorem tax	Total expenditures	(P/F, 12%, n)	Cumulative PWE
Alternative I								
0	10,000	1,000				11,000	1.0000	11,000
1			1,600	227	400	2,227	0.8929	12,989
2			1,680	227	400	2,307	0.7972	14,828
3			1,760	227	400	2,387	0.7118	16,527
4			1,850	227	400	2,477	0.6355	18,101
5	6,500	300	1,940	227	400	9,367	0.5674	23,416
6			2,840	316	660	3,816	0.5066	25,349
7			2,980	316	660	3,956	0.4523	27,138
8			3,130	316	660	4,106	0.4039	28,797
9			3,290	316	660	4,266	0.3606	30,335
10	−1,500		3,450	316	660	2,926	0.3220	31,278
Alternative II								
0	8,000	0				8,000	1.0000	8,000
1			1,300	153	320	1,773	0.8929	9,583
2			1,360	153	320	1,833	0.7972	11,045
3	9,800	600	1,430	153	320	12,303	0.7118	19,802
4			3,000	307	710	4,017	0.6355	22,355
5			3,160	307	710	4,177	0.5674	24,725
6			3,320	307	710	4,337	0.5066	26,923
7			3,480	307	710	4,497	0.4523	28,957
8			3,660	307	710	4,677	0.4039	30,846
9			3,840	307	710	4,867	0.3606	32,597
10	50		4,040	307	710	5,107	0.3220	34,242

FIGURE 16-3
Cumulative Present Worth of Expenditures from Example 12-9

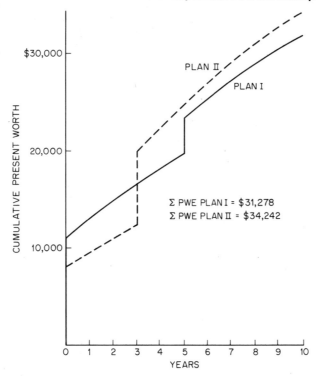

Showing the taxes as $227 for 5 years and $316 for the second 5 years is equivalent to the present worth of $1,465. Alternative II is similar in that the levelized equivalent income tax for the first 3 years is $153, and for the remaining 7 years it is $307.

The expenditures are summarized and the cumulative present worth is found and plotted as in Figure 16-3. Capital expenditures are plotted as if the investment occurred instantaneously. Using rounded factors in the cumulative computations has caused a difference of $1 between the *PWE* conclusion of Example 12-9 and Table 16-1.

When a cumulative *PWE* graph is completed, it shows clearly the differences between plans, as well as the relationship between alternatives in the timing of additions. As Figure 16-3 shows, Plan I will have greater expenditures than Plan II for the first 3 years, but lower expenditures after that. Ultimately, Plan II costs more, but it might be considered if there were a high need to conserve capital during the first 3 years.

It is important to avoid any temptations to make absolute quantity judgments from a graph. The apparent significance of differences between alternatives is sensitive to the scale to which the graph is drawn. Furthermore, the inclusion of expenses that are common to the plans reduce the impact of real differences and may make the plans look very similar. The most useful aspect of the cumulative *PWE* graph is to show the points in time when decisions can be made which have the potential for changing a study outcome. For example, the differences between Plans 1 and 2 could be negligible if after 3 years of Plan 2 the additions could be deferred another year. If accelerated growth were to require the addition in Plan 1 a year earlier than planned, there may be no savings for Plan 1.

Of course, these considerations can be included in the rigorous algebraic analysis, and in the narrative, but they can be appreciated more intuitively from a graph. Plans that have an equivalent present worth of expenditures would probably have different graphs that may show decided advantages in terms of opportunities to re-evaluate growth or to conserve capital over some period.

All the cautions in interpreting *PWE* data apply to interpreting *PWE* graphs. Only the relative merits of plans are shown, and the absolute value of a plan to the firm is not considered at all. A plan could be extremely weak, yet make a beautiful graph.

OTHER AIDS

Tracking (Plan Monitoring)

Tracking, or plan monitoring, is the process of following a decision during its implementation to determine how it correlates with the study model. Obviously, tracking can be advantageous when there is an opportunity for changing the investment decision during its implementation. Testing the decision for sensitivity to various parameters will help determine which parameters should be tracked closely.

Tracking may also be used to assure that the investment decision is implemented as specified so that full economic benefits are realized. For example, if a certain investment is justified because it reduces the current maintenance force by two people within 2 years, tracking may determine if that force reduction is actually achieved. Of course, tracking will not always indicate whether the decision was right or wrong, because it may not be possible to determine which changes in conditions might have occurred if an alternative decision had been made. In addition, Economy study decisions are based upon incremental analysis, and it may not be possible to track these incremental differences, since only one plan is actually implemented. If tracking indicates that some of the factors affecting a plan have changed, further study is required to determine if some other action should be taken because of the changed conditions. Tracking also provides feedback on the validity of study assumptions, so that better data may be used in subsequent studies.

Computer Applications

Computers have become a powerful tool in most fields today, and economics is no exception. The use of mechanized aids allows the analyst to consider many more alternatives under more varied conditions than would be physically possible by manual calculations. Computers make testing results for sensitivity and "what if" models more feasible. They may also increase the degree of detail, and permit a higher level of accuracy to be maintained. There are many computer programs for performing economic analysis, and more are being developed.

These may be either general-purpose programs or programs designed for specific applications. Because of the volume of existing computer aids, and also because programs can be changed and updated to reflect current conditions and objectives, none will be specifically discussed here. In general, however, the analyst should know the tool he is using and apply it only to the situation for which it was intended.

As in all other areas, the output of a computer program is no better than the input data. The analyst must be sure that the values used are the best estimates available under the specific conditions of the study. The analyst must also know all the assumptions made in the program—their significance and how they affect the results. Finally, the analyst must thoroughly understand the model simulated by the program, and be sure that this model is appropriate for the objectives. The computer can then manipulate a great amount of data quickly and accurately and provide results—based on the input data, the assumptions, and the model used.

However, mechanized aids are only tools; they are not a substitute for thinking. These tools can be used as a time-saving device to eliminate the many tedious and repetitive calculations required in a detailed study; but the answers must be analyzed carefully, so they will be interpreted correctly in view of all the relevant factors.

THE RECOMMENDATION

The recommended course of action should be completely and clearly defined. All contingencies of the plan should be identified, and its advantages and disadvantages (both economic and noneconomic) over alternative investments precisely stated.

Furthermore, the course of action recommended must be a result of the consideration of all the relevant environmental factors, and there are many of them. For example, some of these factors are the following:

- Regulatory perspective
- New technology
- Community posture
- Corporate objectives
- Budgetary constraints
- Plant requirements
- Corporate financial indicators
- Service objectives
- Traffic patterns
- Available space
- Fundamental plans

In recommending an alternative, three factors should be considered:

1. The economic indications as determined by the economy study
2. Noneconomic considerations, such as the regulatory perspective and the community posture
3. The analyst's best judgment

It may be desirable to look at the alternatives from more than one perspective. The internal rate of return, the present worth of expenditures, the capital requirements, the risk, the annual cash flows, and so on—all may help in the decision-making process. The analyst and the manager together must use their best judgment of all relevant factors in recommending a plan and making the final decision.

SUMMARY

An economy study should provide the manager with all the information required to make an intelligent investment decision. The study should be presented clearly and concisely with sufficient detail to establish the integrity and validity of the conclusions.

In evaluating the study, then, the manager should be satisfied that the presentation of the study provides the answers to these questions:

1. Why is the study being undertaken?

    - What is the problem?

    - Where is the problem?

    - Why is there a problem?

    - What is the extent of the problem?

2. Can the problem be solved?

3. If so,

    - Why do it?

    - Why do it now?

    - Why do it this way?

4. Is the solution reasonable, and are plans comparable?

5. Are there any reasonable alternatives which were not identified and presented in the study?

6. Was the study method that was used appropriate and applied properly?

7. Are the study data complete, pertinent, and reasonable?

8. Are the study results and recommendations consistent with corporate goals and regulatory constraints?

9. Have computer aids been used appropriately?

10. Have there been any changes in any of the parameters of the study situation which would cause the study to be subject to question?

11. Have the results been tested for sensitivity to uncertainty in parameters?

12. Does the study include a practical plan for monitoring its recommended course of action? Should it?

13. Is there enough documentation to support the assumptions and conclusions?

14. What, if anything, would the reviewer have done differently if the reviewer had performed the study?

15. Is the study complete and clearly written?

Time-Value Factors for Periodic Compounding

This appendix contains tables of time-value factors for periodic compounding. The following nine factors are tabulated for interest rates from 1 to 20%.

- *The Present Worth (P) of a Future Amount (F)*

$$(P/F, i\%, N) = \frac{1}{(1 + i)^N}$$

- *The Future Worth (F) of a Present Amount (P)*

$$(F/P, i\%, N) = (1 + i)^N$$

- *An Annuity (A) from a Present Amount (P)*

$$(A/P, i\%, N) = \frac{i(1 + i)^N}{(1 + i)^N - 1}$$

- *An Annuity (A) for a Future Amount (F)*

$$(A/F, i\%, N) = \frac{i}{(1 + i)^N - 1}$$

- *The Present Worth (P) of an Annuity (A)*

$$(P/A, i\%, N) = \frac{(1 + i)^N - 1}{i(1 + i)^N}$$

- *The Future Worth (F) of an Annuity (A)*

$$(F/A, i\%, N) = \frac{(1 + i)^N - 1}{i}$$

- *An Annuity (A) from a Linear Gradient (G)*

$$(A/G, i\%, N) = \frac{1}{i} - \frac{N}{(1 + i)^N - 1}$$

- *The Present Worth (P) of a Linear Gradient (G)*

$$(P/G, i\%, N) = \frac{1}{i}\left[\frac{(1 + i)^N - 1}{i(1 + i)^N} - \frac{N}{(1 + i)^N}\right]$$

- *The Future Worth (F) of a Linear Gradient (G)*

$$(F/G, i\%, N) = \frac{1}{i}\left[\frac{(1 + i)^N - 1}{i} - N\right]$$

The use of these factors is explained in Chapter 5. Their derivation may be found in the Appendix to Chapter 5, Appendix 5A.

1.00%

ANNUAL COMPOUNDING AT ABOVE EFFECTIVE (i) RATE PER YEAR

FACTORS APPLICABLE TO END OF YEAR AMOUNTS

YR	P/F	F/P	A/P	A/F	P/A	F/A	A/G	P/G	F/G	YR
1	.9901	1.010	1.01000	1.00000	.990	1.000	.000	.000	.00	1
2	.9803	1.020	.50751	.49751	1.970	2.010	.498	.980	1.00	2
3	.9706	1.030	.34002	.33002	2.941	3.030	.993	2.921	3.01	3
4	.9610	1.041	.25628	.24628	3.902	4.060	1.488	5.804	6.04	4
5	.9515	1.051	.20604	.19604	4.853	5.101	1.980	9.610	10.10	5
6	.9420	1.062	.17255	.16255	5.795	6.152	2.471	14.321	15.20	6
7	.9327	1.072	.14863	.13863	6.728	7.214	2.960	19.917	21.35	7
8	.9235	1.083	.13069	.12069	7.652	8.286	3.448	26.381	28.57	8
9	.9143	1.094	.11674	.10674	8.566	9.369	3.934	33.696	36.85	9
10	.9053	1.105	.10558	.09558	9.471	10.462	4.418	41.844	46.22	10
11	.8963	1.116	.09645	.08645	10.368	11.567	4.901	50.807	56.68	11
12	.8874	1.127	.08885	.07885	11.255	12.683	5.381	60.569	68.25	12
13	.8787	1.138	.08241	.07241	12.134	13.809	5.861	71.113	80.93	13
14	.8700	1.149	.07690	.06690	13.004	14.947	6.338	82.422	94.74	14
15	.8613	1.161	.07212	.06212	13.865	16.097	6.814	94.481	109.69	15
16	.8528	1.173	.06794	.05794	14.718	17.258	7.289	107.273	125.79	16
17	.8444	1.184	.06426	.05426	15.562	18.430	7.761	120.783	143.04	17
18	.8360	1.196	.06098	.05098	16.398	19.615	8.232	134.996	161.47	18
19	.8277	1.208	.05805	.04805	17.226	20.811	8.702	149.895	181.09	19
20	.8195	1.220	.05542	.04542	18.046	22.019	9.169	165.466	201.90	20
21	.8114	1.232	.05303	.04303	18.857	23.239	9.635	181.695	223.92	21
22	.8034	1.245	.05086	.04086	19.660	24.472	10.100	198.566	247.16	22
23	.7954	1.257	.04889	.03889	20.456	25.716	10.563	216.066	271.63	23
24	.7876	1.270	.04707	.03707	21.243	26.973	11.024	234.180	297.35	24
25	.7798	1.282	.04541	.03541	22.023	28.243	11.483	252.894	324.32	25
26	.7720	1.295	.04387	.03387	22.795	29.526	11.941	272.196	352.56	26
27	.7644	1.308	.04245	.03245	23.560	30.821	12.397	292.070	382.09	27
28	.7568	1.321	.04112	.03112	24.316	32.129	12.852	312.505	412.91	28
29	.7493	1.335	.03990	.02990	25.066	33.450	13.304	333.486	445.04	29
30	.7419	1.348	.03875	.02875	25.808	34.785	13.756	355.002	478.49	30
31	.7346	1.361	.03768	.02768	26.542	36.133	14.205	377.039	513.27	31
32	.7273	1.375	.03667	.02667	27.270	37.494	14.653	399.586	549.41	32
33	.7201	1.389	.03573	.02573	27.990	38.869	15.099	422.629	586.90	33
34	.7130	1.403	.03484	.02484	28.703	40.258	15.544	446.157	625.77	34
35	.7059	1.417	.03400	.02400	29.409	41.660	15.987	470.158	666.03	35
40	.6717	1.489	.03046	.02046	32.835	48.886	18.178	596.856	888.64	40
45	.6391	1.565	.02771	.01771	36.095	56.481	20.327	733.704	1148.11	45
50	.6080	1.645	.02551	.01551	39.196	64.463	22.436	879.418	1446.32	50
55	.5785	1.729	.02373	.01373	42.147	72.852	24.505	1032.815	1785.25	55
60	.5504	1.817	.02224	.01224	44.955	81.670	26.533	1192.806	2166.97	60
65	.5237	1.909	.02100	.01100	47.627	90.937	28.522	1358.390	2593.66	65
70	.4983	2.007	.01993	.00993	50.169	100.676	30.470	1528.647	3067.63	70
75	.4741	2.109	.01902	.00902	52.587	110.913	32.379	1702.734	3591.28	75
∞	0.0000	∞	.01000	0.00000	100.000	∞	100.000	10000.000	∞	∞

2.00%

ANNUAL COMPOUNDING AT ABOVE EFFECTIVE (i) RATE PER YEAR

FACTORS APPLICABLE TO END OF YEAR AMOUNTS

YR	P/F	F/P	A/P	A/F	P/A	F/A	A/G	P/G	F/G	YR
1	.9804	1.020	1.02000	1.00000	.980	1.000	.000	.000	.00	1
2	.9612	1.040	.51505	.49505	1.942	2.020	.495	.961	1.00	2
3	.9423	1.061	.34675	.32675	2.884	3.060	.987	2.846	3.02	3
4	.9238	1.082	.26262	.24262	3.808	4.122	1.475	5.617	6.08	4
5	.9057	1.104	.21216	.19216	4.713	5.204	1.960	9.240	10.20	5
6	.8880	1.126	.17853	.15853	5.601	6.308	2.442	13.680	15.41	6
7	.8706	1.149	.15451	.13451	6.472	7.434	2.921	18.903	21.71	7
8	.8535	1.172	.13651	.11651	7.325	8.583	3.396	24.878	29.15	8
9	.8368	1.195	.12252	.10252	8.162	9.755	3.868	31.572	37.73	9
10	.8203	1.219	.11133	.09133	8.983	10.950	4.337	38.955	47.49	10
11	.8043	1.243	.10218	.08218	9.787	12.169	4.802	46.998	58.44	11
12	.7885	1.268	.09456	.07456	10.575	13.412	5.264	55.671	70.60	12
13	.7730	1.294	.08812	.06812	11.348	14.680	5.723	64.948	84.02	13
14	.7579	1.319	.08260	.06260	12.106	15.974	6.179	74.800	98.70	14
15	.7430	1.346	.07783	.05783	12.849	17.293	6.631	85.202	114.67	15
16	.7284	1.373	.07365	.05365	13.578	18.639	7.080	96.129	131.96	16
17	.7142	1.400	.06997	.04997	14.292	20.012	7.526	107.555	150.60	17
18	.7002	1.428	.06670	.04670	14.992	21.412	7.968	119.458	170.62	18
19	.6864	1.457	.06378	.04378	15.678	22.841	8.407	131.814	192.03	19
20	.6730	1.486	.06116	.04116	16.351	24.297	8.843	144.600	214.87	20
21	.6598	1.516	.05878	.03878	17.011	25.783	9.276	157.796	239.17	21
22	.6468	1.546	.05663	.03663	17.658	27.299	9.705	171.379	264.95	22
23	.6342	1.577	.05467	.03467	18.292	28.845	10.132	185.331	292.25	23
24	.6217	1.608	.05287	.03287	18.914	30.422	10.555	199.630	321.09	24
25	.6095	1.641	.05122	.03122	19.523	32.030	10.974	214.259	351.51	25
26	.5976	1.673	.04970	.02970	20.121	33.671	11.391	229.199	383.55	26
27	.5859	1.707	.04829	.02829	20.707	35.344	11.804	244.431	417.22	27
28	.5744	1.741	.04699	.02699	21.281	37.051	12.214	259.939	452.56	28
29	.5631	1.776	.04578	.02578	21.844	38.792	12.621	275.706	489.61	29
30	.5521	1.811	.04465	.02465	22.396	40.568	13.025	291.716	528.40	30
31	.5412	1.848	.04360	.02360	22.938	42.379	13.426	307.954	568.97	31
32	.5306	1.885	.04261	.02261	23.468	44.227	13.823	324.403	611.35	32
33	.5202	1.922	.04169	.02169	23.989	46.112	14.217	341.051	655.58	33
34	.5100	1.961	.04082	.02082	24.499	48.034	14.608	357.882	701.69	34
35	.5000	2.000	.04000	.02000	24.999	49.994	14.996	374.883	749.72	35
40	.4529	2.208	.03656	.01656	27.355	60.402	16.889	461.993	1020.10	40
45	.4102	2.438	.03391	.01391	29.490	71.893	18.703	551.565	1344.64	45
50	.3715	2.692	.03182	.01182	31.424	84.579	20.442	642.361	1728.97	50
55	.3365	2.972	.03014	.01014	33.175	98.587	22.106	733.353	2179.33	55
60	.3048	3.281	.02877	.00877	34.761	114.052	23.696	823.698	2702.58	60
65	.2761	3.623	.02763	.00763	36.197	131.126	25.215	912.709	3306.31	65
70	.2500	4.000	.02667	.00667	37.499	149.978	26.663	999.834	3998.90	70
75	.2265	4.416	.02586	.00586	38.677	170.792	28.043	1084.639	4789.59	75
∞	0.0000	∞	.02000	0.00000	50.000	∞	50.000	2500.000	∞	∞

3.00%

ANNUAL COMPOUNDING AT ABOVE EFFECTIVE (i) RATE PER YEAR

FACTORS APPLICABLE TO END OF YEAR AMOUNTS

YR	P/F	F/P	A/P	A/F	P/A	F/A	A/G	P/G	F/G	YR
1	.9709	1.030	1.03000	1.00000	.971	1.000	.000	.000	.00	1
2	.9426	1.061	.52261	.49261	1.913	2.030	.493	.943	1.00	2
3	.9151	1.093	.35353	.32353	2.829	3.091	.980	2.773	3.03	3
4	.8885	1.126	.26903	.23903	3.717	4.184	1.463	5.438	6.12	4
5	.8626	1.159	.21835	.18835	4.580	5.309	1.941	8.889	10.30	5
6	.8375	1.194	.18460	.15460	5.417	6.468	2.414	13.076	15.61	6
7	.8131	1.230	.16051	.13051	6.230	7.662	2.882	17.955	22.08	7
8	.7894	1.267	.14246	.11246	7.020	8.892	3.345	23.481	29.74	8
9	.7664	1.305	.12843	.09843	7.786	10.159	3.803	29.612	38.64	9
10	.7441	1.344	.11723	.08723	8.530	11.464	4.256	36.309	48.80	10
11	.7224	1.384	.10808	.07808	9.253	12.808	4.705	43.533	60.26	11
12	.7014	1.426	.10046	.07046	9.954	14.192	5.148	51.248	73.07	12
13	.6810	1.469	.09403	.06403	10.635	15.618	5.587	59.420	87.26	13
14	.6611	1.513	.08853	.05853	11.296	17.086	6.021	68.014	102.88	14
15	.6419	1.558	.08377	.05377	11.938	18.599	6.450	77.000	119.96	15
16	.6232	1.605	.07961	.04961	12.561	20.157	6.874	86.348	138.56	16
17	.6050	1.653	.07595	.04595	13.166	21.762	7.294	96.028	158.72	17
18	.5874	1.702	.07271	.04271	13.754	23.414	7.708	106.014	180.48	18
19	.5703	1.754	.06981	.03981	14.324	25.117	8.118	116.279	203.90	19
20	.5537	1.806	.06722	.03722	14.877	26.870	8.523	126.799	229.01	20
21	.5375	1.860	.06487	.03487	15.415	28.676	8.923	137.550	255.88	21
22	.5219	1.916	.06275	.03275	15.937	30.537	9.319	148.509	284.56	22
23	.5067	1.974	.06081	.03081	16.444	32.453	9.709	159.657	315.10	23
24	.4919	2.033	.05905	.02905	16.936	34.426	10.095	170.971	347.55	24
25	.4776	2.094	.05743	.02743	17.413	36.459	10.477	182.434	381.98	25
26	.4637	2.157	.05594	.02594	17.877	38.553	10.853	194.026	418.43	26
27	.4502	2.221	.05456	.02456	18.327	40.710	11.226	205.731	456.99	27
28	.4371	2.288	.05329	.02329	18.764	42.931	11.593	217.532	497.70	28
29	.4243	2.357	.05211	.02211	19.188	45.219	11.956	229.414	540.63	29
30	.4120	2.427	.05102	.02102	19.600	47.575	12.314	241.361	585.85	30
31	.4000	2.500	.05000	.02000	20.000	50.003	12.668	253.361	633.42	31
32	.3883	2.575	.04905	.01905	20.389	52.503	13.017	265.399	683.43	32
33	.3770	2.652	.04816	.01816	20.766	55.078	13.362	277.464	735.93	33
34	.3660	2.732	.04732	.01732	21.132	57.730	13.702	289.544	791.01	34
35	.3554	2.814	.04654	.01654	21.487	60.462	14.037	301.627	848.74	35
40	.3066	3.262	.04326	.01326	23.115	75.401	15.650	361.750	1180.04	40
45	.2644	3.782	.04079	.01079	24.519	92.720	17.156	420.632	1590.66	45
50	.2281	4.384	.03887	.00887	25.730	112.797	18.558	477.480	2093.23	50
55	.1968	5.082	.03735	.00735	26.774	136.072	19.860	531.741	2702.39	55
60	.1697	5.892	.03613	.00613	27.676	163.053	21.067	583.053	3435.11	60
65	.1464	6.830	.03515	.00515	28.453	194.333	22.184	631.201	4311.09	65
70	.1263	7.918	.03434	.00434	29.123	230.594	23.215	676.087	5353.14	70
75	.1089	9.179	.03367	.00367	29.702	272.631	24.163	717.698	6587.70	75
∞	0.0000	∞	.03000	0.00000	33.333	∞	33.333	1111.111	∞	∞

4.00%

ANNUAL COMPOUNDING AT ABOVE EFFECTIVE (i) RATE PER YEAR

FACTORS APPLICABLE TO END OF YEAR AMOUNTS

YR	P/F	F/P	A/P	A/F	P/A	F/A	A/G	P/G	F/G	YR
1	.9615	1.040	1.04000	1.00000	.962	1.000	.000	.000	.00	1
2	.9246	1.082	.53020	.49020	1.886	2.040	.490	.925	1.00	2
3	.8890	1.125	.36035	.32035	2.775	3.122	.974	2.703	3.04	3
4	.8548	1.170	.27549	.23549	3.630	4.246	1.451	5.267	6.16	4
5	.8219	1.217	.22463	.18463	4.452	5.416	1.922	8.555	10.41	5
6	.7903	1.265	.19076	.15076	5.242	6.633	2.386	12.506	15.82	6
7	.7599	1.316	.16661	.12661	6.002	7.898	2.843	17.066	22.46	7
8	.7307	1.369	.14853	.10853	6.733	9.214	3.294	22.181	30.36	8
9	.7026	1.423	.13449	.09449	7.435	10.583	3.739	27.801	39.57	9
10	.6756	1.480	.12329	.08329	8.111	12.006	4.177	33.881	50.15	10
11	.6496	1.539	.11415	.07415	8.760	13.486	4.609	40.377	62.16	11
12	.6246	1.601	.10655	.06655	9.385	15.026	5.034	47.248	75.65	12
13	.6006	1.665	.10014	.06014	9.986	16.627	5.453	54.455	90.67	13
14	.5775	1.732	.09467	.05467	10.563	18.292	5.866	61.962	107.30	14
15	.5553	1.801	.08994	.04994	11.118	20.024	6.272	69.735	125.59	15
16	.5339	1.873	.08582	.04582	11.652	21.825	6.672	77.744	145.61	16
17	.5134	1.948	.08220	.04220	12.166	23.698	7.066	85.958	167.44	17
18	.4936	2.026	.07899	.03899	12.659	25.645	7.453	94.350	191.14	18
19	.4746	2.107	.07614	.03614	13.134	27.671	7.834	102.893	216.78	19
20	.4564	2.191	.07358	.03358	13.590	29.778	8.209	111.565	244.45	20
21	.4388	2.279	.07128	.03128	14.029	31.969	8.578	120.341	274.23	21
22	.4220	2.370	.06920	.02920	14.451	34.248	8.941	129.202	306.20	22
23	.4057	2.465	.06731	.02731	14.857	36.618	9.297	138.128	340.45	23
24	.3901	2.563	.06559	.02559	15.247	39.083	9.648	147.101	377.07	24
25	.3751	2.666	.06401	.02401	15.622	41.646	9.993	156.104	416.15	25
26	.3607	2.772	.06257	.02257	15.983	44.312	10.331	165.121	457.79	26
27	.3468	2.883	.06124	.02124	16.330	47.084	10.664	174.138	502.11	27
28	.3335	2.999	.06001	.02001	16.663	49.968	10.991	183.142	549.19	28
29	.3207	3.119	.05888	.01888	16.984	52.966	11.312	192.121	599.16	29
30	.3083	3.243	.05783	.01783	17.292	56.085	11.627	201.062	652.12	30
31	.2965	3.373	.05686	.01686	17.588	59.328	11.937	209.956	708.21	31
32	.2851	3.508	.05595	.01595	17.874	62.701	12.241	218.792	767.54	32
33	.2741	3.648	.05510	.01510	18.148	66.210	12.540	227.563	830.24	33
34	.2636	3.794	.05431	.01431	18.411	69.858	12.832	236.261	896.45	34
35	.2534	3.946	.05358	.01358	18.665	73.652	13.120	244.877	966.31	35
40	.2083	4.801	.05052	.01052	19.793	95.026	14.477	286.530	1375.64	40
45	.1712	5.841	.04826	.00826	20.720	121.029	15.705	325.403	1900.73	45
50	.1407	7.107	.04655	.00655	21.482	152.667	16.812	361.164	2566.68	50
55	.1157	8.646	.04523	.00523	22.109	191.159	17.807	393.689	3403.98	55
60	.0951	10.520	.04420	.00420	22.623	237.991	18.697	422.997	4449.77	60
65	.0781	12.799	.04339	.00339	23.047	294.968	19.491	449.201	5749.21	65
70	.0642	15.572	.04275	.00275	23.395	364.290	20.196	472.479	7357.26	70
75	.0528	18.945	.04223	.00223	23.680	448.631	20.821	493.041	9340.78	75
∞	0.0000	∞	.04000	0.00000	25.000	∞	25.000	625.000	∞	∞

5.00%

ANNUAL COMPOUNDING AT ABOVE EFFECTIVE (i) RATE PER YEAR

FACTORS APPLICABLE TO END OF YEAR AMOUNTS

YR	P/F	F/P	A/P	A/F	P/A	F/A	A/G	P/G	F/G	YR
1	.9524	1.050	1.05000	1.00000	.952	1.000	.000	.000	.00	1
2	.9070	1.103	.53780	.48780	1.859	2.050	.488	.907	1.00	2
3	.8638	1.158	.36721	.31721	2.723	3.153	.967	2.635	3.05	3
4	.8227	1.216	.28201	.23201	3.546	4.310	1.439	5.103	6.20	4
5	.7835	1.276	.23097	.18097	4.329	5.526	1.903	8.237	10.51	5
6	.7462	1.340	.19702	.14702	5.076	6.802	2.358	11.968	16.04	6
7	.7107	1.407	.17282	.12282	5.786	8.142	2.805	16.232	22.84	7
8	.6768	1.477	.15472	.10472	6.463	9.549	3.245	20.970	30.98	8
9	.6446	1.551	.14069	.09069	7.108	11.027	3.676	26.127	40.53	9
10	.6139	1.629	.12950	.07950	7.722	12.578	4.099	31.652	51.56	10
11	.5847	1.710	.12039	.07039	8.306	14.207	4.514	37.499	64.14	11
12	.5568	1.796	.11283	.06283	8.863	15.917	4.922	43.624	78.34	12
13	.5303	1.886	.10646	.05646	9.394	17.713	5.322	49.988	94.26	13
14	.5051	1.980	.10102	.05102	9.899	19.599	5.713	56.554	111.97	14
15	.4810	2.079	.09634	.04634	10.380	21.579	6.097	63.288	131.57	15
16	.4581	2.183	.09227	.04227	10.838	23.657	6.474	70.160	153.15	16
17	.4363	2.292	.08870	.03870	11.274	25.840	6.842	77.140	176.81	17
18	.4155	2.407	.08555	.03555	11.690	28.132	7.203	84.204	202.65	18
19	.3957	2.527	.08275	.03275	12.085	30.539	7.557	91.328	230.78	19
20	.3769	2.653	.08024	.03024	12.462	33.066	7.903	98.488	261.32	20
21	.3589	2.786	.07800	.02800	12.821	35.719	8.242	105.667	294.39	21
22	.3418	2.925	.07597	.02597	13.163	38.505	8.573	112.846	330.10	22
23	.3256	3.072	.07414	.02414	13.489	41.430	8.897	120.009	368.61	23
24	.3101	3.225	.07247	.02247	13.799	44.502	9.214	127.140	410.04	24
25	.2953	3.386	.07095	.02095	14.094	47.727	9.524	134.228	454.54	25
26	.2812	3.556	.06956	.01956	14.375	51.113	9.827	141.259	502.27	26
27	.2678	3.733	.06829	.01829	14.643	54.669	10.122	148.223	553.38	27
28	.2551	3.920	.06712	.01712	14.898	58.403	10.411	155.110	608.05	28
29	.2429	4.116	.06605	.01605	15.141	62.323	10.694	161.913	666.45	29
30	.2314	4.322	.06505	.01505	15.372	66.439	10.969	168.623	728.78	30
31	.2204	4.538	.06413	.01413	15.593	70.761	11.238	175.233	795.22	31
32	.2099	4.765	.06328	.01328	15.803	75.299	11.501	181.739	865.98	32
33	.1999	5.003	.06249	.01249	16.003	80.064	11.757	188.135	941.28	33
34	.1904	5.253	.06176	.01176	16.193	85.067	12.006	194.417	1021.34	34
35	.1813	5.516	.06107	.01107	16.374	90.320	12.250	200.581	1106.41	35
40	.1420	7.040	.05828	.00828	17.159	120.800	13.377	229.545	1616.00	40
45	.1113	8.985	.05626	.00626	17.774	159.700	14.364	255.315	2294.00	45
50	.0872	11.467	.05478	.00478	18.256	209.348	15.223	277.915	3186.96	50
55	.0683	14.636	.05367	.00367	18.633	272.713	15.966	297.510	4354.25	55
60	.0535	18.679	.05283	.00283	18.929	353.584	16.606	314.343	5871.67	60
65	.0419	23.840	.05219	.00219	19.161	456.798	17.154	328.691	7835.96	65
70	.0329	30.426	.05170	.00170	19.343	588.529	17.621	340.841	10370.57	70
75	.0258	38.833	.05132	.00132	19.485	756.654	18.018	351.072	13633.07	75
∞	0.0000	∞	.05000	0.00000	20.000	∞	20.000	400.000	∞	∞

6.00%

ANNUAL COMPOUNDING AT ABOVE EFFECTIVE (i) RATE PER YEAR

FACTORS APPLICABLE TO END OF YEAR AMOUNTS

YR	P/F	F/P	A/P	A/F	P/A	F/A	A/G	P/G	F/G	YR
1	.9434	1.060	1.06000	1.00000	.943	1.000	.000	.000	.00	1
2	.8900	1.124	.54544	.48544	1.833	2.060	.485	.890	1.00	2
3	.8396	1.191	.37411	.31411	2.673	3.184	.961	2.569	3.06	3
4	.7921	1.262	.28859	.22859	3.465	4.375	1.427	4.946	6.24	4
5	.7473	1.338	.23740	.17740	4.212	5.637	1.884	7.935	10.62	5
6	.7050	1.419	.20336	.14336	4.917	6.975	2.330	11.459	16.26	6
7	.6651	1.504	.17914	.11914	5.582	8.394	2.768	15.450	23.23	7
8	.6274	1.594	.16104	.10104	6.210	9.897	3.195	19.842	31.62	8
9	.5919	1.689	.14702	.08702	6.802	11.491	3.613	24.577	41.52	9
10	.5584	1.791	.13587	.07587	7.360	13.181	4.022	29.602	53.01	10
11	.5268	1.898	.12679	.06679	7.887	14.972	4.421	34.870	66.19	11
12	.4970	2.012	.11928	.05928	8.384	16.870	4.811	40.337	81.17	12
13	.4688	2.133	.11296	.05296	8.853	18.882	5.192	45.963	98.04	13
14	.4423	2.261	.10758	.04758	9.295	21.015	5.564	51.713	116.92	14
15	.4173	2.397	.10296	.04296	9.712	23.276	5.926	57.555	137.93	15
16	.3936	2.540	.09895	.03895	10.106	25.673	6.279	63.459	161.21	16
17	.3714	2.693	.09544	.03544	10.477	28.213	6.624	69.401	186.88	17
18	.3503	2.854	.09236	.03236	10.828	30.906	6.960	75.357	215.09	18
19	.3305	3.026	.08962	.02962	11.158	33.760	7.287	81.306	246.00	19
20	.3118	3.207	.08718	.02718	11.470	36.786	7.605	87.230	279.76	20
21	.2942	3.400	.08500	.02500	11.764	39.993	7.915	93.114	316.55	21
22	.2775	3.604	.08305	.02305	12.042	43.392	8.217	98.941	356.54	22
23	.2618	3.820	.08128	.02128	12.303	46.996	8.510	104.701	399.93	23
24	.2470	4.049	.07968	.01968	12.550	50.816	8.795	110.381	446.93	24
25	.2330	4.292	.07823	.01823	12.783	54.865	9.072	115.973	497.74	25
26	.2198	4.549	.07690	.01690	13.003	59.156	9.341	121.468	552.61	26
27	.2074	4.822	.07570	.01570	13.211	63.706	9.603	126.860	611.76	27
28	.1956	5.112	.07459	.01459	13.406	68.528	9.857	132.142	675.47	28
29	.1846	5.418	.07358	.01358	13.591	73.640	10.103	137.310	744.00	29
30	.1741	5.743	.07265	.01265	13.765	79.058	10.342	142.359	817.64	30
31	.1643	6.088	.07179	.01179	13.929	84.802	10.574	147.286	896.69	31
32	.1550	6.453	.07100	.01100	14.084	90.890	10.799	152.090	981.50	32
33	.1462	6.841	.07027	.01027	14.230	97.343	11.017	156.768	1072.39	33
34	.1379	7.251	.06960	.00960	14.368	104.184	11.228	161.319	1169.73	34
35	.1301	7.686	.06897	.00897	14.498	111.435	11.432	165.743	1273.91	35
40	.0972	10.286	.06646	.00646	15.046	154.762	12.359	185.957	1912.70	40
45	.0727	13.765	.06470	.00470	15.456	212.744	13.141	203.110	2795.73	45
50	.0543	18.420	.06344	.00344	15.762	290.336	13.796	217.457	4005.60	50
55	.0406	24.650	.06254	.00254	15.991	394.172	14.341	229.322	5652.87	55
60	.0303	32.988	.06188	.00188	16.161	533.128	14.791	239.043	7885.47	60
65	.0227	44.145	.06139	.00139	16.289	719.083	15.160	246.945	10901.38	65
70	.0169	59.076	.06103	.00103	16.385	967.932	15.461	253.327	14965.54	70
75	.0126	79.057	.06077	.00077	16.456	1300.949	15.706	258.453	20432.48	75
∞	0.0000	∞	.06000	0.00000	16.667	∞	16.667	277.778	∞	∞

7.00%

ANNUAL COMPOUNDING AT ABOVE EFFECTIVE (i) RATE PER YEAR

FACTORS APPLICABLE TO END OF YEAR AMOUNTS

YR	P/F	F/P	A/P	A/F	P/A	F/A	A/G	P/G	F/G	YR
1	.9346	1.070	1.07000	1.00000	.935	1.000	.000	.000	.00	1
2	.8734	1.145	.55309	.48309	1.808	2.070	.483	.873	1.00	2
3	.8163	1.225	.38105	.31105	2.624	3.215	.955	2.506	3.07	3
4	.7629	1.311	.29523	.22523	3.387	4.440	1.416	4.795	6.28	4
5	.7130	1.403	.24389	.17389	4.100	5.751	1.865	7.647	10.72	5
6	.6663	1.501	.20980	.13980	4.767	7.153	2.303	10.978	16.48	6
7	.6227	1.606	.18555	.11555	5.389	8.654	2.730	14.715	23.63	7
8	.5820	1.718	.16747	.09747	5.971	10.260	3.147	18.789	32.28	8
9	.5439	1.838	.15349	.08349	6.515	11.978	3.552	23.140	42.54	9
10	.5083	1.967	.14238	.07238	7.024	13.816	3.946	27.716	54.52	10
11	.4751	2.105	.13336	.06336	7.499	15.784	4.330	32.466	68.34	11
12	.4440	2.252	.12590	.05590	7.943	17.888	4.703	37.351	84.12	12
13	.4150	2.410	.11965	.04965	8.358	20.141	5.065	42.330	102.01	13
14	.3878	2.579	.11434	.04434	8.745	22.550	5.417	47.372	122.15	14
15	.3624	2.759	.10979	.03979	9.108	25.129	5.758	52.446	144.70	15
16	.3387	2.952	.10586	.03586	9.447	27.888	6.090	57.527	169.83	16
17	.3166	3.159	.10243	.03243	9.763	30.840	6.411	62.592	197.72	17
18	.2959	3.380	.09941	.02941	10.059	33.999	6.722	67.622	228.56	18
19	.2765	3.617	.09675	.02675	10.336	37.379	7.024	72.599	262.56	19
20	.2584	3.870	.09439	.02439	10.594	40.995	7.316	77.509	299.94	20
21	.2415	4.141	.09229	.02229	10.836	44.865	7.599	82.339	340.93	21
22	.2257	4.430	.09041	.02041	11.061	49.006	7.872	87.079	385.80	22
23	.2109	4.741	.08871	.01871	11.272	53.436	8.137	91.720	434.80	23
24	.1971	5.072	.08719	.01719	11.469	58.177	8.392	96.255	488.24	24
25	.1842	5.427	.08581	.01581	11.654	63.249	8.639	100.676	546.41	25
26	.1722	5.807	.08456	.01456	11.826	68.676	8.877	104.981	609.66	26
27	.1609	6.214	.08343	.01343	11.987	74.484	9.107	109.166	678.34	27
28	.1504	6.649	.08239	.01239	12.137	80.698	9.329	113.226	752.82	28
29	.1406	7.114	.08145	.01145	12.278	87.347	9.543	117.162	833.52	29
30	.1314	7.612	.08059	.01059	12.409	94.461	9.749	120.972	920.87	30
31	.1228	8.145	.07980	.00980	12.532	102.073	9.947	124.655	1015.33	31
32	.1147	8.715	.07907	.00907	12.647	110.218	10.138	128.212	1117.40	32
33	.1072	9.325	.07841	.00841	12.754	118.933	10.322	131.643	1227.62	33
34	.1002	9.978	.07780	.00780	12.854	128.259	10.499	134.951	1346.55	34
35	.0937	10.677	.07723	.00723	12.948	138.237	10.669	138.135	1474.81	35
40	.0668	14.974	.07501	.00501	13.332	199.635	11.423	152.293	2280.50	40
45	.0476	21.002	.07350	.00350	13.606	285.749	12.036	163.756	3439.28	45
50	.0339	29.457	.07246	.00246	13.801	406.529	12.529	172.905	5093.27	50
55	.0242	41.315	.07174	.00174	13.940	575.929	12.921	180.124	7441.84	55
60	.0173	57.946	.07123	.00123	14.039	813.520	13.232	185.768	10764.58	60
65	.0123	81.273	.07087	.00087	14.110	1146.755	13.476	190.145	15453.65	65
70	.0088	113.989	.07062	.00062	14.160	1614.134	13.666	193.519	22059.06	70
75	.0063	159.876	.07044	.00044	14.196	2269.657	13.814	196.104	31352.25	75
∞	0.0000	∞	.07000	0.00000	14.286	∞	14.286	204.082	∞	∞

8.00%

ANNUAL COMPOUNDING AT ABOVE EFFECTIVE (i) RATE PER YEAR

FACTORS APPLICABLE TO END OF YEAR AMOUNTS

YR	P/F	F/P	A/P	A/F	P/A	F/A	A/G	P/G	F/G	YR
1	.9259	1.080	1.08000	1.00000	.926	1.000	.000	.000	.00	1
2	.8573	1.166	.56077	.48077	1.783	2.080	.481	.857	1.00	2
3	.7938	1.260	.38803	.30803	2.577	3.246	.949	2.445	3.08	3
4	.7350	1.360	.30192	.22192	3.312	4.506	1.404	4.650	6.33	4
5	.6806	1.469	.25046	.17046	3.993	5.867	1.846	7.372	10.83	5
6	.6302	1.587	.21632	.13632	4.623	7.336	2.276	10.523	16.70	6
7	.5835	1.714	.19207	.11207	5.206	8.923	2.694	14.024	24.04	7
8	.5403	1.851	.17401	.09401	5.747	10.637	3.099	17.806	32.96	8
9	.5002	1.999	.16008	.08008	6.247	12.488	3.491	21.808	43.59	9
10	.4632	2.159	.14903	.06903	6.710	14.487	3.871	25.977	56.08	10
11	.4289	2.332	.14008	.06008	7.139	16.645	4.240	30.266	70.57	11
12	.3971	2.518	.13270	.05270	7.536	18.977	4.596	34.634	87.21	12
13	.3677	2.720	.12652	.04652	7.904	21.495	4.940	39.046	106.19	13
14	.3405	2.937	.12130	.04130	8.244	24.215	5.273	43.472	127.69	14
15	.3152	3.172	.11683	.03683	8.559	27.152	5.594	47.886	151.90	15
16	.2919	3.426	.11298	.03298	8.851	30.324	5.905	52.264	179.05	16
17	.2703	3.700	.10963	.02963	9.122	33.750	6.204	56.588	209.38	17
18	.2502	3.996	.10670	.02670	9.372	37.450	6.492	60.843	243.13	18
19	.2317	4.316	.10413	.02413	9.604	41.446	6.770	65.013	280.58	19
20	.2145	4.661	.10185	.02185	9.818	45.762	7.037	69.090	322.02	20
21	.1987	5.034	.09983	.01983	10.017	50.423	7.294	73.063	367.79	21
22	.1839	5.437	.09803	.01803	10.201	55.457	7.541	76.926	418.21	22
23	.1703	5.871	.09642	.01642	10.371	60.893	7.779	80.673	473.67	23
24	.1577	6.341	.09498	.01498	10.529	66.765	8.007	84.300	534.56	24
25	.1460	6.848	.09368	.01368	10.675	73.106	8.225	87.804	601.32	25
26	.1352	7.396	.09251	.01251	10.810	79.954	8.435	91.184	674.43	26
27	.1252	7.988	.09145	.01145	10.935	87.351	8.636	94.439	754.38	27
28	.1159	8.627	.09049	.01049	11.051	95.339	8.829	97.569	841.74	28
29	.1073	9.317	.08962	.00962	11.158	103.966	9.013	100.574	937.07	29
30	.0994	10.063	.08883	.00883	11.258	113.283	9.190	103.456	1041.04	30
31	.0920	10.868	.08811	.00811	11.350	123.346	9.358	106.216	1154.32	31
32	.0852	11.737	.08745	.00745	11.435	134.214	9.520	108.857	1277.67	32
33	.0789	12.676	.08685	.00685	11.514	145.951	9.674	111.382	1411.88	33
34	.0730	13.690	.08630	.00630	11.587	158.627	9.821	113.792	1557.83	34
35	.0676	14.785	.08580	.00580	11.655	172.317	9.961	116.092	1716.46	35
40	.0460	21.725	.08386	.00386	11.925	259.057	10.570	126.042	2738.21	40
45	.0313	31.920	.08259	.00259	12.108	386.506	11.045	133.733	4268.82	45
50	.0213	46.902	.08174	.00174	12.233	573.770	11.411	139.593	6547.13	50
55	.0145	68.914	.08118	.00118	12.319	848.923	11.690	144.006	9924.04	55
60	.0099	101.257	.08080	.00080	12.377	1253.213	11.902	147.300	14915.17	60
65	.0067	148.780	.08054	.00054	12.416	1847.248	12.060	149.739	22278.10	65
70	.0046	218.606	.08037	.00037	12.443	2720.080	12.178	151.533	33126.00	70
75	.0031	321.205	.08025	.00025	12.461	4002.557	12.266	152.845	49094.46	75
∞	0.0000	∞	.08000	0.00000	12.500	∞	12.500	156.250	∞	∞

9.00%

ANNUAL COMPOUNDING AT ABOVE EFFECTIVE (i) RATE PER YEAR

FACTORS APPLICABLE TO END OF YEAR AMOUNTS

YR	P/F	F/P	A/P	A/F	P/A	F/A	A/G	P/G	F/G	YR
1	.9174	1.090	1.09000	1.00000	.917	1.000	.000	.000	.00	1
2	.8417	1.188	.56847	.47847	1.759	2.090	.478	.842	1.00	2
3	.7722	1.295	.39505	.30505	2.531	3.278	.943	2.386	3.09	3
4	.7084	1.412	.30867	.21867	3.240	4.573	1.393	4.511	6.37	4
5	.6499	1.539	.25709	.16709	3.890	5.985	1.828	7.111	10.94	5
6	.5963	1.677	.22292	.13292	4.486	7.523	2.250	10.092	16.93	6
7	.5470	1.828	.19869	.10869	5.033	9.200	2.657	13.375	24.45	7
8	.5019	1.993	.18067	.09067	5.535	11.028	3.051	16.888	33.65	8
9	.4604	2.172	.16680	.07680	5.995	13.021	3.431	20.571	44.68	9
10	.4224	2.367	.15582	.06582	6.418	15.193	3.798	24.373	57.70	10
11	.3875	2.580	.14695	.05695	6.805	17.560	4.151	28.248	72.89	11
12	.3555	2.813	.13965	.04965	7.161	20.141	4.491	32.159	90.45	12
13	.3262	3.066	.13357	.04357	7.487	22.953	4.818	36.073	110.59	13
14	.2992	3.342	.12843	.03843	7.786	26.019	5.133	39.963	133.55	14
15	.2745	3.642	.12406	.03406	8.061	29.361	5.435	43.807	159.57	15
16	.2519	3.970	.12030	.03030	8.313	33.003	5.724	47.585	188.93	16
17	.2311	4.328	.11705	.02705	8.544	36.974	6.002	51.282	221.93	17
18	.2120	4.717	.11421	.02421	8.756	41.301	6.269	54.886	258.90	18
19	.1945	5.142	.11173	.02173	8.950	46.018	6.524	58.387	300.21	19
20	.1784	5.604	.10955	.01955	9.129	51.160	6.767	61.777	346.22	20
21	.1637	6.109	.10762	.01762	9.292	56.765	7.001	65.051	397.38	21
22	.1502	6.659	.10590	.01590	9.442	62.873	7.223	68.205	454.15	22
23	.1378	7.258	.10438	.01438	9.580	69.532	7.436	71.236	517.02	23
24	.1264	7.911	.10302	.01302	9.707	76.790	7.638	74.143	586.55	24
25	.1160	8.623	.10181	.01181	9.823	84.701	7.832	76.926	663.34	25
26	.1064	9.399	.10072	.01072	9.929	93.324	8.016	79.586	748.04	26
27	.0976	10.245	.09973	.00973	10.027	102.723	8.191	82.124	841.37	27
28	.0895	11.167	.09885	.00885	10.116	112.968	8.357	84.542	944.09	28
29	.0822	12.172	.09806	.00806	10.198	124.135	8.515	86.842	1057.06	29
30	.0754	13.268	.09734	.00734	10.274	136.308	8.666	89.028	1181.19	30
31	.0691	14.462	.09669	.00669	10.343	149.575	8.808	91.102	1317.50	31
32	.0634	15.763	.09610	.00610	10.406	164.037	8.944	93.069	1467.08	32
33	.0582	17.182	.09556	.00556	10.464	179.800	9.072	94.931	1631.11	33
34	.0534	18.728	.09508	.00508	10.518	196.982	9.193	96.693	1810.91	34
35	.0490	20.414	.09464	.00464	10.567	215.711	9.308	98.359	2007.90	35
40	.0318	31.409	.09296	.00296	10.757	337.882	9.796	105.376	3309.80	40
45	.0207	48.327	.09190	.00190	10.881	525.859	10.160	110.556	5342.87	45
50	.0134	74.358	.09123	.00123	10.962	815.084	10.430	114.325	8500.93	50
55	.0087	114.408	.09079	.00079	11.014	1260.092	10.626	117.036	13389.91	55
60	.0057	176.031	.09051	.00051	11.048	1944.792	10.768	118.968	20942.13	60
65	.0037	270.846	.09033	.00033	11.070	2998.288	10.870	120.334	32592.09	65
70	.0024	416.730	.09022	.00022	11.084	4619.223	10.943	121.294	50546.92	70
75	.0016	641.191	.09014	.00014	11.094	7113.232	10.994	121.965	78202.58	75
∞	0.0000	∞	.09000	0.00000	11.111	∞	11.111	123.457	∞	∞

10.00%

ANNUAL COMPOUNDING AT ABOVE EFFECTIVE (i) RATE PER YEAR

FACTORS APPLICABLE TO END OF YEAR AMOUNTS

YR	P/F	F/P	A/P	A/F	P/A	F/A	A/G	P/G	F/G	YR
1	.9091	1.100	1.10000	1.00000	.909	1.000	.000	.000	.00	1
2	.8264	1.210	.57619	.47619	1.736	2.100	.476	.826	1.00	2
3	.7513	1.331	.40211	.30211	2.487	3.310	.937	2.329	3.10	3
4	.6830	1.464	.31547	.21547	3.170	4.641	1.381	4.378	6.41	4
5	.6209	1.611	.26380	.16380	3.791	6.105	1.810	6.862	11.05	5
6	.5645	1.772	.22961	.12961	4.355	7.716	2.224	9.684	17.16	6
7	.5132	1.949	.20541	.10541	4.868	9.487	2.622	12.763	24.87	7
8	.4665	2.144	.18744	.08744	5.335	11.436	3.004	16.029	34.36	8
9	.4241	2.358	.17364	.07364	5.759	13.579	3.372	19.421	45.79	9
10	.3855	2.594	.16275	.06275	6.145	15.937	3.725	22.891	59.37	10
11	.3505	2.853	.15396	.05396	6.495	18.531	4.064	26.396	75.31	11
12	.3186	3.138	.14676	.04676	6.814	21.384	4.388	29.901	93.84	12
13	.2897	3.452	.14078	.04078	7.103	24.523	4.699	33.377	115.23	13
14	.2633	3.797	.13575	.03575	7.367	27.975	4.996	36.800	139.75	14
15	.2394	4.177	.13147	.03147	7.606	31.772	5.279	40.152	167.72	15
16	.2176	4.595	.12782	.02782	7.824	35.950	5.549	43.416	199.50	16
17	.1978	5.054	.12466	.02466	8.022	40.545	5.807	46.582	235.45	17
18	.1799	5.560	.12193	.02193	8.201	45.599	6.053	49.640	275.99	18
19	.1635	6.116	.11955	.01955	8.365	51.159	6.286	52.583	321.59	19
20	.1486	6.727	.11746	.01746	8.514	57.275	6.508	55.407	372.75	20
21	.1351	7.400	.11562	.01562	8.649	64.002	6.719	58.110	430.02	21
22	.1228	8.140	.11401	.01401	8.772	71.403	6.919	60.689	494.03	22
23	.1117	8.954	.11257	.01257	8.883	79.543	7.108	63.146	565.43	23
24	.1015	9.850	.11130	.01130	8.985	88.497	7.288	65.481	644.97	24
25	.0923	10.835	.11017	.01017	9.077	98.347	7.458	67.696	733.47	25
26	.0839	11.918	.10916	.00916	9.161	109.182	7.619	69.794	831.82	26
27	.0763	13.110	.10826	.00826	9.237	121.100	7.770	71.777	941.00	27
28	.0693	14.421	.10745	.00745	9.307	134.210	7.914	73.650	1062.10	28
29	.0630	15.863	.10673	.00673	9.370	148.631	8.049	75.415	1196.31	29
30	.0573	17.449	.10608	.00608	9.427	164.494	8.176	77.077	1344.94	30
31	.0521	19.194	.10550	.00550	9.479	181.943	8.296	78.640	1509.43	31
32	.0474	21.114	.10497	.00497	9.526	201.138	8.409	80.108	1691.38	32
33	.0431	23.225	.10450	.00450	9.569	222.252	8.515	81.486	1892.52	33
34	.0391	25.548	.10407	.00407	9.609	245.477	8.615	82.777	2114.77	34
35	.0356	28.102	.10369	.00369	9.644	271.024	8.709	83.987	2360.24	35
40	.0221	45.259	.10226	.00226	9.779	442.593	9.096	88.953	4025.93	40
45	.0137	72.890	.10139	.00139	9.863	718.905	9.374	92.454	6739.05	45
50	.0085	117.391	.10086	.00086	9.915	1163.909	9.570	94.889	11139.09	50
55	.0053	189.059	.10053	.00053	9.947	1880.591	9.708	96.562	18255.91	55
60	.0033	304.482	.10033	.00033	9.967	3034.816	9.802	97.701	29748.16	60
65	.0020	490.371	.10020	.00020	9.980	4893.707	9.867	98.471	48287.07	65
70	.0013	789.747	.10013	.00013	9.987	7887.469	9.911	98.987	78174.69	70
75	.0008	1271.895	.10008	.00008	9.992	12708.953	9.941	99.332	126339.54	75
∞	0.0000	∞	.10000	0.00000	10.000	∞	10.000	100.000	∞	∞

11.00%

ANNUAL COMPOUNDING AT ABOVE EFFECTIVE (i) RATE PER YEAR

FACTORS APPLICABLE TO END OF YEAR AMOUNTS

YR	P/F	F/P	A/P	A/F	P/A	F/A	A/G	P/G	F/G	YR
1	.9009	1.110	1.11000	1.00000	.901	1.000	.000	.000	.00	1
2	.8116	1.232	.58393	.47393	1.713	2.110	.474	.812	1.00	2
3	.7312	1.368	.40921	.29921	2.444	3.342	.931	2.274	3.11	3
4	.6587	1.518	.32233	.21233	3.102	4.710	1.370	4.250	6.45	4
5	.5935	1.685	.27057	.16057	3.696	6.228	1.792	6.624	11.16	5
6	.5346	1.870	.23638	.12638	4.231	7.913	2.198	9.297	17.39	6
7	.4817	2.076	.21222	.10222	4.712	9.783	2.586	12.187	25.30	7
8	.4339	2.305	.19432	.08432	5.146	11.859	2.958	15.225	35.09	8
9	.3909	2.558	.18060	.07060	5.537	14.164	3.314	18.352	46.95	9
10	.3522	2.839	.16980	.05980	5.889	16.722	3.654	21.522	61.11	10
11	.3173	3.152	.16112	.05112	6.207	19.561	3.979	24.695	77.83	11
12	.2858	3.498	.15403	.04403	6.492	22.713	4.288	27.839	97.39	12
13	.2575	3.883	.14815	.03815	6.750	26.212	4.582	30.929	120.11	13
14	.2320	4.310	.14323	.03323	6.982	30.095	4.862	33.945	146.32	14
15	.2090	4.785	.13907	.02907	7.191	34.405	5.127	36.871	176.41	15
16	.1883	5.311	.13552	.02552	7.379	39.190	5.379	39.695	210.82	16
17	.1696	5.895	.13247	.02247	7.549	44.501	5.618	42.409	250.01	17
18	.1528	6.544	.12984	.01984	7.702	50.396	5.844	45.007	294.51	18
19	.1377	7.263	.12756	.01756	7.839	56.939	6.057	47.486	344.90	19
20	.1240	8.062	.12558	.01558	7.963	64.203	6.259	49.842	401.84	20
21	.1117	8.949	.12384	.01384	8.075	72.265	6.449	52.077	466.05	21
22	.1007	9.934	.12231	.01231	8.176	81.214	6.628	54.191	538.31	22
23	.0907	11.026	.12097	.01097	8.266	91.148	6.797	56.186	619.53	23
24	.0817	12.239	.11979	.00979	8.348	102.174	6.956	58.066	710.67	24
25	.0736	13.585	.11874	.00874	8.422	114.413	7.104	59.832	812.85	25
26	.0663	15.080	.11781	.00781	8.488	127.999	7.244	61.490	927.26	26
27	.0597	16.739	.11699	.00699	8.548	143.079	7.375	63.043	1055.26	27
28	.0538	18.580	.11626	.00626	8.602	159.817	7.498	64.497	1198.34	28
29	.0485	20.624	.11561	.00561	8.650	178.397	7.613	65.854	1358.16	29
30	.0437	22.892	.11502	.00502	8.694	199.021	7.721	67.121	1536.55	30
31	.0394	25.410	.11451	.00451	8.733	221.913	7.821	68.302	1735.57	31
32	.0355	28.206	.11404	.00404	8.769	247.324	7.915	69.401	1957.49	32
33	.0319	31.308	.11363	.00363	8.801	275.529	8.002	70.423	2204.81	33
34	.0288	34.752	.11326	.00326	8.829	306.837	8.084	71.372	2480.34	34
35	.0259	38.575	.11293	.00293	8.855	341.590	8.159	72.254	2787.18	35
40	.0154	65.001	.11172	.00172	8.951	581.826	8.466	75.779	4925.69	40
45	.0091	109.530	.11101	.00101	9.008	986.639	8.676	78.155	8560.35	45
50	.0054	184.565	.11060	.00060	9.042	1668.771	8.819	79.734	14716.10	50
55	.0032	311.002	.11035	.00035	9.062	2818.204	8.913	80.771	25120.04	55
60	.0019	524.057	.11021	.00021	9.074	4755.066	8.976	81.446	42682.42	60
65	.0011	883.067	.11012	.00012	9.081	8018.790	9.017	81.882	72307.18	65
70	.0007	1488.019	.11007	.00007	9.085	13518.356	9.044	82.161	122257.78	70
75	.0004	2507.399	.11004	.00004	9.087	22785.444	9.061	82.340	206458.58	75
∞	0.0000	∞	.11000	0.00000	9.091	∞	9.091	82.645	∞	∞

12.00%

ANNUAL COMPOUNDING AT ABOVE EFFECTIVE (i) RATE PER YEAR

FACTORS APPLICABLE TO END OF YEAR AMOUNTS

YR	P/F	F/P	A/P	A/F	P/A	F/A	A/G	P/G	F/G	YR
1	.8929	1.120	1.12000	1.00000	.893	1.000	.000	.000	.00	1
2	.7972	1.254	.59170	.47170	1.690	2.120	.472	.797	1.00	2
3	.7118	1.405	.41635	.29635	2.402	3.374	.925	2.221	3.12	3
4	.6355	1.574	.32923	.20923	3.037	4.779	1.359	4.127	6.49	4
5	.5674	1.762	.27741	.15741	3.605	6.353	1.775	6.397	11.27	5
6	.5066	1.974	.24323	.12323	4.111	8.115	2.172	8.930	17.63	6
7	.4523	2.211	.21912	.09912	4.564	10.089	2.551	11.644	25.74	7
8	.4039	2.476	.20130	.08130	4.968	12.300	2.913	14.471	35.83	8
9	.3606	2.773	.18768	.06768	5.328	14.776	3.257	17.356	48.13	9
10	.3220	3.106	.17698	.05698	5.650	17.549	3.585	20.254	62.91	10
11	.2875	3.479	.16842	.04842	5.938	20.655	3.895	23.129	80.45	11
12	.2567	3.896	.16144	.04144	6.194	24.133	4.190	25.952	101.11	12
13	.2292	4.363	.15568	.03568	6.424	28.029	4.468	28.702	125.24	13
14	.2046	4.887	.15087	.03087	6.628	32.393	4.732	31.362	153.27	14
15	.1827	5.474	.14682	.02682	6.811	37.280	4.980	33.920	185.66	15
16	.1631	6.130	.14339	.02339	6.974	42.753	5.215	36.367	222.94	16
17	.1456	6.866	.14046	.02046	7.120	48.884	5.435	38.697	265.70	17
18	.1300	7.690	.13794	.01794	7.250	55.750	5.643	40.908	314.58	18
19	.1161	8.613	.13576	.01576	7.366	63.440	5.838	42.998	370.33	19
20	.1037	9.646	.13388	.01388	7.469	72.052	6.020	44.968	433.77	20
21	.0926	10.804	.13224	.01224	7.562	81.699	6.191	46.819	505.82	21
22	.0826	12.100	.13081	.01081	7.645	92.503	6.351	48.554	587.52	22
23	.0738	13.552	.12956	.00956	7.718	104.603	6.501	50.178	680.02	23
24	.0659	15.179	.12846	.00846	7.784	118.155	6.641	51.693	784.63	24
25	.0588	17.000	.12750	.00750	7.843	133.334	6.771	53.105	902.78	25
26	.0525	19.040	.12665	.00665	7.896	150.334	6.892	54.418	1036.12	26
27	.0469	21.325	.12590	.00590	7.943	169.374	7.005	55.637	1186.45	27
28	.0419	23.884	.12524	.00524	7.984	190.699	7.110	56.767	1355.82	28
29	.0374	26.750	.12466	.00466	8.022	214.583	7.207	57.814	1546.52	29
30	.0334	29.960	.12414	.00414	8.055	241.333	7.297	58.782	1761.11	30
31	.0298	33.555	.12369	.00369	8.085	271.293	7.381	59.676	2002.44	31
32	.0266	37.582	.12328	.00328	8.112	304.848	7.459	60.501	2273.73	32
33	.0238	42.092	.12292	.00292	8.135	342.429	7.530	61.261	2578.58	33
34	.0212	47.143	.12260	.00260	8.157	384.521	7.596	61.961	2921.01	34
35	.0189	52.800	.12232	.00232	8.176	431.664	7.658	62.605	3305.53	35
40	.0107	93.051	.12130	.00130	8.244	767.091	7.899	65.116	6059.10	40
45	.0061	163.988	.12074	.00074	8.283	1358.230	8.057	66.734	10943.58	45
50	.0035	289.002	.12042	.00042	8.304	2400.018	8.160	67.762	19583.49	50
55	.0020	509.321	.12024	.00024	8.317	4236.005	8.225	68.408	34841.71	55
60	.0011	897.597	.12013	.00013	8.324	7471.641	8.266	68.810	61763.68	60
65	.0006	1581.873	.12008	.00008	8.328	13173.938	8.292	69.058	109241.15	65
70	.0004	2787.800	.12004	.00004	8.330	23223.332	8.308	69.210	192944.44	70
75	.0002	4913.056	.12002	.00002	8.332	40933.799	8.318	69.303	340489.99	75
∞	0.0000	∞	.12000	0.00000	8.333	∞	8.333	69.444	∞	∞

13.00%

ANNUAL COMPOUNDING AT ABOVE EFFECTIVE (i) RATE PER YEAR

FACTORS APPLICABLE TO END OF YEAR AMOUNTS

YR	P/F	F/P	A/P	A/F	P/A	F/A	A/G	P/G	F/G	YR
1	.8850	1.130	1.13000	1.00000	.885	1.000	.000	.000	.00	1
2	.7831	1.277	.59948	.46948	1.668	2.130	.469	.783	1.00	2
3	.6931	1.443	.42352	.29352	2.361	3.407	.919	2.169	3.13	3
4	.6133	1.630	.33619	.20619	2.974	4.850	1.348	4.009	6.54	4
5	.5428	1.842	.28431	.15431	3.517	6.480	1.757	6.180	11.39	5
6	.4803	2.082	.25015	.12015	3.998	8.323	2.147	8.582	17.87	6
7	.4251	2.353	.22611	.09611	4.423	10.405	2.517	11.132	26.19	7
8	.3762	2.658	.20839	.07839	4.799	12.757	2.869	13.765	36.59	8
9	.3329	3.004	.19487	.06487	5.132	15.416	3.201	16.428	49.35	9
10	.2946	3.395	.18429	.05429	5.426	18.420	3.516	19.080	64.77	10
11	.2607	3.836	.17584	.04584	5.687	21.814	3.813	21.687	83.19	11
12	.2307	4.335	.16899	.03899	5.918	25.650	4.094	24.224	105.00	12
13	.2042	4.898	.16335	.03335	6.122	29.985	4.357	26.674	130.65	13
14	.1807	5.535	.15867	.02867	6.302	34.883	4.605	29.023	160.64	14
15	.1599	6.254	.15474	.02474	6.462	40.417	4.837	31.262	195.52	15
16	.1415	7.067	.15143	.02143	6.604	46.672	5.055	33.384	235.94	16
17	.1252	7.986	.14861	.01861	6.729	53.739	5.259	35.388	282.61	17
18	.1108	9.024	.14620	.01620	6.840	61.725	5.449	37.271	336.35	18
19	.0981	10.197	.14413	.01413	6.938	70.749	5.627	39.037	398.07	19
20	.0868	11.523	.14235	.01235	7.025	80.947	5.792	40.685	468.82	20
21	.0768	13.021	.14081	.01081	7.102	92.470	5.945	42.221	549.77	21
22	.0680	14.714	.13948	.00948	7.170	105.491	6.088	43.649	642.24	22
23	.0601	16.627	.13832	.00832	7.230	120.205	6.220	44.972	747.73	23
24	.0532	18.788	.13731	.00731	7.283	136.831	6.343	46.196	867.93	24
25	.0471	21.231	.13643	.00643	7.330	155.620	6.457	47.326	1004.77	25
26	.0417	23.991	.13565	.00565	7.372	176.850	6.561	48.369	1160.39	26
27	.0369	27.109	.13498	.00498	7.409	200.841	6.658	49.328	1337.24	27
28	.0326	30.633	.13439	.00439	7.441	227.950	6.747	50.209	1538.08	28
29	.0289	34.616	.13387	.00387	7.470	258.583	6.830	51.018	1766.03	29
30	.0256	39.116	.13341	.00341	7.496	293.199	6.905	51.759	2024.61	30
31	.0226	44.201	.13301	.00301	7.518	332.315	6.975	52.438	2317.81	31
32	.0200	49.947	.13266	.00266	7.538	376.516	7.039	53.059	2650.12	32
33	.0177	56.440	.13234	.00234	7.556	426.463	7.097	53.626	3026.64	33
34	.0157	63.777	.13207	.00207	7.572	482.903	7.151	54.143	3453.10	34
35	.0139	72.069	.13183	.00183	7.586	546.681	7.200	54.615	3936.01	35
40	.0075	132.782	.13099	.00099	7.634	1013.704	7.389	56.409	7490.03	40
45	.0041	244.641	.13053	.00053	7.661	1874.165	7.508	57.515	14070.50	45
50	.0022	450.736	.13029	.00029	7.675	3459.507	7.581	58.187	26226.98	50
55	.0012	830.452	.13016	.00016	7.683	6380.398	7.626	58.591	48656.91	55
60	.0007	1530.054	.13009	.00009	7.687	11761.950	7.653	58.831	90015.00	60
65	.0004	2819.025	.13005	.00005	7.690	21677.111	7.669	58.973	166247.01	65
70	.0002	5193.870	.13003	.00003	7.691	39945.153	7.679	59.057	306731.95	70
75	.0001	9569.369	.13001	.00001	7.692	73602.835	7.684	59.105	565598.73	75
∞	0.0000	∞	.13000	0.00000	7.692	∞	7.692	59.172	∞	∞

14.00%

ANNUAL COMPOUNDING AT ABOVE EFFECTIVE (i) RATE PER YEAR

FACTORS APPLICABLE TO END OF YEAR AMOUNTS

YR	P/F	F/P	A/P	A/F	P/A	F/A	A/G	P/G	F/G	YR
1	.8772	1.140	1.14000	1.00000	.877	1.000	.000	.000	.00	1
2	.7695	1.300	.60729	.46729	1.647	2.140	.467	.769	1.00	2
3	.6750	1.482	.43073	.29073	2.322	3.440	.913	2.119	3.14	3
4	.5921	1.689	.34320	.20320	2.914	4.921	1.337	3.896	6.58	4
5	.5194	1.925	.29128	.15128	3.433	6.610	1.740	5.973	11.50	5
6	.4556	2.195	.25716	.11716	3.889	8.536	2.122	8.251	18.11	6
7	.3996	2.502	.23319	.09319	4.288	10.730	2.483	10.649	26.65	7
8	.3506	2.853	.21557	.07557	4.639	13.233	2.825	13.103	37.38	8
9	.3075	3.252	.20217	.06217	4.946	16.085	3.146	15.563	50.61	9
10	.2697	3.707	.19171	.05171	5.216	19.337	3.449	17.991	66.69	10
11	.2366	4.226	.18339	.04339	5.453	23.045	3.733	20.357	86.03	11
12	.2076	4.818	.17667	.03667	5.660	27.271	4.000	22.640	109.08	12
13	.1821	5.492	.17116	.03116	5.842	32.089	4.249	24.825	136.35	13
14	.1597	6.261	.16661	.02661	6.002	37.581	4.482	26.901	168.44	14
15	.1401	7.138	.16281	.02281	6.142	43.842	4.699	28.862	206.02	15
16	.1229	8.137	.15962	.01962	6.265	50.980	4.901	30.706	249.86	16
17	.1078	9.276	.15692	.01692	6.373	59.118	5.089	32.430	300.84	17
18	.0946	10.575	.15462	.01462	6.467	68.394	5.263	34.038	359.96	18
19	.0829	12.056	.15266	.01266	6.550	78.969	5.424	35.531	428.35	19
20	.0728	13.743	.15099	.01099	6.623	91.025	5.573	36.914	507.32	20
21	.0638	15.668	.14954	.00954	6.687	104.768	5.711	38.190	598.35	21
22	.0560	17.861	.14830	.00830	6.743	120.436	5.838	39.366	703.11	22
23	.0491	20.362	.14723	.00723	6.792	138.297	5.955	40.446	823.55	23
24	.0431	23.212	.14630	.00630	6.835	158.659	6.062	41.437	961.85	24
25	.0378	26.462	.14550	.00550	6.873	181.871	6.161	42.344	1120.51	25
26	.0331	30.167	.14480	.00480	6.906	208.333	6.251	43.173	1302.38	26
27	.0291	34.390	.14419	.00419	6.935	238.499	6.334	43.929	1510.71	27
28	.0255	39.204	.14366	.00366	6.961	272.889	6.410	44.618	1749.21	28
29	.0224	44.693	.14320	.00320	6.983	312.094	6.479	45.244	2022.10	29
30	.0196	50.950	.14280	.00280	7.003	356.787	6.542	45.813	2334.19	30
31	.0172	58.083	.14245	.00245	7.020	407.737	6.600	46.330	2690.98	31
32	.0151	66.215	.14215	.00215	7.035	465.820	6.652	46.798	3098.72	32
33	.0132	75.485	.14188	.00188	7.048	532.035	6.700	47.222	3564.54	33
34	.0116	86.053	.14165	.00165	7.060	607.520	6.743	47.605	4096.57	34
35	.0102	98.100	.14144	.00144	7.070	693.573	6.782	47.952	4704.09	35
40	.0053	188.884	.14075	.00075	7.105	1342.025	6.930	49.238	9300.18	40
45	.0027	363.679	.14039	.00039	7.123	2590.565	7.019	49.996	18182.61	45
50	.0014	700.233	.14020	.00020	7.133	4994.521	7.071	50.438	35318.01	50
55	.0007	1348.239	.14010	.00010	7.138	9623.135	7.102	50.691	68343.82	55
60	.0004	2595.919	.14005	.00005	7.140	18535.134	7.120	50.836	131965.24	60
65	.0002	4998.220	.14003	.00003	7.141	35694.427	7.130	50.917	254495.91	65
70	.0001	9623.645	.14001	.00001	7.142	68733.181	7.136	50.963	490451.29	70
75	.0001	18529.507	.14001	.00001	7.142	132346.480	7.139	50.989	944796.28	75
∞	0.0000	∞	.14000	0.00000	7.143	∞	7.143	51.020	∞	∞

15.00%

ANNUAL COMPOUNDING AT ABOVE EFFECTIVE (i) RATE PER YEAR

FACTORS APPLICABLE TO END OF YEAR AMOUNTS

YR	P/F	F/P	A/P	A/F	P/A	F/A	A/G	P/G	F/G	YR
1	.8696	1.150	1.15000	1.00000	.870	1.000	.000	.000	.00	1
2	.7561	1.323	.61512	.46512	1.626	2.150	.465	.756	1.00	2
3	.6575	1.521	.43798	.28798	2.283	3.473	.907	2.071	3.15	3
4	.5718	1.749	.35027	.20027	2.855	4.993	1.326	3.786	6.62	4
5	.4972	2.011	.29832	.14832	3.352	6.742	1.723	5.775	11.62	5
6	.4323	2.313	.26424	.11424	3.784	8.754	2.097	7.937	18.36	6
7	.3759	2.660	.24036	.09036	4.160	11.067	2.450	10.192	27.11	7
8	.3269	3.059	.22285	.07285	4.487	13.727	2.781	12.481	38.18	8
9	.2843	3.518	.20957	.05957	4.772	16.786	3.092	14.755	51.91	9
10	.2472	4.046	.19925	.04925	5.019	20.304	3.383	16.979	68.69	10
11	.2149	4.652	.19107	.04107	5.234	24.349	3.655	19.129	89.00	11
12	.1869	5.350	.18448	.03448	5.421	29.002	3.908	21.185	113.34	12
13	.1625	6.153	.17911	.02911	5.583	34.352	4.144	23.135	142.35	13
14	.1413	7.076	.17469	.02469	5.724	40.505	4.362	24.972	176.70	14
15	.1229	8.137	.17102	.02102	5.847	47.580	4.565	26.693	217.20	15
16	.1069	9.358	.16795	.01795	5.954	55.717	4.752	28.296	264.78	16
17	.0929	10.761	.16537	.01537	6.047	65.075	4.925	29.783	320.50	17
18	.0808	12.375	.16319	.01319	6.128	75.836	5.084	31.156	385.58	18
19	.0703	14.232	.16134	.01134	6.198	88.212	5.231	32.421	461.41	19
20	.0611	16.367	.15976	.00976	6.259	102.444	5.365	33.582	549.62	20
21	.0531	18.822	.15842	.00842	6.312	118.810	5.488	34.645	652.07	21
22	.0462	21.645	.15727	.00727	6.359	137.632	5.601	35.615	770.88	22
23	.0402	24.891	.15628	.00628	6.399	159.276	5.704	36.499	908.51	23
24	.0349	28.625	.15543	.00543	6.434	184.168	5.798	37.302	1067.79	24
25	.0304	32.919	.15470	.00470	6.464	212.793	5.883	38.031	1251.95	25
26	.0264	37.857	.15407	.00407	6.491	245.712	5.961	38.692	1464.75	26
27	.0230	43.535	.15353	.00353	6.514	283.569	6.032	39.289	1710.46	27
28	.0200	50.066	.15306	.00306	6.534	327.104	6.096	39.828	1994.03	28
29	.0174	57.575	.15265	.00265	6.551	377.170	6.154	40.315	2321.13	29
30	.0151	66.212	.15230	.00230	6.566	434.745	6.207	40.753	2698.30	30
31	.0131	76.144	.15200	.00200	6.579	500.957	6.254	41.147	3133.05	31
32	.0114	87.565	.15173	.00173	6.591	577.100	6.297	41.501	3634.00	32
33	.0099	100.700	.15150	.00150	6.600	664.666	6.336	41.818	4211.10	33
34	.0086	115.805	.15131	.00131	6.609	765.365	6.371	42.103	4875.77	34
35	.0075	133.176	.15113	.00113	6.617	881.170	6.402	42.359	5641.13	35
40	.0037	267.864	.15056	.00056	6.642	1779.090	6.517	43.283	11593.94	40
45	.0019	538.769	.15028	.00028	6.654	3585.129	6.583	43.805	23600.86	45
50	.0009	1083.657	.15014	.00014	6.661	7217.716	6.620	44.096	47784.78	50
55	.0005	2179.622	.15007	.00007	6.664	14524.148	6.641	44.256	96460.99	55
60	.0002	4383.999	.15003	.00003	6.665	29219.992	6.653	44.343	194399.95	60
65	.0001	8817.788	.15002	.00002	6.666	58778.584	6.659	44.390	391423.89	65
70	.0001	17735.721	.15001	.00001	6.666	118231.470	6.663	44.416	787743.13	70
75	.0000	35672.869	.15000	.00000	6.666	237812.461	6.665	44.429	1584916.41	75
∞	0.0000	∞	.15000	0.00000	6.667	∞	6.667	44.444	∞	∞

16.00%

ANNUAL COMPOUNDING AT ABOVE EFFECTIVE (i) RATE PER YEAR

FACTORS APPLICABLE TO END OF YEAR AMOUNTS

YR	P/F	F/P	A/P	A/F	P/A	F/A	A/G	P/G	F/G	YR
1	.8621	1.160	1.16000	1.00000	.862	1.000	.000	.000	.00	1
2	.7432	1.346	.62296	.46296	1.605	2.160	.463	.743	1.00	2
3	.6407	1.561	.44526	.28526	2.246	3.506	.901	2.024	3.16	3
4	.5523	1.811	.35738	.19738	2.798	5.066	1.316	3.681	6.67	4
5	.4761	2.100	.30541	.14541	3.274	6.877	1.706	5.586	11.73	5
6	.4104	2.436	.27139	.11139	3.685	8.977	2.073	7.638	18.61	6
7	.3538	2.826	.24761	.08761	4.039	11.414	2.417	9.761	27.59	7
8	.3050	3.278	.23022	.07022	4.344	14.240	2.739	11.896	39.00	8
9	.2630	3.803	.21708	.05708	4.607	17.519	3.039	14.000	53.24	9
10	.2267	4.411	.20690	.04690	4.833	21.321	3.319	16.040	70.76	10
11	.1954	5.117	.19886	.03886	5.029	25.733	3.578	17.994	92.08	11
12	.1685	5.936	.19241	.03241	5.197	30.850	3.819	19.847	117.81	12
13	.1452	6.886	.18718	.02718	5.342	36.786	4.041	21.590	148.66	13
14	.1252	7.988	.18290	.02290	5.468	43.672	4.246	23.217	185.45	14
15	.1079	9.266	.17936	.01936	5.575	51.660	4.435	24.728	229.12	15
16	.0930	10.748	.17641	.01641	5.668	60.925	4.609	26.124	280.78	16
17	.0802	12.468	.17395	.01395	5.749	71.673	4.768	27.407	341.71	17
18	.0691	14.463	.17188	.01188	5.818	84.141	4.913	28.583	413.38	18
19	.0596	16.777	.17014	.01014	5.877	98.603	5.046	29.656	497.52	19
20	.0514	19.461	.16867	.00867	5.929	115.380	5.167	30.632	596.12	20
21	.0443	22.574	.16742	.00742	5.973	134.841	5.277	31.518	711.50	21
22	.0382	26.186	.16635	.00635	6.011	157.415	5.377	32.320	846.34	22
23	.0329	30.376	.16545	.00545	6.044	183.601	5.467	33.044	1003.76	23
24	.0284	35.236	.16467	.00467	6.073	213.978	5.549	33.697	1187.36	24
25	.0245	40.874	.16401	.00401	6.097	249.214	5.623	34.284	1401.34	25
26	.0211	47.414	.16345	.00345	6.118	290.088	5.690	34.811	1650.55	26
27	.0182	55.000	.16296	.00296	6.136	337.502	5.750	35.284	1940.64	27
28	.0157	63.800	.16255	.00255	6.152	392.503	5.804	35.707	2278.14	28
29	.0135	74.009	.16219	.00219	6.166	456.303	5.853	36.086	2670.65	29
30	.0116	85.850	.16189	.00189	6.177	530.312	5.896	36.423	3126.95	30
31	.0100	99.586	.16162	.00162	6.187	616.162	5.936	36.725	3657.26	31
32	.0087	115.520	.16140	.00140	6.196	715.747	5.971	36.993	4273.42	32
33	.0075	134.003	.16120	.00120	6.203	831.267	6.002	37.232	4989.17	33
34	.0064	155.443	.16104	.00104	6.210	965.270	6.030	37.444	5820.44	34
35	.0055	180.314	.16089	.00089	6.215	1120.713	6.055	37.633	6785.71	35
40	.0026	378.721	.16042	.00042	6.233	2360.757	6.144	38.299	14504.73	40
45	.0013	795.444	.16020	.00020	6.242	4965.274	6.193	38.660	30751.71	45
50	.0006	1670.704	.16010	.00010	6.246	10435.649	6.220	38.852	64910.31	50
55	.0003	3509.049	.16005	.00005	6.248	21925.305	6.234	38.953	136689.41	55
60	.0001	7370.201	.16002	.00002	6.249	46057.509	6.242	39.006	287484.43	60
65	.0001	15479.941	.16001	.00001	6.250	96743.383	6.246	39.034	604239.89	65
70	.0000	32513.165	.16000	.00000	6.250	203201.032	6.248	39.048	1269568.95	70
75	.0000	68288.756	.16000	.00000	6.250	426798.474	6.249	39.055	2667021.72	75
∞	0.0000	∞	.16000	0.00000	6.250	∞	6.250	39.062	∞	∞

17.00%

ANNUAL COMPOUNDING AT ABOVE EFFECTIVE (i) RATE PER YEAR

FACTORS APPLICABLE TO END OF YEAR AMOUNTS

YR	P/F	F/P	A/P	A/F	P/A	F/A	A/G	P/G	F/G	YR
1	.8547	1.170	1.17000	1.00000	.855	1.000	.000	.000	.00	1
2	.7305	1.369	.63083	.46083	1.585	2.170	.461	.731	1.00	2
3	.6244	1.602	.45257	.28257	2.210	3.539	.896	1.979	3.17	3
4	.5337	1.874	.36453	.19453	2.743	5.141	1.305	3.580	6.71	4
5	.4561	2.192	.31256	.14256	3.199	7.014	1.689	5.405	11.85	5
6	.3898	2.565	.27861	.10861	3.589	9.207	2.049	7.354	18.86	6
7	.3332	3.001	.25495	.08495	3.922	11.772	2.385	9.353	28.07	7
8	.2848	3.511	.23769	.06769	4.207	14.773	2.697	11.346	39.84	8
9	.2434	4.108	.22469	.05469	4.451	18.285	2.987	13.294	54.62	9
10	.2080	4.807	.21466	.04466	4.659	22.393	3.255	15.166	72.90	10
11	.1778	5.624	.20676	.03676	4.836	27.200	3.503	16.944	95.29	11
12	.1520	6.580	.20047	.03047	4.988	32.824	3.732	18.616	122.49	12
13	.1299	7.699	.19538	.02538	5.118	39.404	3.942	20.175	155.32	13
14	.1110	9.007	.19123	.02123	5.229	47.103	4.134	21.618	194.72	14
15	.0949	10.539	.18782	.01782	5.324	56.110	4.310	22.946	241.82	15
16	.0811	12.330	.18500	.01500	5.405	66.649	4.470	24.163	297.93	16
17	.0693	14.426	.18266	.01266	5.475	78.979	4.616	25.272	364.58	17
18	.0592	16.879	.18071	.01071	5.534	93.406	4.749	26.279	443.56	18
19	.0506	19.748	.17907	.00907	5.584	110.285	4.869	27.190	536.97	19
20	.0433	23.106	.17769	.00769	5.628	130.033	4.978	28.013	647.25	20
21	.0370	27.034	.17653	.00653	5.665	153.139	5.076	28.753	777.29	21
22	.0316	31.629	.17555	.00555	5.696	180.172	5.164	29.417	930.42	22
23	.0270	37.006	.17472	.00472	5.723	211.801	5.244	30.011	1110.60	23
24	.0231	43.297	.17402	.00402	5.746	248.808	5.315	30.542	1322.40	24
25	.0197	50.658	.17342	.00342	5.766	292.105	5.379	31.016	1571.21	25
26	.0169	59.270	.17292	.00292	5.783	342.763	5.436	31.438	1863.31	26
27	.0144	69.345	.17249	.00249	5.798	402.032	5.487	31.813	2206.07	27
28	.0123	81.134	.17212	.00212	5.810	471.378	5.533	32.146	2608.10	28
29	.0105	94.927	.17181	.00181	5.820	552.512	5.574	32.441	3079.48	29
30	.0090	111.065	.17154	.00154	5.829	647.439	5.610	32.702	3631.99	30
31	.0077	129.946	.17132	.00132	5.837	758.504	5.642	32.932	4279.43	31
32	.0066	152.036	.17113	.00113	5.844	888.449	5.670	33.136	5037.94	32
33	.0056	177.883	.17096	.00096	5.849	1040.486	5.696	33.316	5926.39	33
34	.0048	208.123	.17082	.00082	5.854	1218.368	5.718	33.475	6966.87	34
35	.0041	243.503	.17070	.00070	5.858	1426.491	5.738	33.614	8185.24	35
40	.0019	533.869	.17032	.00032	5.871	3134.522	5.807	34.097	18203.07	40
45	.0009	1170.479	.17015	.00015	5.877	6879.291	5.844	34.346	40201.71	45
50	.0004	2566.215	.17007	.00007	5.880	15089.502	5.863	34.474	88467.66	50
55	.0002	5626.294	.17003	.00003	5.881	33089.963	5.873	34.538	194323.31	55
60	.0001	12335.356	.17001	.00001	5.882	72555.038	5.877	34.571	426441.40	60
65	.0000	27044.628	.17001	.00001	5.882	159080.166	5.880	34.587	935383.33	65
70	.0000	59293.942	.17000	.00000	5.882	348782.011	5.881	34.595	2051247.12	70
75	.0000	129998.886	.17000	.00000	5.882	764693.446	5.882	34.598	4497755.56	75
∞	0.0000	∞	.17000	0.00000	5.882	∞	5.882	34.602	∞	∞

18.00%

ANNUAL COMPOUNDING AT ABOVE EFFECTIVE (i) RATE PER YEAR

FACTORS APPLICABLE TO END OF YEAR AMOUNTS

YR	P/F	F/P	A/P	A/F	P/A	F/A	A/G	P/G	F/G	YR
1	.8475	1.180	1.18000	1.00000	.847	1.000	.000	.000	.00	1
2	.7182	1.392	.63872	.45872	1.566	2.180	.459	.718	1.00	2
3	.6086	1.643	.45992	.27992	2.174	3.572	.890	1.935	3.18	3
4	.5158	1.939	.37174	.19174	2.690	5.215	1.295	3.483	6.75	4
5	.4371	2.288	.31978	.13978	3.127	7.154	1.673	5.231	11.97	5
6	.3704	2.700	.28591	.10591	3.498	9.442	2.025	7.083	19.12	6
7	.3139	3.185	.26236	.08236	3.812	12.142	2.353	8.967	28.56	7
8	.2660	3.759	.24524	.06524	4.078	15.327	2.656	10.829	40.71	8
9	.2255	4.435	.23239	.05239	4.303	19.086	2.936	12.633	56.03	9
10	.1911	5.234	.22251	.04251	4.494	23.521	3.194	14.352	75.12	10
11	.1619	6.176	.21478	.03478	4.656	28.755	3.430	15.972	98.64	11
12	.1372	7.288	.20863	.02863	4.793	34.931	3.647	17.481	127.39	12
13	.1163	8.599	.20369	.02369	4.910	42.219	3.845	18.877	162.33	13
14	.0985	10.147	.19968	.01968	5.008	50.818	4.025	20.158	204.54	14
15	.0835	11.974	.19640	.01640	5.092	60.965	4.189	21.327	255.36	15
16	.0708	14.129	.19371	.01371	5.162	72.939	4.337	22.389	316.33	16
17	.0600	16.672	.19149	.01149	5.222	87.068	4.471	23.348	389.27	17
18	.0508	19.673	.18964	.00964	5.273	103.740	4.592	24.212	476.33	18
19	.0431	23.214	.18810	.00810	5.316	123.414	4.700	24.988	580.08	19
20	.0365	27.393	.18682	.00682	5.353	146.628	4.798	25.681	703.49	20
21	.0309	32.324	.18575	.00575	5.384	174.021	4.885	26.300	850.12	21
22	.0262	38.142	.18485	.00485	5.410	206.345	4.963	26.851	1024.14	22
23	.0222	45.008	.18409	.00409	5.432	244.487	5.033	27.339	1230.48	23
24	.0188	53.109	.18345	.00345	5.451	289.494	5.095	27.772	1474.97	24
25	.0160	62.669	.18292	.00292	5.467	342.603	5.150	28.155	1764.46	25
26	.0135	73.949	.18247	.00247	5.480	405.272	5.199	28.494	2107.07	26
27	.0115	87.260	.18209	.00209	5.492	479.221	5.243	28.791	2512.34	27
28	.0097	102.967	.18177	.00177	5.502	566.481	5.281	29.054	2991.56	28
29	.0082	121.501	.18149	.00149	5.510	669.447	5.315	29.284	3558.04	29
30	.0070	143.371	.18126	.00126	5.517	790.948	5.345	29.486	4227.49	30
31	.0059	169.177	.18107	.00107	5.523	934.319	5.371	29.664	5018.44	31
32	.0050	199.629	.18091	.00091	5.528	1103.496	5.394	29.819	5952.76	32
33	.0042	235.563	.18077	.00077	5.532	1303.125	5.415	29.955	7056.25	33
34	.0036	277.964	.18065	.00065	5.536	1538.688	5.433	30.074	8359.38	34
35	.0030	327.997	.18055	.00055	5.539	1816.652	5.449	30.177	9898.06	35
40	.0013	750.378	.18024	.00024	5.548	4163.213	5.502	30.527	22906.74	40
45	.0006	1716.684	.18010	.00010	5.552	9531.577	5.529	30.701	52703.21	45
50	.0003	3927.357	.18005	.00005	5.554	21813.093	5.543	30.786	120906.08	50
55	.0001	8984.841	.18002	.00002	5.555	49910.228	5.549	30.827	276973.49	55
60	.0000	20555.140	.18001	.00001	5.555	114189.666	5.553	30.846	634053.70	60
65	.0000	47025.180	.18000	.00000	5.555	261245.446	5.554	30.856	1451002.48	65
70	.0000	107582.221	.18000	.00000	5.556	597673.452	5.555	30.860	3320019.19	70
75	.0000	246122.059	.18000	.00000	5.556	1367339.232	5.555	30.862	7595912.44	75
∞	0.0000	∞	.18000	0.00000	5.556	∞	5.556	30.864	∞	∞

19.00%

ANNUAL COMPOUNDING AT ABOVE EFFECTIVE (i) RATE PER YEAR

FACTORS APPLICABLE TO END OF YEAR AMOUNTS

YR	P/F	F/P	A/P	A/F	P/A	F/A	A/G	P/G	F/G	YR
1	.8403	1.190	1.19000	1.00000	.840	1.000	.000	.000	.00	1
2	.7062	1.416	.64662	.45662	1.547	2.190	.457	.706	1.00	2
3	.5934	1.685	.46731	.27731	2.140	3.606	.885	1.893	3.19	3
4	.4987	2.005	.37899	.18899	2.639	5.291	1.284	3.389	6.80	4
5	.4190	2.386	.32705	.13705	3.058	7.297	1.657	5.065	12.09	5
6	.3521	2.840	.29327	.10327	3.410	9.683	2.002	6.826	19.38	6
7	.2959	3.379	.26985	.07985	3.706	12.523	2.321	8.601	29.07	7
8	.2487	4.021	.25289	.06289	3.954	15.902	2.615	10.342	41.59	8
9	.2090	4.785	.24019	.05019	4.163	19.923	2.886	12.014	57.49	9
10	.1756	5.695	.23047	.04047	4.339	24.709	3.133	13.594	77.42	10
11	.1476	6.777	.22289	.03289	4.486	30.404	3.359	15.070	102.12	11
12	.1240	8.064	.21690	.02690	4.611	37.180	3.564	16.434	132.53	12
13	.1042	9.596	.21210	.02210	4.715	45.244	3.751	17.684	169.71	13
14	.0876	11.420	.20823	.01823	4.802	54.841	3.920	18.823	214.95	14
15	.0736	13.590	.20509	.01509	4.876	66.261	4.072	19.853	269.79	15
16	.0618	16.172	.20252	.01252	4.938	79.850	4.209	20.781	336.05	16
17	.0520	19.244	.20041	.01041	4.990	96.022	4.331	21.612	415.90	17
18	.0437	22.901	.19868	.00868	5.033	115.266	4.441	22.354	511.93	18
19	.0367	27.252	.19724	.00724	5.070	138.166	4.539	23.015	627.19	19
20	.0308	32.429	.19605	.00605	5.101	165.418	4.627	23.601	765.36	20
21	.0259	38.591	.19505	.00505	5.127	197.847	4.705	24.119	930.78	21
22	.0218	45.923	.19423	.00423	5.149	236.438	4.773	24.576	1128.62	22
23	.0183	54.649	.19354	.00354	5.167	282.362	4.834	24.979	1365.06	23
24	.0154	65.032	.19297	.00297	5.182	337.010	4.888	25.333	1647.42	24
25	.0129	77.388	.19249	.00249	5.195	402.042	4.936	25.643	1984.43	25
26	.0109	92.092	.19209	.00209	5.206	479.431	4.978	25.914	2386.48	26
27	.0091	109.589	.19175	.00175	5.215	571.522	5.015	26.151	2865.91	27
28	.0077	130.411	.19147	.00147	5.223	681.112	5.047	26.358	3437.43	28
29	.0064	155.189	.19123	.00123	5.229	811.523	5.075	26.539	4118.54	29
30	.0054	184.675	.19103	.00103	5.235	966.712	5.100	26.696	4930.06	30
31	.0046	219.764	.19087	.00087	5.239	1151.387	5.121	26.832	5896.78	31
32	.0038	261.519	.19073	.00073	5.243	1371.151	5.140	26.951	7048.16	32
33	.0032	311.207	.19061	.00061	5.246	1632.670	5.157	27.054	8419.31	33
34	.0027	370.337	.19051	.00051	5.249	1943.877	5.171	27.143	10051.98	34
35	.0023	440.701	.19043	.00043	5.251	2314.214	5.184	27.220	11995.86	35
40	.0010	1051.667	.19018	.00018	5.258	5529.829	5.225	27.474	28893.84	40
45	.0004	2509.651	.19008	.00008	5.261	13203.424	5.245	27.595	69254.86	45
50	.0002	5988.914	.19003	.00003	5.262	31515.336	5.255	27.652	165607.03	50
55	.0001	14291.666	.19001	.00001	5.263	75214.034	5.259	27.679	395573.86	55
60	.0000	34104.970	.19001	.00001	5.263	179494.580	5.261	27.691	944392.53	60
65	.0000	81386.521	.19000	.00000	5.263	428344.847	5.262	27.696	2254104.47	65
70	.0000	194217.020	.19000	.00000	5.263	1022189.582	5.263	27.699	5379576.75	70
∞	0.0000	∞	.19000	0.00000	5.263	∞	5.263	27.701	∞	∞

20.00%

ANNUAL COMPOUNDING AT ABOVE EFFECTIVE (i) RATE PER YEAR

FACTORS APPLICABLE TO END OF YEAR AMOUNTS

YR	P/F	F/P	A/P	A/F	P/A	F/A	A/G	P/G	F/G	YR
1	.8333	1.200	1.20000	1.00000	.833	1.000	.000	.000	.00	1
2	.6944	1.440	.65455	.45455	1.528	2.200	.455	.694	1.00	2
3	.5787	1.728	.47473	.27473	2.106	3.640	.879	1.852	3.20	3
4	.4823	2.074	.38629	.18629	2.589	5.368	1.274	3.299	6.84	4
5	.4019	2.488	.33438	.13438	2.991	7.442	1.641	4.906	12.21	5
6	.3349	2.986	.30071	.10071	3.326	9.930	1.979	6.581	19.65	6
7	.2791	3.583	.27742	.07742	3.605	12.916	2.290	8.255	29.58	7
8	.2326	4.300	.26061	.06061	3.837	16.499	2.576	9.883	42.50	8
9	.1938	5.160	.24808	.04808	4.031	20.799	2.836	11.434	58.99	9
10	.1615	6.192	.23852	.03852	4.192	25.959	3.074	12.887	79.79	10
11	.1346	7.430	.23110	.03110	4.327	32.150	3.289	14.233	105.75	11
12	.1122	8.916	.22526	.02526	4.439	39.581	3.484	15.467	137.90	12
13	.0935	10.699	.22062	.02062	4.533	48.497	3.660	16.588	177.48	13
14	.0779	12.839	.21689	.01689	4.611	59.196	3.817	17.601	225.98	14
15	.0649	15.407	.21388	.01388	4.675	72.035	3.959	18.509	285.18	15
16	.0541	18.488	.21144	.01144	4.730	87.442	4.085	19.321	357.21	16
17	.0451	22.186	.20944	.00944	4.775	105.931	4.198	20.042	444.65	17
18	.0376	26.623	.20781	.00781	4.812	128.117	4.298	20.680	550.58	18
19	.0313	31.948	.20646	.00646	4.843	154.740	4.386	21.244	678.70	19
20	.0261	38.338	.20536	.00536	4.870	186.688	4.464	21.739	833.44	20
21	.0217	46.005	.20444	.00444	4.891	225.026	4.533	22.174	1020.13	21
22	.0181	55.206	.20369	.00369	4.909	271.031	4.594	22.555	1245.15	22
23	.0151	66.247	.20307	.00307	4.925	326.237	4.647	22.887	1516.18	23
24	.0126	79.497	.20255	.00255	4.937	392.484	4.694	23.176	1842.42	24
25	.0105	95.396	.20212	.00212	4.948	471.981	4.735	23.428	2234.91	25
26	.0087	114.475	.20176	.00176	4.956	567.377	4.771	23.646	2706.89	26
27	.0073	137.371	.20147	.00147	4.964	681.853	4.802	23.835	3274.26	27
28	.0061	164.845	.20122	.00122	4.970	819.223	4.829	23.999	3956.12	28
29	.0051	197.814	.20102	.00102	4.975	984.068	4.853	24.141	4775.34	29
30	.0042	237.376	.20085	.00085	4.979	1181.882	4.873	24.263	5759.41	30
31	.0035	284.852	.20070	.00070	4.982	1419.258	4.891	24.368	6941.29	31
32	.0029	341.822	.20059	.00059	4.985	1704.109	4.906	24.459	8360.55	32
33	.0024	410.186	.20049	.00049	4.988	2045.931	4.919	24.537	10064.66	33
34	.0020	492.224	.20041	.00041	4.990	2456.118	4.931	24.604	12110.59	34
35	.0017	590.668	.20034	.00034	4.992	2948.341	4.941	24.661	14566.71	35
40	.0007	1469.772	.20014	.00014	4.997	7343.858	4.973	24.847	36519.29	40
45	.0003	3657.262	.20005	.00005	4.999	18281.310	4.988	24.932	91181.55	45
50	.0001	9100.438	.20002	.00002	4.999	45497.190	4.995	24.970	227235.95	50
55	.0000	22644.802	.20001	.00001	5.000	113219.009	4.998	24.987	565820.05	55
60	.0000	56347.513	.20000	.00000	5.000	281732.564	4.999	24.994	1408362.83	60
65	.0000	140210.643	.20000	.00000	5.000	701048.213	5.000	24.998	3504916.06	65
70	.0000	348888.946	.20000	.00000	5.000	1744439.733	5.000	24.999	8721848.75	70
∞	0.0000	∞	.20000	0.00000	5.000	∞	5.000	25.000	∞	∞

APPENDIX B

Time-Value Factors for Continuous Compounding

This appendix contains tables of time-value factors for continuous compounding. The following six factors are tabulated for *effective* interest rates from 8 to 15%

- *The Present Worth (P) of a Continuous Future Amount (F̄)*

$$(P/\bar{F}, i\%, N) = \frac{e^r - 1}{re^{rN}}$$

- *The Future Worth (F) of a Continuous Present Amount (P̄)*

$$(F/\bar{P}, i\%, N) = \frac{e^{rN}(e^r - 1)}{re^r}$$

- *The Continuous Annuity (Ā) from a Present Amount (P)*

$$(\bar{A}/P, i\%, N) = \frac{re^{rN}}{e^{rN} - 1}$$

- *A Continuous Annuity (Ā) for a Future Amount (F)*

$$(\bar{A}/F, i\%, N) = \frac{r}{e^{rN} - 1}$$

- *The Present Worth (P) of a Continuous Annuity (Ā)*

$$(P/\bar{A}, i\%, N) = \frac{e^{rN} - 1}{re^{rN}}$$

- *The Future Worth (F) of a Continuous Annuity (Ā)*

$$(F/\bar{A}, i\%, N) = \frac{e^{rN} - 1}{r}$$

The use of these factors is explained in Chapter 5. Their derivation may be found in the appendix to Chapter 5, Appendix 5A.

8.0%

CONTINUOUS COMPOUNDING AT ABOVE EFFECTIVE RATE (i) PER YEAR

NOMINAL OR FORCE OF INTEREST RATE (r) =   7.69610%

YR	P/$\overline{F}$	$\overline{F}$/P	$\overline{A}$/P	$\overline{A}$/F	P/$\overline{A}$	F/$\overline{A}$	YR
1	.9625	1.039	1.03897	.96201	.962	1.039	1
2	.8912	1.123	.53947	.46251	1.854	2.162	2
3	.8252	1.212	.37329	.29633	2.679	3.375	3
4	.7641	1.309	.29045	.21349	3.443	4.684	4
5	.7075	1.414	.24094	.16398	4.150	6.098	5
6	.6551	1.527	.20810	.13114	4.805	7.626	6
7	.6065	1.650	.18478	.10782	5.412	9.275	7
8	.5616	1.781	.16740	.09044	5.974	11.057	8
9	.5200	1.924	.15400	.07704	6.494	12.981	9
10	.4815	2.078	.14337	.06641	6.975	15.059	10
11	.4458	2.244	.13476	.05779	7.421	17.303	11
12	.4128	2.424	.12765	.05069	7.834	19.726	12
13	.3822	2.618	.12172	.04475	8.216	22.344	13
14	.3539	2.827	.11669	.03973	8.570	25.171	14
15	.3277	3.053	.11239	.03543	8.897	28.224	15
16	.3034	3.297	.10869	.03172	9.201	31.522	16
17	.2809	3.561	.10546	.02850	9.482	35.083	17
18	.2601	3.846	.10265	.02569	9.742	38.929	18
19	.2409	4.154	.10017	.02321	9.983	43.083	19
20	.2230	4.486	.09798	.02102	10.206	47.569	20
21	.2065	4.845	.09604	.01908	10.412	52.414	21
22	.1912	5.233	.09431	.01735	10.604	57.647	22
23	.1770	5.651	.09276	.01580	10.781	63.298	23
24	.1639	6.103	.09137	.01441	10.945	69.401	24
25	.1518	6.592	.09012	.01316	11.096	75.993	25
26	.1405	7.119	.08899	.01203	11.237	83.112	26
27	.1301	7.688	.08797	.01101	11.367	90.800	27
28	.1205	8.303	.08705	.01009	11.487	99.103	28
29	.1116	8.968	.08621	.00925	11.599	108.071	29
30	.1033	9.685	.08545	.00849	11.702	117.756	30
31	.0956	10.460	.08476	.00780	11.798	128.216	31
32	.0886	11.297	.08413	.00717	11.887	139.513	32
33	.0820	12.201	.08355	.00659	11.969	151.714	33
34	.0759	13.177	.08303	.00606	12.044	164.890	34
35	.0703	14.231	.08254	.00558	12.115	179.121	35
40	.0478	20.910	.08067	.00371	12.395	269.286	40
45	.0326	30.723	.07945	.00249	12.587	401.768	45
50	.0222	45.142	.07864	.00168	12.717	596.427	50
55	.0151	66.329	.07809	.00113	12.805	882.445	55
60	.0103	97.459	.07773	.00077	12.865	1302.699	60
65	.0070	143.199	.07748	.00052	12.906	1920.190	65
70	.0048	210.406	.07731	.00035	12.934	2827.488	70
75	.0032	309.156	.07720	.00024	12.953	4160.606	75
∞			.07696	.00000	12.994		∞

9.0%

CONTINUOUS COMPOUNDING AT ABOVE EFFECTIVE RATE (i) PER YEAR

NOMINAL OR FORCE OF INTEREST RATE (r) =   8.61777%

YR	P/$\overline{F}$	F/$\overline{P}$	$\overline{A}$/P	$\overline{A}$/F	P/$\overline{A}$	F/$\overline{A}$	YR
1	.9581	1.044	1.04371	.95753	.958	1.044	1
2	.8790	1.138	.54433	.45815	1.837	2.183	2
3	.8064	1.241	.37828	.29210	2.644	3.423	3
4	.7398	1.352	.29556	.20938	3.383	4.776	4
5	.6788	1.474	.24617	.16000	4.062	6.250	5
6	.6227	1.607	.21345	.12727	4.685	7.857	6
7	.5713	1.751	.19025	.10407	5.256	9.609	7
8	.5241	1.909	.17300	.08682	5.780	11.518	8
9	.4808	2.081	.15971	.07354	6.261	13.599	9
10	.4411	2.268	.14920	.06302	6.702	15.867	10
11	.4047	2.472	.14071	.05453	7.107	18.339	11
12	.3713	2.695	.13372	.04754	7.478	21.034	12
13	.3406	2.937	.12789	.04172	7.819	23.971	13
14	.3125	3.202	.12298	.03680	8.131	27.173	14
15	.2867	3.490	.11879	.03261	8.418	30.663	15
16	.2630	3.804	.11519	.02901	8.681	34.467	16
17	.2413	4.146	.11208	.02590	8.923	38.614	17
18	.2214	4.520	.10936	.02318	9.144	43.133	18
19	.2031	4.926	.10699	.02081	9.347	48.060	19
20	.1863	5.370	.10489	.01872	9.533	53.429	20
21	.1710	5.853	.10305	.01687	9.704	59.282	21
22	.1568	6.380	.10141	.01523	9.861	65.662	22
23	.1439	6.954	.09995	.01377	10.005	72.616	23
24	.1320	7.580	.09865	.01247	10.137	80.196	24
25	.1211	8.262	.09748	.01130	10.258	88.458	25
26	.1111	9.006	.09644	.01026	10.369	97.463	26
27	.1019	9.816	.09550	.00932	10.471	107.279	27
28	.0935	10.699	.09465	.00848	10.565	117.979	28
29	.0858	11.662	.09389	.00771	10.651	129.641	29
30	.0787	12.712	.09320	.00702	10.729	142.353	30
31	.0722	13.856	.09258	.00640	10.802	156.209	31
32	.0663	15.103	.09201	.00584	10.868	171.313	32
33	.0608	16.462	.09150	.00533	10.929	187.775	33
34	.0558	17.944	.09104	.00486	10.984	205.719	34
35	.0512	19.559	.09062	.00444	11.035	225.278	35
40	.0332	30.094	.08901	.00283	11.234	352.869	40
45	.0216	46.303	.08800	.00182	11.364	549.183	45
50	.0140	71.244	.08735	.00117	11.448	851.236	50
55	.0091	109.617	.08694	.00076	11.503	1315.982	55
60	.0059	168.660	.08667	.00049	11.538	2031.051	60
65	.0039	259.504	.08650	.00032	11.561	3131.274	65
70	.0025	399.279	.08638	.00021	11.576	4824.103	70
75	.0016	614.340	.08631	.00013	11.586	7428.731	75
∞			.08618	.00000	11.604		∞

10.0%

CONTINUOUS COMPOUNDING AT ABOVE EFFECTIVE RATE (i) PER YEAR

NOMINAL OR FORCE OF INTEREST RATE (r) =   9.53102%

YR	P/$\overline{F}$	$\overline{F}$/P	$\overline{A}$/P	$\overline{A}$/F	P/$\overline{A}$	$\overline{F}$/$\overline{A}$	YR
1	.9538	1.049	1.04841	.95310	.954	1.049	1
2	.8671	1.154	.54917	.45386	1.821	2.203	2
3	.7883	1.270	.38326	.28795	2.609	3.473	3
4	.7166	1.396	.30068	.20537	3.326	4.869	4
5	.6515	1.536	.25143	.15612	3.977	6.406	5
6	.5922	1.690	.21884	.12353	4.570	8.095	6
7	.5384	1.859	.19577	.10046	5.108	9.954	7
8	.4895	2.045	.17865	.08334	5.597	11.999	8
9	.4450	2.249	.16550	.07019	6.042	14.248	9
10	.4045	2.474	.15511	.05980	6.447	16.722	10
11	.3677	2.721	.14674	.05143	6.815	19.443	11
12	.3343	2.994	.13988	.04457	7.149	22.437	12
13	.3039	3.293	.13418	.03887	7.453	25.729	13
14	.2763	3.622	.12938	.03407	7.729	29.352	14
15	.2512	3.984	.12531	.03000	7.980	33.336	15
16	.2283	4.383	.12182	.02651	8.209	37.719	16
17	.2076	4.821	.11882	.02351	8.416	42.540	17
18	.1887	5.303	.11621	.02090	8.605	47.843	18
19	.1716	5.833	.11394	.01863	8.777	53.676	19
20	.1560	6.417	.11195	.01664	8.932	60.093	20
21	.1418	7.059	.11020	.01489	9.074	67.152	21
22	.1289	7.764	.10866	.01335	9.203	74.916	22
23	.1172	8.541	.10729	.01198	9.320	83.457	23
24	.1065	9.395	.10608	.01077	9.427	92.852	24
25	.0968	10.334	.10500	.00969	9.524	103.186	25
26	.0880	11.368	.10404	.00873	9.612	114.554	26
27	.0800	12.505	.10318	.00787	9.692	127.059	27
28	.0728	13.755	.10241	.00710	9.765	140.814	28
29	.0661	15.131	.10172	.00641	9.831	155.944	29
30	.0601	16.644	.10110	.00579	9.891	172.588	30
31	.0547	18.308	.10055	.00524	9.945	190.896	31
32	.0497	20.139	.10005	.00474	9.995	211.035	32
33	.0452	22.153	.09960	.00429	10.040	233.188	33
34	.0411	24.368	.09919	.00388	10.081	257.556	34
35	.0373	26.805	.09883	.00352	10.119	284.360	35
40	.0232	43.169	.09746	.00215	10.260	464.371	40
45	.0144	69.525	.09664	.00133	10.348	754.279	45
50	.0089	111.970	.09613	.00082	10.403	1221.180	50
55	.0055	180.329	.09582	.00051	10.437	1973.128	55
60	.0034	290.422	.09562	.00031	10.458	3184.147	60
65	.0021	467.727	.09550	.00019	10.471	5134.506	65
70	.0013	753.279	.09543	.00012	10.479	8275.579	70
75	.0008	1213.164	.09539	.00007	10.484	13334.309	75
∞			.09531	.00000	10.492		∞

11.0%

CONTINUOUS COMPOUNDING AT ABOVE EFFECTIVE RATE (i) PER YEAR

NOMINAL OR FORCE OF INTEREST RATE (r) = 10.43600%

YR	P/$\overline{F}$	F/$\overline{P}$	$\overline{A}$/P	$\overline{A}$/F	P/$\overline{A}$	F/$\overline{A}$	YR
1	.9496	1.054	1.05309	.94873	.950	1.054	1
2	.8555	1.170	.55399	.44963	1.805	2.224	2
3	.7707	1.299	.38823	.28387	2.576	3.523	3
4	.6943	1.442	.30580	.20144	3.270	4.964	4
5	.6255	1.600	.25670	.15234	3.896	6.564	5
6	.5635	1.776	.22426	.11990	4.459	8.340	6
7	.5077	1.971	.20133	.09697	4.967	10.312	7
8	.4574	2.188	.18436	.08000	5.424	12.500	8
9	.4121	2.429	.17134	.06698	5.836	14.929	9
10	.3712	2.696	.16110	.05674	6.208	17.626	10
11	.3344	2.993	.15286	.04850	6.542	20.619	11
12	.3013	3.322	.14613	.04177	6.843	23.941	12
13	.2714	3.688	.14055	.03619	7.115	27.628	13
14	.2445	4.093	.13588	.03152	7.359	31.721	14
15	.2203	4.543	.13194	.02757	7.579	36.265	15
16	.1985	5.043	.12857	.02421	7.778	41.308	16
17	.1788	5.598	.12568	.02132	7.957	46.906	17
18	.1611	6.214	.12319	.01883	8.118	53.120	18
19	.1451	6.897	.12102	.01666	8.263	60.017	19
20	.1307	7.656	.11914	.01478	8.394	67.673	20
21	.1178	8.498	.11749	.01313	8.511	76.171	21
22	.1061	9.433	.11604	.01168	8.618	85.603	22
23	.0956	10.470	.11477	.01041	8.713	96.074	23
24	.0861	11.622	.11365	.00929	8.799	107.696	24
25	.0776	12.901	.11265	.00829	8.877	120.597	25
26	.0699	14.320	.11177	.00741	8.947	134.916	26
27	.0630	15.895	.11099	.00663	9.010	150.811	27
28	.0567	17.643	.11030	.00594	9.066	168.454	28
29	.0511	19.584	.10968	.00532	9.118	188.038	29
30	.0460	21.738	.10913	.00477	9.164	209.777	30
31	.0415	24.129	.10864	.00428	9.205	233.906	31
32	.0374	26.784	.10820	.00384	9.242	260.690	32
33	.0337	29.730	.10780	.00344	9.276	290.420	33
34	.0303	33.000	.10745	.00309	9.306	323.420	34
35	.0273	36.630	.10714	.00278	9.334	360.050	35
40	.0162	61.724	.10599	.00163	9.435	613.270	40
45	.0096	104.009	.10532	.00096	9.495	1039.960	45
50	.0057	175.261	.10493	.00057	9.530	1758.958	50
55	.0034	295.324	.10470	.00034	9.551	2970.510	55
60	.0020	497.639	.10456	.00020	9.564	5012.047	60
65	.0012	838.551	.10448	.00012	9.571	8452.155	65
70	.0007	1413.006	.10443	.00007	9.576	14248.936	70
75	.0004	2380.998	.10440	.00004	9.578	24016.851	75
∞			.10436	.00000	9.582		∞

12.0%

CONTINUOUS COMPOUNDING AT ABOVE EFFECTIVE RATE (i) PER YEAR

NOMINAL OR FORCE OF INTEREST RATE (r) =   11.33287%

YR	P/$\overline{F}$	F/$\overline{P}$	$\overline{A}$/P	$\overline{A}$/F	P/$\overline{A}$	F/$\overline{A}$	YR
1	.9454	1.059	1.05773	.94441	.945	1.059	1
2	.8441	1.186	.55880	.44547	1.790	2.245	2
3	.7537	1.328	.39320	.27987	2.543	3.573	3
4	.6729	1.488	.31093	.19760	3.216	5.061	4
5	.6008	1.666	.26199	.14866	3.817	6.727	5
6	.5365	1.866	.22970	.11638	4.353	8.593	6
7	.4790	2.090	.20694	.09361	4.832	10.683	7
8	.4277	2.341	.19011	.07678	5.260	13.024	8
9	.3818	2.622	.17725	.06392	5.642	15.645	9
10	.3409	2.936	.16714	.05382	5.983	18.582	10
11	.3044	3.289	.15905	.04572	6.287	21.870	11
12	.2718	3.683	.15246	.03913	6.559	25.554	12
13	.2427	4.125	.14702	.03369	6.802	29.679	13
14	.2167	4.620	.14248	.02915	7.018	34.299	14
15	.1935	5.175	.13866	.02533	7.212	39.474	15
16	.1727	5.796	.13542	.02209	7.385	45.270	16
17	.1542	6.491	.13265	.01932	7.539	51.761	17
18	.1377	7.270	.13027	.01694	7.676	59.032	18
19	.1229	8.143	.12822	.01489	7.799	67.174	19
20	.1098	9.120	.12644	.01311	7.909	76.294	20
21	.0980	10.214	.12489	.01156	8.007	86.508	21
22	.0875	11.440	.12354	.01021	8.095	97.948	22
23	.0781	12.813	.12236	.00903	8.173	110.761	23
24	.0698	14.350	.12132	.00799	8.243	125.111	24
25	.0623	16.072	.12041	.00708	8.305	141.183	25
26	.0556	18.001	.11961	.00628	8.360	159.184	26
27	.0497	20.161	.11890	.00558	8.410	179.345	27
28	.0443	22.580	.11828	.00495	8.454	201.925	28
29	.0396	25.290	.11773	.00440	8.494	227.215	29
30	.0353	28.325	.11724	.00391	8.529	255.539	30
31	.0316	31.724	.11681	.00348	8.561	287.263	31
32	.0282	35.530	.11643	.00310	8.589	322.793	32
33	.0252	39.794	.11609	.00276	8.614	362.587	33
34	.0225	44.569	.11578	.00246	8.637	407.157	34
35	.0201	49.918	.11552	.00219	8.657	457.074	35
40	.0114	87.972	.11456	.00123	8.729	812.248	40
45	.0065	155.037	.11402	.00070	8.770	1438.185	45
50	.0037	273.228	.11372	.00039	8.793	2541.300	50
55	.0021	481.520	.11355	.00022	8.807	4485.366	55
60	.0012	848.603	.11346	.00013	8.814	7911.474	60
65	.0007	1495.529	.11340	.00007	8.818	13949.447	65
70	.0004	2635.633	.11337	.00004	8.821	24590.418	70
75	.0002	4644.886	.11335	.00002	8.822	43343.446	75
∞			.11333	.00000	8.824		∞

13.0%

CONTINUOUS COMPOUNDING AT ABOVE EFFECTIVE RATE (i) PER YEAR

NOMINAL OR FORCE OF INTEREST RATE (r) =   12.22176%

YR	P/$\overline{F}$	F/$\overline{P}$	$\overline{A}$/P	$\overline{A}$/F	P/$\overline{A}$	F/$\overline{A}$	YR
1	.9413	1.064	1.06235	.94014	.941	1.064	1
2	.8330	1.202	.56360	.44138	1.774	2.266	2
3	.7372	1.358	.39817	.27595	2.512	3.624	3
4	.6524	1.535	.31607	.19385	3.164	5.159	4
5	.5773	1.734	.26729	.14508	3.741	6.893	5
6	.5109	1.960	.23518	.11296	4.252	8.853	6
7	.4521	2.215	.21257	.09036	4.704	11.067	7
8	.4001	2.502	.19591	.07369	5.104	13.570	8
9	.3541	2.828	.18320	.06099	5.458	16.397	9
10	.3133	3.195	.17326	.05104	5.772	19.593	10
11	.2773	3.611	.16531	.04310	6.049	23.203	11
12	.2454	4.080	.15887	.03665	6.294	27.283	12
13	.2172	4.611	.15357	.03135	6.512	31.894	13
14	.1922	5.210	.14917	.02695	6.704	37.104	14
15	.1701	5.887	.14548	.02326	6.874	42.991	15
16	.1505	6.653	.14236	.02014	7.024	49.644	16
17	.1332	7.517	.13971	.01749	7.158	57.161	17
18	.1179	8.495	.13745	.01523	7.275	65.656	18
19	.1043	9.599	.13551	.01329	7.380	75.254	19
20	.0923	10.847	.13383	.01161	7.472	86.101	20
21	.0817	12.257	.13238	.01017	7.554	98.358	21
22	.0723	13.850	.13113	.00891	7.626	112.208	22
23	.0640	15.651	.13004	.00782	7.690	127.859	23
24	.0566	17.685	.12909	.00687	7.747	145.544	24
25	.0501	19.984	.12826	.00604	7.797	165.529	25
26	.0443	22.582	.12753	.00532	7.841	188.111	26
27	.0392	25.518	.12690	.00468	7.880	213.629	27
28	.0347	28.835	.12634	.00412	7.915	242.465	28
29	.0307	32.584	.12585	.00364	7.946	275.049	29
30	.0272	36.820	.12542	.00321	7.973	311.869	30
31	.0241	41.607	.12505	.00283	7.997	353.476	31
32	.0213	47.016	.12471	.00250	8.018	400.491	32
33	.0188	53.128	.12442	.00220	8.037	453.619	33
34	.0167	60.034	.12416	.00195	8.054	513.653	34
35	.0148	67.839	.12394	.00172	8.069	581.491	35
40	.0080	124.988	.12315	.00093	8.121	1078.253	40
45	.0043	230.283	.12272	.00050	8.149	1993.505	45
50	.0024	424.281	.12249	.00027	8.164	3679.796	50
55	.0013	781.710	.12236	.00015	8.172	6786.678	55
60	.0007	1440.249	.12230	.00008	8.177	12510.908	60
65	.0004	2653.566	.12226	.00004	8.179	23057.430	65
70	.0002	4889.023	.12224	.00002	8.181	42488.713	70
75	.0001	9007.709	.12223	.00001	8.181	78289.594	75
∞			.12222	.00000	8.182		∞

14.0%

CONTINUOUS COMPOUNDING AT ABOVE EFFECTIVE RATE (i) PER YEAR

NOMINAL OR FORCE OF INTEREST RATE (r) =   13.10283%

YR	$P/\overline{F}$	$F/\overline{P}$	$\overline{A}/P$	$\overline{A}/F$	$P/\overline{A}$	$F/\overline{A}$	YR
1	.9373	1.068	1.06694	.93592	.937	1.068	1
2	.8222	1.218	.56837	.43734	1.759	2.287	2
3	.7212	1.389	.40313	.27210	2.481	3.675	3
4	.6326	1.583	.32121	.19018	3.113	5.258	4
5	.5549	1.805	.27262	.14159	3.668	7.063	5
6	.4868	2.057	.24068	.10965	4.155	9.120	6
7	.4270	2.345	.21825	.08722	4.582	11.465	7
8	.3746	2.674	.20176	.07073	4.956	14.139	8
9	.3286	3.048	.18921	.05818	5.285	17.187	9
10	.2882	3.475	.17943	.04840	5.573	20.661	10
11	.2528	3.961	.17164	.04061	5.826	24.622	11
12	.2218	4.516	.16535	.03432	6.048	29.138	12
13	.1945	5.148	.16019	.02917	6.242	34.286	13
14	.1706	5.868	.15593	.02490	6.413	40.154	14
15	.1497	6.690	.15238	.02135	6.563	46.844	15
16	.1313	7.627	.14939	.01836	6.694	54.471	16
17	.1152	8.694	.14686	.01583	6.809	63.165	17
18	.1010	9.912	.14471	.01368	6.910	73.077	18
19	.0886	11.299	.14288	.01185	6.999	84.376	19
20	.0777	12.881	.14131	.01028	7.077	97.258	20
21	.0682	14.685	.13996	.00893	7.145	111.942	21
22	.0598	16.740	.13880	.00777	7.205	128.682	22
23	.0525	19.084	.13780	.00677	7.257	147.766	23
24	.0460	21.756	.13693	.00590	7.303	169.522	24
25	.0404	24.802	.13617	.00515	7.344	194.324	25
26	.0354	28.274	.13552	.00449	7.379	222.598	26
27	.0311	32.232	.13495	.00392	7.410	254.830	27
28	.0273	36.745	.13446	.00343	7.437	291.574	28
29	.0239	41.889	.13403	.00300	7.461	333.463	29
30	.0210	47.753	.13365	.00262	7.482	381.217	30
31	.0184	54.439	.13332	.00230	7.501	435.655	31
32	.0161	62.060	.13304	.00201	7.517	497.716	32
33	.0142	70.749	.13279	.00176	7.531	568.464	33
34	.0124	80.653	.13257	.00154	7.543	649.118	34
35	.0109	91.945	.13238	.00135	7.554	741.063	35
40	.0057	177.032	.13173	.00070	7.592	1433.916	40
45	.0029	340.860	.13139	.00036	7.611	2767.945	45
50	.0015	656.298	.13122	.00019	7.621	5336.505	50
55	.0008	1263.645	.13113	.00010	7.626	10282.048	55
60	.0004	2433.040	.13108	.00005	7.629	19804.267	60
65	.0002	4684.611	.13105	.00003	7.630	38138.488	65
70	.0001	9019.819	.13104	.00001	7.631	73439.464	70
75	.0001	17366.891	.13104	.00001	7.632	141408.477	75
∞			.13103	.00000	7.632		∞

15.0%

CONTINUOUS COMPOUNDING AT ABOVE EFFECTIVE RATE (i) PER YEAR

NOMINAL OR FORCE OF INTEREST RATE (r) =   13.976619%

YR	P/$\overline{F}$	F/$\overline{P}$	A/P	$\overline{A}$/F	P/$\overline{A}$	F/$\overline{A}$	YR
1	.9333	1.073	1.07151	.93175	.933	1.073	1
2	.8115	1.234	.57313	.43337	1.745	2.307	2
3	.7057	1.419	.40808	.26832	2.450	3.727	3
4	.6136	1.632	.32636	.18660	3.064	5.359	4
5	.5336	1.877	.27795	.13819	3.598	7.236	5
6	.4640	2.159	.24620	.10644	4.062	9.395	6
7	.4035	2.483	.22395	.08419	4.465	11.877	7
8	.3508	2.855	.20764	.06788	4.816	14.732	8
9	.3051	3.283	.19527	.05551	5.121	18.015	9
10	.2653	3.776	.18565	.04589	5.386	21.791	10
11	.2307	4.342	.17803	.03827	5.617	26.133	11
12	.2006	4.993	.17189	.03213	5.818	31.126	12
13	.1744	5.742	.16689	.02712	5.992	36.868	13
14	.1517	6.604	.16277	.02300	6.144	43.472	14
15	.1319	7.594	.15934	.01958	6.276	51.066	15
16	.1147	8.733	.15648	.01672	6.390	59.799	16
17	.0997	10.043	.15408	.01432	6.490	69.842	17
18	.0867	11.550	.15205	.01229	6.577	81.392	18
19	.0754	13.282	.15032	.01056	6.652	94.674	19
20	.0656	15.274	.14886	.00910	6.718	109.948	20
21	.0570	17.565	.14760	.00784	6.775	127.513	21
22	.0496	20.200	.14653	.00677	6.824	147.714	22
23	.0431	23.230	.14561	.00585	6.868	170.944	23
24	.0375	26.715	.14482	.00506	6.905	197.659	24
25	.0326	30.722	.14414	.00438	6.938	228.381	25
26	.0284	35.330	.14355	.00379	6.966	263.711	26
27	.0247	40.630	.14305	.00329	6.991	304.341	27
28	.0214	46.724	.14261	.00285	7.012	351.066	28
29	.0186	53.733	.14223	.00247	7.031	404.799	29
30	.0162	61.793	.14191	.00214	7.047	466.592	30
31	.0141	71.062	.14162	.00186	7.061	537.654	31
32	.0123	81.721	.14138	.00161	7.073	619.375	32
33	.0107	93.980	.14116	.00140	7.084	713.355	33
34	.0093	108.076	.14098	.00122	7.093	821.431	34
35	.0081	124.288	.14082	.00106	7.101	945.719	35
40	.0040	249.987	.14029	.00052	7.128	1909.415	40
45	.0020	502.814	.14002	.00026	7.142	3847.752	45
50	.0010	1011.339	.13989	.00013	7.148	7746.440	50
55	.0005	2034.163	.13983	.00006	7.152	15588.094	55
60	.0002	4091.429	.13979	.00003	7.153	31360.462	60
65	.0001	8229.323	.13978	.00002	7.154	63084.326	65
70	.0001	16552.109	.13977	.00001	7.155	126892.350	70
75	.0000	33292.209	.13977	.00000	7.155	255233.080	75
∞			.13976	.00000	7.155		∞

APPENDIX C

Present Worth of ADR Accelerated Tax Depreciation

(P/D_t) FACTORS

This appendix contains tables of factors to determine the present worth of ADR accelerated tax depreciation. Tables are included for interest rates from 8 to 15%. Four tables are provided for each interest rate. They are:

Depreciation Method	Tax Life
ADR DDB/SYD	80% of N
ADR DDB/SYD	N
ADR 1.5DB/SL	80% of N
ADR 1.5DB/SL	N

The use of these tables is explained in Chapter 9.

TAX DEPRECIATION TABLE

ASSET DEPRECIATION RANGE
ADR HYBRID
DDB WITH SWITCH TO SYD

ADR LIFE 80% OF SERVICE LIFE
8.0% RATE OF RETURN
ANNUAL COMPOUNDING

PRESENT WORTH FACTORS

* N *	0%	10%	GROSS 20%	30%	SALVAGE 40%	50%	60%	70%	80%	90%	100%	*COR ADJ * *+COR% X FACTOR	* N *
3	.91377	.83439	.74938	.66136	.56876	.47617	.38358	.29099	.19839	.10580	.01321	.79383	3
4	.89103	.81753	.73839	.65523	.56950	.48205	.38946	.29687	.20427	.11168	.01909	.73503	4
5	.87086	.80280	.72831	.64893	.56637	.48064	.39490	.30231	.20972	.11713	.02453	.68058	5
6	.85246	.78944	.71815	.64252	.56314	.47996	.39423	.30735	.21476	.12217	.02958	.63017	6
7	.83490	.77655	.70940	.63635	.55909	.47971	.39482	.30908	.21943	.12684	.03424	.58349	7
8	.81827	.76424	.70006	.63012	.55658	.47720	.39608	.31034	.22375	.13116	.03857	.54027	8
9	.80251	.75249	.69163	.62438	.55191	.47571	.39632	.31196	.22623	.13516	.04257	.50025	9
10	.78754	.74122	.68311	.61812	.54831	.47481	.39553	.31376	.22803	.13887	.04627	.46319	10
11	.77259	.72970	.67445	.61243	.54470	.47197	.39520	.31564	.22990	.14230	.04970	.42888	11
12	.75824	.71853	.66609	.60589	.54027	.46979	.39520	.31582	.23178	.14547	.05288	.39711	12
13	.74448	.70771	.65764	.60027	.53658	.46809	.39459	.31603	.23362	.14789	.05582	.36770	13
14	.73126	.69722	.64972	.59388	.53238	.46541	.39326	.31639	.23540	.14967	.05855	.34046	14
15	.71856	.68704	.64152	.58809	.52811	.46275	.39222	.31684	.23711	.15137	.06107	.31524	15
16	.70581	.67662	.63350	.58182	.52423	.46040	.39136	.31736	.23797	.15300	.06340	.29189	16
17	.69353	.66651	.62531	.57576	.51959	.45779	.39031	.31718	.23851	.15453	.06557	.27027	17
18	.68171	.65668	.61752	.56977	.51528	.45474	.38853	.31663	.23907	.15597	.06757	.25025	18
19	.67030	.64713	.60974	.56372	.51121	.45200	.38696	.31617	.23962	.15732	.06942	.23171	19
20	.65928	.63783	.60218	.55798	.50661	.44947	.38556	.31580	.24015	.15859	.07114	.21455	20
21	.64826	.62839	.59447	.55177	.50225	.44613	.38420	.31547	.24067	.15977	.07273	.19866	21
22	.63761	.61922	.58689	.54602	.49788	.44307	.38209	.31505	.24116	.16087	.07420	.18394	22
23	.62733	.61029	.57956	.53996	.49332	.44025	.38016	.31399	.24144	.16189	.07556	.17032	23
24	.61738	.60161	.57223	.53424	.48907	.43707	.37838	.31302	.24122	.16268	.07682	.15770	24
25	.60775	.59314	.56520	.52846	.48463	.43385	.37672	.31211	.24102	.16309	.07799	.14602	25
26	.59814	.58461	.55794	.52267	.48012	.43079	.37453	.31126	.24082	.16347	.07881	.13520	26
27	.58883	.57631	.55096	.51698	.47589	.42771	.37236	.31046	.24064	.16381	.07957	.12519	27
28	.57982	.56823	.54402	.51130	.47135	.42435	.37033	.30928	.24046	.16413	.08027	.11591	28
29	.57109	.56036	.53730	.50584	.46701	.42119	.36841	.30798	.24029	.16442	.08091	.10733	29
30	.56262	.55269	.53067	.50028	.46283	.41820	.36624	.30675	.24012	.16469	.08151	.09938	30
31	.55419	.54499	.52403	.49488	.45835	.41482	.36393	.30557	.23960	.16493	.08206	.09202	31
32	.54602	.53750	.51754	.48938	.45409	.41156	.36174	.30444	.23898	.16514	.08257	.08520	32
33	.53808	.53019	.51115	.48410	.44987	.40846	.35966	.30328	.23837	.16533	.08304	.07889	33
34	.53038	.52308	.50495	.47883	.44559	.40528	.35751	.30179	.23780	.16550	.08347	.07305	34
35	.52290	.51613	.49881	.47367	.44150	.40202	.35516	.30036	.23724	.16565	.08387	.06763	35
36	.51546	.50920	.49274	.46852	.43729	.39886	.35288	.29898	.23669	.16557	.08423	.06262	36
37	.50824	.50244	.48672	.46341	.43312	.39580	.35071	.29765	.23616	.16539	.08457	.05799	37
38	.50122	.49585	.48088	.45847	.42914	.39253	.34858	.29638	.23560	.16520	.08488	.05369	38
39	.49439	.48942	.47513	.45351	.42503	.38940	.34620	.29492	.23476	.16502	.08516	.04971	39
40	.48775	.48314	.46952	.44873	.42106	.38639	.34392	.29338	.23394	.16483	.08542	.04603	40
45	.45650	.45337	.44265	.42540	.40168	.37104	.33276	.28610	.23022	.16389	.08641	.03133	45
50	.42862	.42648	.41803	.40373	.38342	.35642	.32183	.27873	.22586	.16234	.08687	.02132	50
55	.40362	.40217	.39550	.38365	.36625	.34238	.31120	.27139	.22158	.16054	.08669	.01451	55
60	.38113	.38014	.37489	.36507	.35014	.32909	.30084	.26407	.21725	.15873	.08644	.00988	60
65	.36081	.36014	.35599	.34786	.33503	.31647	.29091	.25694	.21275	.15655	.08614	.00672	65
70	.34239	.34193	.33866	.33191	.32090	.30454	.28143	.25003	.20846	.15462	.08583	.00457	70
75	.32564	.32532	.32274	.31714	.30770	.29326	.27237	.24334	.20420	.15238	.08536	.00311	75

TAX DEPRECIATION TABLE

ASSET DEPRECIATION RANGE
ADR HYBRID
DDB WITH SWITCH TO SYD

ADR LIFE 100% OF SERVICE LIFE
8.0% RATE OF RETURN
ANNUAL COMPOUNDING

PRESENT WORTH FACTORS

N	0%	10%	GROSS 20%	30%	SALVAGE 40%	50%	60%	70%	80%	90%	100%	*COR ADJ* *+COR% X* * FACTOR *	N
3	.89601	.81662	.73724	.65221	.56648	.47617	.38358	.29099	.19839	.10580	.01321	.79383	3
4	.87086	.79735	.72287	.64349	.56093	.47519	.38946	.29687	.20427	.11168	.01909	.73503	4
5	.84806	.78000	.70977	.63509	.55570	.47378	.38805	.30231	.20972	.11713	.02453	.68058	5
6	.82680	.76378	.69796	.62627	.55081	.47143	.38851	.30278	.21476	.12217	.02958	.63017	6
7	.80669	.74834	.68691	.61896	.54568	.46865	.38927	.30418	.21845	.12684	.03424	.58349	7
8	.78754	.73352	.67541	.61041	.54060	.46710	.38782	.30606	.22032	.13116	.03857	.54027	8
9	.76923	.71921	.66452	.60289	.53564	.46345	.38722	.30784	.22242	.13516	.04257	.50025	9
10	.75167	.70535	.65412	.59502	.53032	.46081	.38718	.30779	.22460	.13887	.04627	.46319	10
11	.73480	.69191	.64362	.58738	.52544	.45801	.38540	.30811	.22678	.14105	.04970	.42888	11
12	.71856	.67885	.63333	.57990	.51992	.45456	.38403	.30866	.22892	.14319	.05288	.39711	12
13	.70293	.66616	.62342	.57222	.51507	.45176	.38305	.30935	.22997	.14525	.05582	.36770	13
14	.68786	.65381	.61360	.56502	.50943	.44824	.38138	.30886	.23075	.14722	.05855	.34046	14
15	.67332	.64179	.60387	.55738	.50447	.44479	.37945	.30838	.23158	.14909	.06107	.31524	15
16	.65928	.63009	.59444	.55024	.49887	.44174	.37782	.30807	.23242	.15085	.06340	.29189	16
17	.64573	.61871	.58520	.54283	.49373	.43784	.37614	.30786	.23326	.15251	.06557	.27027	17
18	.63264	.60761	.57599	.53574	.48831	.43435	.37381	.30720	.23407	.15406	.06757	.25025	18
19	.61998	.59681	.56706	.52861	.48306	.43085	.37173	.30617	.23418	.15548	.06942	.23171	19
20	.60775	.58629	.55835	.52160	.47778	.42700	.36987	.30526	.23417	.15624	.07114	.21455	20
21	.59591	.57604	.54966	.51473	.47248	.42347	.36739	.30446	.23418	.15695	.07240	.19866	21
22	.58445	.56606	.54122	.50784	.46733	.41970	.36500	.30363	.23421	.15763	.07358	.18394	22
23	.57336	.55633	.53300	.50120	.46203	.41590	.36280	.30222	.23424	.15826	.07467	.17032	23
24	.56262	.54685	.52483	.49445	.45700	.41237	.36041	.30092	.23429	.15886	.07568	.15770	24
25	.55222	.53762	.51688	.48798	.45173	.40844	.35781	.29971	.23387	.15941	.07662	.14602	25
26	.54214	.52862	.50912	.48145	.44677	.40470	.35539	.29858	.23335	.15992	.07749	.13520	26
27	.53237	.51985	.50147	.47513	.44160	.40107	.35306	.29709	.23287	.16039	.07830	.12519	27
28	.52290	.51130	.49398	.46884	.43667	.39719	.35033	.29553	.23241	.16082	.07904	.11591	28
29	.51371	.50297	.48668	.46268	.43167	.39353	.34775	.29406	.23197	.16093	.07973	.10733	29
30	.50479	.49486	.47950	.45663	.42679	.38982	.34531	.29266	.23155	.16094	.08037	.09938	30
31	.49614	.48694	.47246	.45063	.42195	.38605	.34267	.29119	.23083	.16093	.08096	.09202	31
32	.48775	.47923	.46560	.44481	.41714	.38247	.34000	.28946	.23002	.16092	.08150	.08520	32
33	.47959	.47170	.45888	.43899	.41245	.37874	.33746	.28782	.22924	.16089	.08200	.07889	33
34	.47167	.46437	.45227	.43336	.40773	.37510	.33493	.28625	.22850	.16086	.08200	.07305	34
35	.46398	.45722	.44583	.42773	.40318	.37159	.33222	.28475	.22778	.16082	.08246	.06763	35
36	.45650	.45024	.43952	.42227	.39855	.36791	.32963	.28297	.22709	.16076	.08289	.06262	36
37	.44924	.44344	.43333	.41687	.39412	.36438	.32715	.28123	.22621	.16070	.08328	.05799	37
38	.44217	.43680	.42728	.41158	.38963	.36089	.32447	.27956	.22523	.16054	.08364	.05369	38
39	.43530	.43033	.42137	.40638	.38528	.35735	.32186	.27795	.22429	.16020	.08397	.04971	39
40	.42862	.42401	.41556	.40126	.38095	.35395	.31935	.27626	.22338	.15987	.08428	.04603	40
45	.39778	.39464	.38837	.37707	.36028	.33718	.30669	.26764	.21862	.15830	.08440	.03133	45
50	.37071	.36858	.36391	.35498	.34114	.32136	.29451	.25911	.21367	.15629	.08482	.02132	50
55	.34683	.34538	.34191	.33483	.32342	.30650	.28283	.25076	.20866	.15413	.08497	.01451	55
60	.32564	.32465	.32206	.31647	.30703	.29258	.27170	.24267	.20353	.15170	.08496	.00988	60
65	.30673	.30606	.30413	.29971	.29191	.27957	.26115	.23486	.19846	.14935	.08468	.00672	65
70	.28979	.28933	.28789	.28439	.27795	.26739	.25114	.22736	.19352	.14679	.08408	.00457	70
75	.27453	.27422	.27315	.27037	.26505	.25601	.24171	.22016	.18874	.14441	.08350	.00311	75

TAX DEPRECIATION TABLE

ASSET DEPRECIATION RANGE
ADR
1.5DB WITH SWITCH TO STRAIGHT LINE

PRESENT WORTH FACTORS

ADR LIFE   80% OF SERVICE LIFE
8.0% RATE OF RETURN
ANNUAL COMPOUNDING

N	GROSS SALVAGE											*COR ADJ* *+COR% X* * FACTOR*	N
	0%	10%	20%	30%	40%	50%	60%	70%	80%	90%	100%		
3	.89742	.81804	.73490	.64935	.56362	.47617	.38358	.29099	.19839	.10580	.01321	.79383	3
4	.87100	.79749	.71868	.63930	.55878	.47305	.38732	.29687	.20427	.11168	.01909	.73503	4
5	.84677	.77871	.70521	.63143	.55204	.47207	.38633	.30060	.20972	.11713	.02453	.68058	5
6	.82496	.76194	.69283	.62087	.54736	.46821	.38709	.30135	.21476	.12217	.02958	.63017	6
7	.80319	.74484	.67953	.61147	.54005	.46579	.38640	.30296	.21722	.12684	.03424	.58349	7
8	.78222	.72820	.66619	.60241	.53435	.46176	.38524	.30498	.21925	.13116	.03857	.54027	8
9	.76214	.71211	.65397	.59291	.52787	.45861	.38488	.30550	.22147	.13516	.04257	.50025	9
10	.74250	.69618	.64215	.58411	.52178	.45482	.38290	.30566	.22374	.13801	.04627	.46319	10
11	.72412	.68123	.62975	.57473	.51524	.45092	.38136	.30614	.22600	.14027	.04970	.42888	11
12	.70654	.66683	.61803	.56580	.50897	.44716	.37993	.30679	.22745	.14247	.05288	.39711	12
13	.68946	.65269	.60684	.55746	.50283	.44302	.37761	.30608	.22827	.14459	.05582	.36770	13
14	.67278	.63874	.59574	.54838	.49635	.43922	.37572	.30563	.22916	.14661	.05855	.34046	14
15	.65654	.62501	.58445	.53973	.49016	.43468	.37351	.30535	.23008	.14852	.06107	.31524	15
16	.64154	.61235	.57394	.53155	.48380	.43091	.37105	.30481	.23101	.15032	.06340	.29189	16
17	.62688	.59985	.56372	.52312	.47788	.42638	.36889	.30373	.23167	.15201	.06557	.27027	17
18	.61257	.58755	.55337	.51505	.47141	.42226	.36626	.30280	.23170	.15343	.06757	.25025	18
19	.59871	.57554	.54332	.50692	.46545	.41780	.36360	.30201	.23177	.15428	.06933	.23171	19
20	.58545	.56400	.53377	.49907	.45927	.41360	.36124	.30075	.23187	.15509	.07071	.21455	20
21	.57278	.55291	.52435	.49151	.45339	.40931	.35833	.29937	.23199	.15586	.07199	.19866	21
22	.56042	.54203	.51496	.48371	.44732	.40494	.35558	.29811	.23210	.15658	.07319	.18394	22
23	.54850	.53147	.50595	.47628	.44150	.40073	.35295	.29694	.23161	.15726	.07429	.17032	23
24	.53701	.52124	.49728	.46906	.43571	.39637	.34997	.29525	.23115	.15789	.07532	.15770	24
25	.52583	.51123	.48862	.46180	.42999	.39227	.34719	.29367	.23074	.15848	.07628	.14602	25
26	.51510	.50158	.48017	.45478	.42432	.38784	.34435	.29219	.23034	.15902	.07716	.13520	26
27	.50477	.49225	.47207	.44788	.41874	.38375	.34138	.29062	.22997	.15918	.07798	.12519	27
28	.49476	.48317	.46415	.44113	.41333	.37949	.33859	.28881	.22926	.15927	.07874	.11591	28
29	.48502	.47429	.45632	.43458	.40786	.37537	.33565	.28711	.22848	.15934	.07944	.10733	29
30	.47555	.46561	.44870	.42797	.40257	.37120	.33268	.28549	.22773	.15940	.08008	.09938	30
31	.46655	.45735	.44141	.42168	.39729	.36715	.32991	.28366	.22702	.15944	.08068	.09202	3
32	.45779	.44927	.43417	.41549	.39222	.36314	.32692	.28180	.22634	.15947	.08123	.08520	3
33	.44926	.44137	.42712	.40935	.38705	.35909	.32402	.28002	.22549	.15949	.08174	.07889	3
34	.44098	.43368	.42027	.40340	.38209	.35519	.32122	.27832	.22451	.15950	.08221	.07305	3
35	.43301	.42625	.41362	.39757	.37717	.35123	.31826	.27635	.22356	.15937	.08264	.06763	3
36	.42530	.41904	.40709	.39185	.37237	.34744	.31542	.27447	.22265	.15907	.08304	.06262	3
37	.41780	.41200	.40072	.38628	.36759	.34354	.31261	.27266	.22177	.15879	.08333	.05799	3
38	.41054	.40517	.39455	.38077	.36294	.33981	.30972	.27083	.22085	.15850	.08351	.05369	3
39	.40350	.39853	.38850	.37544	.35837	.33607	.30697	.26888	.21975	.15822	.08367	.04971	3
40	.39666	.39205	.38259	.37020	.35386	.33240	.30418	.26700	.21869	.15794	.08381	.04603	4
45	.36540	.36226	.35521	.34561	.33256	.31479	.29063	.25775	.21342	.15618	.08429	.03133	4
50	.33841	.33628	.33099	.32356	.31309	.29838	.27766	.24849	.20797	.15401	.08450	.02132	5
55	.31491	.31346	.30951	.30375	.29537	.28317	.26542	.23959	.20248	.15150	.08446	.01451	5
60	.29433	.29334	.29038	.28592	.27922	.26912	.25391	.23104	.19703	.14906	.08391	.00988	
65	.27621	.27554	.27333	.26988	.26451	.25615	.24311	.22280	.19168	.14632	.08336	.00672	
70	.26016	.25970	.25805	.25537	.25107	.24414	.23299	.21498	.18646	.14370	.08284	.00457	
75	.24586	.24554	.24430	.24224	.23879	.23304	.22348	.20753	.18141	.14106	.08214	.00311	

Annuity from a Present Amount Reflecting Dispersion of Lives

Periodic Compounding
(α/P) FACTORS

This appendix contains tables of annuity from a present amount factors which recognize the effect of dispersion in the retirement of the initial amount. Tables are provided for interest rates from 8 to 15%, and for seven different survivor curves. The seven curves, and the derivation and use of the factors, are covered in Chapter 14.

The following tables do not include the (α/F) factors. To find the (α/F) factor, use

$$(\alpha/F) = (\alpha/P) - i$$

To duplicate these tables of (α/P) factors, it is necessary to know about the original computer program which generated them. The $(\alpha/P)_{\text{negative exponential}}$ factors are computed using the formula given in Chapter 14, and the $(\alpha/P)_{\text{rectangular}}$ factors are just the (A/P) factors. The factors for the Gompertz-Makeham curves (No. 1 through No. 5) are developed using a graphical integration technique. Because these curves are changing rapidly during a single year, if the average life is short, the average survivors in a single year are found as the average of the average survivors for portions of the year. In the program which produces the following tables, each year is divided into an integral number of parts equal to 75 divided by the average life (truncated). Therefore, each part of a year (Δt) is 1/INT(75/avg. life) of a year. The average number of survivors is then

$$\text{Average survivors}_n = \sum_{x=1}^{1/\Delta t} L_{[n-1+(x-1/2)\Delta t]\Delta t}$$

If the average life is greater than 37.5 years, the Δt is 1 year [INT(75/38) = 1] and average survivors is simply

$$\text{Average survivors}_n = L_{n-1/2}$$

8.0%

DISPERSED ANNUITY FACTORS FROM A PRESENT AMOUNT

FACTORS APPLICABLE TO AN INITIAL AMOUNT
HAVING AN AVERAGE LIFE OF (N) YEARS
FOR VARIOUS SURVIVOR CURVES
ANNUALLY COMPOUNDED AT ABOVE EFFECTIVE RATE (i) PER YEAR

N	NEG EXPON	#1	SURVIVOR CURVES #2	#3	#4	#5	RECT LIFE	N
3	0.42650	0.40413	0.39875	0.39469	0.39212	0.39008	0.38803	3
4	0.33987	0.31804	0.31268	0.30860	0.30603	0.30397	0.30192	4
5	0.28790	0.26657	0.26124	0.25715	0.25458	0.25251	0.25046	5
6	0.25325	0.23241	0.22712	0.22302	0.22045	0.21838	0.21632	6
7	0.22850	0.20814	0.20288	0.19878	0.19621	0.19413	0.19207	7
8	0.20994	0.19003	0.18482	0.18072	0.17815	0.17608	0.17401	8
9	0.19550	0.17604	0.17088	0.16678	0.16422	0.16214	0.16008	9
10	0.18395	0.16492	0.15981	0.15571	0.15316	0.15109	0.14903	10
11	0.17450	0.15589	0.15082	0.14674	0.14420	0.14213	0.14008	11
12	0.16662	0.14841	0.14340	0.13933	0.13681	0.13474	0.13270	12
13	0.15996	0.14214	0.13719	0.13313	0.13062	0.12856	0.12652	13
14	0.15425	0.13680	0.13191	0.12787	0.12537	0.12332	0.12130	14
15	0.14930	0.13221	0.12738	0.12336	0.12088	0.11884	0.11683	15
16	0.14497	0.12823	0.12346	0.11947	0.11701	0.11498	0.11298	16
17	0.14115	0.12475	0.12004	0.11607	0.11363	0.11161	0.10963	17
18	0.13775	0.12168	0.11703	0.11309	0.11068	0.10867	0.10670	18
19	0.13471	0.11896	0.11438	0.11046	0.10807	0.10608	0.10413	19
20	0.13197	0.11653	0.11201	0.10812	0.10576	0.10378	0.10185	20
21	0.12950	0.11436	0.10990	0.10604	0.10370	0.10174	0.09983	21
22	0.12725	0.11239	0.10800	0.10417	0.10186	0.09992	0.09803	22
23	0.12520	0.11062	0.10629	0.10249	0.10021	0.09829	0.09642	23
24	0.12331	0.10900	0.10473	0.10098	0.09872	0.09682	0.09498	24
25	0.12158	0.10753	0.10333	0.09960	0.09738	0.09550	0.09368	25
26	0.11998	0.10619	0.10204	0.09836	0.09616	0.09430	0.09251	26
27	0.11850	0.10495	0.10087	0.09722	0.09505	0.09322	0.09145	27
28	0.11712	0.10382	0.09979	0.09618	0.09404	0.09223	0.09049	28
29	0.11584	0.10277	0.09880	0.09523	0.09312	0.09134	0.08962	29
30	0.11465	0.10180	0.09789	0.09436	0.09228	0.09052	0.08883	30
31	0.11353	0.10090	0.09705	0.09355	0.09151	0.08977	0.08811	31
32	0.11248	0.10006	0.09627	0.09281	0.09080	0.08908	0.08745	32
33	0.11150	0.09928	0.09555	0.09213	0.09015	0.08846	0.08685	33
34	0.11057	0.09855	0.09488	0.09149	0.08954	0.08788	0.08630	34
35	0.10970	0.09787	0.09425	0.09091	0.08899	0.08735	0.08580	35
40	0.10599	0.09505	0.09169	0.08853	0.08677	0.08526	0.08386	40
45	0.10310	0.09293	0.08981	0.08684	0.08522	0.08384	0.08259	45
50	0.10079	0.09130	0.08839	0.08559	0.08412	0.08286	0.08174	50
55	0.09890	0.09000	0.08729	0.08465	0.08331	0.08216	0.08118	55
60	0.09732	0.08896	0.08642	0.08393	0.08270	0.08166	0.08080	60
65	0.09599	0.08810	0.08571	0.08337	0.08225	0.08130	0.08054	65
70	0.09485	0.08738	0.08513	0.08292	0.08189	0.08103	0.08037	70
75	0.09386	0.08677	0.08465	0.08255	0.08161	0.08083	0.08025	75
∞	0.08000	0.08000	0.08000	0.08000	0.08000	0.08000	0.08000	∞

FOR DISPERSED ANNUITY FACTOR FOR A FUTURE AMOUNT, SUBTRACT I

9.0%

DISPERSED ANNUITY FACTORS FROM A PRESENT AMOUNT

FACTORS APPLICABLE TO AN INITIAL AMOUNT
HAVING AN AVERAGE LIFE OF (N) YEARS
FOR VARIOUS SURVIVOR CURVES
ANNUALLY COMPOUNDED AT ABOVE EFFECTIVE RATE (i) PER YEAR

N	NEG EXPON	#1	SURVIVOR CURVES #2	#3	#4	#5	RECT LIFE	N
3	0.43812	0.41318	0.40713	0.40256	0.39966	0.39736	0.39505	3
4	0.35109	0.32680	0.32079	0.31619	0.31330	0.31098	0.30867	4
5	0.29887	0.27521	0.26924	0.26463	0.26173	0.25941	0.25709	5
6	0.26406	0.24101	0.23508	0.23047	0.22757	0.22524	0.22292	6
7	0.23919	0.21672	0.21085	0.20624	0.20335	0.20101	0.19869	7
8	0.22054	0.19863	0.19282	0.18821	0.18533	0.18299	0.18067	8
9	0.20604	0.18467	0.17892	0.17432	0.17145	0.16911	0.16680	9
10	0.19444	0.17359	0.16790	0.16331	0.16046	0.15813	0.15582	10
11	0.18494	0.16459	0.15898	0.15440	0.15157	0.14924	0.14695	11
12	0.17703	0.15717	0.15162	0.14707	0.14425	0.14194	0.13965	12
13	0.17033	0.15094	0.14547	0.14094	0.13814	0.13584	0.13357	13
14	0.16460	0.14564	0.14025	0.13575	0.13298	0.13069	0.12843	14
15	0.15962	0.14110	0.13579	0.13132	0.12857	0.12630	0.12406	15
16	0.15527	0.13717	0.13193	0.12749	0.12477	0.12252	0.12030	16
17	0.15143	0.13373	0.12857	0.12417	0.12148	0.11924	0.11705	17
18	0.14802	0.13070	0.12562	0.12126	0.11860	0.11638	0.11421	18
19	0.14497	0.12802	0.12302	0.11869	0.11607	0.11388	0.11173	19
20	0.14222	0.12563	0.12071	0.11642	0.11384	0.11166	0.10955	20
21	0.13973	0.12349	0.11865	0.11441	0.11185	0.10971	0.10762	21
22	0.13747	0.12156	0.11680	0.11260	0.11009	0.10796	0.10590	22
23	0.13541	0.11982	0.11514	0.11098	0.10850	0.10641	0.10438	23
24	0.13351	0.11824	0.11364	0.10953	0.10708	0.10502	0.10302	24
25	0.13177	0.11680	0.11227	0.10821	0.10581	0.10377	0.10181	25
26	0.13017	0.11549	0.11103	0.10702	0.10465	0.10264	0.10072	26
27	0.12868	0.11428	0.10990	0.10593	0.10361	0.10163	0.09973	27
28	0.12730	0.11317	0.10887	0.10494	0.10266	0.10071	0.09885	28
29	0.12601	0.11214	0.10791	0.10404	0.10179	0.09988	0.09806	29
30	0.12481	0.11120	0.10704	0.10321	0.10101	0.09912	0.09734	30
31	0.12369	0.11032	0.10623	0.10245	0.10029	0.09844	0.09669	31
32	0.12264	0.10950	0.10548	0.10175	0.09963	0.09781	0.09610	32
33	0.12165	0.10875	0.10479	0.10111	0.09902	0.09724	0.09556	33
34	0.12072	0.10804	0.10415	0.10052	0.09847	0.09671	0.09508	34
35	0.11984	0.10738	0.10355	0.09997	0.09796	0.09624	0.09464	35
40	0.11611	0.10464	0.10111	0.09776	0.09594	0.09438	0.09296	40
45	0.11321	0.10259	0.09933	0.09620	0.09455	0.09314	0.09190	45
50	0.11089	0.10100	0.09799	0.09506	0.09357	0.09231	0.09123	50
55	0.10899	0.09975	0.09695	0.09421	0.09286	0.09173	0.09079	55
60	0.10741	0.09874	0.09613	0.09355	0.09234	0.09132	0.09051	60
65	0.10607	0.09791	0.09546	0.09304	0.09194	0.09103	0.09033	65
70	0.10492	0.09721	0.09492	0.09264	0.09164	0.09082	0.09022	70
75	0.10392	0.09663	0.09447	0.09231	0.09140	0.09066	0.09014	75
∞	0.09000	0.09000	0.09000	0.09000	0.09000	0.09000	0.09000	∞

FOR DISPERSED ANNUITY FACTOR FOR A FUTURE AMOUNT, SUBTRACT I

10.0%

DISPERSED ANNUITY FACTORS FROM A PRESENT AMOUNT

FACTORS APPLICABLE TO AN INITIAL AMOUNT
HAVING AN AVERAGE LIFE OF (N) YEARS
FOR VARIOUS SURVIVOR CURVES
ANNUALLY COMPOUNDED AT ABOVE EFFECTIVE RATE (i) PER YEAR

N	NEG EXPON	#1	#2	#3	#4	#5	RECT LIFE	N
3	0.44974	0.42226	0.41555	0.41046	0.40724	0.40468	0.4021ı	3
4	0.36230	0.33561	0.32895	0.32384	0.32062	0.31804	0.31547	4
5	0.30984	0.28391	0.27731	0.27218	0.26896	0.26637	0.26380	5
6	0.27487	0.24967	0.24312	0.23799	0.23478	0.23218	0.22961	6
7	0.24989	0.22538	0.21891	0.21378	0.21058	0.20798	0.20541	7
8	0.23115	0.20732	0.20092	0.19580	0.19261	0.19001	0.18744	8
9	0.21658	0.19339	0.18707	0.18197	0.17879	0.17620	0.17364	9
10	0.20492	0.18235	0.17611	0.17103	0.16788	0.16530	0.16275	10
11	0.19538	0.17340	0.16725	0.16220	0.15907	0.15650	0.15396	11
12	0.18743	0.16602	0.15996	0.15494	0.15184	0.14928	0.14676	12
13	0.18071	0.15983	0.15388	0.14888	0.14581	0.14328	0.14078	13
14	0.17494	0.15459	0.14873	0.14377	0.14074	0.13822	0.13575	14
15	0.16995	0.15009	0.14433	0.13942	0.13642	0.13392	0.13147	15
16	0.16558	0.14620	0.14054	0.13567	0.13271	0.13024	0.12782	16
17	0.16172	0.14281	0.13724	0.13242	0.12950	0.12705	0.12466	17
18	0.15829	0.13982	0.13435	0.12958	0.12670	0.12429	0.12193	18
19	0.15522	0.13718	0.13181	0.12709	0.12425	0.12187	0.11955	19
20	0.15246	0.13483	0.12955	0.12489	0.12210	0.11974	0.11746	20
21	0.14996	0.13273	0.12754	0.12294	0.12019	0.11787	0.11562	21
22	0.14769	0.13083	0.12575	0.12119	0.11850	0.11621	0.11401	22
23	0.14562	0.12912	0.12413	0.11964	0.11699	0.11474	0.11257	23
24	0.14372	0.12757	0.12267	0.11824	0.11563	0.11342	0.11130	24
25	0.14197	0.12616	0.12135	0.11698	0.11442	0.11225	0.11017	25
26	0.14035	0.12488	0.12016	0.11584	0.11333	0.11119	0.10916	26
27	0.13886	0.12369	0.11906	0.11480	0.11235	0.11025	0.10826	27
28	0.13747	0.12261	0.11806	0.11386	0.11146	0.10940	0.10745	28
29	0.13618	0.12161	0.11715	0.11301	0.11065	0.10863	0.10673	29
30	0.13497	0.12068	0.11630	0.11222	0.10991	0.10793	0.10608	30
31	0.13385	0.11983	0.11553	0.11151	0.10924	0.10730	0.10550	31
32	0.13279	0.11903	0.11481	0.11085	0.10863	0.10673	0.10497	32
33	0.13179	0.11829	0.11415	0.11024	0.10808	0.10622	0.10450	33
34	0.13086	0.11760	0.11353	0.10969	0.10757	0.10574	0.10407	34
35	0.12998	0.11696	0.11296	0.10917	0.10710	0.10532	0.10369	35
40	0.12623	0.11430	0.11063	0.10712	0.10526	0.10367	0.10226	40
45	0.12332	0.11230	0.10894	0.10567	0.10401	0.10260	0.10139	45
50	0.12098	0.11076	0.10766	0.10462	0.10314	0.10189	0.10086	50
55	0.11908	0.10955	0.10668	0.10384	0.10252	0.10141	0.10053	55
60	0.11749	0.10857	0.10590	0.10325	0.10206	0.10107	0.10033	60
65	0.11614	0.10776	0.10527	0.10279	0.10171	0.10084	0.10020	65
70	0.11499	0.10708	0.10475	0.10242	0.10145	0.10067	0.10013	70
75	0.11399	0.10651	0.10432	0.10212	0.10124	0.10054	0.10008	75
∞	0.10000	0.10000	0.10000	0.10000	0.10000	0.10000	0.10000	∞

FOR DISPERSED ANNUITY FACTOR FOR A FUTURE AMOUNT, SUBTRACT I

11.0%

DISPERSED ANNUITY FACTORS FROM A PRESENT AMOUNT

FACTORS APPLICABLE TO AN INITIAL AMOUNT
HAVING AN AVERAGE LIFE OF (N) YEARS
FOR VARIOUS SURVIVOR CURVES
ANNUALLY COMPOUNDED AT ABOVE EFFECTIVE RATE (i) PER YEAR

N	NEG EXPON	#1	SURVIVOR CURVES #2	#3	#4	#5	RECT LIFE	N
3	0.46135	0.43137	0.42401	0.41840	0.41486	0.41203	0.40921	3
4	0.37351	0.34448	0.33717	0.33154	0.32800	0.32516	0.32233	4
5	0.32081	0.29267	0.28544	0.27979	0.27626	0.27340	0.27057	5
6	0.28567	0.25839	0.25124	0.24560	0.24207	0.23921	0.23638	6
7	0.26058	0.23412	0.22705	0.22142	0.21790	0.21505	0.21222	7
8	0.24176	0.21608	0.20911	0.20349	0.19999	0.19714	0.19432	8
9	0.22712	0.20219	0.19532	0.18972	0.18625	0.18341	0.18060	9
10	0.21540	0.19119	0.18443	0.17886	0.17542	0.17259	0.16980	10
11	0.20582	0.18229	0.17564	0.17011	0.16670	0.16389	0.16112	11
12	0.19784	0.17496	0.16842	0.16293	0.15956	0.15677	0.15403	12
13	0.19108	0.16882	0.16240	0.15696	0.15363	0.15087	0.14815	13
14	0.18529	0.16363	0.15732	0.15193	0.14865	0.14591	0.14323	14
15	0.18027	0.15918	0.15299	0.14766	0.14442	0.14171	0.13907	15
16	0.17588	0.15533	0.14926	0.14399	0.14080	0.13813	0.13552	16
17	0.17200	0.15198	0.14603	0.14081	0.13768	0.13504	0.13247	17
18	0.16856	0.14903	0.14320	0.13805	0.13497	0.13237	0.12984	18
19	0.16548	0.14643	0.14071	0.13563	0.13260	0.13004	0.12756	19
20	0.16270	0.14412	0.13851	0.13350	0.13052	0.12801	0.12558	20
21	0.16019	0.14205	0.13655	0.13161	0.12869	0.12622	0.12384	21
22	0.15791	0.14019	0.13481	0.12993	0.12707	0.12464	0.12231	22
23	0.15583	0.13851	0.13324	0.12843	0.12563	0.12325	0.12097	23
24	0.15392	0.13699	0.13182	0.12709	0.12435	0.12201	0.11979	24
25	0.15216	0.13560	0.13054	0.12589	0.12320	0.12091	0.11874	25
26	0.15054	0.13434	0.12939	0.12480	0.12217	0.11993	0.11781	26
27	0.14904	0.13319	0.12833	0.12381	0.12125	0.11905	0.11699	27
28	0.14764	0.13212	0.12736	0.12292	0.12041	0.11826	0.11626	28
29	0.14635	0.13115	0.12648	0.12211	0.11966	0.11756	0.11561	29
30	0.14513	0.13024	0.12567	0.12137	0.11897	0.11692	0.11502	30
31	0.14400	0.12941	0.12492	0.12069	0.11835	0.11635	0.11451	31
32	0.14294	0.12863	0.12423	0.12007	0.11779	0.11583	0.11404	32
33	0.14194	0.12791	0.12360	0.11950	0.11727	0.11536	0.11363	33
34	0.14100	0.12723	0.12301	0.11898	0.11680	0.11494	0.11326	34
35	0.14012	0.12660	0.12246	0.11849	0.11637	0.11455	0.11293	35
40	0.13635	0.12401	0.12023	0.11657	0.11470	0.11310	0.11172	40
45	0.13342	0.12206	0.11861	0.11523	0.11358	0.11217	0.11101	45
50	0.13108	0.12057	0.11739	0.11427	0.11280	0.11157	0.11060	50
55	0.12916	0.11938	0.11645	0.11355	0.11225	0.11117	0.11035	55
60	0.12757	0.11842	0.11571	0.11300	0.11184	0.11089	0.11021	60
65	0.12622	0.11764	0.11510	0.11258	0.11154	0.11070	0.11012	65
70	0.12506	0.11698	0.11461	0.11224	0.11130	0.11056	0.11007	70
75	0.12405	0.11642	0.11420	0.11197	0.11112	0.11046	0.11004	75
∞	0.11000	0.11000	0.11000	0.11000	0.11000	0.11000	0.11000	∞

FOR DISPERSED ANNUITY FACTOR FOR A FUTURE AMOUNT, SUBTRACT I

12.0%

DISPERSED ANNUITY FACTORS FROM A PRESENT AMOUNT

FACTORS APPLICABLE TO AN INITIAL AMOUNT
HAVING AN AVERAGE LIFE OF (N) YEARS
FOR VARIOUS SURVIVOR CURVES
ANNUALLY COMPOUNDED AT ABOVE EFFECTIVE RATE (i) PER YEAR

N	NEG EXPON	#1	SURVIVOR CURVES #2	#3	#4	#5	RECT LIFE	N
3	0.47296	0.44052	0.43251	0.42638	0.42252	0.41943	0.41635	3
4	0.38472	0.35338	0.34544	0.33929	0.33543	0.33232	0.32923	4
5	0.33177	0.30149	0.29363	0.28747	0.28361	0.28050	0.27741	5
6	0.29648	0.26718	0.25943	0.25328	0.24943	0.24632	0.24323	6
7	0.27127	0.24292	0.23527	0.22913	0.22531	0.22220	0.21912	7
8	0.25236	0.22491	0.21738	0.21127	0.20747	0.20437	0.20130	8
9	0.23765	0.21106	0.20366	0.19758	0.19381	0.19073	0.18768	9
10	0.22589	0.20011	0.19284	0.18680	0.18308	0.18001	0.17698	10
11	0.21626	0.19126	0.18412	0.17813	0.17445	0.17141	0.16842	11
12	0.20824	0.18398	0.17698	0.17104	0.16741	0.16440	0.16144	12
13	0.20145	0.17790	0.17103	0.16516	0.16158	0.15860	0.15568	13
14	0.19563	0.17275	0.16602	0.16022	0.15670	0.15375	0.15087	14
15	0.19059	0.16834	0.16176	0.15603	0.15256	0.14966	0.14682	15
16	0.18618	0.16454	0.15810	0.15244	0.14904	0.14617	0.14339	16
17	0.18229	0.16123	0.15492	0.14934	0.14600	0.14319	0.14046	17
18	0.17883	0.15833	0.15215	0.14665	0.14338	0.14061	0.13794	18
19	0.17573	0.15576	0.14972	0.14430	0.14110	0.13838	0.13576	19
20	0.17294	0.15348	0.14758	0.14224	0.13910	0.13644	0.13388	20
21	0.17042	0.15144	0.14567	0.14042	0.13735	0.13474	0.13224	21
22	0.16813	0.14962	0.14397	0.13880	0.13580	0.13324	0.13081	22
23	0.16604	0.14797	0.14244	0.13736	0.13443	0.13193	0.12956	23
24	0.16412	0.14647	0.14107	0.13607	0.13321	0.13077	0.12846	24
25	0.16235	0.14511	0.13983	0.13492	0.13213	0.12974	0.12750	25
26	0.16073	0.14388	0.13871	0.13388	0.13116	0.12883	0.12665	26
27	0.15922	0.14275	0.13769	0.13294	0.13029	0.12801	0.12590	27
28	0.15782	0.14170	0.13676	0.13209	0.12951	0.12729	0.12524	28
29	0.15651	0.14075	0.13590	0.13132	0.12880	0.12664	0.12466	29
30	0.15530	0.13986	0.13512	0.13062	0.12816	0.12606	0.12414	30
31	0.15416	0.13904	0.13440	0.12998	0.12759	0.12554	0.12369	31
32	0.15309	0.13828	0.13374	0.12939	0.12707	0.12507	0.12328	32
33	0.15209	0.13757	0.13313	0.12886	0.12659	0.12465	0.12292	33
34	0.15114	0.13691	0.13256	0.12836	0.12616	0.12427	0.12260	34
35	0.15025	0.13630	0.13203	0.12791	0.12576	0.12393	0.12232	35
40	0.14647	0.13376	0.12989	0.12611	0.12424	0.12264	0.12130	40
45	0.14353	0.13186	0.12834	0.12487	0.12322	0.12184	0.12074	45
50	0.14118	0.13040	0.12717	0.12397	0.12252	0.12133	0.12042	50
55	0.13925	0.12924	0.12627	0.12331	0.12203	0.12099	0.12024	55
60	0.13765	0.12831	0.12555	0.12280	0.12167	0.12076	0.12013	60
65	0.13629	0.12754	0.12497	0.12241	0.12140	0.12060	0.12008	65
70	0.13513	0.12689	0.12450	0.12210	0.12119	0.12048	0.12004	70
75	0.13412	0.12634	0.12410	0.12185	0.12103	0.12040	0.12002	75
∞	0.12000	0.12000	0.12000	0.12000	0.12000	0.12000	0.12000	∞

FOR DISPERSED ANNUITY FACTOR FOR A FUTURE AMOUNT, SUBTRACT I

13.0%

DISPERSED ANNUITY FACTORS FROM A PRESENT AMOUNT

FACTORS APPLICABLE TO AN INITIAL AMOUNT
HAVING AN AVERAGE LIFE OF (N) YEARS
FOR VARIOUS SURVIVOR CURVES
ANNUALLY COMPOUNDED AT ABOVE EFFECTIVE RATE (i) PER YEAR

N	NEG EXPON	#1	#2	SURVIVOR CURVES #3	#4	#5	RECT LIFE	N
3	0.48456	0.44971	0.44104	0.43440	0.43021	0.42686	0.42352	3
4	0.39592	0.36234	0.35376	0.34709	0.34291	0.33954	0.33619	4
5	0.34274	0.31035	0.30188	0.29521	0.29104	0.28766	0.28431	5
6	0.30728	0.27603	0.26768	0.26103	0.25687	0.25350	0.25015	6
7	0.28195	0.25179	0.24357	0.23693	0.23281	0.22944	0.22611	7
8	0.26296	0.23381	0.22574	0.21914	0.21505	0.21170	0.20839	8
9	0.24819	0.22001	0.21208	0.20553	0.20148	0.19815	0.19487	9
10	0.23637	0.20911	0.20134	0.19484	0.19084	0.18754	0.18429	10
11	0.22670	0.20031	0.19270	0.18627	0.18232	0.17905	0.17584	11
12	0.21864	0.19308	0.18563	0.17927	0.17539	0.17215	0.16899	12
13	0.21182	0.18704	0.17976	0.17347	0.16966	0.16647	0.16335	13
14	0.20598	0.18194	0.17482	0.16862	0.16487	0.16173	0.15867	14
15	0.20091	0.17759	0.17063	0.16451	0.16084	0.15775	0.15474	15
16	0.19648	0.17383	0.16703	0.16101	0.15741	0.15437	0.15143	16
17	0.19257	0.17056	0.16392	0.15799	0.15447	0.15149	0.14861	17
18	0.18909	0.16769	0.16121	0.15537	0.15193	0.14901	0.14620	18
19	0.18598	0.16516	0.15883	0.15309	0.14973	0.14687	0.14413	19
20	0.18318	0.16291	0.15674	0.15110	0.14782	0.14502	0.14235	20
21	0.18065	0.16091	0.15488	0.14934	0.14614	0.14341	0.14081	21
22	0.17835	0.15911	0.15322	0.14778	0.14466	0.14200	0.13948	22
23	0.17625	0.15749	0.15174	0.14640	0.14336	0.14076	0.13832	23
24	0.17432	0.15602	0.15040	0.14516	0.14221	0.13967	0.13731	24
25	0.17255	0.15468	0.14920	0.14406	0.14118	0.13871	0.13643	25
26	0.17091	0.15347	0.14812	0.14307	0.14027	0.13787	0.13565	26
27	0.16940	0.15236	0.14713	0.14217	0.13945	0.13712	0.13498	27
28	0.16799	0.15134	0.14622	0.14136	0.13872	0.13645	0.13439	28
29	0.16668	0.15040	0.14540	0.14063	0.13806	0.13585	0.13387	29
30	0.16546	0.14953	0.14464	0.13997	0.13747	0.13533	0.13341	30
31	0.16431	0.14872	0.14395	0.13936	0.13693	0.13485	0.13301	31
32	0.16324	0.14798	0.14331	0.13881	0.13645	0.13443	0.13266	32
33	0.16223	0.14729	0.14272	0.13830	0.13601	0.13405	0.13234	33
34	0.16128	0.14664	0.14217	0.13783	0.13561	0.13371	0.13207	34
35	0.16039	0.14604	0.14166	0.13741	0.13525	0.13340	0.13183	35
40	0.15659	0.14356	0.13960	0.13572	0.13385	0.13227	0.13099	40
45	0.15364	0.14170	0.13811	0.13456	0.13293	0.13158	0.13053	45
50	0.15127	0.14026	0.13698	0.13372	0.13230	0.13114	0.13029	50
55	0.14934	0.13913	0.13611	0.13310	0.13185	0.13085	0.13016	55
60	0.14773	0.13821	0.13542	0.13263	0.13153	0.13065	0.13009	60
65	0.14636	0.13745	0.13486	0.13227	0.13129	0.13052	0.13005	65
70	0.14520	0.13682	0.13440	0.13198	0.13110	0.13042	0.13003	70
75	0.14418	0.13628	0.13402	0.13175	0.13095	0.13035	0.13001	75
∞	0.13000	0.13000	0.13000	0.13000	0.13000	0.13000	0.13000	∞

FOR DISPERSED ANNUITY FACTOR FOR A FUTURE AMOUNT, SUBTRACT I

Engineering Economy

14.0%

DISPERSED ANNUITY FACTORS FROM A PRESENT AMOUNT

FACTORS APPLICABLE TO AN INITIAL AMOUNT
HAVING AN AVERAGE LIFE OF (N) YEARS
FOR VARIOUS SURVIVOR CURVES
ANNUALLY COMPOUNDED AT ABOVE EFFECTIVE RATE (i) PER YEAR

N	NEG EXPON	#1	SURVIVOR CURVES #2	#3	#4	#5	RECT LIFE	N
3	0.49616	0.45893	0.44961	0.44245	0.43794	0.43433	0.43073	3
4	0.40712	0.37133	0.36212	0.35495	0.35044	0.34681	0.34320	4
5	0.35369	0.31927	0.31019	0.30301	0.29852	0.29489	0.29128	5
6	0.31808	0.28494	0.27601	0.26885	0.26438	0.26075	0.25716	6
7	0.29264	0.26071	0.25194	0.24481	0.24039	0.23677	0.23319	7
8	0.27356	0.24278	0.23418	0.22710	0.22272	0.21912	0.21557	8
9	0.25872	0.22902	0.22060	0.21358	0.20925	0.20568	0.20217	9
10	0.24685	0.21817	0.20993	0.20298	0.19872	0.19519	0.19171	10
11	0.23713	0.20942	0.20137	0.19450	0.19031	0.18682	0.18339	11
12	0.22904	0.20224	0.19438	0.18760	0.18348	0.18004	0.17667	12
13	0.22219	0.19626	0.18858	0.18189	0.17786	0.17447	0.17116	13
14	0.21632	0.19121	0.18372	0.17713	0.17317	0.16985	0.16661	14
15	0.21123	0.18689	0.17959	0.17311	0.16924	0.16597	0.16281	15
16	0.20678	0.18318	0.17605	0.16968	0.16590	0.16270	0.15962	16
17	0.20285	0.17994	0.17300	0.16674	0.16305	0.15992	0.15692	17
18	0.19936	0.17711	0.17035	0.16420	0.16061	0.15755	0.15462	18
19	0.19624	0.17462	0.16802	0.16199	0.15849	0.15550	0.15266	19
20	0.19342	0.17240	0.16598	0.16006	0.15665	0.15374	0.15099	20
21	0.19088	0.17043	0.16416	0.15836	0.15505	0.15221	0.14954	21
22	0.18857	0.16866	0.16255	0.15686	0.15364	0.15088	0.14830	22
23	0.18646	0.16706	0.16111	0.15553	0.15240	0.14972	0.14723	23
24	0.18452	0.16561	0.15981	0.15435	0.15131	0.14871	0.14630	24
25	0.18274	0.16430	0.15864	0.15329	0.15034	0.14781	0.14550	25
26	0.18110	0.16311	0.15759	0.15234	0.14948	0.14703	0.14480	26
27	0.17957	0.16202	0.15663	0.15149	0.14872	0.14634	0.14419	27
28	0.17816	0.16101	0.15576	0.15072	0.14803	0.14573	0.14366	28
29	0.17684	0.16009	0.15496	0.15002	0.14742	0.14518	0.14320	29
30	0.17562	0.15924	0.15423	0.14939	0.14686	0.14470	0.14280	30
31	0.17447	0.15845	0.15355	0.14882	0.14637	0.14427	0.14245	31
32	0.17339	0.15772	0.15293	0.14829	0.14592	0.14389	0.14215	32
33	0.17238	0.15704	0.15236	0.14781	0.14551	0.14355	0.14188	33
34	0.17143	0.15640	0.15183	0.14738	0.14514	0.14325	0.14165	34
35	0.17053	0.15581	0.15134	0.14697	0.14481	0.14297	0.14144	35
40	0.16671	0.15338	0.14935	0.14539	0.14353	0.14197	0.14075	40
45	0.16374	0.15155	0.14791	0.14429	0.14269	0.14137	0.14039	45
50	0.16137	0.15015	0.14682	0.14351	0.14212	0.14099	0.14020	50
55	0.15943	0.14903	0.14598	0.14293	0.14171	0.14074	0.14010	55
60	0.15781	0.14813	0.14531	0.14249	0.14142	0.14058	0.14005	60
65	0.15644	0.14739	0.14477	0.14215	0.14119	0.14046	0.14003	65
70	0.15526	0.14676	0.14433	0.14188	0.14102	0.14038	0.14001	70
75	0.15425	0.14623	0.14395	0.14166	0.14089	0.14032	0.14001	75
∞	0.14000	0.14000	0.14000	0.14000	0.14000	0.14000	0.14000	∞

FOR DISPERSED ANNUITY FACTOR FOR A FUTURE AMOUNT, SUBTRACT I

15.0%

DISPERSED ANNUITY FACTORS FROM A PRESENT AMOUNT

FACTORS APPLICABLE TO AN INITIAL AMOUNT
HAVING AN AVERAGE LIFE OF (N) YEARS
FOR VARIOUS SURVIVOR CURVES
ANNUALLY COMPOUNDED AT ABOVE EFFECTIVE RATE (i) PER YEAR

N	NEG EXPON	#1	SURVIVOR CURVES #2	#3	#4	#5	RECT LIFE	N
3	0.50775	0.46818	0.45822	0.45054	0.44571	0.44184	0.43798	3
4	0.41831	0.38037	0.37054	0.36284	0.35802	0.35413	0.35027	4
5	0.36465	0.32823	0.31856	0.31087	0.30607	0.30218	0.29832	5
6	0.32888	0.29390	0.28439	0.27674	0.27197	0.26808	0.26424	6
7	0.30332	0.26970	0.26038	0.25277	0.24805	0.24418	0.24036	7
8	0.28416	0.25180	0.24269	0.23514	0.23047	0.22664	0.22285	8
9	0.26925	0.23810	0.22919	0.22171	0.21711	0.21331	0.20957	9
10	0.25733	0.22730	0.21860	0.21121	0.20669	0.20294	0.19925	10
11	0.24757	0.21860	0.21012	0.20283	0.19839	0.19469	0.19107	11
12	0.23944	0.21147	0.20321	0.19602	0.19168	0.18803	0.18448	12
13	0.23256	0.20554	0.19748	0.19041	0.18616	0.18259	0.17911	13
14	0.22666	0.20053	0.19269	0.18574	0.18159	0.17808	0.17469	14
15	0.22155	0.19626	0.18863	0.18180	0.17775	0.17433	0.17102	15
16	0.21708	0.19258	0.18516	0.17846	0.17451	0.17117	0.16795	16
17	0.21313	0.18939	0.18216	0.17559	0.17176	0.16849	0.16537	17
18	0.20963	0.18659	0.17956	0.17312	0.16940	0.16621	0.16319	18
19	0.20649	0.18413	0.17729	0.17098	0.16736	0.16426	0.16134	19
20	0.20366	0.18195	0.17529	0.16911	0.16560	0.16259	0.15976	20
21	0.20111	0.18000	0.17352	0.16747	0.16407	0.16114	0.15842	21
22	0.19878	0.17825	0.17195	0.16603	0.16273	0.15989	0.15727	22
23	0.19666	0.17668	0.17054	0.16475	0.16155	0.15880	0.15628	23
24	0.19472	0.17526	0.16928	0.16362	0.16052	0.15786	0.15543	24
25	0.19293	0.17397	0.16815	0.16260	0.15960	0.15703	0.15470	25
26	0.19128	0.17279	0.16712	0.16170	0.15879	0.15630	0.15407	26
27	0.18975	0.17172	0.16619	0.16088	0.15807	0.15566	0.15353	27
28	0.18833	0.17073	0.16534	0.16015	0.15743	0.15510	0.15306	28
29	0.18701	0.16982	0.16457	0.15949	0.15686	0.15461	0.15265	29
30	0.18578	0.16898	0.16386	0.15889	0.15634	0.15417	0.15230	30
31	0.18462	0.16821	0.16321	0.15834	0.15588	0.15378	0.15200	31
32	0.18354	0.16749	0.16261	0.15784	0.15546	0.15344	0.15173	32
33	0.18252	0.16682	0.16205	0.15739	0.15509	0.15313	0.15150	33
34	0.18157	0.16620	0.16154	0.15697	0.15474	0.15286	0.15131	34
35	0.18066	0.16562	0.16106	0.15659	0.15443	0.15261	0.15113	35
40	0.17683	0.16323	0.15914	0.15510	0.15326	0.15173	0.15056	40
45	0.17385	0.16143	0.15774	0.15406	0.15249	0.15120	0.15028	45
50	0.17147	0.16005	0.15669	0.15333	0.15196	0.15087	0.15014	50
55	0.16951	0.15895	0.15587	0.15278	0.15159	0.15066	0.15007	55
60	0.16789	0.15806	0.15522	0.15237	0.15132	0.15051	0.15003	60
65	0.16651	0.15733	0.15470	0.15205	0.15112	0.15041	0.15002	65
70	0.16533	0.15672	0.15426	0.15179	0.15096	0.15034	0.15001	70
75	0.16431	0.15619	0.15390	0.15159	0.15084	0.15029	0.15000	75
∞	0.15000	0.15000	0.15000	0.15000	0.15000	0.15000	0.15000	∞

FOR DISPERSED ANNUITY FACTOR FOR A FUTURE AMOUNT, SUBTRACT I

Annuity from a Present Amount Reflecting Dispersion of Lives

Continuous Compounding
$(\bar{a}/P)$ FACTORS

This appendix contains tables of continuous annuity from present amount factors which recognize the effect of dispersion in the continuous retirement of the initial amount. Tables are provided for *effective* interest rates from 8 to 15%. These tables are similar to the periodic tables in Appendix F, except they provide the rate per year of a *continuous* annuity which will provide for capital repayment and return of an initial amount which is retiring continuously throughout its total period of existence.

The following tables do not include the $(\bar{a}/F)$ factors. To find the $(\bar{a}/F)$ factor, use

$$(\bar{a}/F) = (\bar{a}/P) - r$$

The computer program which develops these continuous factors is similar to the periodic factor program. See Appendix D for a discussion of the program details.

8.0%
DISPERSED ANNUITY FACTORS FROM A PRESENT AMOUNT

FACTORS APPLICABLE TO AN INITIAL AMOUNT
HAVING AN AVERAGE LIFE OF (N) YEARS
FOR VARIOUS SURVIVOR CURVES
CONTINUOUS COMPOUNDING AT ABOVE EFFECTIVE RATE (i) PER YEAR
NOMINAL OR FORCE OF INTEREST RATE (r) =   7.6961 %

N	NEG EXPON	#1	#2	#3	#4	#5	RECT LIFE	N
3	0.41029	0.38878	0.38361	0.37970	0.37723	0.37526	0.37329	3
4	0.32696	0.30596	0.30080	0.29688	0.29440	0.29242	0.29045	4
5	0.27696	0.25644	0.25132	0.24738	0.24491	0.24292	0.24094	5
6	0.24363	0.22358	0.21849	0.21455	0.21207	0.21008	0.20810	6
7	0.21982	0.20023	0.19518	0.19123	0.18875	0.18676	0.18478	7
8	0.20196	0.18282	0.17780	0.17386	0.17138	0.16939	0.16740	8
9	0.18807	0.16935	0.16439	0.16044	0.15798	0.15598	0.15400	9
10	0.17696	0.15866	0.15374	0.14980	0.14734	0.14535	0.14337	10
11	0.16787	0.14997	0.14509	0.14117	0.13872	0.13673	0.13476	11
12	0.16029	0.14278	0.13796	0.13404	0.13161	0.12962	0.12765	12
13	0.15388	0.13674	0.13198	0.12807	0.12566	0.12367	0.12172	13
14	0.14839	0.13160	0.12690	0.12301	0.12061	0.11864	0.11669	14
15	0.14363	0.12719	0.12254	0.11868	0.11629	0.11433	0.11239	15
16	0.13946	0.12336	0.11877	0.11493	0.11256	0.11061	0.10869	16
17	0.13578	0.12001	0.11548	0.11166	0.10932	0.10737	0.10546	17
18	0.13252	0.11706	0.11259	0.10879	0.10647	0.10454	0.10265	18
19	0.12959	0.11444	0.11003	0.10626	0.10396	0.10205	0.10017	19
20	0.12696	0.11211	0.10776	0.10402	0.10174	0.09984	0.09798	20
21	0.12458	0.11001	0.10572	0.10201	0.09976	0.09788	0.09604	21
22	0.12242	0.10812	0.10390	0.10022	0.09799	0.09613	0.09431	22
23	0.12044	0.10642	0.10225	0.09860	0.09640	0.09456	0.09276	23
24	0.11863	0.10486	0.10076	0.09714	0.09497	0.09315	0.09137	24
25	0.11696	0.10345	0.09940	0.09582	0.09368	0.09187	0.09012	25
26	0.11542	0.10216	0.09817	0.09462	0.09251	0.09072	0.08899	26
27	0.11400	0.10097	0.09704	0.09353	0.09144	0.08968	0.08797	27
28	0.11268	0.09987	0.09600	0.09253	0.09047	0.08873	0.08705	28
29	0.11144	0.09886	0.09505	0.09161	0.08959	0.08787	0.08621	29
30	0.11029	0.09793	0.09417	0.09077	0.08878	0.08708	0.08545	30
31	0.10922	0.09706	0.09336	0.09000	0.08803	0.08636	0.08476	31
32	0.10821	0.09626	0.09261	0.08928	0.08735	0.08570	0.08413	32
33	0.10726	0.09551	0.09192	0.08863	0.08672	0.08510	0.08355	33
34	0.10637	0.09481	0.09127	0.08802	0.08614	0.08454	0.08303	34
35	0.10553	0.09415	0.09067	0.08745	0.08561	0.08403	0.08254	35
40	0.10196	0.09144	0.08821	0.08517	0.08347	0.08202	0.08067	40
45	0.09918	0.08940	0.08640	0.08354	0.08199	0.08066	0.07945	45
50	0.09696	0.08783	0.08503	0.08234	0.08092	0.07971	0.07864	50
55	0.09514	0.08659	0.08397	0.08144	0.08014	0.07904	0.07809	55
60	0.09363	0.08558	0.08313	0.08074	0.07956	0.07856	0.07773	60
65	0.09235	0.08475	0.08246	0.08020	0.07912	0.07821	0.07748	65
70	0.09125	0.08406	0.08190	0.07977	0.07878	0.07796	0.07731	70
75	0.09029	0.08348	0.08144	0.07942	0.07851	0.07776	0.07720	75
∞	0.07696	0.07696	0.07696	0.07696	0.07696	0.07696	0.07696	∞

FOR DISPERSED ANNUITY FACTOR FOR A FUTURE AMOUNT, SUBTRACT R

9.0%
DISPERSED ANNUITY FACTORS FROM A PRESENT AMOUNT

FACTORS APPLICABLE TO AN INITIAL AMOUNT
HAVING AN AVERAGE LIFE OF (N) YEARS
FOR VARIOUS SURVIVOR CURVES
CONTINUOUS COMPOUNDING AT ABOVE EFFECTIVE RATE (i) PER YEAR
NOMINAL OR FORCE OF INTEREST RATE (r) =   8.61777 %

N	NEG EXPON	#1	#2	#3	#4	#5	RECT LIFE	N
3	0.41951	0.39563	0.38984	0.38546	0.38269	0.38048	0.37828	3
4	0.33618	0.31292	0.30717	0.30276	0.29999	0.29777	0.29556	4
5	0.28618	0.26352	0.25781	0.25339	0.25062	0.24839	0.24617	5
6	0.25284	0.23077	0.22510	0.22068	0.21791	0.21567	0.21345	6
7	0.22903	0.20752	0.20190	0.19748	0.19471	0.19247	0.19025	7
8	0.21118	0.19020	0.18463	0.18022	0.17746	0.17522	0.17300	8
9	0.19729	0.17683	0.17132	0.16691	0.16416	0.16193	0.15971	9
10	0.18618	0.16621	0.16077	0.15638	0.15364	0.15141	0.14920	10
11	0.17709	0.15760	0.15223	0.14785	0.14513	0.14290	0.14071	11
12	0.16951	0.15049	0.14518	0.14082	0.13812	0.13591	0.13372	12
13	0.16310	0.14453	0.13929	0.13495	0.13227	0.13007	0.12789	13
14	0.15761	0.13946	0.13430	0.12998	0.12733	0.12514	0.12298	14
15	0.15284	0.13511	0.13002	0.12574	0.12311	0.12093	0.11879	15
16	0.14868	0.13134	0.12633	0.12208	0.11947	0.11731	0.11519	16
17	0.14500	0.12805	0.12311	0.11889	0.11632	0.11418	0.11208	17
18	0.14173	0.12515	0.12029	0.11611	0.11356	0.11144	0.10936	18
19	0.13881	0.12258	0.11780	0.11365	0.11114	0.10904	0.10699	19
20	0.13618	0.12030	0.11559	0.11148	0.10900	0.10692	0.10489	20
21	0.13380	0.11825	0.11361	0.10955	0.10710	0.10505	0.10305	21
22	0.13163	0.11640	0.11184	0.10782	0.10541	0.10338	0.10141	22
23	0.12966	0.11473	0.11025	0.10627	0.10390	0.10189	0.09995	23
24	0.12784	0.11322	0.10881	0.10487	0.10254	0.10056	0.09865	24
25	0.12618	0.11184	0.10751	0.10361	0.10131	0.09936	0.09748	25
26	0.12464	0.11058	0.10632	0.10247	0.10021	0.09829	0.09644	26
27	0.12321	0.10943	0.10523	0.10143	0.09921	0.09731	0.09550	27
28	0.12189	0.10836	0.10424	0.10049	0.09830	0.09643	0.09465	28
29	0.12066	0.10738	0.10333	0.09962	0.09747	0.09564	0.09389	29
30	0.11951	0.10647	0.10249	0.09883	0.09672	0.09491	0.09320	30
31	0.11844	0.10563	0.10172	0.09810	0.09603	0.09425	0.09258	31
32	0.11743	0.10485	0.10100	0.09743	0.09540	0.09365	0.09201	32
33	0.11648	0.10413	0.10034	0.09682	0.09482	0.09311	0.09150	33
34	0.11559	0.10345	0.09973	0.09625	0.09429	0.09261	0.09104	34
35	0.11475	0.10282	0.09915	0.09572	0.09380	0.09215	0.09062	35
40	0.11118	0.10019	0.09682	0.09361	0.09186	0.09037	0.08901	40
45	0.10840	0.09823	0.09511	0.09211	0.09054	0.08919	0.08800	45
50	0.10618	0.09671	0.09383	0.09102	0.08960	0.08839	0.08735	50
55	0.10436	0.09552	0.09283	0.09020	0.08892	0.08783	0.08694	55
60	0.10284	0.09455	0.09204	0.08958	0.08842	0.08744	0.08667	60
65	0.10156	0.09375	0.09141	0.08909	0.08804	0.08716	0.08650	65
70	0.10046	0.09309	0.09089	0.08870	0.08775	0.08696	0.08638	70
75	0.09951	0.09252	0.09045	0.08839	0.08752	0.08681	0.08631	75
∞	0.08618	0.08618	0.08618	0.08618	0.08618	0.08618	0.08618	∞

FOR DISPERSED ANNUITY FACTOR FOR A FUTURE AMOUNT, SUBTRACT R

10.0%
DISPERSED ANNUITY FACTORS FROM A PRESENT AMOUNT

FACTORS APPLICABLE TO AN INITIAL AMOUNT
HAVING AN AVERAGE LIFE OF (N) YEARS
FOR VARIOUS SURVIVOR CURVES
CONTINUOUS COMPOUNDING AT ABOVE EFFECTIVE RATE (i) PER YEAR
NOMINAL OR FORCE OF INTEREST RATE (r) = 9.53102 %

N	NEG EXPON	#1	SURVIVOR CURVES #2	#3	#4	#5	RECT LIFE	N
3	0.42864	0.40245	0.39606	0.39121	0.38814	0.38570	0.38326	3
4	0.34531	0.31987	0.31353	0.30865	0.30558	0.30313	0.30068	4
5	0.29531	0.27060	0.26430	0.25942	0.25635	0.25388	0.25143	5
6	0.26198	0.23796	0.23172	0.22683	0.22377	0.22130	0.21884	6
7	0.23817	0.21481	0.20864	0.20376	0.20070	0.19823	0.19577	7
8	0.22031	0.19759	0.19150	0.18662	0.18358	0.18110	0.17865	8
9	0.20642	0.18432	0.17830	0.17343	0.17041	0.16794	0.16550	9
10	0.19531	0.17379	0.16785	0.16301	0.16000	0.15754	0.15511	10
11	0.18622	0.16527	0.15941	0.15459	0.15161	0.14916	0.14674	11
12	0.17864	0.15823	0.15246	0.14767	0.14472	0.14228	0.13988	12
13	0.17223	0.15234	0.14666	0.14190	0.13898	0.13656	0.13418	13
14	0.16674	0.14734	0.14175	0.13703	0.13414	0.13174	0.12938	14
15	0.16198	0.14306	0.13756	0.13288	0.13002	0.12764	0.12531	15
16	0.15781	0.13935	0.13395	0.12931	0.12648	0.12413	0.12182	16
17	0.15413	0.13611	0.13080	0.12621	0.12342	0.12109	0.11882	17
18	0.15087	0.13326	0.12805	0.12350	0.12076	0.11846	0.11621	18
19	0.14794	0.13075	0.12563	0.12113	0.11843	0.11615	0.11394	19
20	0.14531	0.12851	0.12348	0.11903	0.11637	0.11413	0.11195	20
21	0.14293	0.12650	0.12156	0.11717	0.11455	0.11234	0.11020	21
22	0.14076	0.12470	0.11985	0.11551	0.11294	0.11076	0.10866	22
23	0.13879	0.12307	0.11831	0.11403	0.11150	0.10936	0.10729	23
24	0.13698	0.12159	0.11692	0.11269	0.11021	0.10810	0.10608	24
25	0.13531	0.12025	0.11566	0.11149	0.10906	0.10698	0.10500	25
26	0.13377	0.11902	0.11452	0.11041	0.10802	0.10598	0.10404	26
27	0.13235	0.11789	0.11348	0.10942	0.10708	0.10508	0.10318	27
28	0.13102	0.11686	0.11253	0.10852	0.10623	0.10427	0.10241	28
29	0.12979	0.11591	0.11165	0.10771	0.10546	0.10353	0.10172	29
30	0.12864	0.11502	0.11085	0.10696	0.10476	0.10287	0.10110	30
31	0.12757	0.11421	0.11011	0.10628	0.10412	0.10227	0.10055	31
32	0.12656	0.11345	0.10943	0.10565	0.10354	0.10173	0.10005	32
33	0.12561	0.11274	0.10879	0.10507	0.10301	0.10123	0.09960	33
34	0.12472	0.11209	0.10821	0.10454	0.10252	0.10079	0.09919	34
35	0.12388	0.11147	0.10766	0.10405	0.10207	0.10038	0.09883	35
40	0.12031	0.10893	0.10545	0.10209	0.10032	0.09881	0.09746	40
45	0.11753	0.10703	0.10383	0.10072	0.09913	0.09779	0.09664	45
50	0.11531	0.10557	0.10261	0.09972	0.09831	0.09711	0.09613	50
55	0.11349	0.10441	0.10167	0.09897	0.09771	0.09665	0.09582	55
60	0.11198	0.10347	0.10093	0.09841	0.09727	0.09633	0.09562	60
65	0.11069	0.10270	0.10033	0.09797	0.09694	0.09611	0.09550	65
70	0.10960	0.10206	0.09984	0.09762	0.09669	0.09595	0.09543	70
75	0.10864	0.10152	0.09942	0.09733	0.09650	0.09583	0.09539	75
∞	0.09531	0.09531	0.09531	0.09531	0.09531	0.09531	0.09531	∞

FOR DISPERSED ANNUITY FACTOR FOR A FUTURE AMOUNT, SUBTRACT R

11.0%
DISPERSED ANNUITY FACTORS FROM A PRESENT AMOUNT

FACTORS APPLICABLE TO AN INITIAL AMOUNT
HAVING AN AVERAGE LIFE OF (N) YEARS
FOR VARIOUS SURVIVOR CURVES
CONTINUOUS COMPOUNDING AT ABOVE EFFECTIVE RATE (i) PER YEAR
NOMINAL OR FORCE OF INTEREST RATE (r) = 10.436 %

N	NEG EXPON	#1	#2	#3	#4	#5	RECT LIFE	N
3	0.43769	0.40925	0.40227	0.39695	0.39359	0.39091	0.38823	3
4	0.35436	0.32681	0.31988	0.31454	0.31118	0.30848	0.30580	4
5	0.30436	0.27767	0.27080	0.26545	0.26209	0.25939	0.25670	5
6	0.27103	0.24515	0.23836	0.23300	0.22965	0.22695	0.22426	6
7	0.24722	0.22211	0.21541	0.21006	0.20673	0.20402	0.20133	7
8	0.22936	0.20500	0.19838	0.19305	0.18974	0.18703	0.18436	8
9	0.21547	0.19182	0.18530	0.17999	0.17670	0.17401	0.17134	9
10	0.20436	0.18139	0.17497	0.16969	0.16642	0.16374	0.16110	10
11	0.19527	0.17294	0.16663	0.16139	0.15815	0.15549	0.15286	11
12	0.18769	0.16599	0.15978	0.15458	0.15138	0.14873	0.14613	12
13	0.18128	0.16017	0.15407	0.14891	0.14575	0.14313	0.14055	13
14	0.17579	0.15524	0.14925	0.14414	0.14103	0.13843	0.13588	14
15	0.17103	0.15102	0.14515	0.14009	0.13701	0.13445	0.13194	15
16	0.16686	0.14737	0.14161	0.13661	0.13358	0.13104	0.12857	16
17	0.16318	0.14419	0.13854	0.13359	0.13062	0.12812	0.12568	17
18	0.15992	0.14139	0.13586	0.13097	0.12805	0.12558	0.12319	18
19	0.15699	0.13892	0.13350	0.12868	0.12580	0.12337	0.12102	19
20	0.15436	0.13673	0.13141	0.12665	0.12383	0.12144	0.11914	20
21	0.15198	0.13476	0.12955	0.12486	0.12210	0.11975	0.11749	21
22	0.14981	0.13300	0.12789	0.12327	0.12056	0.11825	0.11604	22
23	0.14784	0.13141	0.12641	0.12185	0.11919	0.11693	0.11477	23
24	0.14603	0.12996	0.12506	0.12058	0.11798	0.11576	0.11365	24
25	0.14436	0.12865	0.12385	0.11943	0.11689	0.11471	0.11265	25
26	0.14282	0.12746	0.12275	0.11840	0.11591	0.11378	0.11177	26
27	0.14140	0.12636	0.12175	0.11746	0.11503	0.11295	0.11099	27
28	0.14007	0.12535	0.12083	0.11662	0.11424	0.11220	0.11030	28
29	0.13884	0.12442	0.12000	0.11585	0.11352	0.11153	0.10968	29
30	0.13769	0.12356	0.11923	0.11514	0.11287	0.11093	0.10913	30
31	0.13662	0.12277	0.11852	0.11450	0.11228	0.11038	0.10864	31
32	0.13561	0.12203	0.11787	0.11391	0.11175	0.10989	0.10820	32
33	0.13466	0.12135	0.11726	0.11337	0.11126	0.10945	0.10780	33
34	0.13377	0.12071	0.11670	0.11288	0.11081	0.10905	0.10745	34
35	0.13293	0.12011	0.11618	0.11242	0.11041	0.10868	0.10711	35
40	0.12936	0.11765	0.11407	0.11060	0.10882	0.10730	0.10599	40
45	0.12658	0.11580	0.11253	0.10933	0.10775	0.10642	0.10532	45
50	0.12436	0.11438	0.11137	0.10841	0.10702	0.10585	0.10493	50
55	0.12254	0.11326	0.11048	0.10773	0.10649	0.10547	0.10470	55
60	0.12103	0.11235	0.10977	0.10721	0.10611	0.10521	0.10456	60
65	0.11974	0.11160	0.10920	0.10681	0.10582	0.10502	0.10448	65
70	0.11865	0.11098	0.10873	0.10649	0.10560	0.10489	0.10443	70
75	0.11769	0.11045	0.10834	0.10623	0.10542	0.10479	0.10440	75
∞	0.10436	0.10436	0.10436	0.10436	0.10436	0.10436	0.10436	∞

FOR DISPERSED ANNUITY FACTOR FOR A FUTURE AMOUNT, SUBTRACT R

12.0%
DISPERSED ANNUITY FACTORS FROM A PRESENT AMOUNT

FACTORS APPLICABLE TO AN INITIAL AMOUNT
HAVING AN AVERAGE LIFE OF (N) YEARS
FOR VARIOUS SURVIVOR CURVES
CONTINUOUS COMPOUNDING AT ABOVE EFFECTIVE RATE (i) PER YEAR
NOMINAL OR FORCE OF INTEREST RATE (r) =   11.3329 %

N	NEG EXPON	#1	#2	#3	#4	#5	RECT LIFE	N
3	0.44666	0.41603	0.40846	0.40268	0.39903	0.39611	0.39320	3
4	0.36333	0.33374	0.32623	0.32043	0.31678	0.31385	0.31093	4
5	0.31333	0.28472	0.27730	0.27149	0.26785	0.26491	0.26199	5
6	0.28000	0.25233	0.24500	0.23919	0.23556	0.23262	0.22970	6
7	0.25619	0.22941	0.22219	0.21640	0.21278	0.20985	0.20694	7
8	0.23833	0.21241	0.20530	0.19952	0.19594	0.19301	0.19011	8
9	0.22444	0.19933	0.19233	0.18659	0.18304	0.18012	0.17725	9
10	0.21333	0.18899	0.18212	0.17642	0.17290	0.17000	0.16714	10
11	0.20424	0.18063	0.17389	0.16823	0.16475	0.16188	0.15905	11
12	0.19666	0.17375	0.16714	0.16153	0.15810	0.15526	0.15246	12
13	0.19025	0.16801	0.16152	0.15598	0.15260	0.14978	0.14702	13
14	0.18476	0.16314	0.15679	0.15131	0.14798	0.14520	0.14248	14
15	0.18000	0.15899	0.15277	0.14735	0.14408	0.14134	0.13866	15
16	0.17583	0.15540	0.14931	0.14396	0.14075	0.13805	0.13542	16
17	0.17215	0.15227	0.14631	0.14104	0.13789	0.13523	0.13265	17
18	0.16888	0.14952	0.14370	0.13850	0.13541	0.13280	0.13027	18
19	0.16596	0.14710	0.14140	0.13628	0.13325	0.13069	0.12822	19
20	0.16333	0.14495	0.13937	0.13433	0.13137	0.12885	0.12644	20
21	0.16095	0.14302	0.13757	0.13261	0.12971	0.12725	0.12489	21
22	0.15878	0.14130	0.13596	0.13108	0.12825	0.12584	0.12354	22
23	0.15681	0.13974	0.13452	0.12972	0.12696	0.12459	0.12236	23
24	0.15500	0.13833	0.13323	0.12851	0.12581	0.12350	0.12132	24
25	0.15333	0.13705	0.13206	0.12742	0.12478	0.12253	0.12041	25
26	0.15179	0.13588	0.13100	0.12644	0.12387	0.12166	0.11961	26
27	0.15037	0.13481	0.13003	0.12555	0.12305	0.12090	0.11890	27
28	0.14904	0.13383	0.12915	0.12475	0.12231	0.12021	0.11828	28
29	0.14781	0.13292	0.12835	0.12402	0.12164	0.11960	0.11773	29
30	0.14666	0.13208	0.12761	0.12336	0.12104	0.11905	0.11724	30
31	0.14559	0.13131	0.12693	0.12275	0.12050	0.11856	0.11681	31
32	0.14458	0.13059	0.12630	0.12220	0.12000	0.11812	0.11643	32
33	0.14363	0.12992	0.12572	0.12169	0.11955	0.11772	0.11609	33
34	0.14274	0.12930	0.12519	0.12123	0.11914	0.11736	0.11578	34
35	0.14190	0.12872	0.12469	0.12080	0.11877	0.11704	0.11552	35
40	0.13833	0.12633	0.12267	0.11910	0.11733	0.11582	0.11456	40
45	0.13555	0.12453	0.12120	0.11793	0.11637	0.11507	0.11402	45
50	0.13333	0.12315	0.12010	0.11708	0.11571	0.11458	0.11372	50
55	0.13151	0.12206	0.11925	0.11645	0.11525	0.11426	0.11355	55
60	0.13000	0.12117	0.11857	0.11597	0.11490	0.11404	0.11346	60
65	0.12871	0.12044	0.11803	0.11560	0.11465	0.11389	0.11340	65
70	0.12761	0.11984	0.11758	0.11531	0.11445	0.11378	0.11337	70
75	0.12666	0.11932	0.11720	0.11507	0.11430	0.11370	0.11335	75
∞	0.11333	0.11333	0.11333	0.11333	0.11333	0.11333	0.11333	∞

FOR DISPERSED ANNUITY FACTOR FOR A FUTURE AMOUNT, SUBTRACT R

13.0%
DISPERSED ANNUITY FACTORS FROM A PRESENT AMOUNT

FACTORS APPLICABLE TO AN INITIAL AMOUNT
HAVING AN AVERAGE LIFE OF (N) YEARS
FOR VARIOUS SURVIVOR CURVES,
CONTINUOUS COMPOUNDING AT ABOVE EFFECTIVE RATE (i) PER YEAR
NOMINAL OR FORCE OF INTEREST RATE (r) =   12.2218 %

N	NEG EXPON	#1	SURVIVOR CURVES #2	#3	#4	#5	RECT LIFE	N
3	0.45555	0.42279	0.41464	0.40839	0.40446	0.40131	0.39817	3
4	0.37222	0.34065	0.33258	0.32632	0.32238	0.31922	0.31607	4
5	0.32222	0.29177	0.28381	0.27754	0.27362	0.27044	0.26729	5
6	0.28888	0.25951	0.25166	0.24540	0.24149	0.23832	0.23518	6
7	0.26507	0.23671	0.22899	0.22275	0.21887	0.21571	0.21257	7
8	0.24722	0.21982	0.21223	0.20602	0.20218	0.19903	0.19591	8
9	0.23333	0.20684	0.19939	0.19323	0.18942	0.18629	0.18320	9
10	0.22222	0.19659	0.18929	0.18318	0.17942	0.17631	0.17326	10
11	0.21313	0.18832	0.18117	0.17512	0.17141	0.16834	0.16531	11
12	0.20555	0.18152	0.17452	0.16854	0.16489	0.16185	0.15887	12
13	0.19914	0.17585	0.16900	0.16309	0.15950	0.15650	0.15357	13
14	0.19365	0.17105	0.16436	0.15853	0.15500	0.15205	0.14917	14
15	0.18888	0.16695	0.16041	0.15466	0.15121	0.14830	0.14548	15
16	0.18472	0.16342	0.15703	0.15137	0.14798	0.14513	0.14236	16
17	0.18104	0.16034	0.15411	0.14853	0.14522	0.14242	0.13971	17
18	0.17777	0.15765	0.15156	0.14607	0.14284	0.14009	0.13745	18
19	0.17485	0.15527	0.14932	0.14393	0.14077	0.13808	0.13551	19
20	0.17222	0.15316	0.14735	0.14205	0.13897	0.13634	0.13383	20
21	0.16984	0.15127	0.14561	0.14040	0.13739	0.13482	0.13238	21
22	0.16767	0.14958	0.14405	0.13893	0.13600	0.13350	0.13113	22
23	0.16570	0.14806	0.14265	0.13763	0.13478	0.13233	0.13004	23
24	0.16388	0.14668	0.14140	0.13647	0.13369	0.13131	0.12909	24
25	0.16222	0.14542	0.14027	0.13543	0.13273	0.13041	0.12826	25
26	0.16068	0.14428	0.13925	0.13450	0.13187	0.12961	0.12753	26
27	0.15925	0.14324	0.13832	0.13366	0.13110	0.12891	0.12690	27
28	0.15793	0.14228	0.13747	0.13290	0.13041	0.12828	0.12634	28
29	0.15670	0.14139	0.13670	0.13221	0.12980	0.12772	0.12585	29
30	0.15555	0.14058	0.13598	0.13159	0.12924	0.12722	0.12542	30
31	0.15448	0.13982	0.13533	0.13102	0.12874	0.12678	0.12505	31
32	0.15347	0.13912	0.13473	0.13050	0.12828	0.12638	0.12471	32
33	0.15252	0.13847	0.13417	0.13002	0.12787	0.12603	0.12442	33
34	0.15163	0.13786	0.13366	0.12958	0.12749	0.12571	0.12416	34
35	0.15079	0.13730	0.13318	0.12918	0.12715	0.12542	0.12394	35
40	0.14722	0.13496	0.13125	0.12760	0.12584	0.12435	0.12315	40
45	0.14444	0.13321	0.12984	0.12650	0.12497	0.12370	0.12272	45
50	0.14222	0.13187	0.12878	0.12572	0.12438	0.12329	0.12249	50
55	0.14040	0.13080	0.12796	0.12513	0.12396	0.12301	0.12236	55
60	0.13888	0.12994	0.12731	0.12469	0.12365	0.12283	0.12230	60
65	0.13760	0.12922	0.12679	0.12435	0.12343	0.12271	0.12226	65
70	0.13650	0.12863	0.12636	0.12408	0.12325	0.12261	0.12224	70
75	0.13555	0.12812	0.12600	0.12386	0.12311	0.12255	0.12223	75
∞	0.12222	0.12222	0.12222	0.12222	0.12222	0.12222	0.12222	∞

FOR DISPERSED ANNUITY FACTOR FOR A FUTURE AMOUNT, SUBTRACT R

14.0%
DISPERSED ANNUITY FACTORS FROM A PRESENT AMOUNT

FACTORS APPLICABLE TO AN INITIAL AMOUNT
HAVING AN AVERAGE LIFE OF (N) YEARS
FOR VARIOUS SURVIVOR CURVES
CONTINUOUS COMPOUNDING AT ABOVE EFFECTIVE RATE (i) PER YEAR
NOMINAL OR FORCE OF INTEREST RATE (r) =   13.1028 %

N	NEG EXPON	#1	#2	#3	#4	#5	RECT LIFE	N
3	0.46436	0.42952	0.42080	0.41410	0.40988	0.40650	0.40313	3
4	0.38103	0.34754	0.33892	0.33220	0.32798	0.32459	0.32121	4
5	0.33103	0.29881	0.29031	0.28359	0.27939	0.27599	0.27262	5
6	0.29769	0.26668	0.25832	0.25162	0.24744	0.24404	0.24068	6
7	0.27389	0.24401	0.23580	0.22913	0.22498	0.22160	0.21825	7
8	0.25603	0.22722	0.21917	0.21254	0.20844	0.20508	0.20176	8
9	0.24214	0.21434	0.20646	0.19989	0.19584	0.19250	0.18921	9
10	0.23103	0.20419	0.19648	0.18997	0.18598	0.18268	0.17943	10
11	0.22194	0.19600	0.18846	0.18204	0.17811	0.17485	0.17164	11
12	0.21436	0.18928	0.18192	0.17557	0.17172	0.16850	0.16535	12
13	0.20795	0.18368	0.17650	0.17024	0.16646	0.16329	0.16019	13
14	0.20246	0.17895	0.17194	0.16578	0.16208	0.15896	0.15593	14
15	0.19769	0.17492	0.16808	0.16201	0.15839	0.15534	0.15238	15
16	0.19353	0.17144	0.16477	0.15881	0.15527	0.15228	0.14939	16
17	0.18985	0.16841	0.16192	0.15606	0.15261	0.14968	0.14686	17
18	0.18658	0.16576	0.15943	0.15368	0.15031	0.14745	0.14471	18
19	0.18366	0.16343	0.15726	0.15161	0.14833	0.14554	0.14288	19
20	0.18103	0.16136	0.15534	0.14980	0.14661	0.14389	0.14131	20
21	0.17865	0.15951	0.15364	0.14821	0.14511	0.14246	0.13996	21
22	0.17648	0.15785	0.15213	0.14681	0.14380	0.14121	0.13880	22
23	0.17451	0.15635	0.15078	0.14556	0.14264	0.14013	0.13780	23
24	0.17269	0.15500	0.14957	0.14446	0.14161	0.13918	0.13693	24
25	0.17103	0.15377	0.14848	0.14347	0.14071	0.13834	0.13617	25
26	0.16949	0.15266	0.14749	0.14258	0.13990	0.13761	0.13552	26
27	0.16807	0.15164	0.14659	0.14178	0.13919	0.13696	0.13495	27
28	0.16674	0.15070	0.14577	0.14106	0.13854	0.13639	0.13446	28
29	0.16551	0.14983	0.14503	0.14041	0.13797	0.13588	0.13403	29
30	0.16436	0.14903	0.14434	0.13982	0.13745	0.13543	0.13365	30
31	0.16329	0.14830	0.14371	0.13928	0.13699	0.13503	0.13332	31
32	0.16228	0.14761	0.14313	0.13879	0.13657	0.13467	0.13304	32
33	0.16133	0.14697	0.14260	0.13834	0.13619	0.13435	0.13279	33
34	0.16044	0.14638	0.14210	0.13793	0.13584	0.13407	0.13257	34
35	0.15960	0.14583	0.14164	0.13755	0.13553	0.13381	0.13238	35
40	0.15603	0.14355	0.13978	0.13607	0.13433	0.13287	0.13173	40
45	0.15325	0.14184	0.13843	0.13504	0.13354	0.13231	0.13139	45
50	0.15103	0.14052	0.13741	0.13431	0.13301	0.13195	0.13122	50
55	0.14921	0.13948	0.13663	0.13377	0.13263	0.13172	0.13113	55
60	0.14769	0.13864	0.13600	0.13336	0.13235	0.13157	0.13108	60
65	0.14641	0.13794	0.13549	0.13304	0.13215	0.13146	0.13105	65
70	0.14531	0.13736	0.13508	0.13279	0.13199	0.13138	0.13104	70
75	0.14436	0.13686	0.13473	0.13258	0.13186	0.13132	0.13104	75
∞	0.13103	0.13103	0.13103	0.13103	0.13103	0.13103	0.13103	∞

FOR DISPERSED ANNUITY FACTOR FOR A FUTURE AMOUNT, SUBTRACT R

15.0%
DISPERSED ANNUITY FACTORS FROM A PRESENT AMOUNT

FACTORS APPLICABLE TO AN INITIAL AMOUNT
HAVING AN AVERAGE LIFE OF (N) YEARS
FOR VARIOUS SURVIVOR CURVES
CONTINUOUS COMPOUNDING AT ABOVE EFFECTIVE RATE (i) PER YEAR
NOMINAL OR FORCE OF INTEREST RATE (r) = 13.9762 %

N	NEG EXPON	#1	#2	#3	#4	#5	RECT LIFE	N
3	0.47310	0.43622	0.42694	0.41979	0.41529	0.41168	0.40808	3
4	0.38976	0.35441	0.34525	0.33808	0.33358	0.32996	0.32636	4
5	0.33976	0.30583	0.29682	0.28965	0.28518	0.28155	0.27795	5
6	0.30643	0.27384	0.26498	0.25785	0.25341	0.24979	0.24620	6
7	0.28262	0.25129	0.24261	0.23552	0.23112	0.22752	0.22395	7
8	0.26476	0.23462	0.22612	0.21909	0.21474	0.21117	0.20764	8
9	0.25087	0.22184	0.21354	0.20658	0.20230	0.19875	0.19527	9
10	0.23976	0.21178	0.20368	0.19680	0.19258	0.18909	0.18565	10
11	0.23067	0.20368	0.19578	0.18899	0.18485	0.18140	0.17803	11
12	0.22310	0.19704	0.18934	0.18264	0.17859	0.17520	0.17189	12
13	0.21669	0.19151	0.18401	0.17742	0.17346	0.17013	0.16689	13
14	0.21119	0.18684	0.17954	0.17306	0.16919	0.16593	0.16277	14
15	0.20643	0.18286	0.17575	0.16939	0.16562	0.16243	0.15934	15
16	0.20226	0.17944	0.17252	0.16628	0.16260	0.15948	0.15648	16
17	0.19859	0.17646	0.16973	0.16361	0.16003	0.15699	0.15408	17
18	0.19532	0.17386	0.16731	0.16131	0.15783	0.15487	0.15205	18
19	0.19239	0.17156	0.16519	0.15931	0.15594	0.15305	0.15032	19
20	0.18976	0.16953	0.16333	0.15757	0.15430	0.15149	0.14886	20
21	0.18738	0.16771	0.16168	0.15604	0.15287	0.15014	0.14760	21
22	0.18522	0.16609	0.16021	0.15470	0.15162	0.14898	0.14653	22
23	0.18324	0.16462	0.15890	0.15351	0.15052	0.14796	0.14561	23
24	0.18143	0.16329	0.15773	0.15245	0.14956	0.14708	0.14482	24
25	0.17976	0.16209	0.15667	0.15150	0.14871	0.14631	0.14414	25
26	0.17822	0.16100	0.15572	0.15066	0.14795	0.14563	0.14355	26
27	0.17680	0.16000	0.15485	0.14990	0.14728	0.14504	0.14305	27
28	0.17548	0.15908	0.15406	0.14922	0.14669	0.14452	0.14261	28
29	0.17424	0.15823	0.15334	0.14860	0.14615	0.14405	0.14223	29
30	0.17310	0.15745	0.15267	0.14804	0.14567	0.14365	0.14191	30
31	0.17202	0.15673	0.15207	0.14753	0.14524	0.14329	0.14162	31
32	0.17101	0.15606	0.15151	0.14707	0.14485	0.14296	0.14138	32
33	0.17006	0.15544	0.15099	0.14665	0.14450	0.14268	0.14116	33
34	0.16917	0.15485	0.15051	0.14626	0.14418	0.14242	0.14098	34
35	0.16833	0.15431	0.15007	0.14590	0.14389	0.14220	0.14082	35
40	0.16476	0.15209	0.14828	0.14451	0.14280	0.14137	0.14029	40
45	0.16198	0.15041	0.14697	0.14355	0.14208	0.14088	0.14002	45
50	0.15976	0.14912	0.14599	0.14286	0.14159	0.14057	0.13989	50
55	0.15794	0.14810	0.14523	0.14235	0.14125	0.14037	0.13983	55
60	0.15643	0.14727	0.14463	0.14197	0.14099	0.14024	0.13979	60
65	0.15515	0.14659	0.14414	0.14167	0.14080	0.14015	0.13978	65
70	0.15405	0.14602	0.14373	0.14143	0.14066	0.14008	0.13977	70
75	0.15310	0.14553	0.14339	0.14124	0.14054	0.14003	0.13977	75
∞	0.13976	0.13976	0.13976	0.13976	0.13976	0.13976	0.13976	∞

FOR DISPERSED ANNUITY FACTOR FOR A FUTURE AMOUNT, SUBTRACT R

Index